THE LOGIC OF INFINITY

Few mathematical results capture the imagination like Georg Cantor's ground-breaking work on infinity in the late nineteenth century. This opened the door to an intricate axiomatic theory of sets, which was born in the decades that followed.

Written for the motivated novice, this book provides an overview of key ideas in set theory, bridging the gap between technical accounts of mathematical foundations and popular accounts of logic. Readers will learn of the formal construction of the classical number systems, from the natural numbers to the real numbers and beyond, and see how set theory has evolved to analyse such deep questions as the status of the Continuum Hypothesis and the Axiom of Choice. Remarks and digressions introduce the reader to some of the philosophical aspects of the subject and to adjacent mathematical topics. The rich, annotated bibliography encourages the dedicated reader to delve into what is now a vast literature.

BARNABY SHEPPARD is a freelance writer. He has previously held positions at Lancaster University, the University of Durham and University College Dublin.

THE LOGIC OF INFINITY

BARNABY SHEPPARD

CAMBRIDGE
UNIVERSITY PRESS

CAMBRIDGE
UNIVERSITY PRESS

University Printing House, Cambridge CB2 8BS, United Kingdom

Cambridge University Press is part of the University of Cambridge.

It furthers the University's mission by disseminating knowledge in the pursuit of education, learning and research at the highest international levels of excellence.

www.cambridge.org
Information on this title: www.cambridge.org/9781107678668

© Barnaby Sheppard 2014

First published 2014
First paperback edition 2015

A catalogue record for this publication is available from the British Library

ISBN 978-1-107-67866-8 Paperback

Audience

This book is to be carefully enclosed in a time machine and sent to a much younger version of myself, to whom it is addressed: a hopeless daydreamer already intoxicated with mathematics, but perhaps not yet fully entangled in the wheels of a university education. Having taught myself all the mathematics I could find in a pre-internet era, I was aware that a whole universe of new ideas was out there waiting to be discovered. I would have opened this book, read parts of it in a peculiar order, and eventually completed all of it after filling a few notebooks with mathematical sketches and questions. It would have captured my imagination and made me want to learn more, so I can only hope that there are others who might also find something of interest herein.

Contents

Preface   page   xi

Synopsis   xvii

1 Introduction   1
  1.1   Primitive notions   1
  1.2   Natural numbers   19
  1.3   Equivalence classes and order   34
  1.4   Integers   38
  1.5   Rational numbers   44
  1.6   Real numbers   48
  1.7   Limits and continuity   75
  1.8   Complex numbers   106
  1.9   Algebraic numbers   127
  1.10 Higher infinities   132
  1.11 From order types to ordinal numbers   135
  1.12 Cardinal numbers   149
  1.13 A finite Universe   154
  1.14 Three curious axioms   171
  1.15 A theory of sets   177

2 Logical foundations   185
  2.1   Language   185
  2.2   Well-formed formulas   203
  2.3   Axioms and rules of inference   206
  2.4   The axiomatic method   218
  2.5   Equality, substitution and extensionality   233

3 Avoiding Russell's paradox   239
  3.1   Russell's paradox and some of its relatives   239
  3.2   Introducing classes   247
  3.3   Set hierarchies   250

4   Further axioms   **255**
      4.1   Pairs   255
      4.2   Unions   259
      4.3   Power sets   260
      4.4   Replacement   262
      4.5   Regularity   267

5   Relations and order   **273**
      5.1   Cartesian products and relations   273
      5.2   Foundational relations   277
      5.3   Isomorphism invariance   280

6   Ordinal numbers and the Axiom of Infinity   **283**
      6.1   Ordinal numbers   283
      6.2   The Axiom of Infinity   288
      6.3   Transfinite induction   292
      6.4   Transfinite recursion   294
      6.5   Rank   298

7   Infinite arithmetic   **303**
      7.1   Basic operations   303
      7.2   Exponentiation and normal form   307

8   Cardinal numbers   **319**
      8.1   Cardinal theorems and Cantor's paradox   319
      8.2   Finite and infinite sets   322
      8.3   Perspectives on cardinal numbers   325
      8.4   Cofinality and inaccessible cardinals   327

9   The Axiom of Choice and the Continuum Hypothesis   **335**
      9.1   The Axiom of Choice   335
      9.2   Well-Ordering, trichotomy and Zorn's Lemma   348
      9.3   The Continuum Hypothesis   353

10 Models   **365**
      10.1 Satisfaction and restriction   365
      10.2 General models   368
      10.3 The Reflection Principle and absoluteness   375
      10.4 Standard transitive models of ZF   378

11 From Gödel to Cohen   **383**
      11.1 The constructible universe   383
      11.2 Limitations of inner models   389
      11.3 Gödel numbering   391
      11.4 Arithmetization   398
      11.5 A sketch of forcing   400
      11.6 The evolution of forcing   407

A Peano Arithmetic   **411**
   A.1 The axioms   411
   A.2 Some familiar results   412
   A.3 Exponentiation   414
   A.4 Second-order Peano Arithmetic   414
   A.5 Weaker theories   415

B Zermelo–Fraenkel set theory   **417**

C Gödel's Incompleteness Theorems   **419**
   C.1 The basic idea   420
   C.2 Negation incompleteness   422
   C.3 Consistency   425
   C.4 A further abstraction   426

Bibliography   **429**

Index   **462**

Preface

*No other question has ever moved so profoundly the spirit of man;
no other idea has so fruitfully stimulated his intellect; yet no other
concept stands in greater need of clarification than that of the infinite.*

– DAVID HILBERT[1]

In the later years of the nineteenth century Georg Cantor discovered that there
are different sizes of infinity. What had begun as the study of a concrete problem
concerning the convergence of trigonometric series had turned into something far
more profound. Further investigation led Cantor to a new theory of the infinite
and the mathematical community stood with a mixture of bewilderment and
disbelief before an unfamiliar universe. This theory was refined and developed,
and continues to this day as axiomatic set theory. The theory of sets is a
body of work that I feel should be known to a much wider audience than the
mathematicians and mathematical philosophers who hold it in such high esteem.

This book is about the concepts which lie at the logical foundations of mathematics, including the rigorous notions of infinity born in the ground-breaking
work of Cantor. It has been my intention to make the heroes of the book the
ideas themselves rather than the multitude of mathematicians who developed
them, so the treatment is deliberately sparse on biographical information, and
on lengthy technical arguments. The aim is to convey the big ideas, to advertise
the theory, leaving the enthused to seek further details in what is now a vast
literature. This omission of technicalities is not to be interpreted as an act of
laziness – in fact the effort required to hold back from giving rigorous justification for some of the items discussed herein was significant, and I have often
buckled under the pressure, giving an outline of a proof or providing some technical details in the remarks at the end of each section. There are many books
on the subject which flesh out these logical arguments at full length, some of
which are listed in the bibliography. If an amateur reader is drawn to at least
one of these as a result of reading this book then I would consider the effort of
writing it worthwhile.

I cannot pretend that it is possible to write a truly 'popular' account of set
theory. The subject needs a certain amount of mathematical maturity to appreciate. Added to this is the difficulty that many of the conceptually challenging

---

[1]'Über das Unendliche', *Mathematische Annalen*, **95**, 161–90, (1926).

parts of the subject appear at the very beginning – the student is forced to detach himself from a comfortable but naive position – and there is little that can be done about this. A technical book on the same subject can afford to start tersely with a list of the symbols to be used together with the axioms, assuming that the motivation is already known to the reader. It can be argued that this is why axiomatic set theory is not as widely known as some other branches of mathematics. It is necessary to have amassed some experience with mathematics in order to get a firm grasp of what the abstract theory is talking about – set theory would not have been born at all without the force of centuries of mathematical research behind it, for this body of work informs the choice of axioms and the choice of basic definitions that need to be made. For this reason the introductory chapter is significantly longer than any other.

Mathematics is a gargantuan subject, bigger than most are aware. In his *Letters to a Young Mathematician*[1] Ian Stewart estimates, based on reasonable assumptions drawn from the output of the *Mathematical Reviews*, that approximately one million pages of new mathematics is published every year. One mathematician could easily spend a thousand lifetimes navigating the subject but still feel that the surface had barely been scratched; each proof or insight leads to a legion of new problems and areas to explore. Yet, despite its size, there are few regions in the expanse of mathematics that could accurately be described as remote. Two apparently different branches of the subject are never more than a few theorems away from one another, indeed some of the most interesting parts of mathematics are precisely those that forge such deep connections between seemingly distinct fields. It is a direct result of this combination of dense connectivity and huge size that in any account of the subject the temptation to wander into adjacent territory is enormous. In order to keep the book at a reasonable length I have had to be particularly disciplined, at times brutal, in controlling digressions.

Although the general public is, for the most part, rather shy of mathematics (and the sciences in general) and, more worryingly, unaware of the *existence* of mathematics beyond calculus, or even arithmetic, I have been pleasantly surprised at the genuine interest that some non-mathematicians have shown in the subject. In fact, without the encouragement of these individuals this book would most likely never have been written. Mathematics is a huge, endlessly fascinating and deeply creative subject which is alive and very well indeed; it grows significantly by the day. It is not very well advertised in the media, but it nevertheless seems to have captured the imagination of many amateurs.

Set theory has a dual role in mathematics. It can be viewed as merely another branch of the subject, just as any other, with its own peculiarities and diversions, yet it has a second role, that of a foundation of mathematics itself. This foundational role makes set theory a particularly ripe source for philosophical discussion. Each of the notions we encounter in mathematics can be modelled by particular sets or classes of objects, and consequently each part of mathematics may be 'embedded' within set theory; it is taken for granted

---

[1]Stewart [**202**].

that the problems of mathematics can be so expressed in set theoretic terms. We spend much of the introduction describing various classical structures, indicating how they may be translated into the new rigorous language.

There is great power in the set theoretic description of the mathematical universe. As we shall see in detail, one of Cantor's questions was, naively: *'how many points are there in a line?'*. This had little more than vague philosophical meaning before the arrival of set theory, but with set theory the question can be made precise. The punch-line in this case is that the standard theory of sets – a theory which is powerful enough to sensibly formulate the question in the first place – is not strong enough to give a definite answer, it only gives an infinite number of possible answers, any one of which can be adopted without causing contradiction. The solution remains undecidable; the theory does not know the size of the beast it has created! This flexibility in the solution is arguably far more intriguing than any definite answer would have been – it transfers our focus from a dull fixed single universe of truths to a rich multiverse, one consistent universe for each alternative answer.

Cantor's research was not initiated by lofty thoughts on the nature of the infinite. He was led to the theory through work on far more sober problems. Such humble origins are typical of grand discoveries; successful abstract theories are rarely invented just for the sake of it, there is usually a very concrete and, dare I say it, practical problem lurking at the theory's genesis. In turn, solutions to practical problems often come as corollaries of highly abstract bodies of work studied just for the love of knowledge. This lesson from history increasingly falls on deaf ears when it comes to the funding and practice of science; far-reaching research with no *immediate* practical use is always in danger of being overlooked in favour of the sort of uninspired (and uninspiring) drudgery which is worked out today only to be used tomorrow. We will meet some of the original motivating problems that helped to ignite the theory of the infinite.

Many people, when they first encounter the proof of Cantor's Theorem, which implies an infinitude of sizes of infinity, or the special case that the set of real numbers is not equipollent with the set of natural numbers, probably share a sense of injustice at being kept in the dark about such a remarkable result. This is a fairly simple proposition with an easy proof, but it is nevertheless extraordinary. *Why wasn't I told this earlier?*, *Why doesn't everyone know about this?*, might be typical reactions. That said, there is no point in introducing set theory to anyone before they have acquired an intuitive grasp of some basic mathematical ideas. I have assumed of the reader a confidence in dealing with 'abstract symbolism', for want of a better term. An important point which needs to be emphasized early on is that infinite objects generally do not exhibit the familiar combinatorial behaviour of finite objects, and nor should we have any reason to expect them to.

The language of set theory has been widely adopted; almost every branch of mathematics is written, with varying degrees of precision, in set theoretic terms. The main notions of set theory, including cardinal and ordinal numbers and the various incarnations of transfinite induction and recursion, are familiar landmarks in the landscape of mathematics.

Although some mathematicians with research interests in set theory will undoubtedly peek at the book out of curiosity, they are far from the intended audience and I'm afraid they will almost certainly find that their favourite idea, to which they have devoted a life's work, is condensed to a short sentence, or is most likely not here at all. An homuncular researcher has been perched on my shoulder throughout the writing, poking my ear with a little pitchfork whenever I have left an avenue unexplored or truncated an idea. I can only apologize if the reader feels that I have unduly neglected something.

The idea to write this book was partly triggered by a casual rereading, some time in 2004 or 2005, of a dog-eared second-hand copy of Takeuti and Zaring's *Introduction to Axiomatic Set Theory*.[1] Takeuti and Zaring's book is based on lectures delivered by Takeuti in the 1960s, not so long after Cohen's invention of forcing (one of the final subjects of this book). It is written largely in logical symbols, the occasional English sentence or paragraph providing some direction. I wondered if it was possible to write an accessible account of this opaque-looking material, and related matter, for a wider audience who did not want to become too bogged down in the symbol-crunching details. Other books have admittedly covered similar ground but none seem to discuss all of the topics I wanted to include. After an inexcusably long period of procrastination I decided that the project might be worthwhile and set to work on what turned out to be a surprisingly challenging task. Much later, in fact several years later than intended, after repeatedly retrieving abandoned copies of the manuscript out of oblivion, and after indulging in countless side-projects and distractions, a book has emerged – nothing like the book I had imagined in the early days, but nonetheless something approximating it.

I'd like to take this opportunity to thank Roger Astley at Cambridge University Press, who first showed an interest in the book, and also Clare Dennison, Charlotte Thomas and the rest of the production team for their expert handling of the project. Frances Nex copy-edited the manuscript with painstaking precision, and her suggestions greatly improved the text. If any errors remain, they are a result of my subsequent meddling. I am also most grateful to the staff of The British Library, who never failed to hunt down even the most obscure reference that I could throw at them.

REMARKS

1. Each section of the book is followed by a remark or three. This gives me an opportunity to sketch any technical details, comment on peripheral topics and digress a little from the main theme without undue disruption.

2. It must be very surprising to the layman to learn that different sizes of infinity were discovered in the late nineteenth century, as if they were glimpsed through a telescope or found growing in a forgotten Petri dish. Our task is to

---
[1] Takeuti and Zaring [**207**].

describe exactly what it was that Cantor found and along the way perhaps indirectly clarify what mathematicians mean when they say something exists, and indeed clarify what the objects of mathematics are. Broadly speaking the approach taken by most mathematicians who are not constrained by the peculiar demands of physics or computer science is to equate coexistence with coconsistency. Mathematics is then viewed as a free and liberating game of the imagination restricted only by the need for logical consistency. In fact it is generally only philosophers of mathematics who worry about the ontology of mathematical objects – mathematicians are, quite rightly, primarily interested in producing new mathematics.

3. I have already mentioned that this book can hardly be accurately classified as a popular exposition of the subject. There is a tendency in some popular science writing to focus on the 'human story', or to create one if one cannot be found, discarding the scientific details in favour of fairly shallow journalistic accounts of the private lives of the people involved. When done well, this can certainly be diverting, if not educational, but it seems to be used as an excuse to avoid talking about the hard (and therefore interesting) stuff. If you finish the book or article thinking that the lives of the scientists and mathematicians are more interesting than the science and mathematics they discovered, then something is clearly amiss, the book can be argued to have missed its target. On the other hand, to write a serious biography is a very demanding task involving time consuming historical research. To have included such material over and above the mathematics would, I feel, have obscured the content, so my focus is entirely on the subject matter. By way of balance I have added some biographical works to the bibliography.

4. One question which is guaranteed to make a pure mathematician's blood boil is *'what are the applications?'*, or the closely related *'what is it for?'*. In a more tired moment, caught off guard, there is a temptation to reply by quoting Benjamin Franklin, or was it Michael Faraday, who (probably never) said *'what use is a newborn baby?'*. The implication here is that the pure mathematics of today is the physics of tomorrow, which in turn is the everyday technology of the next century. Although history provides enough examples to assert that this is generally a fairly accurate portrait of the relationship between mathematics and the so-called real world, it is absolutely *not* the reason why most pure mathematicians do mathematics. Mathematicians are not a species of altruists hoping to improve the quality of life of their great-great grandchildren. The reason for doing pure mathematics is simply that it is paralyzingly fascinating and seems to mine deeply into the difficult questions of the Universe. That it happens to have useful spin-offs somewhere in the future is beside the point. Analyze the motivation behind the question and flinch at it. It is very difficult to predict which parts of current mathematics will find a use in tomorrow's physics, but one can confidently formulate the following two laws of mathematical utility: (i) as soon as a piece of pure mathematics is described as useless, it is destined to become extremely useful in some branch of applied mathematics or physics;

and (ii) as soon as a piece of mathematics is hailed by its creator as being ground-breaking and full of potential applications, it is destined to vanish without a trace.

5. One of the dangers of learning about Cantor's work (or any piece of pure mathematics, for that matter) without an adequate mathematical and logical background, and without the philosophical maturity that comes with it, is that there is a tendency to take the defined terms at face value and in turn react aggressively against all that is counterintuitive. That 'counterintuitive' does not mean 'false' is one of pure mathematics' prize exhibits, but it takes some effort to reach these results with clarity. We have also discovered that 'seemingly obvious' does not mean 'true' in mathematics. This phenomenon is not confined to mathematics: the physical world is also wildly counterintuitive at its extremes.

6. Mathematics was practiced for millennia without set theory. Many mathematicians continue to produce beautiful results without knowing any set theoretic axioms, and perhaps only having a vague acquaintance with the results of axiomatic set theory. Despite it acting as a foundation of mathematics, one can do mathematics without knowing axiomatic set theory. The point of set theory is not to provide a new way of finding mathematical results. Producing mathematics is, as it always has been, a matter of hard creative effort. Visual and conceptual reasoning play a critical role.

Synopsis

Proof is the idol before whom the pure mathematician tortures himself.

– Sir Arthur Eddington[1]

The vast majority of all humanly describable logical truths are, if presented without adequate preparation, either counterintuitive or beyond intuitive judgement. This is not a feature of an inherently peculiar Universe but merely a reminder of our limited cognitive ability; we have a thin grasp of large abstract structures. Fortunately we can gain access to the far reaches of such alien territory by building long strings of logical inferences, developing a new intuition as we do so. A proof of a theorem describes one such path through the darkness. Another important aspect of mathematics more closely resembles the empirical sciences, where features of a mathematical landscape are revealed experimentally, through the design of algorithms and meta-algorithms. In this book I assemble a miniature collage of a part of mathematics; an initial fragment of a huge body of work known as axiomatic set theory. The ambition of the book is a humble one – my intention is simply to present a snapshot of some of the basic themes and ideas of the theory.

Despite the impression given by the impatient practices of the media, it is not possible to faithfully condense into one convenient soundbite the details of any significant idea. One cannot hope to explain the rules of chess in six syllables, and it would be equally absurd to expect a short accessible account of set theory to be anything more than a fleeting glimpse of the whole. Perhaps the answer is not to attempt such a thing at all. My experiment was to see if something brief, coherent, yet still accurate, could be forged out of a difficult literature. I do not know if I have been successful, but I don't want the book to be an apology for omissions.

Writing prose about mathematics is notoriously difficult; the subject simply doesn't lend itself to a wordy treatment. At its worst an expository article can be a cold and off-putting list of facts punctuated by indigestibly dense passages of definitions. Technical texts do not provoke such *ennui* because the reader is an active participant in the theory, investing a lot of time on each assertion, carefully exploring adjacent ideas in an absorbing stream of consciousness journey, so that any list-like character of the original text is disguised by the slow

---

[1]Eddington [**60**]. Chapter 15.

motion of digestion and digression. The message is that passive reading benefits nobody. Many technical mathematical texts are designed to be read more than once, or are to be dipped into, rarely being read from cover to cover (at least not in the order of pagination). I have, naturally, written the text so that it can be read in sequence, but I will be neither surprised nor insulted if the reader skips back and forth between chapters and there may even be some benefit in doing this.

In any foundational investigation one inevitably feels an overpowering sense that 'everything I thought I knew is wrong', which is always a healthy development, even if it is a little disorientating at first. To those who never decamp from the foothills of such work, however, the proceedings can seem like a pointless exercise in pedantry, or even a reinvention of the wheel in obscure terms. I have tried to avoid giving this false impression, although there are some necessary preliminaries concerning familiar number systems which need to be covered before moving ahead to less intuitive matters.

Mathematicians who dip their toes into foundational philosophical matters often become the victims of a form of digressional paralysis, which may well be a pleasant state of mind for philosophers but is a bit of a hindrance for the mathematician who would rather produce concrete results than chase rainbows. Like a participant in Zeno's race, not being able to start the journey, the traveller must first argue about the sense of 'existence' and then argue about the sense of 'sense'. There are many statements herein which could easily be dismantled and turned into a full-blown dialogue between philosophers, logicians, mathematicians and physicists in a devil's staircase of counteropinions. I have tried, for the sake of brevity and sanity, not to indulge too frequently in this microscopic analysis of each and every assertion. Stones left unturned in good quantity make fine fodder for debate in any case.

Although the ideas discussed in this book are now well-established in mainstream mathematics, in each generation since Cantor there have always been a few individuals who have argued, for various philosophical reasons, that mathematics should confine itself to finite sets. These we shall call Finitists. Despite my occasional sympathies with some of their arguments, I cannot claim to be a Finitist. I'm certainly a 'Physical Finitist' to some extent, by which I mean that I am doubtful of the existence of infinite physical aggregates, however I do not extend this material limitation to the Platonic constructions of mathematics (the set of natural numbers, the set of real numbers, the class of all groups, etc.), which seem perfectly sound as *ideas* and which are well-captured in the formalized language of set theory. However, I cannot pretend to be a strong Platonist either (although my feelings often waver), so my objection is not based on a quasi-mystical belief that there is a mathematical infinity 'out there' somewhere. In practice I find it most convenient to gravitate towards a type of Formalism, and insofar as this paints a picture of mathematics as a kind of elaborate game[1] played with a finite collection of symbols, this makes me a Finitist in a certain Hilbertian sense – but I have no difficulty in *imag-*

---

[1] If it is a game, then Nature seems to be playing by the same rules!

ining all manner of infinite sets, and it is from such imaginings that a lot of interesting and very useful mathematics springs, not from staring at or manipulating strings of logical symbols.[1] I am not alone in being so fickle as to my philosophical leanings; ask a logician and a physicist to comment on the quote attributed to Bertrand Russell *'any universe of objects...'* which opens Section 11.1 of Chapter 11 and witness the slow dynamics of shifting opinions and perspectives as the argument unfolds.

The history of mathematical ideas is often very difficult to untangle, and because ideas evolve gradually over a long period of time it is impossible to draw an exact boundary between a given theory and its offspring. It is consequently a painful task to credit some developments to a small number of authors. On occasion I have had to resort to the less than satisfactory device of mentioning only a few of the prominent names. On those occasions when I label an idea with a date it is because that is the year that the idea, possibly in its final form, was published. Behind each date is a thick volume of information and behind every bold statement is a host of qualifications which, if included in their entirety, would render the present book an unpalatable paper brick. In an ideal world it would be possible to attach a meta-commentary to the text, extending the dimensions of the book. A skillfully constructed website might be able to achieve this, but then there would be a lack of cohesion and permanence, and a slimmer chance that all will be read.

I have indulged in 'strategic repetition', a luxury which is uniquely available in writing at this less than technical level. I have done this both to reinforce ideas and to view the same objects from different perspectives. I hope the reader will forgive me for telling them the same thing, albeit from a different point of view, on more than one occasion. For example, I point out several times the equivalence of the Well-Ordering Theorem and the Axiom of Choice, and, under the influence of the Axiom of Choice, the equivalence of different ways of looking at cardinal and ordinal numbers. I also repeat the fact of the absence of atoms in ZF. I wish to interpret these repetitions – in particular the material which is recast in more precise terms from its intuitive base in the introduction – simply as motifs revisited against a clearer background.

There are one or two occasions when I commit the logical *faux pas* of mentioning a concept before it is properly defined. When this happens, the reader should simply log the intuitive idea as a proto-definition, ready for future rigorous clarification. The logically sound alternative turns a slightly turbulent prose into a tense pile-up of preliminaries awaiting resolution, leaving the reader to agonize over where it is all leading. There is also a sizeable conceptual distance between some of the simple ideas covered in the earlier part of the introduction and the more technical matter towards the end of the book.

---

[1]Formalism is sometimes unfairly painted as merely a naive way of avoiding philosophical discussion (more precisely as a tactic used to dodge ontological commitment). Even more unfairly it is presented as being sterile and, perhaps worst of all, as being associated in some way with the notion that mathematics is invented and not discovered. Some fashions come and go. The ever growing literature on competing philosophies of mathematics could fill a library and plausible arguments can be found for almost any imaginable position.

In the first section we meet the core idea of cardinality together with associated terminology and notation. Once sufficient motivation has been generated the formal constructions of the classical number systems are described. It should already be clear at this stage that the reader is expected to have an intuitive familiarity with these systems before viewing them in these formal terms. The point of the exercise is to look at these structures anew and to attempt to put them all on the same logical footing. Once the hurdle of constructing the natural numbers has been cleared the rest is relatively smooth going thanks to some simple algebraic machinery. Although the introduction only covers the basics we can afford to give a fairly detailed exposition. The passage from the natural numbers to the complex numbers is described in its entirety, while the developments of later chapters necessarily become progressively more fragmented. The main goal of the first part of the introduction is to describe the rigorous construction of the real numbers (the complex numbers being a simple extension). This gives us an opportunity to view the continuum from various different angles – it appears as the class of all Dedekind cuts; certain Cauchy sequence equivalence classes; and synthetically as a complete ordered field. Some interesting bijections are presented and, most importantly, via Cantor's well-known diagonal argument, the strict cardinal difference between the set of natural numbers and the set of real numbers is demonstrated. The algebraic numbers are the last of the classical number types to be discussed.

Having introduced the cardinal infinite, the limiting processes associated with calculus are described; this material is a quick sketch of the theory that students usually meet early on in their study of analysis. The operations of calculus are presented in non-mysterious terms, removing the need for undefined 'infinitesimals'. In terms of condensing information into almost nothing the latter in particular imitates a whale being forced through the eye of a needle. The history of calculus is lengthy and deserves a multi-volume treatment. This is equally true, as I stress (squirming at my own self censorship), of the later treatment of complex numbers.

The 'order type' beginnings of ordinal numbers are detailed and Cantor's approach to the ordinal building process is described. This is material which is to be viewed again from various different perspectives (all equivalent assuming the Axiom of Choice) later in the book. The ordinal notion is revealed to have its roots in analysis via iterative processes which fail to stabilize after 'ω' many iterations. The Russell–Whitehead realization of cardinal numbers is repeated and the cardinal arithmetic operations are described – again these are to be revisited later.

I have indulged in a little fantasy in the section on the Finite Universe for the purpose of defending a finitistic view of the physical Universe (not in itself a controversial viewpoint but a useful contrast to the infinite implied in purely mathematical structures). Following this a brief tribute to Zeno's paradoxes is given.

The Axiom of Choice, the Continuum Hypothesis and its generalization are central to the text. In the introduction we meet them briefly, setting the scene for future discussion. The introduction ends with an impression of what is to come.

Chapter 2 through to the end might be regarded as one long endnote, describing in more formal terms what is needed to place the ideas of the introduction on a sound logical foundation. What do we assume at the outset, what are the primitive terms and sentences of the language? What are the logical axioms and rules of inference? At this point we adopt Zermelo–Fraenkel set theory (ZF) as our principal theory and all future statements are to be interpreted within this biased framework. Here we also meet the first non-logical axiom of ZF, Extensionality.

Russell's paradox must be regarded as a major catalyst in the development of modern set theory, although it is of course not the only one. The origin of the paradox is spelled out and some of its relatives are described. We try to identify what needs to be avoided in order to resolve the paradox and mention some strategies. Classes are introduced as defined terms and the formal development of this theory is outlined. Set hierarchies are discussed in general and we get a first glimpse of the von Neumann set hierarchy.

Further non-logical axioms appear: Pairing, Unions, Powers, Replacement and Regularity. We return to the idea of order in a closer look at cartesian products and relations, foundational relations, and the all important notion of well-ordering which underpins ordinal numbers. Ordinal numbers reappear in a new disguise and we meet the last of the non-logical axioms of ZF, the Axiom of Infinity. A Formalist approach is adopted to avoid interpretative controversy. Having opened the gates, via the last axiom, to the huge hierarchy $\mathcal{O}n$ of all ordinal numbers we introduce the powerful methods of transfinite induction and transfinite recursion. The arithmetic of ordinals is described and we return to the von Neumann hierarchy to look at the elementary notion of rank. Having reconstructed ordinal numbers in purely set theoretic terms we revisit cardinal numbers, again stressing different perspectives and exhibiting different definitions of 'finite'. We also look at the upper cardinal boundaries imposed by the axioms of ZF and the Axiom of Choice and examine new axioms needed to break through these inaccessible heights.

We are reacquainted with the ZF-undecidable Axiom of Choice and the Continuum Hypothesis. Some well-known equivalents of the former are described and the Banach–Tarski paradox makes an appearance, a result which is sometimes (misguidedly) given as a reason to reject the Axiom of Choice. Alternatives to the Axiom of Choice are discussed.

The theory of models is introduced, a subject which removes a lot of mystery from set theory, giving us something tangible to work with when proving metamathematical statements and, in particular, allowing us to gain access to independence results. The final chapter, necessarily sketchy, attempts to describe how independence results are proved by modifying existing models. The focus is on the Axiom of Constructibility. First we meet Gödel's constructible universe, a model of ZF in which the Axiom of Choice and the Continuum Hypothesis both hold. Finally we indicate some of the classical ideas that were used to generate a model in which Constructibility fails. The latter material is intended to be little more than a 'six syllable rules of chess' account of the topic, although more detail is given than most popular texts would dare. The book must end here, for further details would demand a more thorough technical grounding.

I have included three appendices. Appendix A briefly describes Peano Arithmetic, the first-order formalization of the arithmetic of natural numbers. Appendix B collects together the axioms of ZF which are discussed in more detail in the text. Appendix C is my attempt to give a postage stamp sized account of Gödel's Incompleteness Theorems. This should be regarded as 'impressionistic' at best and for the purposes of the main text of the book it suffices to know that in Peano Arithmetic and Zermelo–Fraenkel set theory (and in similar theories) there exist arithmetical statements which cannot be proved or disproved, and that such theories cannot prove their own consistency.

If more details on set theory are sought then it is time to mine the bibliography for items of interest; I have provided a little summary of the contents of each listed item. I urge the reader to at least browse through it to get an impression of the size of the subject and to see what little I have been able to cover here. The bibliography is broad, ranging from popular texts readable by the layman to fiendishly technical monographs demanding many years of intense study. Due to the small readership a lot of these items are now regrettably out of print, so the collector must be prepared to hunt for an affordable second hand copy or head for a good library. I must add that there is a torrent of current research on the subject which is not even hinted at either in the text or in the bibliography.

Among many other texts, I should record the influence of Michael Potter's *Set Theory and its Philosophy*,[1] which I discovered only part way through the first draft of this book, and, to a lesser but still significant extent, Mary Tiles' *The Philosophy of Set Theory*.[2] Takeuti and Zaring's *Introduction to Axiomatic Set Theory*[3] was, as mentioned in the preface, the original motivating text, and its presence, as well as that of Cohen's pioneering exposition of his method,[4] can still be felt in some of the more technical sections in the latter half of the book. Raymond Smullyan's *Gödel's Incompleteness Theorems*[5] provided a useful template for Appendix C. Even if the content of the many texts that I consulted when writing this book is not reflected herein, the sole fact of their existence provided some inspiration to continue writing.

Just one last word is in order concerning the quotations at the beginning of each section. In writing this book I have discovered something I have long suspected to be true: most famous quotations are either incorrect, misattributed or fictitious. I have been careful to give fuller quotes where possible, setting them in their proper context, and in most cases I have given the primary source of the quote.

---

[1] Potter [**170**].
[2] Tiles [**212**].
[3] Takeuti and Zaring [**207**].
[4] Cohen [**34**].
[5] Smullyan [**198**].

REMARKS

1. I think I ought to clarify the simple distinction between 'counterintuitive' and 'beyond intuitive' as an addendum to my opening paragraph. Both rely on a plastic intuitive base of results entirely dependent on the experience of the individual. A true result is counterintuitive to an individual if it seems to contradict his intuitive grounding. For example, the Banach–Tarski paradox, which we will meet later (see Subsection 9.1.5), is counterintuitive to anyone who thinks of geometry in purely physical terms. Through careful study or an appropriate change of perspective a counterintuitive result can in time become intuitive, the intuitive ground now suitably remoulded. A result or conjecture that is beyond intuitive judgment is one that really doesn't seem to relate to the intuitive base of results at all, so it is difficult to make any call as to whether it might be true or not. To find examples of such statements go to a conference or open a book in a branch of mathematics you know nothing about. Find a conjecture that the sub-community agrees on: 'it seems likely that every Hausdorff quasi-stratum is a Galois regular connector' and there is often no hint as to where this conjecture might have come from, even after you learn what Hausdorff quasi-strata and Galois regular connectors are.[1] Spend a few years studying it and you start to get an idea of why the conjecture was made, and eventually it might even become intuitively obvious to you, and lead to further conjectures. Of course there are statements which are beyond anyone's grasp, and may remain so forever; since we can only ever understand finitely many things, and since in any symbolic logic one can generate a potentially infinite number of implications, most statements will fall into the latter inaccessible class.

2. Studying technical mathematics often seems to involve more writing than reading. One sentence can trigger a good few pages of notes (drawings, diagrams, misproofs, proofs, dead ends), first in an attempt to fully understand what the sentence says, and then a series of playful modifications of the assumptions, moving off on various related threads, extending the results in some direction.

3. Axioms in the Euclidean sense are intended to be 'self-evident truths'. Modern axioms need not be so obvious. They are chosen for other reasons, often as distillations of observed phenomena in existing theories. The emphasis is more on the consequences of the new assumptions rather than any immediate intuitive plausibility of the assumptions themselves. An axiom, as it is meant in the modern sense, is just an assumption whose consequences one can explore.

---

[1] To put the reader at ease, these two terms are (as far as I know) my own inventions.

4. Mentioning a concept before giving its formal definition is not quite the criminal act that some would pretend it to be. In his sometimes controversial article *The Pernicious Influence of Mathematics Upon Philosophy*[1] Gian-Carlo Rota stresses the distinction between mathematical and philosophical enquiry, mathematics starting with a definition and philosophy ending with one. Even within mathematics definitions and theorems are mutually influential.

5. I should stress the importance of having an intuitive understanding of the common number systems before looking at any formal definitions. As just mentioned, the formal definitions we make are made only after thoroughly examining the underlying intuitive ideas, and they are not made with the intention of explaining what a number really is, whatever that might mean.

---

[1]See Rota [**178**].

1

Introduction

1.1 Primitive notions

The word 'definition' has come to have a dangerously reassuring sound, owing no doubt to its frequent occurrence in logical and mathematical writings.

– WILLARD VAN ORMAN QUINE[1]

1.1.1 Definitions – avoiding circularity

Some elementary observations have profound corollaries. Here is an example. Suppose finitely many points are distributed in some space and each point is joined to a number of other points by arrows, forming a complex directed network. We choose a point at random and trace a path, following the direction of the arrows, spoilt for choice at each turn. No matter how skillfully we traverse the network, and no matter how large the network is, we are forced at some stage to return to a point we have already visited. Every road eventually becomes part of a loop, in fact many loops.

A dictionary is a familiar example of such a network. Represent each word by a point and connect it via outwardly pointing arrows to each of the words used in its definition. We see that a dictionary is a dense minefield of circular definitions. In practice it is desirable to make these loops as large as possible, but this is a tactic knowingly founded on denial. The union of all such loops forms the core of the artificial language world of the dictionary, every word therein definable in terms of the loop members. So not only does this language core contain circular definitions, it is made of them! (In an appropriate tribute to the problem, the dictionary on my desk has some very short loops in the region containing the words *meaning* and *sense*.)

Any idealistic attempt to populate a dictionary with a finite number of words defined finitely and exclusively in terms of one another leads either to

---

[1] 'Two dogmas of Empiricism', in Quine [**172**].

the humiliation and dissatisfaction of circularity or, what is the better of the two options, to the nomination of some notions as 'primitive'. These primitive elements by nature cannot be defined in terms of other words, yet they form the building blocks of all other entries in the dictionary.

No dictionary of natural language is organized in this way, and nor should we expect it to be. Lexicographers skillfully try to evade circularity by using certain observable phenomena, the sensory and emotional meat of daily existence, and abstractions thereof, as their primitive reference points. As the observables are given fairly crude finitistic word descriptions such efforts will always fail for the reason just given, but the points at which a definition struggles to avoid circularity signpost those areas where the dictionary begins to step outside the boundaries of its original function; it has never pretended to be, and never will be, a self-contained account of the meaning of all things, and it is not a philosophical text. Perhaps it is not surprising that extremist philosophies adopting the position that 'all is language' tend to fall into a curious form of nihilism, turning all philosophical discussion into petty exercises in obscurantist creative writing.

By a long process of generalization and analogy, natural language reaches for high concepts far removed from the concrete stuff that lies at its foothills. This process is intuitive – new concepts are described as and when they are needed. Of course our dictionary example is just a convenient somewhat artificial one – however, the same finitistic obstruction applies to any attempt to chase meaning. How is it that a finite universe can harbour sense?

Most accounts of mathematics rely on an intuitive base of instantly recognizable Platonic objects, for example the classical number systems and various geometric notions. That these are 'instantly recognizable' (however this may be interpreted) is perhaps surprising given that none of them exactly match the comparatively crude real worldly things that helped to inspire them; but this act of abstraction, an effortless ability to simplify by assigning an ideal object to a large number of perceived objects is, to our great benefit, the way we have evolved to 'make sense of what we sense' – we simply couldn't think without it. However if we want to describe these familiar structures without deferring to vague and unreliable intuitions we have no choice but to embrace the 'primitive element' approach. An unworkable alternative which is only marginally better than circularity is to admit an infinite regress of definitions – a bottomless pit where each notion is defined in terms of lower notions.

Stretching the analogy to breaking point, by fixing a set of primitive elements we create an artificial platform spanning a cross-section of the bottomless pit. From this base we can look up (what properties follow from our basic assumptions?) and we can look down (what, at a deeper level, is capable of describing all of the properties we have chosen?). In a sense the two directions represent different aspects of mathematical versus logical enquiry, although we mustn't take this naive picture too seriously.

The crucial question is: what should we take to be the primitive elements of all of mathematics? As nineteenth century ideas moved into the twentieth, it was generally agreed that the primitive notions should be axioms governing 'aggregates' together with a means of describing or isolating objects within such aggregates (sets/classes and the idea of membership). This was the birth of set theory.

REMARKS

1. There is a natural way to modify directed networks which relieves them of all loops. The basic idea is in essence to repeatedly contract each minimal loop to a single vertex. However, in order not to lose too much information, we have to be careful not to combine two adjacent loops in one operation. To this end we repeat the following simple process. Suppose our directed network has labelled vertices. To each loop of minimal size we associate a new vertex whose label is the set of all labels of the vertices of the loop it represents. We then delete all vertices belonging to these minimal loops (which means we also delete any edges that are deprived of one or two end vertices). Next we connect the newly created vertices to the rest of the network in the natural way: the outgoing edges are the outgoing edges of the original loop vertices (in the case where minimal cycles are joined to one another by an edge, or share edges or vertices, some of these edges will be joined to one of the new vertices); and the incoming edges are the incoming edges of the original loop vertices. We then repeat the construction on the new network. Eventually all loops will vanish and we will be left with a tree with some set-labelled vertices. In extreme cases we might end up with just one vertex with a complicated label of nested sets. Performing this construction on the network associated with a good sized dictionary would create an amusing toy model for what might be called a lexicographic universe; one gets something of a burning curiosity to see this process in action (at least to determine the first few stages – what are the initial few minimal cycles?).

2. The common practice of mathematical abstraction is this: Take some intuitive object. Determine some properties of this object (express these properties in set theoretic/algebraic terms). Then consider the class of all objects satisfying these properties. Of course this class will include a model of the intuitive object you started with, but it might include a host of other objects too, possibly some surprises. The extremal objects in this set might reveal something new about the original object. To what extent can the original object be isolated within this larger class of models; is there a further set of properties which completely characterizes it among its cousins? This is a simple idea, but it is remarkable how powerful it can be. We will see some examples of this later.

A chain of definitions must be an infinite regress, an eventual cycle or it must terminate in an undefined primitive notion. Rejecting the first two options we must choose the primitive notions of the language of mathematics. Can all of mathematics be described in terms of a theory of aggregates?

1.1.2   Intuition and its dangers

Fed by constant sensory bombardment from the outset, and through total immersion in our environment, we develop an intuitive, albeit often misguided, grasp of some fundamental notions. Exposed to a tidal wave of examples we build soft conceptual structures on softer foundations and form a picture of the physical and logical universe that is childishly simplistic yet, provided we don't question our ideas too closely, free from alarms and surprises. Such a thin grasp of the world about us is adequate for the vast majority of human activities. Once the sensory parameters are shifted, perhaps by space (the weirdness of the subatomic world, or at the other extreme of magnification, clusters of galaxies mere specks of dust) or by time (superslow motion, all living creatures now at a geological pace, or superfast where we watch continents drift and observe the evolution of species) we are immediately alienated, the intuitions moulded by our mesoscopic upbringing of little use.

Without the ability to quickly construct a rough understanding of some notions we would be helpless. This basic intelligence is essential in order to gain a foothold in any subject; whether it is by forming a vague picture or by drawing a rough analogy, we need some foundation on which to hang our thoughts. Later enquiry may change the initial picture, sometimes drastically, but nevertheless the ability to concoct an immediate (fuzzy) impression when presented with a new concept is crucial.

The reality is that we, short-lived impatient biological beasts, work on unfamiliar complicated stimuli from the 'top down', dissecting pieces as required. We tend to stop further scrutiny when we feel satisfied that the new entity has been explained in terms of ideas we think we already understand. What qualifies as 'satisfaction' is a matter of taste and experience. If, on the other hand, we are working from the bottom up, building the foundations of a subject, we are most content when the primitive assumptions are few in number, self-evident and consistent to the best of our knowledge. A system of assumptions is consistent if from it we cannot deduce both a statement and its negation. The qualification 'to the best of our knowledge' may seem like an unsatisfactory appeal to the arbitrary limits of human mathematical ability, however, as Gödel famously demonstrated (see Appendix C), there are many systems for which it is impossible to prove consistency without appealing to principles outside the system.

It can be very rewarding to analyze notions that we take for granted. Such analysis often uncovers surprises and we realize what little we understood in the first place. Sometimes what we pretend to be obvious is far from clear, and through the microscope we perceive magnificent strangeness. It is partly the *apparent* surreal quality of some results in mathematics which initially draws many people to it. These exotic ideas, when they first appear, seem to live far from the utilitarian regions of science, yet they have a curious habit of barging their way to the forefront of physics.

Many are drawn to mathematics simply for the pleasure of total immersion in a consistent but otherworldly universe. Stanisław Ulam acknowledges this escapist aspect of the subject in his autobiography *Adventures of a Mathematician*.[1]

Eventually, in the mature stages of such critical investigation, we come to regard the newly uncovered world as normal. All too often, mimicking the progress of science itself, we must abandon or more often demote to crude approximations and special cases our preconceptions and gradually develop a new intuition based on a more refined point of view. The ability, and more importantly, the willingness, to constantly question one's beliefs and understanding seems essential for any sort of intellectual progress.

REMARKS

1. On reflection it is quite alarming how many of our intuitions are based on pure guesswork, or even prejudice, founded on almost nothing at all. Even the most elementary facts about the physical world can be counterintuitive at first; here I am especially thinking of the pre-Galilean misconception that heavier bodies fall faster than lighter bodies. Of course a well designed experiment will swiftly correct this (simply ensure that both objects experience the same air resistance). But there is no need to climb the steps of the Tower of Pisa; a moment's thought can help to shatter the false belief: imagine two bodies falling and joining one another, or the reverse, a single body breaking up into two smaller pieces as it falls. The notion of a body momentarily changing its rate of acceleration as it splits or coalesces is clearly unnatural. In particular consider the moment when two large bodies touch at a point; are we to believe that the united body will suddenly start to plummet even faster owing to this tiny point of contact, and then equally suddenly decelerate as the contact is broken? Alternatively consider connecting a light and heavy body together with a length of string. Is the lighter body somehow expected to reduce the speed of descent of the heavier body, as if it were a parachute, even though their combined weight exceeds the weight of each component part? From these considerations alone one would conjecture that all bodies experiencing the same air resistance, and in particular all bodies in a vacuum, fall at the same rate, and experiment verifies this. Finding the right way to think about something is the difficult part.

---

[1]Ulam [**217**], p. 120. Ulam compares some mathematical practices with drug use, or with absorption in a game of chess, which some mathematicians embrace as a means of avoiding the events of a world from which they wish to escape!

2. From a certain extremely reductionist viewpoint, what Eugene Wigner famously described as 'the unreasonable effectiveness of mathematics in the natural sciences'[1] is not quite as unreasonable as it might seem at first glance. That is to say, if one imagines a bottom-up description of the physical Universe, starting with some primitive elements governed by some simple relations, then there is no surprise at all that the emergent macroscopic world that grows out of this will be a mathematical one. From this point of view every mathematical result, no matter how esoteric, that is describable in the underlying 'logic' of the Universe has the potential to describe something physical. There is perhaps still some room for mystery in that many of the macroscopic mathematical relations in the Universe are so surprisingly tractable (inverse square laws, beautiful symmetry and so forth) and so elegantly related to one another. It is remarkable that the symbolically expressed imaginative fantasies of mathematics can have such concrete applications. If the Universe is not mathematical, what else could it possibly be?[2]

Intuition is reliable only in the limited environment in which it has evolved. Unable to abandon its prejudices completely, we must constantly question what appears to be obvious, often revealing conceptual problems and hidden paradoxes. One intuitive notion which is ultimately paradoxical is that of arbitrary collections.

1.1.3   Arbitrary collections

One notion that we take for granted, to the point of blissful ignorance, has been mentioned already: finiteness. To give a sound definition of finiteness is a surprisingly sticky problem, but it is not intractable. We shall come to its formal definition later, but for now we will have to settle for the intuitive idea and take on trust that it can be formalized.

Finiteness is a property of certain arbitrary collections of objects. Without reference to the sophisticated notion of 'number' or 'counting' how might we compare such aggregates? We shall give an answer to this shortly. The objects we have in mind, in this naive introduction at least, are free to be anything we care to imagine, physical objects or, most often, abstract notions. We deliberately delay mentioning 'numbers' for reasons which will soon become clear; we will model numbers as certain specific sets of objects. By an *element* of a given collection we mean one of the objects in the collection.

---

[1] *Communications on Pure and Applied Mathematics*, **13**, 1–14 (1960). A copy of the article is easily found online.

[2] There is of course a huge literature on this subject. For an interesting take on the relationship between mathematics and the empirical sciences from a *fictionalist* point of view, see Leng [**137**].

It would be liberating to have at our disposal a list of synonyms for the word 'collection'. However, as is often the case in mathematics, the most attractive possible labels have very particular meanings attached to them already ('group', 'category', 'set' and 'class', for example, are all taken). One can be fairly confident that any alternative to 'aggregate' has acquired a technical definition in some branch of mathematics. This can cause interpretative difficulties in certain expositions, but in practice context dictates meaning.

We will be seeing much talk of 'sets' and 'classes'. Roughly speaking, a set is a 'well-behaved' collection formed from other sets according to certain rules. Each finite collection is a set, as are many infinite collections, including the basic number systems we are about to introduce, with the exception of the collection of all ordinal numbers and the collection of all cardinal numbers.[1] Most set theories include an axiom stating that a collection modelling the natural numbers is a set, or one can prove this from more general axioms, and further axioms give rise to infinitely many other infinite sets. A set theory with no infinite sets is obviously possible, but such a setting is clearly not the right environment for the subject of this book, and to adopt such an unnecessarily restrictive theory would also mean having to reject vast amounts of beautiful (and very useful) mathematics.

We can think of classes as collections of objects which share a common property. All sets are classes, but there are classes which are not sets. These sprawling monsters, called *proper classes*, are banished from the safer world of sets because their presence gives rise to unpleasant consequences: the paradoxes. If we were to allow proper classes to be sets then the theory would face an internal contradiction and come crashing down, collapsing under the weight of its own ambition. In fact, because of this delicate divide between sets and proper classes, the notion of 'set' is rather subtle and deserves more discussion; 'set' is treated as a primitive term in set theory, just as 'line' and 'point' are primitive terms in axiomatic geometry. Exactly what is meant by a set – that is, its principal interpretation; how it differs from the intuitive notion of a collection – is very difficult to answer briefly, and it is this difficulty that makes the subject of set theory peculiar at its outset. Most other branches of mathematics can define their central concepts early in the development and with relative ease; group theory begins (along with motivating examples) with the definition of a group; topology begins (again with motivating examples) with the definition of a topological space; measure theory soon gives the definition of a measure and so on. Set theory, on the other hand, starts with 'set variables', 'set' being a primitive term loaded with intuitive interpretation, the *naive* theory of which turns out to be fundamentally flawed. In more explicit terms, which will be clarified later, most mathematical theories grow from a plethora of examples, the common features of which dictate the axioms of the underlying theory (each example comprising a *model* of the theory). The historical development of set theory did not fully conform to this familiar pattern; the axioms came before

---

[1]When such statements are made I am tacitly making reference to the set theory known as Zermelo–Fraenkel set theory (ZF), which we will meet later, and which is the main theory considered in this book.

all genuine models, only the successes and failures of naive set theory serving as a guide. Every abstract theory begins, like set theory, with a set of constants together with variables, relation symbols and axioms (see Section 2.3), but it is quite rare to establish such a system without having a collection of models, even vague intuitive ones, in mind.

It was the twentieth century clarification of the notions of theory and model which was to bring the question 'what is a set?' into sharper focus. Any meaningful answer to the question must be postponed until an explicit model of the theory has been exhibited, and once this is done the answer is fairly banal (a set, in a given model, is simply a member of the model).[1] Prior to the construction of concrete models of set theory, at the theory-building stage, the question should instead be 'what properties do we want sets to have?'. Our answer to this question is to present a short list of properties (the non-logical axioms) which correspond to the properties we imagine intuitive sets – mathematical aggregates – to have, based on our prior experience with mathematics.

Much of the early development of set theory concerns itself with the task of determining when a class is and when it is not a set, or at least in which ways sets can be combined to form other sets. I will use 'set' and 'class' quite casually in this introduction with the hope that the reader understands that not all classes are sets, and that the scope of some notions such as size (i.e. 'number of elements') is restricted to sets and has no meaning for proper classes. At the centre of set theory is the powerful idea of regarding as a single entity any collection of objects.

We exhibit small finite sets by listing their elements inside braces, e.g. $\{a, b, 4, 7\}$. Larger finite sets with an easily discernible pattern are listed in the informal manner $\{0, 2, 4, \ldots, 96, 98, 100\}$, and a similar conceit can be used for some infinite sets. We can be more precise and write $\{n : n$ is an even integer no less than 0 and no greater than 100$\}$ for the latter set. More generally we use $\{x : \phi\}$, meaning 'the class of all x satisfying the property ϕ', where ϕ is some statement. A casual use of this simple device leads to profound difficulties which will be discussed later.

In the presentation of a set the order in which its elements are displayed is of no consequence, and any repetition of an element is redundant. It is important to understand the distinction between a and $\{a\}$; the latter is a set comprising a single element, namely the set a, while the former is the set a itself, which may have many elements.

---

[1]Analogously it is meaningless to ask, when presented with the theory of groups (see Subsection 2.4.1), 'what is an element of a group?'. This is a question for an individual *model* of group theory, i.e. a particular group, to answer, and the answer, depending on the model, could be that it is a permutation, a rotation or translation in space, an invertible matrix, an integer, a function, or indeed any other set theoretically describable mathematical object. In axiomatic theories of geometry we postulate the properties of objects suggestively called 'points' and 'lines', but in describing models of the theory one needn't take these elements to be the familiar points and lines as they are understood in the conventional sense. It can be advantageous to depart from the usual intuitive objects which motivated the axioms. Indeed, this kind of imaginative departure was the conceptual leap required to produce concrete models of non-Euclidean geometries (see Subsection 2.4.2).

REMARKS

1. Jumping ahead of ourselves a little, we can give a precise set theoretic definition of a natural number, and so we can collect all such numbers together to form a class. Do we want this collection to be a set, or should it remain a proper class? After a century of very close scrutiny no contradiction has been shown to follow from the assumption that the class of all natural numbers is a set, so it would seem to be needlessly restrictive not to allow it to be a set. We shall come back to this later.

2. It should be stressed that ZF is just one among many approaches to set theory. Some alternatives closely resemble ZF, others are wildly different. Indeed, there are approaches to the foundations of mathematics that are not 'set theoretic' at all. However, I think ZF is perhaps the most natural place to begin; and once its ideas have been absorbed it is easy to consider variations and deviations from the norm.

It is surprisingly difficult to rigorously define finiteness. In studying this problem we begin to ask what it means for two sets to have the same number of objects. What is 'number'?

1.1.4   Equipollence

What do we mean when we say that two sets are of the same 'size'? If these sets are finite then the notion is something we seem to be able to cope with at a very early stage in our cognitive development. We might proceed as follows. Let us assume that we have been presented with two finite sets of objects. We then simultaneously take one object from each set, continuing to remove pairs until at least one of the sets has been exhausted. If at this terminal stage there still remain objects in front of us then the pair of sets we started with clearly had a different 'number of elements'. Essentially we have just paired off objects, that is, we say that two sets are of the same size if we can pair off the elements of the first set with the elements of the second set in a one-to-one fashion.

As this is such a crucial concept we need to introduce a little terminology in order to make the notion precise.[1] A *function* between two sets A and B is a mapping associating each element of A with some element of B. If we call the function f then the statement that f maps the element a of A to the element b

---

[1] At this stage we are denoting arbitrary sets by upper case roman letters. Later we adopt the convention that sets are to be denoted by lower case letters and classes by upper case letters.

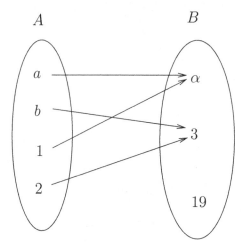

Figure 1.1   A simple-minded depiction of a function mapping the finite set $A = \{a, b, 1, 2\}$ to the finite set $B = \{\alpha, 3, 19\}$, the details of the mapping illustrated by a collection of arrows indicating which elements of B are associated with each element of A.

of B is denoted $f(a) = b$. One might regard, and in fact later *define*, a function f from A to B as a collection **F** of ordered pairs $(x, f(x))$ where each element of the set A appears as the first entry of exactly one of the ordered pairs in **F** and where $f(x)$ is in B for all x in A. The notation $f : A \to B$ is used as a shorthand for the statement 'f is a function from A to B'. We call A the *domain* of f and B the *codomain* of f. The notion of function is one of the most basic and widespread in mathematics.

It is commonplace to picture a function between arbitrary finite sets of objects as a collection of arrows joining elements of the domain to their targets in the codomain (as in Figure 1.1). If the function maps x to y and we wish to suppress the name of the function itself we often use the notation $x \mapsto y$.

We need to consider certain types of function. A function $f : A \to B$ is *injective* (or *one-to-one*) if no two different elements of A are mapped to the same element of B, that is to say $f(x) = f(y)$ implies $x = y$. The function illustrated in Figure 1.1 is not injective because both elements a and 1 of A map to the same element of B, namely α. A function is *surjective* (or *onto*) if each element of B is the image of at least one element of A, that is for every b in B there exists an a in A such that $f(a) = b$. The function illustrated in Figure 1.1 is not surjective because no element of A maps to the element 19 of B. A function is *bijective*, or is a *bijection*, if it is both injective and surjective.

In the case when A and B are finite sets the intuitive statement that A and B have the same number of elements is perfectly captured by the rigorous statement that there exists a bijection $A \to B$. We extend this particular manifestation of number to all sets, finite or otherwise, and say that sets C and D have the same *cardinality* if there exists a bijection $C \to D$. We interpret this correspondence as meaning that C and D have the same 'number' of elements,

A	B	!	"	G	9	7	f	p	*	)	(	l	v	+	=	{	}	j
:	;	@	'	O	H	¡	,	¿	.	?	/	q	a	Z	w	s	x	b

Figure 1.2 As the symbols have been arranged conveniently in pairs we can see instantaneously, without having to count and without assigning a label to the common quantity, that the two rows in the table above have the same number of elements. The equipollence of these two classes is a far more primitive and immediately recognizable characteristic than their shared 'nineteenness'.

even though we may not be able to attach one of the familiar natural number labels $1, 2, 3, \ldots$ to this shared magnitude. Exactly what labels we attach to these infinite sets will be discussed later. If there exists a bijection between two sets we will also say that they are *equipollent* (see Figure 1.2). We think of A as 'strictly smaller than' B if there is an injection from A into B but no bijection A onto B. It is important to stress that we do not need to exhibit an explicit bijection between two sets in order to show that they have the same cardinality, we need only prove that such a bijection exists. There are various ways of proving such existence in practice; the Cantor–Bernstein Theorem, which we shall meet later, often plays a role.

We introduce the notation $x \in A$ to mean x is an element of A and $x \notin A$ to mean x is not an element of A. We say that A is a *subset* of B, $A \subseteq B$ in shorthand, if $x \in A$ implies $x \in B$, that is, if all elements of A are also elements of B. A is a *proper subset* of B, denoted $A \subset B$, if A is a subset of B but is not equal to B, so that there exists at least one element of B not in A. It is not unusual to find $A \subset B$ in place of $A \subseteq B$ in some parts of the mathematical literature, in which case an alternative notation is used to indicate that A is a proper subset of B. The relation of being a subset provides a very crude means of comparison of magnitudes. It is a poor substitute for cardinality, after all, no one would deny that $\{0, 1, 2\}$ has the same number of elements as $\{88, 87, 512\}$, yet neither is a subset of the other; in terms of subsets these two sets are incomparable.

REMARK

When I first learnt about equipollence it was described in terms of a remote tribe who only had words for 'one', 'two' and 'many'. In preparation for a day of hunting the tribe's spear-maker had to craft a new spear for each huntsman. He takes an empty bowl and asks each of the hunters to put a stone in the bowl. Then, with the hunters free to go, the spear-maker begins work; he makes a spear, picks a stone out of the bowl, makes another spear, picks another stone out of the bowl, and continues until the bowl is empty. Although there is some primitive sense of counting here there is no need to give the unknown quantity of spears a name. (There are many variations of this story.)

Such restricted 'innumerate' languages are not fictions invented for the sake of illustration. Studies of the Pirahã tribe of remote northwestern Brazil indicated that they only had words for 'one', 'two' and 'many', or perhaps only 'few' and 'many'. But the truth is a little stranger. A follow-up study revealed that the word for 'one' was actually used for quantities between one and four; and 'two' could be used for five or six. So they get by with only three degrees of quantity: a few, a little and a lot.

The relative size of two sets is measured in terms of equipollence, a one-to-one pairing off of the elements of each set. This gives us a way of comparing sets, but it still leaves open the problem of defining finiteness.

1.1.5   Finiteness

I have freely used the terms finite and infinite expecting the reader to have an intuitive grasp of these notions. Clearly we wish infinite to mean 'not finite', but then how are we to define 'finite'? A natural approach, and one that is encoded within the theory of sets we are about to describe, is to say that a set A is finite either if it has no elements at all or if there exists a natural number n such that A is equipollent to $\{0, 1, 2, \ldots, n\}$. It will be convenient to call this property 'ordinary finiteness'. This is a fair definition, but to be comfortable with it we need to know precisely what we mean by 'natural number', and that is to come (we assume an intuitive familiarity with the set of natural numbers below, for the purposes of an example).

There are alternative definitions of finiteness which make no mention of natural numbers. The equivalence of these 'number free' definitions follows if we are permitted to assume the truth of a certain far-reaching principle called the *Axiom of Choice*.[1] We will have much to say about Choice later; briefly it asserts that for any collection $\mathcal{C}$ of sets there exists a set with exactly one element in common with each set in $\mathcal{C}$.[2]

If a set A is ordinary finite then there does not exist a bijection from A onto a proper subset of A – a property known as *Dedekind finiteness*. The Axiom of Choice is needed to prove that every Dedekind finite set is ordinary finite. Contraposing the Choice-free implication[3] we see that a set A is ordinary *infinite* if there exists a bijection from A onto a proper subset of itself. This property of proper self-injection may seem absurd to someone who has not encountered the phenomenon before, but any such surprise is only symptomatic of the fact that we live in a finite world subject to the relatively sober properties of finite sets, infinite collections being inevitably unfamiliar.

---

[1] I will occasionally use the capitalized 'Choice' as an abbreviation for 'Axiom of Choice' to avoid a cluttered prose, and future axioms will be similarly abbreviated.

[2] It is commonplace to state that the collection $\mathcal{C}$ is pairwise disjoint (i.e. no two sets in $\mathcal{C}$ have an element in common), so that the newly formed set is imagined as a selection of elements from a possibly infinite collection of buckets where exactly one element has been taken from each bucket. However, this assumption of pairwise disjointness is not necessary (see Subsection 1.14.2).

[3] The contrapositive of an implication 'ϕ implies ψ' being the equivalent 'not ψ implies not ϕ'.

We illustrate the idea with the simplest of all infinite sets, the natural numbers $\mathbb{N} = \{0, 1, 2, 3, \ldots\}$. It can be shown that any subset of $\mathbb{N}$ is either equipollent to $\mathbb{N}$ or is finite and it is in this sense that the set of natural numbers, or, equally so, anything equipollent to it, can be regarded as the 'simplest' or 'smallest' example of an infinite set (up to equipollence). The function mapping n to $2n$ is a bijection from $\mathbb{N}$ to the set of even natural numbers $\{0, 2, 4, 6, \ldots\}$, and the latter is a proper subset of $\mathbb{N}$. The functions $f(n) = n+1$ and $g(n) = n^2$ are just two more of infinitely many examples of injective functions $\mathbb{N} \to \mathbb{N}$ mapping $\mathbb{N}$ onto a proper subset of $\mathbb{N}$.

The last example, the bijection from the natural numbers to the set of squares, was remarked upon as a curiosity by Galileo[1] and was undoubtedly noticed many thousands of times before him.

Later we will also meet Tarski's definition of finiteness, which can be proved to be equivalent to ordinary finiteness without appealing to the Axiom of Choice.[2]

REMARKS

1. One obvious remark that is still worth making is that any Dedekind infinite set will have an infinite chain of nested infinite subsets: a Dedekind infinite set A_0 is, by definition, equipollent to a proper subset A_1 of A_0. But A_1 is also Dedekind infinite (simply apply the same bijection again, this time restricted to A_1), so it is equipollent to a proper subset A_2 of A_1, and so on, so we obtain a sequence (A_n) of infinite subsets of A_0 each successor a proper subset of its predecessor. Thus there is no such thing as a 'minimal' Dedekind infinite set.

2. It would be inappropriate to write a book about infinity without mentioning the famous example of Hilbert's Hotel. We imagine an infinite hotel with rooms numbered by the natural numbers 1, 2, 3, and so on. Hilbert's Hotel has the following peculiar property: even if every room is occupied, there is always room for more people.

   Should someone arrive when all the rooms are taken, all the hotel manager needs to do is to instruct all his guests to move to the next room in sequence; guests in room n move to room $n+1$. Thus (admittedly at the inconvenience of infinitely many people) room 1 becomes free and our visitor can stay.

---

[1]This observation appears in Galileo's *Discorsi e dimostrazioni matematiche, intorno a due nuove scienze* (Discourses and Mathematical Demonstrations Relating to Two New Sciences), 1638. An English translation (1974) by Stillman Drake is available under the title *Two New Sciences* [**77**]. The equipollence of the set of squares and the set of natural numbers is discussed, although obviously not in these modern terms, in an imaginary dialogue between the two characters Simplicio and Salviati. The 'two new sciences' of the title concern the strength of materials and the motion of objects.

[2]Nor does Tarski finiteness refer to the notion of equipollence – given these properties it is surprising that Tarski finiteness is not used more often as a definition of finiteness in basic set theory texts.

Now suppose k visitors arrive. This time the manager asks everyone to add k to their room number and move to that room, thus freeing rooms 1 to k.

Finally, if an infinite sequence of people arrives, let's call them 'person 1', 'person 2', 'person 3', and so on, then there is still room for all of the new arrivals: the instruction for the present guests is to double their room number and move to that room. This leaves all the odd numbered rooms vacant, so 'person k' can move into room $2k - 1$.

3. The explicit observation that infinite collections exhibit properties which are profoundly different from those of finite collections is sometimes credited to Gregory of Rimini (the fourteenth century scholar).

In trying to define finiteness we seek a set theoretic property that distinguishes between intuitively finite sets and intuitively infinite sets. One such property is the existence of a proper self-injection. A set which does not have this property is Dedekind finite. A further assumption, the Axiom of Choice, is required to ensure that this definition coincides with various other proposed definitions of finiteness. The existence of a proper self-injection is our first glimpse of the peculiar behaviour of infinite sets. At the other extreme of the cardinal spectrum is a similarly misunderstood set, the empty set.

1.1.6   Nothing

Early set theoretic conceptual difficulties were not confined to matters of infinity; there were problems too with the notion of the empty set, the set containing no objects at all, denoted $\emptyset$.[1] The initial reticence of nineteenth century authors to admit the empty set as a meaningful object has parallels with the reluctance to accept the number zero in mathematics centuries earlier. One can also argue that the difficulties arose from distinct ways of thinking of collections, either as sets or as fusions.[2] Modern mathematicians have become so accustomed to thinking of aggregates in terms of sets, with its all-important notion of membership, that the fusions of old, where membership is replaced by the weaker relation of being a subfusion, are nearly forgotten. It is very difficult for us to appreciate these days, but the distinction between inclusion and membership was not clarified until the end of the nineteenth century. Fusions are rarely found in modern mathematics, but there is strong evidence that some of the early set theorists and analysts of the nineteenth century were thinking in terms of fusions, not sets. For example, in his important work *Was sind und was sollen*

---

[1] The symbol for the empty set is the Danish/Norwegian Ø, introduced by André Weil, and is not to be confused with the Greek ϕ.

[2] The distinction between collections and fusions is discussed in Potter [**170**] (Chapter 2).

die Zahlen?[1] Dedekind avoided the empty set, and the same symbol was used for membership and inclusion, both indicating a fusion-theoretic framework.[2]

In the theory we generally refer to in this book (Zermelo–Fraenkel set theory, occasionally digressing to fragments of other theories) sets are not free to be 'anything', in fact they are all constructed ultimately from the empty set; no explicit items derived from the physical world are admitted. One could introduce such real-life artefacts by including 'atoms' (also called *individuals* or *urelements*) in the theory if desired, but it is unnecessary to do so.

Remark

Formal mereological systems make use of the relation of parthood and the axioms generally reflect the properties one would expect if a part of a whole is interpreted as a subset. An atom is an object with no proper parts. A natural antisymmetry condition on parthood ensures that finite mereological systems will always have atoms, but an infinite system need not, unless a special axiom is introduced to force it to have one. A classical example of an atomless fusion is given by the line, regarded as a fusion of line segments: any line segment can be divided into subsegments indefinitely. The notion of a line as a set of points is a relatively modern one which emerged hand in hand with the development of set theory. Classically one could always specify a point on a line, but since there always exists a new point between any given pair of points it was perhaps thought absurd that one could saturate the entire line by points. We will soon see how such a saturation can be realized.

The notion of the empty set posed some difficulties for early set theorists, especially those who thought in terms of fusions, where the subset relation plays the fundamental role in place of membership. Some set theories build all of their sets from the foundation of the empty set, others assume a wider foundation of atoms. In the course of developing a set theory we begin to ask what it is we are trying to describe. What is a set and in what sense does a set exist?

1.1.7   Sets

Set theory grew out of the work of the nineteenth century mathematicians who sought a rigorous foundation for the calculus. The sets Cantor and his contemporaries originally considered were the sets occurring in classical mathematics, such as sets of integers and real numbers. Anyone with experience in mathematics has an intuitive grasp of such sets, despite their abstract nature, largely

---

[1] Dedekind [**43**].

[2] The philosophical term for the study of objects in terms of their parts in this way, roughly speaking the logic of sets with respect to inclusion, is 'mereology'. Although it hasn't made much of an impact within mathematics itself, mereology does have some influence in related disciplines.

owing to a familiarity with the physical objects or physical phenomena they were originally born to model.

To study the theory of aggregates in more generality it is necessary to clarify exactly what is meant by a set. In the early years of the theory this problem was not an issue at all and no one would have thought to raise the question – a naive set is simply an unrestricted collection of objects – but once it was discovered that naive set theory had internal contradictions (we come to this crucial issue later) the question of exactly what a set is moved from being an unnecessary or unthinkable issue to being a matter of great urgency.

In his later work, Cantor considered sets with arbitrary elements, not necessarily sets of numbers or functions. We mustn't get carried away and assume that the use of such arbitrary classes arrived in the nineteenth century, the idea is an ancient and natural one, nevertheless the rigorous treatment of sets, in particular the treatment of infinite collections such as the set of all natural numbers as *single entities*, and all the surprises of the consequent theory, is a relatively recent development.

Philosophical discussion of mathematics inevitably gets tied up in ontological matters and the subject of Platonism raises itself – we have already commented on the question of whether infinite sets exist or not. Much has been written on the subject of Platonism in mathematics but, preferring to stick to mathematical issues, I will not be honouring this scholarly body of work with a thorough account here. I will have occasion to mention Platonism again but there will be no in-depth analysis.[1] The strict Platonist regards the abstract objects of mathematics (real numbers, geometric objects, Hilbert spaces, functors, groups, homomorphisms, sets, functions,...) as having an existence quite independent of the minds of those individuals who occasionally cast an eye over such matters. It is sometimes difficult to resist this ancient picture of an ideal universe of ideas, especially when in the midst of trying to prove something; there is a feeling when doing research that a Platonic realm is slowly being uncovered, not being created – the theorems we publish seem like (poorly translated) partial observations seen through fog, not fresh inventions. It is easy to picture the class of all consequences of a given set of axioms as a rigid infinite body of results waiting to be discovered (Figure 1.3) and there is surely no controversy in claiming that this web of relative truths exists in some sense, much in the same way that the set of all grammatically correct sentences, or the set of all primes of the form $n^2 + 1$, exists.

A strict Formalist regards mathematics purely as an absorbing exercise in determining the geography of such webs of consequences, treating the interpretation as of secondary importance, while the Platonist goes further, suggesting that any intelligible natural statement, say, about the arithmetic of natural numbers, ought to be either true or false and that the phenomenon of incompleteness, where the union of all provable statements and their negations fails to saturate all statements, simply highlights the failure of the underlying set of

---

[1]A discussion can be found, for example, in Balaguer [**8**]. See also the series of conversations between Jean-Pierre Changeux and Alain Connes [**32**].

axioms to capture the Platonic truths of the Universe.

REMARKS

1. As soon as we are given a set of axioms and a collection of rules of inference then, provided everything is suitably recursively enumerable (i.e. provided there is an algorithm which can generate the class of provable statements), we can in principle begin the never ending task of cataloguing all provable sentences. As a means of finding new mathematics this is by and large hopelessly ineffective – one might as well try to create a poem by writing out all possible word combinations from a dictionary.

   Given an arbitrary statement, as we well know, it may be very difficult or even impossible to determine whether it is provable or not; just take any long unsolved conjecture as an example. There are some rare 'decidable' theories for which there exists an effective algorithm which will tell whether a given statement is provable or not, but even then there is no guarantee that the algorithm will terminate in a realistic time; what use is an oracle that takes longer to speak than the life of the Universe? It is still presently more efficient to be creative; think about the mathematics intuitively, try a few ideas, build on the work of your ancestors, and hopefully come up with a proof or a disproof.

   There is a school of thought that one should only admit collections which are algorithmically generated in some fashion, where we can only speak of an element of such a collection if we can prove that it will indeed eventually be generated. Adopting this philosophy we would perhaps think of the class of provable statements dynamically as a growing tree of results, but not as a completed infinite tree.

2. The class of all grammatically correct sentences is somewhat similar to the case above, but the rules of grammar are not the clear cut exact rules of logic. Nevertheless one can imagine generating ever longer grammatically correct sentences. This time, though, it is relatively easy to recognize a grammatically correct sentence. As for the class being infinite; it will include sentences of the form 'The first five positive integers are: one, two, three, four, five', and one can of course replace 'five' by an arbitrarily high number.

3. Hardy and Littlewood conjectured that there are infinitely many primes of the form $n^2 + 1$, indeed they conjectured that the number of such primes less than n is asymptotic to $c \cdot \frac{\sqrt{n}}{\ln(n)}$ where the constant c, approximately 1.3727, is explicitly given by a certain infinite product, but, numerical verification aside, there is little progress on the problem. Here, then, is one of many examples of a set which we can happily describe in concrete mathematical terms ('the set of all primes of the form $n^2 + 1$') and yet have no idea whether it is finite or not. We can at least begin to list its elements $\{2, 5, 17, 37, \ldots\}$. The above conjecture is a particular case of the Bunyakovsky conjecture

(1857) which states that if f is an irreducible polynomial of degree at least two with integer coefficients, positive leading coefficient, and if the greatest common divisor of the set of integers $\{f(n) : n \in \mathbb{N}\}$ is d, then there are infinitely many primes of the form $\frac{f(n)}{d}$.

Some intuitive sets seem to have a less tangible existence than others, but it is difficult to draw a clear boundary between these different levels of concreteness. Naively we might expect to be able to include everything, to associate a set with any given property (precisely the set of all things having that property), but by a clever choice of property this leads to a contradiction. One of the tasks of set theory is to exclude these paradox-generating predicates and to describe the stable construction of new sets from old. Mathematics has a long history of creating concrete models of notions which at their birth were difficult abstract ideas.

1.1.8   A classical example

Even a cursory glance at the history of mathematics reveals one crisis of existence after another. The mathematicians of the sixteenth century, for example, famously chanced upon the enormous utility of the concept of $\sqrt{-1}$ but were much troubled by it. It is easy to resolve their difficulties from our comfortable modern perspective. Is there a real number with square equal to -1? No. Can the field of real numbers be embedded in a larger structure in which *the embedded image of* -1 has a square root? Yes. Does there exist such an extension with pleasant algebraic properties, i.e. is there a field extension of the real numbers which admits a square root of -1? Yes, and the smallest such extension is the field of complex numbers, which has a natural planar representation, the general theory of which is deeply geometric in nature.[1]

This construction is more than just a convenient flight of fancy – it turns out that the field of complex numbers, despite its initial appearance as a suspicious piece of man-made trickery, is in a sense far more physically relevant (more closely tied to reality?) than the field of real numbers. In fact the term 'real number' was invented *after* the discovery of its complex extension as a means of distinguishing between the two types of number. The terminology, in retrospect, is unfortunate. The concrete representation of $\sqrt{-1}$ either as a $\frac{\pi}{2}$-radian anticlockwise rotation of the plane about the origin or as a point in the plane neatly conceals its troubled history. The conceptual crisis faced by the sixteenth century mathematicians is clear: the other 'new numbers' of history: zero; negative numbers; irrational numbers (all of these will be formally

---

[1]We see more of this field, briefly, in Section 1.8 – and the precise meaning of 'field' will be revealed in due course.

introduced shortly) are at least interpretable as a magnitude of some sort, or as a directed length, whereas $\sqrt{-1}$ seemed, at first, to come from another realm entirely. The comparable modern difficulties with infinite sets might be said to have been similarly tamed by Formalist logic.

To return to our comments at the beginning of this section, carefully avoiding an infinite regress of definitions, the primitive elements of the theory to be described herein are the notions of set and membership. Logical axioms, rules of inference and some well-chosen set theoretic axioms then form the basis of a theory capable of describing a large part of mathematics in a unified fashion.

REMARK

The technique used to accommodate such new numbers as irrational numbers and complex numbers in the mathematical universe is now familiar to us, but we mustn't forget how challenging it must have been to deal with these new ideas at their infancy. Broadly speaking, if we want something from some structure K to satisfy a property ϕ, and if it turns out that no element of K has this property (or if we are unable to prove that such an element exists), then it is sometimes possible to define an extension of the structure, K^+, which is logically consistent and which is designed to allow objects with the desired property ϕ. One then identifies K with its embedding in K^+. We will see several examples of this throughout the book.

The so-called 'imaginary numbers' were successfully used long before they were properly defined as elements of a concrete field extension of the real numbers. The axiomatic description of a theory generally appears only in its mature stages, after many of its properties have been informally explored. Perhaps the longest duration between the usage and formalization of a notion is that of the natural numbers.

1.2   Natural numbers

> *Mathematics is the gate and key of the sciences... Neglect of mathematics works injury to all knowledge, since he who is ignorant of it cannot know the other sciences or the things of this world.*
>
> – ROGER BACON[1]

1.2.1   A familiar collection

Later we will present some primitive notions from which a model of the natural numbers $\{0, 1, 2, 3, \ldots\}$ can be constructed, but for now it will suffice to exhibit some of the basic properties of this familiar set as we intuitively understand

---

[1]Bacon [**7**].

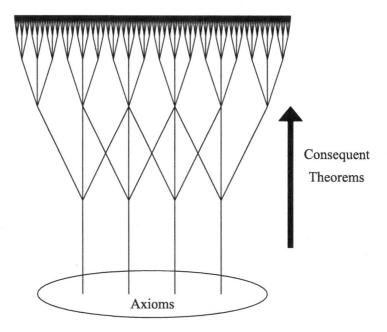

Figure 1.3 From a given base of axioms we might picture the class of all consequent theorems as an infinite outgrowth of statements (the illustration above is a highly stylized representation of this). If the system is consistent then for any statement ϕ it cannot be the
case that both ϕ and its negation are theorems of the system. It can happen, however, that
neither ϕ nor its negation is a theorem of the system (in Zermelo–Fraenkel set theory the
Axiom of Choice and the Continuum Hypothesis are two such undecidable statements). The
strict Platonist insists that in such cases, when the undecidable sentence concerns something
they regard as being set in Platonic stone (perhaps a statement about the arithmetic of the
natural numbers), there is a wider notion of truth which decides, non-arbitrarily, which of ϕ
or its negation are 'true', and that consequently the set of initial axioms has failed to fully
capture their reality. Formalists will readily admit that under any interpretation of set theory
one of ϕ or its negation must be true, and so there are true but not provable statements, but
they do not go as far as favouring one over the other – they are interested only in relative
truths and happily embrace alternative models.

it. It should be noted that, for various reasons, some mathematicians prefer to exclude zero from this set, but I will include it for reasons which will become apparent in the later discussion of ordinal numbers – in any case its inclusion or exclusion makes nothing more than a cosmetic difference to the constructions to come. If we were to exclude zero here, it would make an appearance, in a lightly disguised form, in the construction of the integers later. It is from the natural numbers that we build, using various algebraic constructions, models of the other familiar number systems of mathematics.

As might be expected of a structure which seems to be so fundamental, the natural numbers themselves are comparatively rather difficult to define. We rarely hear the question 'what is 3?', and if it is asked, it tends to be at a very early and impressionable age. Instead of receiving anything approaching a satisfactory answer, we are bombarded with pictures designed to give sufficiently many examples from which the general concept can be abstracted. Memories of this early brainwashing must have a tighter hold on us than we care to imagine; given an urgent prompt to think of several objects we invariably start to blurt out the names of fruit ('three apples') and similarly easy to comprehend items.[1]

This 'explanation by example' is a fairly harmless form of circularity; we are only able to determine whether there are 'three' objects in a given collection by comparison with other collections which, through well-meaning indoctrination, we already know to have 'three' elements. A typical blueprint for comparison in this case is the set comprising the vocalization, in whatever your native language may be, 'one, two, three'. If a set is seen to be equipollent to this set of utterances then we are happy to announce that it has 'three' elements. Our ability to abstract the concept of number with such unthinking ease from our physical environment is extraordinary; the acquisition of a 'number sense', coupled with language, appears to be a critical event in our intellectual evolution.

Thus 'threeness' might be captured as 'the' property shared by all sets, and only those sets, equipollent to the set $\{0, 1, 2\}$. The inverted commas around 'the' suggest some doubt about the uniqueness of such a property. If ϕ and ψ are properties of sets which are true only of sets which are equipollent to $\{0, 1, 2\}$, then within their set theoretic scope, ϕ and ψ are equivalent, i.e. a set satisfies ϕ if and only if it satisfies ψ. What is more important in the present construction is that 'three' is viewed as a unique class of sets, while its defining 'threeness' property can be expressed in many materially different (but logically equivalent) ways. In this scheme the number zero corresponds to the empty set $\emptyset$; the number 1 corresponds to the class of all single element sets; the number two corresponds to the class of all pairs, and so on. The identification of a natural number as a class of sets in this way is certainly a fruitful approach to the formalization of arithmetic but it is arguably rather far removed from any 'natural' assignment, and there is something a little unsatisfactory about it. In order to make the idea sound we must, of course, clarify exactly what the class of *all* sets is in the first place. We come to this mode of thinking in more detail

---

[1] On the other hand one must ask why these particular objects are chosen as canonical representations of natural numbers in the first place.

later.

The definition of the 'number' of a set as the set of all sets equipollent to that set is favoured by Bertrand Russell in his *The Principles of Mathematics*.[1]

> *Mathematically, a number is nothing but a class of similar classes: this definition allows the deduction of all the usual properties of numbers, whether finite or infinite, and is the only one (so far as I know) which is possible in terms of the fundamental concepts of general logic.*

One must bear in mind that this was published in 1903, in set theory's infancy.

Nothing could be more intuitively familiar in mathematics than the natural numbers $0, 1, 2, 3, \ldots$. *But how do we formally define them? Can we isolate some properties of the natural numbers that are sufficient to characterize them?*

1.2.2   Induction

If we try to extract the essential properties of the set of natural numbers based, at least, on our intuitive experience of it, we find ourselves drawn one way or another to the notion of induction. This is commonly introduced in simple problems such as the following.

Consider the infinite sequence of equations:

$$
\begin{aligned}
1 &= 1 \\
1 + 2 &= 3 \\
1 + 2 + 3 &= 6 \\
1 + 2 + 3 + 4 &= 10 \\
&\vdots
\end{aligned}
$$

The task is to find a formula for the general sum of the form $1 + 2 + 3 + \cdots + n$ and to prove that the conjectured formula is correct for all n. Let P_1 be the statement '$1 = 1$', P_2 the statement '$1 + 2 = 3$', P_3 the statement '$1 + 2 + 3 = 6$', and so on. We suspect, after a little thought or experimentation, that P_n, the statement '$1 + 2 + 3 \cdots + n = \frac{1}{2}n(n+1)$', holds for all n in $\{1, 2, 3, 4, 5, \ldots\}$.[2]

---

[1] See Russell [**181**], Chapter XI, Section 111.

[2] The trick is to notice that the sum of the first and last terms is equal to the sum of the second and penultimate terms which is equal to the sum of the third and antepenultimate terms, and so on.

We see that P_1 is (trivially) true. Let us assume that P_n is true, then a little algebraic manipulation:

$$
\begin{aligned}
\text{if } 1 + 2 + \cdots + n &= \frac{1}{2}n(n+1), \text{ then} \\
1 + 2 + \cdots n + (n+1) &= \frac{1}{2}n(n+1) + (n+1) \\
&= (n+1)\left[\frac{1}{2}n + 1\right] \\
&= \frac{1}{2}(n+1)(n+2)
\end{aligned}
$$

reveals that P_{n+1} is true. So, since P_1 is true, so is P_2, and hence P_3 is true, and surfing this infinite cascade of implications we conclude that, for any given $k = 1, 2, 3, \ldots$, P_k is true.

Related to this problem are the following exercises in induction:

$$
\begin{aligned}
1^2 + 2^2 + \cdots + n^2 &= \frac{1}{6}n(n+1)(2n+1) \\
1^3 + 2^3 + \cdots + n^3 &= \frac{1}{4}n^2(n+1)^2 \\
1^4 + 2^4 + \cdots + n^4 &= \frac{1}{30}n(n+1)(2n+1)(3n^2+3n-1).
\end{aligned}
$$

Notice that $1^3 + 2^3 + \cdots + n^3 = (1 + 2 + \cdots + n)^2$. For $k \geq 1$, $1^k + 2^k + \cdots + n^k$ is a polynomial of degree $k+1$ in n, the coefficients of which are well understood.

The method of proof we are using here is the principle of (finite) induction. Generally one wishes to prove an infinite sequence of propositions $P_0, P_1, P_2, \ldots$ (the starting point is of no importance – in the above example we happened to start at P_1). We prove that the first proposition, P_0, is true and we prove that, in general, P_n implies P_{n+1}. The principle of induction tells us that we may conclude from these two assertions that all P_i from P_0 onward are true.

There are several equivalent formulations of induction. For example the so-called *strong induction principle* (also called *complete induction*) is as follows: If P_n is true whenever P_i is true for all $i < n$, then P_n is true for all n.

Another version of induction is the *least number principle*: Every non-empty subset of natural numbers has a least member (where we are taking 'least' to mean least in the sense of the usual ordering $0 < 1 < 2 < \cdots$). This is simply the statement that $<$ is a well-ordering (see later). Its equivalence to induction is proved as follows. Suppose a subset A of the natural numbers has no minimal element. Let P_n be the property 'for all $m < n$, m is not in A'. P_0 is vacuously true (there are no natural numbers less than zero). If P_n is true then P_{n+1} must also be true, for otherwise n would be a minimal element of A. By induction P_n is true for all n and we conclude that A is empty. We have proved that if a subset of the natural numbers has no minimal element, then it is empty.

The contrapositive is precisely the least number principle. To prove that the least number principle implies induction suppose P_n is a property of the natural number n such that P_0 is true and P_n implies P_{n+1} for all n. Let A be the set of all n for which P_n is false. If A is non-empty it possesses, by the least number principle, a minimal element m (and since P_0 is true, we have $m > 0$). But then P_{m-1} is true and P_m is false, contradicting the inductive step assumption. Hence A is empty, i.e. P_n is true for all n.

A version of induction was alluded to, although not explicitly stated, in 1654 by Pascal in the proof of one of the results in his *Traité du Triangle Arithmetique*.[1] Francesco Maurolico used induction to prove that the sum of the first n odd numbers is equal to n^2 in his *Arithmeticorum libri duo* (1575). There are still older examples where some form of induction is clearly being used but no explicit statement of the principle is outlined, most notably in the work of Abu Bakr ibn Muhammad ibn al Husayn al-Karaji in the tenth century and Ibn Yahya al-Maghribi al-Samaw'al and the Indian mathematician Bhaskara, both working in the twelfth century. These early inductions were often employed to prove theorems concerning binomial coefficients.

There is no evidence that the ancient Greeks knew of induction in its modern form, but something akin to the least number principle appears in Euclid's *Elements*, for example. Euclid uses another form of the same principle: there does not exist an infinite strictly decreasing sequence of natural numbers.[2] Proposition 31 of Book VII proves that every composite number has a prime divisor '*Any composite number is measured by some prime number*' and it is in this proof that Euclid uses his version of induction. In Heath's translation the proof is as follows (the principle in question is in bold):

> *Let A be a composite number; I say that A is measured by some prime number. For, since A is composite, some number will measure it. Let a number measure it, and let it be B. Now, if B is prime, what was enjoined will have been done. But if it is composite, some number will measure it. Let a composite number measure it, and let it be C. The, since C measures B, and B measures A, therefore C also measures A. And, if C is prime, what was enjoined will have been done. But if it is composite, some number will measure it. Thus, if the investigation be continued in this way, some prime number will be found which will measure the number before it, which will also measure A. For, if it is not found,* **an infinite series of numbers will measure the number A, each of which is less than the other: which is impossible in numbers.** *Therefore some prime number will be found which will measure the one before it, which will also measure A. Therefore any composite number is measured by some prime number.*

---

[1] An English translation can be found in Smith [**197**].
[2] We will use this in a proof that $\sqrt{2}$ is irrational in Subsection 1.6.1.

REMARKS

1. The sequence of triangular numbers $1, 3, 6, 10, \ldots$ is a fairly innocent example, but like all simple examples one can use it as a creative launch pad, ask some novel questions, and before long formulate some non-trivial conjectures. Such investigations famously led Carl Friedrich Gauss to write in his diary in 1796 'EYPHKA! num$= \triangle + \triangle + \triangle$'. This was a reference to his discovery that every natural number is expressible as the sum of at most three triangular numbers. Earlier, in 1770, Joseph Louis Lagrange had proved the analogous and much more well-known result that every natural number is expressible as the sum of at most four squares. The natural generalization, that every natural number is expressible as the sum of at most n n-gonal numbers, had been one of Pierre de Fermat's unsolved conjectures. (Informally, k is an n-gonal number if one can uniformly arrange k dots to form a filled regular polygon with n sides. Explicitly the mth n-gonal number is $\frac{1}{2}(m^2(n-2) - m(n-4))$.) Augustin-Louis Cauchy finally proved the general case in 1813.

2. $1^k + 2^k + \cdots + n^k$ is a polynomial of degree $k+1$ in n where the coefficient of n^{k+1-m} is $\frac{(-1)^m}{k+1} \binom{k+1}{m} B_m$. Here B_m is the mth Bernoulli number. B_m can be defined in many ways. Perhaps the shortest description is given by the series expansion of $\frac{x}{e^x - 1}$:

$$\frac{x}{e^x - 1} = \sum_{m=0}^{\infty} B_m \frac{x^m}{m!}.$$

Be warned that there are two definitions of Bernoulli numbers; these differ only in the sign of B_1. Using the definition given above we have $B_1 = -\frac{1}{2}$. If we had adopted $B_1 = \frac{1}{2}$ instead, then, since $B_m = 0$ for odd $m \geq 3$, there would be no need for the $(-1)^m$ in the above expression, but an alternative definition of the sequence (B_m) would be required. The sequence of Bernoulli numbers starting with B_0 begins: $1, -\frac{1}{2}, \frac{1}{6}, 0, -\frac{1}{30}, 0, \frac{1}{42}, 0, -\frac{1}{30}, 0, \ldots$ Note in particular that the sum of the coefficients a_i of the polynomial expression $a_{k+1}n^{k+1} + a_k n^k + \cdots + a_1 n$ for $1^k + 2^k + \cdots + n^k$ is always equal to 1 (simply set $n = 1$). (See a good elementary number theory text for full details.)

3. The observation that $1^3 + 2^3 + \cdots + n^3 = (1 + 2 + \cdots + n)^2$ can be found in several ancient texts and is sometimes named after the Neopythagorean mathematician Nichomachus (c. 100AD). Various geometric interpretations of the result exist.

4. We should observe that the fact that the sum of the first n odd numbers is equal to n^2 is geometrically evident: consider an $n \times n$ square of dots and picture it as a dot together with a set of ever larger adjoined L-shaped configurations of dots, so at the top right, say, we find a single dot, this is surrounded below and to the left by a configuration of three dots which in turn is surrounded below and to the left by a configuration of five dots, and so on.

5. Induction is merely a method of proving a conjectured infinite sequence of propositions. It may be that each individual proposition in the sequence is intuitively 'obvious' (then we are halfway to proving all of them, only a little formalization and a proof of the inductive step remains). Alternatively the conjecture may have come as a complete surprise, in which case the induction may be able to shed light on why it is true, or provide some alternative perspective.

6. Euclid's treatment of infinite sets is very subtle and often beautifully clever. His most famous proof, which many mathematicians cite as a personal favourite, is the proof that there are infinitely many primes (Book IX, Proposition 20 of *The Elements*). This makes use of the result cited earlier that every composite number has a prime divisor. Since the original proof is considerably more elegant than some of the misremembered versions which can be found in the modern literature it is worth revisiting.

Give me any collection of primes, says Euclid, then I can find a prime that is not in the collection. The proof is wonderfully simple. Suppose we have n primes $p_1, p_2, \ldots, p_n$. Consider the number $N = p_1 p_2 \ldots p_n + 1$. Either N is prime (in which case we are done, since N is clearly greater than all of the original primes) or it is composite, in which case it has a prime divisor, but since it is visibly the case that none of the original primes $p_1, p_2, \ldots, p_n$ divide N such a prime divisor cannot be in the original collection.

Modern accounts often tend to turn this into a slightly cumbersome proof by contradiction: assume there are finitely many primes, take the product of all primes and add 1, observe that this new number is not divisible by any of the assumed complete list of primes, conclude that it is therefore prime, a contradiction.

Euclid's proof suggests a natural algorithm for generating primes: from an initial finite sequence of primes define the next term in the sequence to be, say, the smallest prime divisor of one plus the product of all the previous terms in the sequence (alternatively we could take the largest prime divisor). So starting with 2 the next term is simply $2 + 1 = 3$. Next we calculate $2 \cdot 3 + 1 = 7$, which happens to be prime. Next we have $2 \cdot 3 \cdot 7 + 1 = 43$, which again is prime. The next term in the sequence is the smallest prime divisor of the composite number $2 \cdot 3 \cdot 7 \cdot 43 + 1 = 1807$, namely 13, and so on. The following six terms are: 5, 6 221 671, 38 709 183 810 571, 139, 2801, 11. This sequence is known as the *Euclid–Mullin sequence*. Not much is known about its behaviour. In particular it is not known if every prime eventually appears. Of course there is nothing to stop us from beginning with a different prime, or indeed an arbitrary set of primes. For instance, if we start with 5 then the sequence begins 5, 2, 11, 3, 331, 19, 199, 53, 21 888 927 391, 29 833, 101, 71, 23, 311, 7,..., and if we start with 11 then the sequence begins 11, 2, 23, 3, 7, 10 627, 433, 17, 13, 10 805 892 983 887, 73,...

The principle of mathematical induction took a few centuries to be formally stated, but the basic technique has been understood in some form since antiquity. The principle of induction is one of the ingredients which characterizes the natural numbers.

1.2.3   Peano's Postulates

The set of natural numbers, as we like to conceive of it, satisfies a small number of basic properties, including an induction principle, which are strong enough to completely characterize it. These postulates are named after Giuseppe Peano who published his axiomatic treatment in 1889.[1] Although Peano credited these axioms to Dedekind (Peano's approach is a 'self sufficient' axiomatization of Dedekind's earlier set theoretic construction[2]), the name 'Peano's Postulates' has now become widespread.

We postulate the existence of a collection N of objects and an object called the **first** together with a primitive notion 'successor' (a function) satisfying the following.

  (i) The **first** is in N.

 (ii) For all $i \in N$ the successor of i is in N.

(iii) For all $i \in N$ the successor of i is not the **first**.

(iv) For all $i, j \in N$ the successor of i and the successor of j are equal if and only if $i = j$.

 (v) (Induction) If a collection of objects A includes the **first** and the successor of each element of A then $N \subseteq A$.

These postulates uniquely characterize the natural numbers in the following sense. If (N, f, s) and (N', f', s') satisfy the postulates, where N, N' are the underlying sets, f, f' are the first elements, and s, s' are the successor functions, then one can find a bijection $\phi : N \to N'$ such that $\phi(f) = f'$ and $\phi(s(i)) = s'(\phi(i))$, i.e. ϕ maps the successor of i to the successor of $\phi(i)$. In other words, the two systems are structurally identical, differing only in the arbitrary labels assigned to their elements.

If we are presented with a collection of objects and a function which, together, we suspect form a copy of the natural numbers and associated successor then to prove that we are indeed dealing with the natural numbers it is simply a matter of demonstrating that the said structure satisfies Peano's Postulates.

---

[1] An English translation *The principles of arithmetic, presented by a new method* can be found in van Heijenoort [**219**].

[2] This appears in Dedekind [**43**].

When we come to it, this may be of considerable philosophical satisfaction to readers who might be suspicious of the principle of induction: starting with natural but more primitive notions we can define a class N (rather than simply postulate its existence) and within this framework induction, and all of the other Peano Postulates, become *theorems*.

The correspondence ϕ between (N, f, s) and (N', f', s') just described is an example of an *isomorphism*. Informally an isomorphism is a 'structure preserving bijection'. A more explicit definition, in particular cases, will be given later as the need arises. The bijection ϕ preserves the successor function and hence all of the arithmetic operations (to be formally defined shortly), that is, we have $\phi(m + n) = \phi(m) + \phi(n)$ and $\phi(mn) = \phi(m)\phi(n)$ for all $n, m \in N$.

Mathematicians usually regard isomorphic structures as identical. The two structures in question may well lie in different parts of mathematics, but such is the beauty of the subject: mathematics is often described as the art and science of finding analogies between analogies.[1] We say that, up to isomorphism, the natural numbers are unique, that is, any two structures satisfying the Peano Postulates are isomorphic.

We give a very brief account of the formal (first-order) theory now known as Peano Arithmetic (PA) in Appendix A. There the distinction between second-order PA, which is essentially what we are discussing here, and first-order PA is underlined. The uniqueness, up to isomorphism, of models of the natural numbers is a fact of second-order PA but *not* of first-order PA. The general notion of a first-order theory and how it differs from a second-order theory is an important one, and we shall come to it in more detail later.

The method of definition just described: to list some postulates, or axioms, to show that, up to isomorphism, there can exist at most one system satisfying the given properties, and then to provide a concrete model of such a system, is very powerful and far-reaching, and one can find it in use throughout modern mathematics. We have given a set of postulates (Peano's Postulates), and we know that any two systems satisfying the postulates are isomorphic, so all that remains is to provide a concrete example of a set (together with a nominated first element and a successor function) which satisfies them. In the absence of such a set the consequent theory could be vacuous; one can happily prove very impressive theorems of the form 'If X satisfies ϕ then X satisfies ψ' but all the hard work may be in vain if it turns out that no X satisfies ϕ. The only model of the natural numbers we have at this stage is the intuitive one. A concrete model constructed from the empty set using a set theoretic successor function will be given later.

---

[1] *'A mathematician is a person who can find analogies between theorems; a better mathematician is one who can see analogies between proofs; and the best mathematician can notice analogies between theories. One can imagine that the ultimate mathematician is one who can see analogies between analogies.'* – attributed to Stefan Banach.

REMARK

Isomorphisms are a mathematical realization of a *perfect analogy* or, if you prefer, a perfect translation between two languages. Indeed, the analogy is so perfect that it often becomes redundant to regard two isomorphic structures as distinct. A kind of mathematical paradise is described by an isomorphism between two structures, especially when it can be used to translate difficult problems in one subject into tractable problems in another. A concrete and very practical example of this (one of many) is the translation of difficult differential equations into algebraic equations via the Laplace transform.

The Peano Postulates are five axioms which describe the properties of the natural numbers as a set together with the notion of a successor. Any two sets satisfying these axioms are isomorphic, i.e. they are identical except for the labels we attach to their elements. Using the successor operator we can recursively define the arithmetic operations on the natural numbers.

1.2.4   Arithmetic and algebraic properties

Let us suppose that we have a set $\mathbb{N}$ with successor function s and first element 0 satisfying the Peano Postulates. We define addition recursively by:

$$\begin{aligned} n + 0 &= n \\ n + s(m) &= s(n + m). \end{aligned}$$

The elements of $\mathbb{N}$ are (omitting parentheses for clarity) $0, s0, ss0, sss0, \ldots$. For example, we evaluate the sum $sss0 + ss0$ by applying the recursion equations repeatedly as follows:

$$\begin{aligned} sss0 + ss0 &= s(sss0 + s0) \\ &= ss(sss0 + 0) \\ &= sssss0. \end{aligned}$$

One can prove the following properties (here we begin a list of ten algebraic properties which will run through to Section 1.5):

(1) *Commutativity of addition*: For all $m, n \in \mathbb{N}$, $n + m = m + n$.
(2) *Associativity of addition*: For all $m, n, k \in \mathbb{N}$, $(n + m) + k = n + (m + k)$.

No rules of inference or logical assumptions have yet been stated, so by 'prove' we mean the usual informal but 'formalizable' sense of proof used in most mathematical expositions. The formal proof of the commutativity of addition within the framework of Peano Arithmetic is outlined in Appendix A.

Multiplication is defined recursively by:

$$n \cdot 0 = 0$$
$$n \cdot s(m) = n \cdot m + n.$$

For example (again omitting parentheses for clarity) the product of $sss0$ and $ss0$ is evaluated as follows (the formal justification for the last of these four equalities is spelled out in Appendix A):

$$sss0 \cdot ss0 = sss0 \cdot s0 + sss0$$
$$= sss0 \cdot 0 + sss0 + sss0$$
$$= 0 + sss0 + sss0$$
$$= sss0 + sss0$$

and by the addition recursion equations we obtain the result $ssssss0$.

Multiplication is also commutative and associative in the following sense:

(3) *Commutativity of multiplication*: For all $m, n \in \mathbb{N}$, $m \cdot n = n \cdot m$.
(4) *Associativity of multiplication*: For all $m, n, k \in \mathbb{N}$, $(n \cdot m) \cdot k = n \cdot (m \cdot k)$.

One can also prove the following property, which incorporates both addition and multiplication:

(5) *Distributivity*: For all $m, n, k \in \mathbb{N}$, $n \cdot (m + k) = n \cdot m + n \cdot k$.

Present in $\mathbb{N}$ is the additive identity 0:

(6) *Additive identity*: There exists an element in $\mathbb{N}$, namely 0, with $n + 0 = n$ for all $n \in \mathbb{N}$,

and the multiplicative identity $s(0)$ (usually denoted 1) (we have $n \cdot 1 = n \cdot s(0) = n \cdot 0 + n = 0 + n = n$):

(7) *Multiplicative identity*: There exists an element in $\mathbb{N}$, namely 1, with $1 \cdot n = n$ for all $n \in \mathbb{N}$.

The additive and multiplicative identities are unique: if z is an additive identity, then $z = z + 0 = 0$ and similarly if z is a multiplicative identity then $z = z \cdot 1 = 1$. We also have the following property:

(8) *No zero divisors*: If $n \cdot m = 0$ then $n = 0$ or $m = 0$.

Finally we define exponentiation by:

$$n^0 = 1$$
$$n^{s(m)} = n^m \cdot n$$

and it is an easy but rather tedious task to prove that the following properties of powers hold:

$$n^a \cdot n^b \;=\; n^{a+b}$$
$$(n^a)^b \;=\; n^{a \cdot b}.$$

In a rare example of philosophical unity there is almost total agreement that Peano's Postulates are the 'right' ones; that they characterize exactly the properties of a set of objects that had been studied, in the absence of any axiomatic development, for thousands of years. The natural numbers, in this introduction at least, form our starting point: we build on and dissect this structure with increasing difficulty and abstraction.

It must be added that most mathematical activity does not involve this kind of meticulous logical dissection and axiomatization of familiar Platonic notions. Mathematicians are generally content with the soundness of the logical concepts they work with and set about exploring and building structures in a fairly intuitive way while we might imagine logicians scrutinizing the logical machinery behind the scenes. But this telescopic-microscopic cartoon depiction of mathematical activity is a gross oversimplification which wildly misrepresents both modes of investigation; in the naked reality of research the distinction between the work of logicians and mathematicians is not clear cut, indeed it is as ill-defined as the boundary between any two branches of mathematics. There are many misconceptions about the nature of research in logic.[1]

REMARK

The elementary properties listed in this section are all so familiar as to be regarded as obvious, and many can be 'seen at a glance' just from geometric considerations. For example, commutativity of multiplication can be viewed in terms of an $m \times n$ grid. Clearly this has the same number of squares as an $n \times m$ grid (simply rotate it), hence $mn = nm$. But to make this observation is to entirely miss the point; obvious though these properties may be, they follow from more primitive assumptions. Other algebraic theories in mathematics use such arithmetic and algebraic rules as axioms, but here we have the luxury of a more primitive pre-structure from which the rules follow. It is a general scientific principle that we should aim to describe a phenomenon using as few hypotheses as possible. As well as asking what few principles might generate an observed set of laws, we can also consider the reverse and ask what are the consequences of a fixed set of assumptions (to study the outgrowth of consequences as a microscopist might observe the growth of an organism).

---

[1] J. Donald Monk remarks in the introduction to his *Mathematical Logic* [149] that the novice first approaching a text on mathematical logic, expecting discussions in the philosophy of mathematics, will often be surprised at the substantial mathematical background assumed of the reader.

> *The natural numbers and the two operations of addition and multiplication sat-*
> *isfy a number of properties that we shall see are common to a wide range of*
> *number systems. These new number systems are built from the natural numbers*
> *using various set theoretic constructions. It is only in the last century or two*
> *that the intuitive system of natural numbers has been given a sure logical footing.*

1.2.5   The evolution of number

Prehistoric notched bones and cave paintings give evidence of the original practical use of natural numbers. We can guess that they recorded such things as
the passing of days, lunar cycles, or perhaps hunting success. They may even
have been used as counting tools. One of the most famous examples, dated as
being about 20 000 years old, is the Ishango bone, a baboon fibula adorned with
numerous tally marks discovered in what is now the Uganda–Congo border in
1960. Older notched bones and musical instruments (for example a flute made
from a vulture bone) have been found in Europe and Africa which have been
dated in excess of 30 000 years. Cave paintings including geometric patterns are
more than twice as old; the cave paintings were as ancient to the bone carvers
as the bones are to us.

The concept of number has evolved over history from its humble beginnings in the form of natural numbers, through integers, rational numbers, real
numbers and complex numbers to Cantor's species of infinite number (ordinals
and cardinals), and beyond. Major shifts in the concept of number are not
the exclusive preserve of modern mathematics. The ancient Greek mathematician Eudoxus' theory of proportion, described in Book V of Euclid's *Elements*,
presents us with a rigorous geometric treatment of irrational quantities which
is a clear prelude to the development of the real numbers from the rational
numbers as given by Dedekind (to be presented in Subsection 1.6.3).

The precise characterization of the integers, rational numbers and real numbers was a core concern for the pioneers of the foundations of mathematics. This
work was partly motivated by an attempt to find a satisfactory answer to the
question 'What is a number?'. However, the development of set theory was not
directly inspired by this question, instead it came from more practical considerations, including Cantor's study of trigonometric series (see Subsection 1.11.7)
and related topics of analysis that were in vogue around 1870. Cantor showed
thin interest in using set theory to model the natural numbers, a problem which
was first taken up by Frege and Dedekind.

Reacting to these new developments, Leopold Kronecker famously proclaim-
ed *'God created the integers. All else is the work of man.'*[1] This was not intended

---

[1] Kronecker apparently said this in a lecture in Berlin in 1886. It appears in print as '*Die ganzen Zahlen hat der liebe Gott gemacht, alles andere ist Menschenwerk*' on page 19 of H. Weber's memorial article 'Leopold Kronecker', *Jahresberichte der Deutschen Mathematiker Vereinigung*, vol ii, 5–31, (1893).

to be a theological announcement, but neatly underlined his philosophical leanings. The idea of natural numbers as primitive may have been usurped by set theory, but the question of which notions we take as primitive will always cause controversy in some quarters. Assuming we have created the natural numbers, by whatever means, we can go on to construct ever more sophisticated types of number, but in order to do this we need some further machinery.

REMARKS

1. It is very difficult to think of numbers in the same way as our ancestors, or, rather, it is difficult to intuit from such scant evidence how abstract their notion of number was. Even in relatively recent times our understanding of certain classes of number has changed, so to make a judgement on Paleolithic numeracy is a fairly hopeless task. Representing the passing of time by a collection of carved notches is already a huge conceptual leap that must not be underestimated. We can perhaps infer that our ancestors had already come to appreciate that the concept of quantity described by these notches is a universal one, applicable not only to numbers of days passed but also to numbers of animals hunted, numbers of mountains on the horizon, numbers of stars in a constellation and so on, so that these diverse collections share an important abstract feature. (Archaeological findings increasingly show that prehistoric man was far more sophisticated than popular gossip would have us believe.)

2. The Greeks did not develop a theory of rational numbers as an extension of natural numbers in the way that is familiar to us now. They did much work on 'ratios', but the operations on ratios described in Euclid's *Elements* are somewhat convoluted. A ratio was regarded as a type of relationship between two wholes and not, as we are accustomed to, a number within a linear collection in which we may embed the natural numbers.

3. Leopold Kronecker was one of the earliest proponents of what one might call constructivism. Indeed he felt so strongly about this approach to mathematics that he tried to prevent Cantor's work from being published in the prestigious *Crelle's Journal* and as a result of his conflict with Cantor and his rejection of some of the beautiful (non-constructive) work in analysis at the time he created a further rift between himself and his former friend Karl Weierstrass. This late nineteenth century divide between the sort of conceptual reasoning used by Cantor and the explicit symbolic computational approach of Kronecker helped to fuel Hilbert's programme, which might be viewed as an attempt to reconcile these two aspects of mathematics.

Natural numbers are taken for granted. Indeed they are such familiar objects that for most of human history no one thought it necessary to try to define them in terms of more primitive logical notions. Once the natural numbers are defined, models of the remaining classical number systems can be constructed with relative ease.

1.3   Equivalence classes and order

Those who assert that the mathematical sciences say nothing of the beautiful or the good are in error. For these sciences say and prove a great deal about them; if they do not expressly mention them, but prove attributes which are their results or definitions, it is not true that they tell us nothing about them. The chief forms of beauty are order and symmetry and definiteness, which the mathematical sciences demonstrate in a special degree.

– ARISTOTLE[1]

1.3.1   Relations

The *cartesian product* of two sets S and T, denoted $S \times T$, is the set of all ordered pairs (a, b) with $a \in S$, $b \in T$. A *relation* on a set S is simply a subset R of $S \times S$. An element $a \in S$ is said to be *R-related* to an element $b \in S$ if $(a, b) \in R$, and in such circumstances we write aRb.

This is an easy definition to make, but its motivation may be lost in abstraction. Prototypical examples of relations are the orders $<, \leq, \geq, >$ on $\mathbb{N}$ and membership and subset relations $\in, \subset, \subseteq$ defined on classes of sets.

We are asked, therefore, to think of the relation $\leq$ on $\mathbb{N}$ as a *set*,

$$L = \{(n, m) : n, m \in \mathbb{N} \text{ and there exists an } x \in \mathbb{N} \text{ such that } m = n + x\}$$

and $n \leq m$ (which we could, by the above convention, write as nLm) is an abbreviation for $(n, m) \in L$ ($n < m$ means $n \leq m$ and $n \neq m$). The x appearing in the condition determining the set L is, if it exists, uniquely determined by the ordered pair (n, m) and will henceforth be denoted $m - n$.

In this scheme, the relation of equality on $\mathbb{N}$ is simply the set

$$\{(n, n) : n \in \mathbb{N}\}.$$

Using the example of equality as a prototype we say that a relation R on a set S is an *equivalence relation* if it has the following three properties:

Reflexive: for all $x \in S$, xRx;
Symmetric: for all $x, y \in S$, if xRy then yRx; and
Transitive: for all $x, y, z \in S$, if xRy and yRz then xRz.

---

[1] *Metaphysics* (XIII) [**3, 6**].

Equivalence relations appear throughout mathematics. As we shall see, the formal constructions of number systems beyond $\mathbb{N}$ rely critically on the notion of equivalence (and on order).

It is very unlikely that anyone, regardless of the extent of their logical or mathematical training, would be compelled to think of the relations of common parlance in these formal terms (as some sort of static globally viewed aggregate of pairs). Instead we tend to think of a relation simply as a criterion. For instance it would be eccentric to regard the relation 'is the son of' on the world's population as a set of pairs of people (incidentally, this relation fails all three conditions for equivalence – it is not reflexive, symmetric or transitive). The point is, in set theory *every* concept must be encoded as a class of objects, even if its intuitive character is far from set-like.

The *union* of two sets A, B, denoted $A \cup B$, is the collection of all elements in A or in B (possibly in both). The *intersection* of two sets, denoted $A \cap B$, is the set of all elements in both A and B. If A and B have no elements in common, i.e. if $A \cap B = \emptyset$, then we say that A and B are *disjoint*. A *partition* of a set X is a collection $\mathbf{P}$ of subsets of X with union X having the property that any pair of distinct sets in $\mathbf{P}$ are disjoint (so intuitively a partition of X is a slicing of X into non-overlapping pieces).

Finding an equivalence relation R on a set S is equivalent to specifying a partition of S: given an equivalence relation R on S, S is naturally partitioned by R into so-called *equivalence classes* $\{\langle x \rangle : x \in S\}$, where $\langle x \rangle = \{y : yRx\}$ (so aRb if and only if $\langle a \rangle = \langle b \rangle$). Conversely any partition of S determines an equivalence relation R on S by declaring xRy if and only if x and y belong to the same set in the partition.

REMARK

Equipollence is an equivalence relation on the class of all sets. That is, we say two sets A and B are related if there exists a bijection from A to B (that this is indeed an equivalence relation is an easy exercise). The equivalence classes induced by equipollence are then precisely the classical cardinal numbers. If we restrict ourselves to Dedekind finite sets then the collection of equivalence classes forms a model of the natural numbers (once the arithmetic operations are appropriately defined).

Although we tend to think of relations as criteria to be satisfied between two objects, in a set theoretic foundation everything must be describable as a set, or at least as a class, so a relation is simply a collection of ordered pairs (x, y). We say x is related to y precisely when (x, y) is in this collection. Equivalence relations will be used to construct most of the number systems to come.

1.3.2   An example

The equivalence relation of equality on $\mathbb{N}$ does not make for a very interesting example: the equivalence classes of $(\mathbb{N}, =)$ are all single element sets of the form $\{(n, n)\}$.

A more interesting example on $\mathbb{N}$ is obtained by declaring n to be related to m if the difference between n and m is a natural multiple of, say, 5. (By 'difference' we mean $m - n$ if $n \leq m$ or $n - m$ if $m < n$.) It is easy to prove that this relation is indeed an equivalence relation. It should be noted that it is somewhat more convenient, and more common, to introduce this example in the context of integers, but we are handicapped by the fact that we haven't yet 'invented' the set of integers! This initial restricted environment, forcing us to reinvent intuitive ideas from scratch, is an unavoidable feature of the early development of any foundational theory.

The equivalence relation just defined partitions the set $\mathbb{N}$ into five equivalence classes which we shall denote by $\bar{0}, \bar{1}, \bar{2}, \bar{3}$ and $\bar{4}$:

$$\bar{0} = \{0, 5, 10, 15, 20, \ldots\};$$
$$\bar{1} = \{1, 6, 11, 16, 21, \ldots\};$$
$$\bar{2} = \{2, 7, 12, 17, 22, \ldots\};$$
$$\bar{3} = \{3, 8, 13, 18, 23, \ldots\};$$
$$\bar{4} = \{4, 9, 14, 19, 24, \ldots\}.$$

Suppose we wish to define, in a consistent fashion, addition and multiplication on the finite set $\{\bar{0}, \bar{1}, \bar{2}, \bar{3}, \bar{4}\}$ of equivalence classes. The most natural way to define $\bar{x} + \bar{y}$ is to select one element of $\bar{x}$, n say, and one element of $\bar{y}$, m say, and define $\bar{x} + \bar{y}$ to be the equivalence class containing $n + m$. There is a worry that the result might depend on the particular choices of n and m, in which case the definition would fall apart, however it is easy to show that this sum $\bar{x} + \bar{y}$ is independent of the choices of n and m, so the operation is well-defined. Multiplication is defined in a similar fashion: $\bar{x}\bar{y}$ is the equivalence class containing nm where n and m are arbitrary elements of $\bar{x}$ and $\bar{y}$ respectively.

As we have a small finite number of elements we can present the operations of addition and multiplication by exhibiting all possible sums and products in tabulated form. This, as we knew all along, is a formalized version of the 'clock arithmetic' that defines the sum or product of two numbers to be the remainder of that value upon division by five.

+	$\bar{0}$	$\bar{1}$	$\bar{2}$	$\bar{3}$	$\bar{4}$
$\bar{0}$	$\bar{0}$	$\bar{1}$	$\bar{2}$	$\bar{3}$	$\bar{4}$
$\bar{1}$	$\bar{1}$	$\bar{2}$	$\bar{3}$	$\bar{4}$	$\bar{0}$
$\bar{2}$	$\bar{2}$	$\bar{3}$	$\bar{4}$	$\bar{0}$	$\bar{1}$
$\bar{3}$	$\bar{3}$	$\bar{4}$	$\bar{0}$	$\bar{1}$	$\bar{2}$
$\bar{4}$	$\bar{4}$	$\bar{0}$	$\bar{1}$	$\bar{2}$	$\bar{3}$

$\cdot$	$\bar{0}$	$\bar{1}$	$\bar{2}$	$\bar{3}$	$\bar{4}$
$\bar{0}$	$\bar{0}$	$\bar{0}$	$\bar{0}$	$\bar{0}$	$\bar{0}$
$\bar{1}$	$\bar{0}$	$\bar{1}$	$\bar{2}$	$\bar{3}$	$\bar{4}$
$\bar{2}$	$\bar{0}$	$\bar{2}$	$\bar{4}$	$\bar{1}$	$\bar{3}$
$\bar{3}$	$\bar{0}$	$\bar{3}$	$\bar{1}$	$\bar{4}$	$\bar{2}$
$\bar{4}$	$\bar{0}$	$\bar{4}$	$\bar{3}$	$\bar{2}$	$\bar{1}$

Such a luxurious all-encompassing presentation is not always available. For example, if the collection of equivalence classes is a large finite, or infinite,

set it will either be impractical or impossible to define operations exhaustively
by exhibiting all possible cases. In such circumstances one must ensure first
that the proposed operation (defined on sets of equivalence classes in terms of
arbitrary elements thereof) is well-defined in the sense outlined above, that is,
the results of the operation must be independent of the choice of representing
elements.

The algebraic structure we have just described, denoted $\mathbb{Z}_5$, is an example of
a (finite commutative) ring. Replacing 5 by an arbitrary natural number $n \geq 2$,
imitating the construction given above, we generate the ring $\mathbb{Z}_n$ of n elements.

Remark

The ring $\mathbb{Z}_n$ is a field (see Section 1.5) if and only if n is prime. More generally
$\bar{a}(\neq \bar{0})$ in $\mathbb{Z}_n$ has a multiplicative inverse if and only if the greatest common
divisor of a and n is 1. We should also note that there are many more finite
fields than those of the form $\mathbb{Z}_p$ (and more generally that the class of finite
rings is far richer than these simple examples suggest); for each prime p and
each natural $n \geq 1$ there is a field of order p^n, unique up to isomorphism, and
conversely every finite field has order p^n for some prime p and $n \geq 1$.

---

*By defining simple equivalence relations on the natural numbers we can gener-
ate further algebraic structures. We shall see later that by defining a certain
equivalence relation on the cartesian product $\mathbb{N} \times \mathbb{N}$ we will be able to construct
the integers. The set of natural numbers also provides another useful prototype,
that of a well-ordered set.*

1.3.3   Order

The notion of order, coloured by various levels of strictness, permeates mathematics. Perhaps the simplest highly structured order is $<$ on $\mathbb{N}$. Salient features
of this order may be abstracted and lead us toward a precise theory of the infinite (in the guise of ordinal numbers). At the looser end of the order spectrum
we find such orders as $\subseteq$ on, say, the set of all subsets of $\mathbb{N}$. In the latter example it is not even the case that all pairs of elements are comparable: neither of
$\{0, 1\}$ and $\{0, 2\}$ is a subset of the other, for instance.

At its bare minimum a (partial) order is a transitive relation R which is also
antisymmetric, that is, if xRy and yRx then $x = y$. It may be the case, as for $<$,
for example, that the condition of antisymmetry is never met, in which case the
property is vacuously satisfied. In this book we will tend to use a fairly strict
definition of order, insisting in particular that all distinct pairs of elements are
comparable, and we shall refer to this as a linear order or total order.

REMARK

A partial order is normally defined as a *reflexive*, transitive and antisymmetric relation. We prefer to exclude the condition of reflexivity because we want to use set membership as an order relation, but the omission is not a drastic one. Any transitive antisymmetric relation R on a set X can be extended to a transitive, antisymmetric and reflexive relation simply by adding to the relation all ordered pairs (x, x), $x \in X$. Conversely, by removing all such pairs from a transitive, antisymmetric and reflexive relation we obtain a relation which continues to be transitive and antisymmetric but which is non-reflexive.

Modelling the familiar orders we naturally encounter in mathematics we define a partial order to be a transitive antisymmetric relation. We often also assume that every pair of elements are comparable, yielding a total, or linear, order. We now have enough machinery to construct the integers.

1.4   Integers

I think I have already said somewhere that mathematics is the art of giving the same name to different things. It is enough that these things, though differing in matter, should be similar in form, to permit of their being, so to speak, run in the same mould. When language has been well chosen, one is astonished to find that all demonstrations made for a known object apply immediately to many new objects: nothing requires to be changed, not even the terms, since the names have become the same.

– JULES HENRI POINCARÉ[1]

1.4.1   The construction of $\mathbb{Z}$

We continue to build the basic number systems of mathematics. The first step, a fairly meagre extension of the natural numbers, is the set of integers. I use meagre in the sense that we need comparatively little work to construct the integers and that the extension is 'small' in that each element of the newly formed class may be identified with either a natural number or with '$-n$' for some natural number n. Of course, historically this is a significant step – the negative numbers were regarded with suspicion for some considerable time, as was zero. This is true, at least, of European mathematics, which did not fully embrace the notion of negative numbers until the seventeenth century. Elsewhere, in Chinese, Indian and Arabic mathematics, the idea had been accepted centuries earlier.

---

[1]Poincaré [**167**]. Chapter II (The Future of Mathematics).

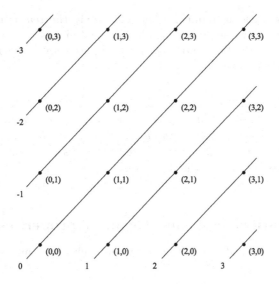

Figure 1.4 The set of integers modelled as equivalence classes of pairs of natural numbers (indicated by the diagonals). The integer -1 is then formally the class of ordered pairs $\{(0,1),(1,2),(2,3),(3,4),\ldots\}$.

The motivation is as follows. Confined, as we presently are, to the realm of natural numbers, the expression $m - n$ has no meaning if $m < n$. If we were to try to attach a meaning to '$0 - 1$', then, based on the arithmetic properties we wish the new entities to have, we would like the same meaning to be attached to '$1 - 2$', '$2 - 3$', and so on. We are led to consider the class of all ordered pairs (n, m) where n and m are natural numbers.

We define an equivalence relation on the cartesian product $\mathbb{N} \times \mathbb{N}$ by declaring (a, b) and (c, d) to be equivalent if and only if $a + d = b + c$. (The relation is therefore a subset of $(\mathbb{N} \times \mathbb{N}) \times (\mathbb{N} \times \mathbb{N})$.) Denote the equivalence class containing (a, b) by $\langle a, b \rangle$, so for example $\langle 0, 1 \rangle$ is the set $\{(0,1),(1,2),(2,3),(3,4),\ldots\}$. We have in mind that $\langle a, b \rangle$ is '$a - b$', so $\langle 0, 1 \rangle$ is a concrete realization of the newly invented number '-1'. These equivalence classes might be pictured as in Figure 1.4. The collection of all such equivalence classes then forms the set, $\mathbb{Z}$, of integers.

REMARK

Here we see our first example of a structural extension designed to accommodate solutions of a class of equations. Explicitly, an equation of the form $x + a = 0$ for $a > 0$ has no solutions in $\mathbb{N}$ so we embed $\mathbb{N}$ in a conservatively larger structure with compatible algebraic properties, namely $\mathbb{Z}$, in which such equations do have solutions. Similarly, the forcing of solutions to such equations as $2x = 1$, $x^2 = 2$ and $x^2 + 1 = 0$ induces the construction of the rational numbers, field extensions of the rational numbers, and ultimately the real numbers and the complex numbers, as we shall see. When expressed in this conveniently modern way, it seems so simple, but hindsight glosses over several centuries of slow and difficult mathematical development.

The integers $\ldots, -3, -2, -1, 0, 1, 2, 3, \ldots$ *are formally defined as a set of equivalence classes on* $\mathbb{N} \times \mathbb{N}$ *and one regards the natural numbers as a subset of this class via a natural embedding. Arithmetic operations are defined on this class which extend those of the natural numbers.*

1.4.2   Operations on $\mathbb{Z}$ and algebraic properties

Addition and multiplication on the equivalence classes comprising the set of integers are defined by:

$$\langle a, b \rangle + \langle c, d \rangle = \langle a + c, b + d \rangle$$
$$\langle a, b \rangle \langle c, d \rangle = \langle ac + bd, ad + bc \rangle.$$

These are well-defined operations, i.e. the defining equations above are independent of the choice of representative elements of each equivalence class. If we use the shorthand $-m$ for $\langle 0, m \rangle$ and identify $\langle n, 0 \rangle$ with the natural number n we see that the set of integers is precisely $\{\ldots, -5, -4, -3, -2, -1, 0, 1, 2, 3, 4, \ldots\}$ and that the restriction of our operations of addition and multiplication to the subclass $\{0, 1, 2, 3, \ldots\}$ coincide with the usual operations on the natural numbers.

It is not only the arithmetical operations that extend to the integers but also their properties such as distributivity, commutativity and associativity. Indeed it is the insistence on preservation of distributivity that motivates the definition of multiplication in the first place and this in turn gives us the identity $(-1)(-1) = 1$. The proof is easy: $0 = (-1)0 = (-1)(1 + (-1)) = (-1)1 + (-1)(-1) = -1 + (-1)(-1)$, so, making use of associativity of addition, we have $1 = (-1)(-1)$.

We tend to say casually that the natural numbers form a subset of the integers. Strictly speaking a *copy* of the natural numbers, namely the set $\{\langle n, 0 \rangle : n \in \mathbb{N}\}$, is a subset of $\mathbb{Z}$, not $\mathbb{N}$ itself, but to make this sort of distinction would become increasingly cumbersome as the theory develops, so there is no harm in making this identification at the outset, and we shall continue this trend whenever we embed one model of a number system in another.

Subtraction is now a well-defined operation on the extended system $\mathbb{Z}$: $n - m = n + (-m) = n + (-1)m$. The ordering of the natural numbers extends to the integers by declaring $n \leq m$ if and only if $m - n \in \mathbb{N}$. As well as the properties (1) to (8) of Section 1.2 (i.e. (1) to (8) hold with $\mathbb{Z}$ in place of $\mathbb{N}$), $\mathbb{Z}$ also has the following property:

(9) *Additive inverse*: For all $n \in \mathbb{Z}$ there exists a unique $m \in \mathbb{Z}$ with $n + m = 0$. (m is, of course, equal to $-n$.)

REMARK

We could have adopted an anachronistic presentation in which the integers are defined first and then the natural numbers are defined as a distinguished subset. Of course this is a perfectly valid alternative, but what I want to emphasize here is the construction of ever more elaborate systems built from a relatively simple base, each system embedded in the next. The only point at which we will be retracing our steps to construct a natural subsystem of an already constructed system will be when we define the algebraic numbers later.

We define operations of addition and multiplication on the set of equivalence classes comprising the integers, making sure that the definition is independent of the representative elements of each class. The set of integers is clearly larger than the set of natural numbers in the sense of inclusion, but what about its cardinality?

1.4.3   The cardinality of $\mathbb{Z}$

In the naive sense of subsets, the set of integers is larger than the set of natural numbers. However, the two sets have the same cardinality, as can be seen via the bijection $\mathbb{N}$ to $\mathbb{Z}$ given by:

$$
\begin{aligned}
0 &\mapsto 0, \\
1 &\mapsto -1, \\
2 &\mapsto 1, \\
3 &\mapsto -2, \\
4 &\mapsto 2, \\
&\cdots
\end{aligned}
$$

That is, we can list the integers as follows: $0, -1, 1, -2, 2, -3, 3, -4, 4, -5, 5, \ldots$. Sets which are equipollent to $\mathbb{N}$ are said to be *countable*.

Consequently, as a pure set of objects, ignoring arithmetical and ordering properties, $\mathbb{Z}$ is no more interesting than $\mathbb{N}$. If we set our sights on building a set with cardinality higher than that of $\mathbb{N}$ our next faltering step may be to consider the cartesian product $\mathbb{Z} \times \mathbb{Z}$, but this attempt fails as we see by considering the spiral depicted in Figure 1.5.

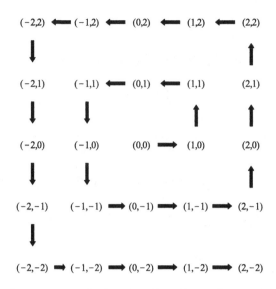

Figure 1.5 The first 25 terms of a bijection $\mathbb{N} \to \mathbb{Z} \times \mathbb{Z}$ illustrated by an anticlockwise directed spiral.

This gives us the bijection

$$
\begin{aligned}
0 &\mapsto (0,0) \\
1 &\mapsto (1,0) \\
2 &\mapsto (1,1) \\
3 &\mapsto (0,1) \\
4 &\mapsto (-1,1) \\
5 &\mapsto (-1,0) \\
&\cdots
\end{aligned}
$$

It follows from the preceding result that if A is any countable set, then A^2 is also countable. We can go on to prove that A^n (i.e. the set of all ordered n-tuples of elements of A) is countable for any $n \in \{1, 2, 3, \ldots\}$ (first observe that A^{n+1} is trivially equipollent to $A \times A^n$ and proceed by induction).

A corollary of the countability of $\mathbb{Z} \times \mathbb{Z}$ is the potentially surprising fact that the set of rational numbers (to be introduced shortly) is countable. It is evident that to break free from countability, if this is possible, we need to employ more sophisticated constructions.

REMARKS

1. If we write $\mathbb{N}$ in the following diagonal pattern

```
54   ...
44   53   ...
35   43   52   ...
27   34   42   51   ...
20   26   33   41   50   ...
14   19   25   32   40   49   ...
 9   13   18   24   31   39   48   ...
 5    8   12   17   23   30   38   47   ...
 2    4    7   11   16   22   29   37   46   ...
 0    1    3    6   10   15   21   28   36   45
```

then we see immediately that $\mathbb{N}$ is equipollent to $\mathbb{N} \times \mathbb{N}$, and furthermore we have an explicit expression for the bijection $\mathbb{N} \times \mathbb{N} \to \mathbb{N}$:

$$f(m,n) = \frac{1}{2}(m+n+1)(m+n) + n.$$

Using this one can derive explicit expressions for bijections $\mathbb{N}^N \to \mathbb{N}$. For example, to get an expression for a bijection $\mathbb{N}^3 \to \mathbb{N}$ first consider the bijection $\mathbb{N}^3 \to \mathbb{N}^2$ given by $(k,m,n) \mapsto (k, f(m,n))$ and compose that with f yielding the bijection $(k,m,n) \mapsto f(k, f(m,n))$. More generally we can define a bijection $f_N : \mathbb{N}^N \to \mathbb{N}$ recursively by

$$f_N(n_1, \ldots, n_N) = \begin{cases} \frac{1}{2}(n_1 + n_2 + 1)(n_1 + n_2) + n_2 & \text{if } N = 2, \\ f_2(n_1, g_{N-1}(n_2, \ldots, n_N)) & \text{if } N > 2. \end{cases}$$

2. Here we see that a sequence of Hilbert Hotels (whether arranged in one street or in an infinite block) is no more capacious than one of its members.

The set of integers, and indeed the set of n-tuples of integers, has the same cardinality as the set of natural numbers – it is 'countable'. A corollary of this is that the set of rational numbers is also countable.

1.5   Rational numbers

> *Our minds are finite, and yet even in these circumstances of finitude*
> *we are surrounded by possibilities that are infinite, and the purpose*
> *of life is to grasp as much as we can out of that infinitude.*
> – ALFRED NORTH WHITEHEAD[1]

1.5.1   The construction of $\mathbb{Q}$

If $np = m$, where n, p and m are integers with $p \neq 0$, then we write $n = \frac{m}{p}$.
We would like to extend the scope of the definition of $\frac{m}{p}$ to all pairs of integers
(m, p) with $p \neq 0$. This motivates the following construction.

Let us define an equivalence relation on the cartesian product $\mathbb{Z} \times \mathbb{Z}^*$, where
$\mathbb{Z}^*$ denotes the set of all non-zero integers. We declare (a, b) and (c, d) to be
equivalent if and only if $ad = bc$. The set of all equivalence classes under this
relation is the class of rational numbers, denoted $\mathbb{Q}$. Two examples of such
equivalence classes are

$$\{\dots, (-2, -4), (-1, -2), (1, 2), (2, 4), (3, 6), (4, 8), \dots\}$$

and

$$\{\dots, (-2, -6), (-1, -3), (1, 3), (2, 6), (3, 9), \dots\}.$$

We agree to denote the equivalence class containing (a, b) by $\frac{a}{b}$. Thus the
first equivalence class above may be labelled $\frac{1}{2}$, or $\frac{2}{4}$, etc., while the second
equivalence class may be labelled $\frac{1}{3}$, or $\frac{2}{6}$, etc. The fact that $\frac{1}{2}$ and $\frac{2}{4}$, or more
generally that $\frac{a}{b}$ and $\frac{ka}{kb}$, for $k \neq 0$, denote the same object captures the simple
idea of 'cancelling common factors' used unthinkingly in daily arithmetic.

This is a standard construction. Of course very few people habitually think
of $\frac{1}{2}$ as an infinite collection of pairs of integers, this formal approach is not
of immediate use in everyday mathematics, but when it comes to working on
foundations we need to explain precisely what a rational number is in terms
of more primitive notions, and the construction just described, or something
similar, is necessary.

The set $\{\frac{n}{1} : n \in \mathbb{Z}\}$ is a copy of the integers embedded in $\mathbb{Q}$. Consequently
we agree to write n in place of $\frac{n}{1}$ and 0 in place of $\frac{0}{m}$ (for any $m \in \mathbb{Z}^*$). When
we say that the integers form a subset of $\mathbb{Q}$ we understand that we are referring
to the copy of $\mathbb{Z}$ embedded in $\mathbb{Q}$ in this way.

REMARK

Although we are accustomed to thinking of rational numbers as a dense set of
points on a line, the definition given here indicates a more natural geometric
interpretation. Regarding the integer lattice $\mathbb{Z}^2$ as a subset of the plane, each

---

[1]Whitehead [**226**]. Chapter 21, June 28, 1941.

rational number corresponds to a non-vertical line passing through the origin $(0,0)$ and some other lattice point (and hence passing through infinitely many lattice points). We will return to this picture later.

The rational numbers are constructed as a collection of equivalence classes of certain pairs of integers. Embedded in this class is a copy of the integers. We define algebraic operations on the rational numbers extending those of the integers.

1.5.2   Operations on $\mathbb{Q}$ and algebraic properties

By defining

$$\frac{a}{b} + \frac{c}{d} = \frac{ad + bc}{bd}; \text{ and}$$
$$\frac{a}{b}\frac{c}{d} = \frac{ac}{bd},$$

we obtain well-defined operations of addition and multiplication on $\mathbb{Q}$ which extend those defined on $\mathbb{Z}$. A rational number $\frac{a}{b}$ is positive if a and b are either both positive or both negative. The order $<$ of $\mathbb{Z}$ extends to $\mathbb{Q}$ by declaring $\frac{a}{b} < \frac{c}{d}$ if and only if $\frac{c}{d} - \frac{a}{b}$ $(= \frac{bc-ad}{bd})$ is positive.

The rational numbers inherit all of the properties (1) to (9) of $\mathbb{Z}$ plus the following:

(10) *Multiplicative inverse*: For every non-zero $q \in \mathbb{Q}$ there exists a unique $r \in \mathbb{Q}$ with $qr = 1$. (Clearly r is equal to $\frac{1}{q}$.)

The 'no zero divisors' condition (8) is rendered redundant by the much stronger condition (10), for if $xy = 0$ then either $x = 0$ or x has a multiplicative inverse, z, in which case $y = z0 = 0$.

Any set with at least two elements together with operations of addition and multiplication satisfying properties (1) to (10) ((8) being redundant) is called a *field*. Up to isomorphism, $\mathbb{Q}$ is the smallest *ordered* field.

For non-zero $x \in \mathbb{Q}$ and $n \in \mathbb{N}$ we define $x^{-n} = \frac{1}{x^n}$ (so x^{-1} is the multiplicative inverse of x) and

$$x^0 = 1$$
$$x^n = x^{n-1}x.$$

This extends to integer exponents our previous definition of exponentiation.

The extensions of number systems we are describing here are all as conservative as possible. $\mathbb{Z}$ is the smallest extension of $\mathbb{N}$ with the property that every element has an additive inverse and such that the familiar arithmetical properties continue to hold, and $\mathbb{Q}$ is the smallest field containing $\mathbb{Z}$. Indeed, if F is a field containing (a copy of) $\mathbb{Z}$ then, being a field, it will contain the multiplicative inverse of each non-zero integer and the products of those inverses with all integers. So F contains a copy of $\mathbb{Q}$.

Natural arithmetical operations are defined on the rational numbers, again being careful to ensure that these definitions are independent of the representative elements one chooses from the equivalence classes that form $\mathbb{Q}$. In terms of inclusion the set of rational numbers is much bigger than the set of integers: with respect to the usual ordering there are infinitely many rational numbers between any two distinct integers (and infinitely many rational numbers between any two distinct rational numbers). However, from the viewpoint of cardinality the set of rational numbers is still small.

1.5.3   The cardinality of $\mathbb{Q}$

If one pictures the integers as an infinite set of equally spaced collinear points arranged from left to right according to the natural order then, in the familiar fashion, one can attach geometric meaning to each rational number as a point on the same line. In this scheme $\frac{1}{2}$ is to be associated with the point mid-way between 0 and 1, $\frac{1}{3}$ and $\frac{2}{3}$ describe the points dividing the interval between 0 and 1 into three equal parts, and so on. The picture one obtains of $\mathbb{Q}$ is of a dense row of points; between any two different rational numbers we find infinitely many others.

It comes as a surprise, when viewing the rational numbers as a densely packed array of points in this way, that there is a bijection from $\mathbb{N}$ to $\mathbb{Q}$. There are various ways of demonstrating this – as mentioned before it is a corollary of the fact that $\mathbb{Z} \times \mathbb{Z}$ is countable. Each $q \in \mathbb{Q}$ is an equivalence class of ordered pairs $(a, b) \in \mathbb{Z} \times \mathbb{Z}$. Map each non-zero q to the unique representative (a, b) of q with the smallest non-negative a (and if $q = 0$, map it to $(0, 1)$). This defines a natural injection $\mathbb{Q} \to \mathbb{Z} \times \mathbb{Z}$ and so the infinite set $\mathbb{Q}$ is equipollent to a subset of the countable set $\mathbb{Z} \times \mathbb{Z}$. An infinite subset of a countable set is countable.

The following explicit enumeration of $\mathbb{Q}$ is particularly appealing. We first arrange the positive rational numbers in a grid.

$$
\begin{array}{cccccc}
\frac{1}{1} & \frac{1}{2} & \frac{1}{3} & \frac{1}{4} & \frac{1}{5} & \frac{1}{6} \quad \cdots \\
\frac{2}{1} & \frac{2}{2} & \frac{2}{3} & \frac{2}{4} & \frac{2}{5} \\
\frac{3}{1} & \frac{3}{2} & \frac{3}{3} & \frac{3}{4} \\
\frac{4}{1} & \frac{4}{2} & \frac{4}{3} \\
\frac{5}{1} & \frac{5}{2} \\
\frac{6}{1} \\
\vdots
\end{array}
$$

Suppose we traverse the grid in the diagonal directions shown, starting each new diagonal from the top row, reading these initial elements from left to right, skipping any repeated equivalence classes as we go, then we obtain a listing of the positive rational numbers (where repeated terms are parenthesized):

$$
\frac{1}{1}, \frac{1}{2}, \frac{2}{1}, \frac{1}{3}, \left(\frac{2}{2}\right), \frac{3}{1}, \frac{1}{4}, \frac{2}{3}, \frac{3}{2}, \frac{4}{1}, \frac{1}{5}, \left(\frac{2}{4}\right), \left(\frac{3}{3}\right), \left(\frac{4}{2}\right), \frac{5}{1}, \ldots
$$

that is,

$$
\frac{1}{1}, \frac{1}{2}, \frac{2}{1}, \frac{1}{3}, \frac{3}{1}, \frac{1}{4}, \frac{2}{3}, \frac{3}{2}, \frac{4}{1}, \frac{1}{5}, \frac{5}{1}, \ldots
$$

and the listing of all of $\mathbb{Q}$ follows easily by alternation of signs and the insertion of an initial 0:

$$
0, \frac{1}{1}, -\frac{1}{1}, \frac{1}{2}, -\frac{1}{2}, \frac{2}{1}, -\frac{2}{1}, \frac{1}{3}, -\frac{1}{3}, \frac{3}{1}, -\frac{3}{1}, \frac{1}{4}, -\frac{1}{4}, \frac{2}{3}, -\frac{2}{3}, \frac{3}{2}, -\frac{3}{2}, \frac{4}{1}, -\frac{4}{1},
$$

$$
\frac{1}{5}, -\frac{1}{5}, \frac{5}{1}, -\frac{5}{1}, \frac{6}{1}, -\frac{6}{1}, \frac{5}{2}, -\frac{5}{2}, \frac{4}{3}, -\frac{4}{3}, \frac{3}{4}, -\frac{3}{4}, \frac{2}{5}, -\frac{2}{5}, \frac{1}{6}, -\frac{1}{6}, \ldots
$$

This is one of the early surprises in the theory. The surprise in this case is present only if we view $\mathbb{Q}$ as a dense set of points on a line. If we picture $\mathbb{Q}$ instead as a (subset of a) grid of numbers then the bijection loses almost all of its counterintuitive impact. Since $\mathbb{Q}$ is countable so is $\mathbb{Q}^n$ for all n in $\{1, 2, 3, \ldots\}$. Again, whether this is to be considered as surprising or not depends on whether one is inclined to think of, say, $\mathbb{Q}^3$, as a dense set of points in space, or otherwise.

REMARK

An alternative proof of the countability of $\mathbb{Q}$ is this. We'll say $\frac{a}{b}$ is *reduced* if b is positive and a and b have greatest common divisor 1 (we agree to represent 0 as $\frac{0}{1}$) and we'll call $|a| + b$ the *length* of $\frac{a}{b}$. For each positive integer k there are only finitely many reduced $\frac{a}{b}$ with length k (so for example the set of reduced rational numbers of length 5 is $\{-\frac{4}{1}, -\frac{3}{2}, -\frac{2}{3}, -\frac{1}{4}, \frac{1}{4}, \frac{2}{3}, \frac{3}{2}, \frac{4}{1}\}$). Then we simply list the rational numbers of length 1, then the rational numbers of length 2, the rational numbers of length 3, and so on.

The set of rational numbers is countable, as is the set of n-tuples of rational numbers for any $n = 1, 2, 3, \ldots$. Owing to its density this is, at first, counter-intuitive. To break free from countability we need to employ more sophisticated methods.

1.6 Real numbers

The so-called Pythagoreans, who were the first to take up mathematics, not only advanced this subject, but saturated with it, they fancied that the principles of mathematics were the principles of all things.

– ARISTOTLE[1]

1.6.1 Irrational numbers

By Pythagoras' Theorem the diagonal length d of a square with sides of length 1 must satisfy the equation $d^2 = 2$ (Figure 1.6). Unfortunately there is no rational number with square equal to 2. Put another way, no matter which unit of length we choose, it is impossible to construct a square with side and diagonal both a natural multiple of that unit – the side and diagonal are 'incommensurable', a fact which, according to some modern commentators, caused considerable distress to Pythagoras and his contemporaries, who would have regarded it as evidence of an unharmonious Universe.

The proof that no rational number has square 2 is well-known. The more commonly seen argument follows similar lines to the one given below but is expressed in terms of common factors of the numerator and denominator, a digression into elementary number theory which I wish to avoid. Instead, as forewarned in Subsection 1.2.2, I appeal to an ancient form of induction.

We assume that there is a positive rational number $\frac{a}{b}$ with square equal to 2, deduce a contradiction, and hence conclude that no such rational number exists. We have $a^2 = 2b^2$, that is, a^2 is even, hence a itself is even. So $a = 2a_1$ for some natural a_1. Then $4a_1^2 = 2b^2$, hence $2a_1^2 = b^2$. This implies b is even, so $b = 2b_1$ for natural b_1. This brings us to $a_1^2 = 2b_1^2$, back where we started with a_1 and b_1 in place of a and b, respectively. Repeating the argument we infer the existence of an infinite strictly descending sequence of natural numbers $\cdots a_4 < a_3 < a_2 < a_1 < a$, and this is the contradiction we were looking for.

The proof of the incommensurability of the side and the diagonal of a square appears in Euclid's *Elements* Book X (everything of this flavour being ingeniously expressed in Eudoxus' language of proportion – recall that the Greeks

---

[1] *Metaphysics* 1–5 [**5, 6**].

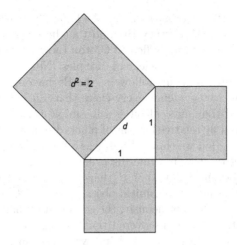

Figure 1.6 A simple theorem with (for the ancient Greeks) disturbing consequences. Pythagoras' Theorem tells us that the diagonal length d of a square with sides of length 1 satisfies the equation $d^2 = 2$ (in this case one can see this almost immediately: simply split the larger square, corner to opposite corner, into four equal triangular pieces and observe that any two of these can be moved to fill one of the smaller squares). However, there is no rational d with this property. Consequently, no matter which basic unit of length u we adopt, the diagonal and side of the square cannot both be natural multiples of u. Eudoxus of Cnidus gave a theory of proportion in the fourth century BC, anticipating the development of real numbers, which was able to cope with such 'incommensurable distances'.

did not regard ratios as an extension of the concept of natural number). The discovery predates Euclid by at least two centuries and is associated with some fanciful tales. The legend is that one of the later Pythagoreans, Hippasus of Metapontum, discovered the incommensurables, perhaps by studying ratios of lengths in a pentagram, the symbol of the Pythagoreans, and as a punishment for revealing this forbidden secret he was drowned at sea. There are various elaborations of the story, in some accounts the secret he revealed was not the existence of incommensurables but the construction of the dodecahedron – time has erased the exact nature of his crime.

REMARKS

1. If we return to the picture of the rational numbers as the set of non-vertical lines passing through the origin and another point in the integer lattice $\mathbb{Z}^2$, the existence of irrational numbers tells us that there exists a line in the plane which passes through the origin but misses all the other integer lattice points.

2. Little is known of Pythagoras the man. We know that he was active around the first half of the sixth century BC and that he started life on the Greek island of Samos, but eventually fled to Croton (now Crotone, in the south of Italy), where he founded his society of disciples. Nevertheless, his influence, or at least the influence of his followers, the Pythagoreans, is considerable. The Pythagoreans saw evidence everywhere that 'all is number'. For example, they observed that musical pitches in simple fractional relation to one another (as realized by relative lengths of plucked strings, blown pipes, struck resonant objects – what we think of today as a ratio of frequencies) produced an effect pleasing to the ear while discordant intervals correspond to more complex ratios. Legend has it that Pythagoras first chanced upon this phenomenon when passing a blacksmith, observing that when some anvils were struck in sequence, or simultaneously, the combination had a musical quality. On further investigation he discovered that the sizes of the anvils responsible for the most concordant musical intervals were in simple ratio to one another.[1] For instance the simple ratios $\frac{1}{1}$, $\frac{2}{1}$, $\frac{3}{2}$, $\frac{5}{4}$ (unison, octave, perfect fifth and major third in just temperament) yield concordant intervals.

3. The subject of musical scales is an interesting application of elementary number theory which has absorbed the time of many mathematicians over the centuries (I confess that I have spent perhaps too long on such things myself). It is amusing to imagine what Pythagoras would have made of the equal tempered division of the octave into twelve semitones of irrational ratio $\sqrt[12]{2}$, a compromise widely adopted in the late eighteenth century to facilitate concordant modulation from one key to another, as was demanded by ever more elaborate and adventurous musical compositions.

There is another less well-explored musical setting for irrational numbers. Suppose we were to set two regular pulses running simultaneously at the frequency of once every a and b seconds respectively. If the ratio $\frac{b}{a}$ is rational then the birhythm so produced would be periodic. If the ratio $\frac{b}{a}$ is irrational then the rhythm would never repeat, although we would discern a quasi-repeating structure restarting on those occasions when the time between the two pulses is so small as to be imperceptible. We cannot reproduce irrational timings with total accuracy, but this hasn't stopped composers experimenting with the idea.[2]

Let me expand a little on what I mean by a quasi-repeating irrational birhythm by focusing on a special case. Suppose we set one regular pulse ticking at one beat per second (this merely serves to illustrate our purpose, the unit of time is irrelevant) and a second regular pulse sounding every $\sqrt{2}$ seconds. We start both pulses simultaneously. Since $\sqrt{2}$ is irrational, the initial beat will be the only true simultaneous event. Let us suppose that

---

[1]Later explorers of the mathematics of harmony made use of a 'monochord' – a single string stretched over a soundbox with a moveable bridge.

[2]Some of Conlon Nancarrow's player piano studies make use of multi-voiced strata with (a close approximation to) irrationally related tempos.

we aren't too good at discriminating between two close beats, so that two beats that are less than a tenth of a second apart will be perceived as being simultaneous. What are the intervals between perceived synchronous beats? The sequence begins:

$$
\begin{array}{ccccccccc}
7 & 10 & 7 & 10 & 7 & 7 & 3 & 7 & \\
  &   & 7 & 10 & 7 & 7 & 3 & 7 & \\
7 & 10 & 7 & 10 & 7 & 7 & 3 & 7 & \\
  &   & 7 & 10 & 7 & 7 & 3 & 7 & \\
7 & 10 & 7 & 10 & 7 & 7 & 3 & 7 & \\
  &   & 7 & 10 & 7 & 7 & 3 & 7 & \\
  &   & 7 & 10 & 7 & 7 & 3 & 7 & \\
7 & 10 & 7 & 10 & 7 & 7 & 3 & 7 & \\
  &   & 7 & 10 & 7 & 7 & 3 & 7 & \\
7 & 10 & 7 & 10 & 7 & 7 & 3 & 7 & \dots
\end{array}
$$

i.e. the rhythm seems to restart after 7 seconds, then again after a further 10 seconds, and so on. There are various ways of dividing this sequence into recognizable subpatterns. In the presentation above it looks as if the sequence is made up of two patterns, of length 6 and 8 respectively, $(7, 10, 7, 7, 3, 7)$ and $(7, 10, 7, 10, 7, 7, 3, 7)$, however a little later we encounter a new pattern of length 13: $(7, 10, 7, 10, 7, 7, 10, 7, 10, 7, 7, 3, 7)$. All of these patterns end with the familiar $(7, 10, 7, 7, 3, 7)$. Inspecting the order in which these three patterns occur we find that they begin to appear in patterns of length 11, 28 and 40, these 'metapatterns' in turn occur in patterns of length 6, 8 and 13, again, which occur in patterns of length 11, 28 and 40, and so on, alternating between the last two groups. The general 'not quite periodic' character is evident at all scales: patterns, metapatterns, metametapatterns... Our choice of a tenth of a second as the threshold is immaterial; whichever choice we make we find a similar phenomenon, for example if we chose one twentieth of a second instead we would find patterns of intervals of 7, 17 and 24 seconds. A threshold of one hundredth of a second yields patterns of 41, 58 and 99. All we are doing here is finding those m such that there exists an n with $|\alpha - \frac{m}{n}| < \varepsilon \frac{1}{n}$, where ε is the threshold and, in this case, $\alpha = \sqrt{2}$. I leave the reader to study this piece of Diophantine approximation more thoroughly.

It has long been known that there are geometric magnitudes which are irrational. In order to algebraicize geometry we need a number system that is able to label these magnitudes, filling in the missing gaps that are left by the system of rational numbers.

1.6.2   Completing the discontinuum

Ideally we would like to construct a number system which is a model of all geometric magnitudes, something that does not falter at the first hurdle, as $\mathbb{Q}$ does. If we were to extend the ordered set $\mathbb{Q}$ to accommodate the irrational length exhibited above (let us follow tradition and label it $\sqrt{2}$) then intuitive geometry alone indicates the position that it ought to occupy in the extended order: if $X = \{x \in \mathbb{Q} : x > 0 \text{ and } x^2 > 2\}$ then for all rational $x \notin X$ and $y \in X$ we should have $x < \sqrt{2} < y$.

We could go further and extend rational addition and multiplication to a larger system (the smallest field extension of $\mathbb{Q}$ containing $\sqrt{2}$)

$$\mathbb{Q}[\sqrt{2}] = \{a + b\sqrt{2} : a, b \in \mathbb{Q}\}$$

with addition and multiplication defined by:

$$(a + b\sqrt{2}) + (c + d\sqrt{2}) = a + c + (b + d)\sqrt{2}$$
$$(a + b\sqrt{2})(c + d\sqrt{2}) = ac + 2bd + (ad + bc)\sqrt{2}$$

(the multiplicative inverse of $a + b\sqrt{2}$ being $\frac{a}{a^2 - 2b^2} - \frac{b}{a^2 - 2b^2}\sqrt{2}$).

Departing from the obvious geometric interpretation the symbol $\sqrt{2}$ can be excised completely from this presentation – this purely algebraic alternative embeds $\mathbb{Q}$ in $\mathbb{Q} \times \mathbb{Q}$ via $x \mapsto (x, 0)$ and defines addition and multiplication on $\mathbb{Q} \times \mathbb{Q}$ by:

$$(a, b) + (c, d) = (a + c, b + d)$$
$$(a, b)(c, d) = (ac + 2bd, ad + bc).$$

The element $(0, 1)$ has the property $(0, 1)^2 = (2, 0)$ (as $(2, 0)$ is the embedded image of 2 in $\mathbb{Q} \times \mathbb{Q}$, $(0, 1)$ is the newly adjoined '$\sqrt{2}$'). We say (a, b) is positive if one of the following holds:

(i) both a and b are positive; or

(ii) a is negative, b is positive and $a^2 < 2b^2$; or

(iii) b is negative, a is positive and $2b^2 < a^2$.

We order $(a, b) < (c, d)$ if and only if $(c - a, d - b)$ is positive.

This resolves the particular problem of attaching a tangible algebraically manipulable object to the purely geometric $\sqrt{2}$, however further investigation reveals infinitely many other irrational magnitudes (viewed geometrically the linear ensemble formed by $\mathbb{Q}$ is densely punctuated by point gaps – plugging just one of these missing points and its arithmetic combinations does little to complete the 'discontinuum'). We need a system which fills *all* the gaps in some sense. The so-called real numbers are designed with this in mind.

REMARK

The theory of algebraic field extensions of $\mathbb{Q}$, a small glimpse of which we have just seen, and the general theory of fields is an extremely rich subject which we cannot possibly do justice to here. We only note that the principle of extension we have just described continues the theme rehearsed earlier: when a structure does not admit elements with some property (in this case $\mathbb{Q}$ does not admit elements x with the property $x^2 = 2$) we extend the structure to a well-defined superstructure that does admit such elements.

One can fill individual gaps in the discontinuum presented by the set of rational numbers in a consistent algebraic fashion, but we need a system that at once fills all of the gaps.

1.6.3   Dedekind's construction

Mathematicians use the real numbers in analytical geometry to model the line, the set of ordered pairs of real numbers to model the plane, and ordered triples, quadruples, etc. of real numbers to model spaces and more general manifolds of higher dimension.

A formal definition of 'manifold' can be found in any text on (differential) topology/geometry. The simplest underfoot example is the sphere, the surface of a ball; every point on the sphere has a neighbourhood which can be continuously and invertibly deformed into a patch of the plane – it is locally 'plane-like', a so-called 2-manifold. Another example of a 2-manifold is the torus, i.e. the surface of a perfect ring doughnut (see Figure 1.7). One similarly defines 3-manifolds as objects which locally resemble three-dimensional space, and so on. More generally a manifold is a topological object which locally resembles some space (not necessarily Euclidean) – other properties are also sometimes imposed on the structure, orientability and different degrees of smoothness for example.

Such notions pervade physics, so we can hardly fail to point out the importance of the analysis of real numbers in applied mathematics. Whether structures formed with real numbers are appropriate models or in any sense the 'right' models for physical ideas is a fascinating subject which has attracted much discussion; the old joke goes that physical reality provides a reasonably good approximation to mathematics. This is a curious phenomenon. Usually measurements are clumsy block-like approximations of a relatively 'smooth' reality, however grainy natured physical reality is a crude substitute for the dense infinite divisibility of real number-derived structures. 'Approximation' is the wrong word for this silk-thread description of rope, but there seems to be no good alternative. There is no doubt that the set of real numbers is a far more exotic creature than is commonly realized. Dedekind, for example, was willing

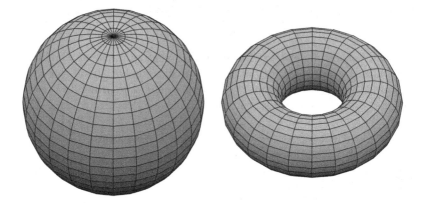

Figure 1.7 The sphere and the torus are just two of infinitely many examples of 2-manifolds. Each point on a 2-manifold has a neighbourhood which can be continuously and invertibly deformed into a patch of the two-dimensional plane. Informally speaking, these objects are 'locally two-dimensional'. One can formalize this idea and study arbitrary manifolds – topological objects which locally resemble some fixed space with some metric structure. The field of real numbers underpins the description of all such manifolds.

to accept that physical space might not be modelled faithfully by real numbers.[1]

Given this geometric motivation – we are trying to model a gapless line, a continuum – how can we go about defining the real numbers in terms of rational numbers? One popular method is to use *Dedekind cuts*. The intuitive idea is to 'pinpoint' a position on the (horizontal) line that we imagine the rational numbers to lie on by specifying a partition of $\mathbb{Q}$, a *cut*, into two sets, one falling to the left of the point and the other to the right. This imaginary line is a geometric crutch on which we lean purely for convenience; the details of the construction must rely solely on the algebraic properties of $\mathbb{Q}$ and its subsets, making no reference to external intuitive geometric ideas. Note that we take great care not to assume the existence of the very thing we are trying to construct; a defect that tended to spoil all naive pre-Weierstrassian limit-based accounts of the theory of real numbers.[2]

We define a *cut* to be an ordered pair of non-empty *sets* of rational numbers (A, B) having the properties:

(i) $A \cup B = \mathbb{Q}$.

(ii) $A \cap B = \emptyset$.

(iii) If $x \in A$ and $y \in B$ then $x < y$.

(iv) A has no largest element.

---

[1] Dedekind, R., *Stetigkeit und irrationale Zahlen*, Braunschweig: Vieweg (1872). Translated in Ewald [**66**].

[2] By Weierstrassian I mean the rigorous (epsilon-delta) developments in analysis which appeared in the second half of the nineteenth century – of course mathematicians other than Weierstrass were involved, but it is convenient to have a label.

We could, in place of (iv), use *B has no smallest element*. This would make only a cosmetic difference to the resulting theory.

Observe that there is some redundancy in the definition. Once one of the sets A or B has been given, the other is determined, so the definition could have been expressed in terms of one set of rational numbers alone, indeed this was the course of action taken by Giuseppe Peano and Bertrand Russell, for example. In fact, when it comes to defining operations on cuts (in particular multiplication) it is far more convenient to take this approach. So rather than considering both subsets of $\mathbb{Q}$ comprising a cut we shall focus on, say, the left-hand sets of Dedekind cuts and, to avoid any confusion, we call these *Dedekind sets*. Thus a Dedekind set D shall be a non-empty subset of $\mathbb{Q}$ such that:

(i) D is not equal to $\mathbb{Q}$;

(ii) if a rational number q is in D then all rational numbers less than q are in D; and

(iii) D has no largest element.

We define the real numbers, denoted by $\mathbb{R}$, to be the set of all Dedekind sets. Once again we stress that in doing so we have formed a model of our intuitive idea of the continuum in terms of sets. So $\sqrt{2}$, for example, which we typically view as a point somewhere on a line, or as a length, is now formally realized as the set $\{q \in \mathbb{Q} : q < 0 \text{ or } q^2 < 2\}$.[1]

Each rational number q corresponds to the Dedekind set

$$\{x \in \mathbb{Q} : x < q\},$$

and we write q in its place, so in particular 0 is the Dedekind set

$$\{x \in \mathbb{Q} : x < 0\}.$$

By making this identification we embed $\mathbb{Q}$ in $\mathbb{R}$ and we call the elements of $\mathbb{R}$ not in this rational embedding, such as $\sqrt{2}$ for example, the *irrational* numbers. The ordering of $\mathbb{Q}$ extends to $\mathbb{R}$ by declaring $A \leq B$ if and only if $A \subseteq B$. All that remains is to extend the operations of addition and multiplication to $\mathbb{R}$.

REMARKS

1. Attempts to classify manifolds have unleashed a tidal wave of sophisticated mathematics and there are many difficult open problems in the field. It is too vast a subject to discuss here but, with no apology for using undefined jargon, it should be pointed out that the two examples of compact orientable

---

[1] If we had not adopted the Peano–Russell convention for cuts, it would be an *ordered pair* of sets, however, we later indicate how an ordered pair of sets can be represented as a single set – see Section 4.1.

2-manifolds we have given here (the sphere and the torus) are good representatives of that class since any connected compact orientable 2-manifold is homeomorphic to either a sphere or a connected sum (essentially a gluing together) of a finite number of tori. Put another way, connected compact orientable 2-manifolds look like spheres with some number of 'handles' attached (possibly none). If we remove the orientability condition we admit a further class described as connected sums of a finite number of real projective planes (see Figure 2.7 in Subsection 2.4.2). So in the case of connected compact 2-manifolds there is a complete classification, up to homeomorphism. We should add that a connected 1-manifold is homeomorphic to either a line or a circle, and a 0-manifold is, of course, a point. Once we reach $n = 3$ and beyond things get very complicated, very interesting and sometimes very strange. (There are some further technical conditions that are imposed on manifolds – they are second countable and Hausdorff – these conditions exclude some pathological examples such as the 'long line', which we will meet later.)

2. On one hand, Dedekind's construction appeals to our geometric intuition, it fills the gaps we imagine to be left by $\mathbb{Q}$ essentially by defining one new object for each gap and one new object for each rational number. But on the other hand there isn't yet an intuition that the class of cuts is very much larger than $\mathbb{Q}$. One can be misled by forgetting the dense nature of $\mathbb{Q}$ – between any two distinct rational numbers there are infinitely many rational numbers – the word 'gap' is too suggestive of a positive interval, but even if we call these gaps points we still don't quite get a grasp on the size of this new collection of objects; at the moment it is still a little mysterious.

The class of Dedekind cuts, or equivalently, Dedekind sets, is designed to model the continuum. It comprises certain subsets of the rational numbers and contains an embedded copy of the rational numbers. We can define arithmetic operations on the class of Dedekind sets which extend those of the rational numbers.

1.6.4 Operations on Dedekind sets

Continuing to work with Dedekind sets rather than cuts, addition of Dedekind sets A and B is defined by

$$A + B = \{a + b : a \in A \text{ and } b \in B\}.$$

Given any Dedekind set A there is a unique Dedekind set '$-A$' such that $A + (-A) = 0$, and a Dedekind set A is said to be non-negative if $0 \leq A$, i.e. if $\{q \in \mathbb{Q} : q < 0\} \subseteq A$.

Multiplication is best defined first for non-negative Dedekind sets by[1]

$$AB = \{q \in \mathbb{Q} : \text{ there exist } a \in A \text{ and } b \in B \text{ such that } a, b \geq 0 \text{ and } q < ab\},$$

and is then extended to all Dedekind sets via the identifications $A(-B) = (-B)A = -(AB)$ and $(-A)(-B) = AB$.

Given a non-zero Dedekind set A, there is a unique Dedekind set A^{-1} such that $AA^{-1} = 1$, where 1 denotes the set $\{q \in \mathbb{Q} : q < 1\}$.

From here on we shall habitually refer to Dedekind sets as real numbers.

Each subset X of $\mathbb{R}$ which is bounded above (i.e. such that one can find a real number s with $x \leq s$ for all $x \in X$) has a least upper bound. This property we call (*Dedekind*) *completeness*. The notion of completeness is of the utmost importance in analysis and underlines the most striking conceptual difference between the set of real numbers and the set of rational numbers. The example $X = \{x \in \mathbb{Q} : x^2 < 2\}$ reveals that $\mathbb{Q}$ is not complete in this sense: X has an upper bound but no least upper bound in $\mathbb{Q}$.

Like $\mathbb{Q}$, the set $\mathbb{R}$ together with the operations just defined satisfies all of the properties (1) to (10) (i.e. it is a field), it is complete in the sense described above, and the order satisfies the following:

(i) for all $x, y, z \in \mathbb{R}$, if $x < y$ then $x + z < y + z$; and

(ii) for all $x, y \in \mathbb{R}$, if $0 < x$ and $0 < y$, then $0 < xy$.

We summarize this by saying that $\mathbb{R}$ is a *complete ordered field*.

Suppose F is an ordered field, i.e. a field with total order $<$ satisfying (i) and (ii) above, with F in place of $\mathbb{R}$. We say that a subset K of F is *inductive* if $0 \in K$ and if $x + 1 \in K$ whenever $x \in K$ (so in particular F itself is inductive). Let N be the intersection of all inductive subsets of F, then N is the smallest inductive subset of F and, defining the successor of n to be $n + 1$, N satisfies Peano's Postulates – it is a model of the natural numbers. The smallest subfield Q of F is isomorphic to $\mathbb{Q}$ (and of course is the field generated by N). For future reference we note that the statement that F is a complete ordered field is equivalent to the statement that F is an ordered field, Q is order dense in F (i.e. for all $x < y$ in F there exists a $q \in Q$ with $x < q < y$) and every Cauchy sequence in F (defined shortly) converges in F.

Up to isomorphism there is only one complete ordered field. More precisely, given any two complete ordered fields R_1 and R_2, there exists a (unique) field isomorphism from R_1 to R_2, that is, a bijection $\phi : R_1 \to R_2$ satisfying $\phi(x+y) = \phi(x) + \phi(y)$ and $\phi(xy) = \phi(x)\phi(y)$ for all $x, y \in R_1$, permitting us to think of R_1 and R_2 as essentially the same object.

---

[1]One cannot use this definition for arbitrary Dedekind sets because the product of, say, the Dedekind sets $-1 = \{q \in \mathbb{Q} : q < -1\}$ and $1 = \{q \in \mathbb{Q} : q < 1\}$ would be $\mathbb{Q}$, which is not a Dedekind set.

REMARKS

1. $-A$ is defined to be

$$\{q \in \mathbb{Q} : \text{ there exists an } a \notin A \text{ such that } q < -a\}.$$

   A^{-1} is defined for positive A to be

$$\{q \in \mathbb{Q} : \text{ there exists an } a \notin A \text{ such that } q < \frac{1}{a}\},$$

   and for negative A is defined as $-(-A)^{-1}$.

2. The proof that $\mathbb{R}$ is complete is particularly easy with this Dedekind set definition. If a set X of real numbers (i.e. a set of Dedekind sets) is bounded above then $\cup_{A \in X} A$ is the least upper bound of X. The details are left to the reader.

3. Our choice of Dedekind set as the left set in a Dedekind cut is one of convenience. Had we chosen the right sets as Dedekind sets then the definition of addition would have been the same but multiplication would have to be defined differently and the ordering would be given by the superset relation (not the subset relation). In either case the definition of multiplication is somewhat awkward and must be defined for positive sets first.

4. Let F be a field with multiplicative identity 1. If $1 + 1 + \cdots + 1 = 0$ (n summands) and is non-zero for any smaller number of summands then we say F has characteristic n (one can easily prove that n must be prime), otherwise we say F has characteristic 0. Ordered fields must have characteristic zero (again an easy exercise) and since all finite fields have prime characteristic we conclude that all ordered fields are infinite.

5. Let us make a comment that includes some material from future chapters (readers who are reading this book for the first time may wish to skip it). The uniqueness being described here, analogous to the uniqueness of models of second-order Peano Arithmetic, is likewise a feature of the *second-order theory* of $\mathbb{R}$, the second-order component in this case being the completeness axiom: *for all subsets X of $\mathbb{R}$, if X has an upper bound then X has a least upper bound in $\mathbb{R}$.* If we replace the completeness axiom with its first-order axiom schema (an infinite collection of axioms, one axiom for each subset A of $\mathbb{R}$ defined by some first-order property) *if A has an upper bound then there is a least upper bound of A in $\mathbb{R}$* then, like any first-order theory with an infinite model, the resulting structure has (infinitely many) non-standard models.

   The fact that the first-order theory is so much weaker than the second-order theory should not be too much of a surprise. In the second-order theory we can talk about an arbitrary subset of $\mathbb{R}$ while the first-order theory is only able to isolate those subsets of $\mathbb{R}$ that are definable in first-order terms (and there are only countably many of those). Likewise in second-order Peano Arithmetic the induction axiom covers arbitrary subsets of $\mathbb{N}$ while the first-order schema covers only that countable class of subsets of $\mathbb{N}$ defined by first-order properties. The distinction between first- and second-order theories will be clarified later (see Subsection 2.3.4).

6. The uniqueness result grants us freedom to construct the complete ordered field in any way we please; whatever the nature of the construction, provided it satisfies the axioms, it will be isomorphic to the field of real numbers.

7. For such constants as π or e or $\log 72$ one needn't go back to 'first principles' and exhibit them as Dedekind sets in order to be sure of their rightful status as real numbers. Now that we have shown that the collection of all Dedekind sets forms a complete ordered field we can follow a well-established route and prove all the stock theorems of real analysis. Using this powerful machinery we are able to identify real numbers via evaluations and zeros of analytic functions and their inverses, definite integrals and all manner of other analytical gadgets.

The class of Dedekind sets together with natural operations of addition and multiplication, and a natural order, is a complete real ordered field. Up to isomorphism there is only one complete real ordered field, and there are several natural alternative ways of describing it concretely.

1.6.5 An alternative construction of $\mathbb{R}$: Cauchy sequences

Another concrete construction of the real numbers (i.e. another model of the complete ordered field) which is more analytic in nature and is in some respects also more constructive is obtained by considering certain equivalence classes of so-called Cauchy sequences in $\mathbb{Q}$. The basic idea is due to Cantor.

A sequence of rational numbers is a collection of rational numbers indexed by the natural numbers: $a_0, a_1, a_2, a_3, \dots$. In other words a sequence is a function $f : \mathbb{N} \to \mathbb{Q}$, where we write a_i in place of $f(i)$. (We can, of course, speak of a sequence in any set S, meaning a function $\mathbb{N} \to S$.)

A sequence (a_n) in $\mathbb{Q}$ is a *Cauchy sequence* if for any rational number $\varepsilon > 0$ there exists a natural number N such that for all $m, n \geq N$ we have $|a_n - a_m| < \varepsilon$. Here, $|x|$, the absolute value of x, is the value of the function mapping x to x if $x \geq 0$ and x to $-x$ if $x < 0$. Note in particular that every constant sequence is a Cauchy sequence.

A sequence of rational numbers (a_n) is a *null sequence* if for any rational number $\varepsilon > 0$ there exists an N such that, for all $n \geq N$, $|a_n| < \varepsilon$. This is a formal way of saying that a_n tends to 0 as n increases. All null sequences are Cauchy sequences.

We say that two Cauchy sequences (a_n) and (b_n) are equivalent if their difference, the sequence $(a_n - b_n)$, is a null sequence. Let $\langle a_n \rangle$ be the equivalence class containing the sequence (a_n) and define:

$$\langle a_n \rangle + \langle b_n \rangle = \langle a_n + b_n \rangle$$
$$\langle a_n \rangle \langle b_n \rangle = \langle a_n b_n \rangle.$$

It is easily shown that these operations are well-defined and that the additive identity (zero) is precisely the equivalence class comprising all null sequences.

A natural order on the set of all equivalence classes is given by $0 < \langle b_n \rangle$ if and only if (b_n) is eventually positive, that is, if there is a natural number N and a $\gamma > 0$ such that for all $n \geq N$, $\gamma < b_n$. We then define $\langle a_n \rangle < \langle b_n \rangle$ if $0 < \langle b_n - a_n \rangle$.

The set of all equivalence classes of Cauchy sequences in $\mathbb{Q}$ with the above operations turns out to be a complete ordered field and hence is a copy of $\mathbb{R}$. The equivalence classes of constant rational number sequences collectively form the embedded image of $\mathbb{Q}$ in this model of $\mathbb{R}$. This is arguably more satisfactory than the Dedekind construction since multiplication is easily described – there is no need to define multiplication for the positive real numbers before extending the definition to all of $\mathbb{R}$. On the other hand it is much easier to see why $\mathbb{R}$ is Dedekind complete using Dedekind sets.

REMARKS

1. Readers familiar with a little ring theory will notice that the null sequences not only form an ideal in the ring of Cauchy sequences but a *maximal* ideal, and so the construction described above, simply a quotient ring, must be a field. Alternatively one must examine the field axioms directly. If $\langle a_n \rangle$ is non-zero then for some N all terms in the tail sequence $(a_n)_{n \geq N}$ are non-zero, so for an arbitrary choice of $b_1, \ldots, b_{N-1}$ and setting $b_n = \frac{1}{a_n}$ for $n \geq N$ we see that $\langle b_n \rangle$ is the multiplicative inverse of $\langle a_n \rangle$. Of the remaining properties of a complete ordered field it is only establishing Dedekind completeness which requires some analytical effort.

2. In the more general setting of an ordered field F with 'rational' subfield Q (the smallest subfield of F, see Subsection 1.6.4 above) the definition of a Cauchy sequence is exactly the same ($|x|$ still makes sense). Indeed, one way to prove that all complete ordered fields are isomorphic is to prove that each is isomorphic to the above-described field of equivalence classes of Cauchy sequences. Suppose F is a complete ordered field. The subfield Q is isomorphic to $\mathbb{Q}$ ($q \in Q$ corresponds to $q' \in \mathbb{Q}$, say). Our isomorphism ϕ from F to the field of real numbers (in Cauchy class form) is constructed in the obvious way: by completeness, for each $x \in F$ there is a Cauchy

sequence (q_n) in Q converging to x. Let $\phi(x)$ be the equivalence class $\langle q'_n \rangle$ of (q'_n) modulo null sequences in $\mathbb{Q}$. One then shows that ϕ is a well-defined bijective homomorphism.

Modelling the complete ordered field as certain equivalence classes of Cauchy sequences is sometimes more intuitive than the Dedekind instruction, and the arithmetical operations are more elegantly defined. One of the ways of labelling the elements of the complete ordered field is so familiar we take it for granted.

1.6.6   A familiar representation

Familiar to all of us, and something I have deliberately avoided until now, is the decimal representation of real numbers. Indeed many people habitually treat this representation as an identification, for this tends to be the way (with a sprinkling of geometry), for good or for bad, that we first meet real numbers. For completeness let us review this well-trodden ground.

The idea, as every reader was taught at an early age, is as follows. A finite expression $a_k \ldots a_0.b_1 \ldots b_s$, where the a_i and b_i are digits in

$$\{0, 1, 2, 3, 4, 5, 6, 7, 8, 9\}$$

is a shorthand for the rational number

$$a_k 10^k + a_{k-1} 10^{k-1} + \cdots + a_1 10^1 + a_0 + b_1 10^{-1} + \cdots + b_s 10^{-s}.$$

The 'non-terminating' expression $a_k \ldots a_0.b_1 b_2 b_3 b_4 \ldots$ is understood to mean the limit of the sequence of rational numbers:

$$a_k \ldots a_0,$$
$$a_k \ldots a_0.b_1,$$
$$a_k \ldots a_0.b_1 b_2,$$
$$a_k \ldots a_0.b_1 b_2 b_3,$$
$$a_k \ldots a_0.b_1 b_2 b_3 b_4,$$
$$\ldots$$

A formal definition of the limit of a sequence of real numbers will be given in Subsection 1.7.2 (I assume the vast majority of readers are already familiar with this) – it suffices to say for now that this sequence of rational numbers gets 'closer and closer' to a unique real number.

There is a non-uniqueness issue to be addressed. The sequence

$$0.9, 0.99, 0.999, 0.9999, \ldots$$

converges, as can easily be shown, to 1.[1] This means we have two distinct ways of representing the real number 1, both as $1.\dot{0}$ and as $0.\dot{9}$, where the dot over the digit denotes infinite recurrence.[2] Any number, except zero, which can be represented by a terminating decimal expression suffers from the same malady. Nevertheless the notation is a successful one: each real number has a decimal representation and every decimal expression represents a unique real number.

Needless to say, there is nothing significant about the number 10 here. It almost certainly became the convention because it is the number of fingers possessed by most human beings. Another choice of base $b \geq 2$ will do just as well, in which case recurrence of the digit $b - 1$ is the source of non-uniqueness. Other bases have been used in some form in various cultures (2, 12, 20, 60 being the most widely known), but even the Babylonian base 60 system (c. 3000BC), which we inherited for angular and time measure, did not use sixty individual symbols, instead employing two basic cuneiform molecules from which 59 characters were formed (there was no symbol for zero). There was still a 10ness to the system, despite its standard positional base 60 form: the two building blocks from which the characters were formed represented units and tens. There are some practical advantages to using 60 as a base – it has an abundance of divisors.[3] The decimal notation we have become accustomed to today is a relatively recent invention, being a simplification of a more cumbersome notation made popular in Western Europe by Simon Stevin in the late sixteenth century.

This lack of uniqueness is something of an irritation, but it is nothing new, after all, each rational number has infinitely many alternative names so the fact that *some* real numbers have two different representations in this particular notation is a comparatively mild difficulty to overcome. Usually we simply insist that our representations do not end in a string of 9s. The avoidance of a tail of 9s leads to some slightly awkward tricks in the coming constructions, but this is a small price to pay for the results obtained.

It is possible to start with the decimal representation alone as a basis for the definition of the real numbers, to define addition and multiplication in a purely combinatorial fashion and prove that the resulting structure is a complete ordered field and hence a copy of $\mathbb{R}$ (however this is spectacularly inelegant and is rarely carried out in practice). One of our reasons for introducing this sometimes cumbersome notation here, besides its basic familiarity, is that it can

---

[1]This is, of course, intuitively immediate. Using the formal definition of sequential convergence to be introduced in Subsection 1.7.2 and the Archimedean property of $\mathbb{R}$ (see the remarks to Subsection 1.7.11) the proof is elementary. Let $a_n = \frac{10^n - 1}{10^n}$, i.e. $0.9\ldots9$ with n 9s. Let $\varepsilon > 0$. By the Archimedean property of $\mathbb{R}$ there exists an integer $N > 0$ such that $\frac{1}{N} < \varepsilon$. Let $n \geq N$. Then

$$|a_n - 1| = \left| \frac{10^n - 1}{10^n} - 1 \right| = \left| -\frac{1}{10^n} \right| = \frac{1}{10^n} \leq \frac{1}{10^N} < \frac{1}{N} < \varepsilon.$$

The proof is complete: $0.\dot{9} = \lim_{n \to \infty} a_n = 1$.

[2]The convention is, of course, to omit recurring 0s entirely and to think of the expression as terminating at the digit preceding the first of the recurring 0s.

[3]12 and 20 are also abundant in the technical sense of the term, n being abundant if the sum of its proper divisors exceeds n.

be used to prove the existence or non-existence of bijections between certain sets.

By the familiar algorithm for division we see that every rational number has a decimal representation which is eventually periodic, that is, the string of digits after the decimal point is of the form

$$.a_1 \ldots a_n b_1 \ldots b_m b_1 \ldots b_m b_1 \ldots b_m b_1 \ldots b_m \ldots$$

which we notate as $.a_1 \ldots a_n \dot{b}_1 \ldots \dot{b}_m$. There may be no initial string of a_is, i.e. n may be 0, for example $\frac{1}{3} = 0.\dot{3}$, and the recurring digit may be zero, i.e. $m = 1$ and $b_1 = 0$, for example $\frac{1}{2} = 0.5$. Terminating decimals (i.e. those that end in a string of zeros) represent precisely the rational numbers $\frac{a}{b}$ where b is of the form $2^k 5^t$, $k, t \in \mathbb{N}$ (and such numbers, with the exception of zero, also have an alternative decimal representation ending in a tail of 9s). Rational numbers not of this terminating form together with irrational numbers have a unique decimal representation. A real number is irrational if and only if its decimal expansion is not eventually periodic. One can exploit this fact to manufacture some rather artificial looking irrational numbers; one simply writes down an aperiodic decimal expansion, for example $0.110\,100\,010\,000\,000\,100\,000\,000\,000\,000\,010\ldots$, where the 1s appear in the 2^kth position, $k = 0, 1, 2, \ldots$.

REMARKS

1. The notions of abundant, perfect and deficient numbers mentioned briefly above might be argued to be remnants of mystical beliefs of our ancestors, however there is a little more mathematical substance to the idea. For $n \in \{1, 2, 3, 4, \ldots\}$ we define $s(n)$ to be the sum of the proper divisors of n, so for example $s(1) = 0$, $s(10) = 1 + 2 + 5 = 8$, $s(12) = 1 + 2 + 3 + 4 + 6 = 16$. If $s(n) < n$, n is *deficient*, if $s(n) = n$, n is *perfect*, and if $s(n) > n$, n is *abundant*. For large N the proportion of abundant numbers in the set $\{1, 2, 3, \ldots, N\}$ is a little under $\frac{1}{4}$. The sequence of perfect numbers begins: $6, 28, 496, 8128, 33\,550\,336, \ldots$ At the time of writing only 48 perfect numbers have been found. No odd perfect numbers are known and it is conjectured that there are none.

   Perfect numbers might have been dismissed as numerological curiosities if it wasn't for Euclid's proof that $2^{p-1}(2^p - 1)$ is an even perfect number whenever $2^p - 1$ is prime and (two thousand years later) Euler's proof that the converse is also true: every even perfect number is of the form $2^{p-1}(2^p - 1)$ for prime $2^p - 1$. It is easy prove that $2^p - 1$ can only be prime if p itself is prime.[1] A prime of the form $2^p - 1$ is called a *Mersenne prime*. So there is a one to one correspondence between Mersenne primes and even perfect numbers. It is not known if there are infinitely many Mersenne primes; every few years a

---

[1] Not all numbers of the form $2^p - 1$, with p prime, are prime. The smallest counterexample is $2^{11} - 1$, which is the product of 23 and 89.

new one is discovered by brute computational force, which in turn yields a new even perfect number.

One more word on $s(n)$. Since perfect numbers are the fixed points of the function s there is some interest in the dynamical system on $\mathbb{N}$ induced by the action of s (let us define $s(0) = 0$). For each n we study the so-called *aliquot sequence* of iterates $n, s(n), s(s(n)), s(s(s(n))), \ldots$. One of two things will happen: either the sequence will eventually get caught in a loop (possibly of length one), or it will be unbounded. Two numbers making a loop of length two are known as *amicable pairs* and have been known since antiquity; the smallest amicable pair is $\{220, 284\}$. Numbers appearing in cycles of length three or more are called *sociable numbers*. No cycles of length three are known. The smallest sociable number is $12\,496$, which yields a cycle of length five: $12\,496, 14\,288, 15\,472, 14\,536, 14\,264$. It is conjectured that all aliquot sequences are eventually cyclic, however the ultimate behaviour of most sequences is still unknown, the smallest of which begins at 276 (more than 1500 terms in the sequence have been calculated at the time of writing and there is no sign of stabilization).

2. Defining real numbers by decimals, as mentioned earlier, is inelegant to the extreme. Let us lightly sketch from a safe distance what is involved. Right at the outset we need to make some awkward identifications. By a digit we mean an element of the set $\{0, 1, 2, 3, 4, 5, 6, 7, 8, 9\}$. A real number will be defined as an ordered pair comprising a finite sequence of digits followed by an infinite sequence of digits. Leading zeros in the finite sequence need to be ignored (this is our first awkward identification) and secondly a number with a tail of 9s in the infinite sequence will, this time by definition, have to be identified with the correct number having a tail of zeros. From this ugly beginning we need to define arithmetical operations. (These are performed by considering sequences of truncated decimals.) An ordering is set up, the field axioms are verified, completeness is proved and so on. Although possibly of some (masochistic?) value, I feel that the alternatives we have given are far better.

Representing real numbers as decimal expansions has its uses, but, like the representation of rational numbers as a quotient of two integers, the representation is generally not unique. Just as $\frac{4}{8}$ and $\frac{1}{2}$ represent the same rational number, $0.4\dot{9}$ and 0.5 represent the same real number. An older representation does not suffer from this problem.

1.6.7   An interlude: continued fractions

I cannot leave this discussion without mentioning the most ancient of all representations of real numbers, continued fractions. A (regular) continued fraction, a natural construction which emerges from the Euclidean division algorithm, is an expression of the form

$$a_0 + \cfrac{1}{a_1 + \cfrac{1}{a_2 + \cdots}}$$

possibly with infinitely many terms $a_0, a_1, a_2, \ldots$. A more compact way of writing this expression is $[a_0; a_1, a_2, \ldots]$. Setting $b_n = [a_0; a_1, a_2, a_3, \ldots, a_n]$ we say the continued fraction converges to α, and write $\alpha = [a_0; a_1, a_2, \ldots]$, if the sequence (b_n) converges to α. It turns out that an infinite continued fraction $[a_0; a_1, a_2, \ldots]$ converges if and only if the series $\sum_{n=0}^{\infty} a_n$ diverges.[1]

We assume now that $a_1, a_2, \ldots$ are all natural numbers and a_0 is an integer. Then for each real number α there exist unique $a_0, a_1, a_2, \ldots$ (finitely many if α is rational and infinitely many if α is irrational) such that $\alpha = [a_0; a_1, a_2, \ldots]$. Conversely every continued fraction $[a_0; a_1, a_2, \ldots]$ of the given type determines a unique real number, so this representation certainly has its advantages.

Continued fractions have many pleasant properties.[2] For example, Joseph Lagrange proved in 1770 that every solution of a quadratic equation $x^2 + bx + c = 0$ with rational coefficients has an eventually periodic continued fraction and conversely that every eventually periodic continued fraction is the solution of such an equation. Denoting the recurring cycle by dots over the digits in imitation of the decimal convention (so $[a; \dot{b}, \dot{c}]$ means $[a; b, c, b, c, b, c, \ldots]$) we have the following simple examples:

$$
\begin{aligned}
\sqrt{2} &= [1; \dot{2}] \\
\sqrt{3} &= [1; \dot{1}, \dot{2}] \\
\sqrt{5} &= [2; \dot{4}] \\
\sqrt{6} &= [2; \dot{2}, \dot{4}] \\
\sqrt{7} &= [2; \dot{1}, 1, 1, \dot{4}] \\
\sqrt{8} &= [2; \dot{1}, \dot{4}] \\
\sqrt{10} &= [3; \dot{6}].
\end{aligned}
$$

The continued fraction $[1; \dot{1}] = \frac{1}{2}(1 + \sqrt{5})$ (the golden ratio) is the very same continued fraction that Hippasus was said to have found when inspecting a ratio of lengths in the pentagram, leading him to the conclusion that the lengths are incommensurable.

Johann Lambert's 1766 proof of the irrationality of π (first conjectured by Aristotle; a technical gap in Lambert's proof was resolved by Adrien-Marie Legendre in 1806) was based on the following (non-regular) continued fraction

---

[1] A rigorous definition of sequential and series convergence is given later in Section 1.7.
[2] For a short introduction see Khinchin [124].

for $\tan z$:

$$\tan z = \cfrac{z}{1 - \cfrac{z^2}{3 - \cfrac{z^2}{5 - \cfrac{z^2}{7 - \cdots}}}}$$

He deduced that $\tan z$ is irrational for all non-zero rational z and in particular, since $\tan \frac{\pi}{4} = 1$, π is irrational.

The reciprocal $\frac{4}{\pi}$ has an interesting (non-regular) continued fraction discovered by William Brouncker in 1656:

$$\frac{4}{\pi} = \cfrac{1}{1 + \cfrac{1^2}{2 + \cfrac{3^2}{2 + \cfrac{5^2}{2 + \cfrac{7^2}{2 + \cdots}}}}}$$

The regular continued fraction of π begins $[3; 7, 15, 1, 292, \ldots]$. By contrast, the regular continued fraction of e has a discernible pattern:[1]

$$e = [2; 1, 2, 1, 1, 4, 1, 1, 6, 1, 1, 8, 1, 1, 10, 1, 1, 12, \ldots].$$

REMARKS

1. The connection between the Euclidean division algorithm and continued fractions is this. To find the continued fraction of the rational number $\frac{a}{b}$ perform the division algorithm to obtain the unique natural numbers m_i and r_i with $r_1 < b$ and $r_{i+1} < r_i$

$$
\begin{aligned}
a &= m_1 b + r_1 \\
b &= m_2 r_1 + r_2 \\
r_1 &= m_3 r_2 + r_3 \\
&\vdots \\
r_n &= m_{n+2} r_{n+1} + 0.
\end{aligned}
$$

Then the coefficients $m_1, m_2, \ldots, m_{n+2}$ give the desired expression $\frac{a}{b} = [m_1; m_2, m_3, \ldots, m_{n+2}]$. To find the continued fraction of an arbitrary real number a we first find the integer part a_0 and fractional part b_0 of a, so $a = a_0 + b_0$, then we find the integer part a_1 and fractional part b_1 of $\frac{1}{b_0}$, then we find the integer part a_2 and fractional part b_2 of $\frac{1}{b_1}$, then the integer part a_3 and fractional part b_3 of $\frac{1}{b_2}$, and so on. We have $a = [a_0; a_1, a_2, \ldots]$.

2. If we call a rational number $\frac{a}{b}$ a *best approximation* of a real number α if $|b\alpha - a| < |b'\alpha - a'|$ whenever $0 < a' \le a$, then each truncated continued fraction of α is a best approximation. Thus four best approximations of $\sqrt{2}$ are given by $[1; 2]$, $[1; 2, 2]$, $[1; 2, 2, 2]$ and $[1; 2, 2, 2, 2]$, i.e. $\frac{3}{2}$, $\frac{7}{5}$, $\frac{17}{12}$ and $\frac{41}{29}$.

---

[1] More historical items on this and other numerical topics included in this introduction can be found in Ebbinghaus, Hermes, Hirzebruch *et al.* [59].

3. The measure theory of continued fractions contains the following surprising result. The geometric mean of the continued fraction coefficients of a real number is almost always equal to a constant

$$K_0 = 2.685\,452\,001\,065\,306\ldots$$

called *Khinchin's constant*. That is, except for a measure zero[1] set of real numbers, the limit of the geometric means $(a_1 \ldots a_n)^{\frac{1}{n}}$ of the first n terms in the continued fraction of an arbitrary real number $[a_1; a_2, a_3, \ldots]$ is K_0. In case the reader wants to check this claim numerically he should be warned that the (measure zero) set of real numbers which do *not* satisfy this property includes almost all of the numbers we have mentioned in this section: rational numbers, solutions of quadratic equations with rational coefficients and e. Numerical computations suggest that π (and Khinchin's constant itself!) have Khinchin's property.

Continued fractions exhibit some interesting properties, but it is difficult to do arithmetic with them, so in this respect the decimal expansion is more practical. Now that we have constructed the field of real numbers in various different ways we can look at its cardinality.

1.6.8 Intervals and their cardinality

Intervals of real numbers are used throughout mathematics. The compact notation (a, b) is shorthand for the set of real numbers x satisfying $a < x < b$. Some texts use $]a, b[$ for an open interval, partly to avoid confusion with the ordered pair (a, b), but this seems to have fallen into disuse. Square brackets are used to include end points, so $[a, b)$ is the set of $x \in \mathbb{R}$ with $a \le x < b$; $(a, b]$ the set of $x \in \mathbb{R}$ with $a < x \le b$, and $[a, b]$ the set of $x \in \mathbb{R}$ with $a \le x \le b$. We use (a, ∞) as a shorthand for the set of real numbers strictly greater than a; $(-\infty, a)$ for the set of real numbers strictly less than a; and $[a, \infty)$ and $(-\infty, a]$ for the same intervals except with the point a included. $(-\infty, \infty)$ is an alternative notation for $\mathbb{R}$. We can exhibit a number of interesting bijections between these intervals.

1. If $a < b$ and $\alpha \in (0, \infty)$ then the map $x \mapsto \alpha x$ (a contraction or expansion) is easily seen to be a bijection from the interval (a, b) to the interval $(\alpha a, \alpha b)$.

2. The real line has the same cardinality as the interval $(-1, 1)$. We can best demonstrate this and other such correspondences with a picture. When we do this it should be clear that the geometry can be removed entirely by expressing the illustrated projections as functions, so that the bijection can be given by a purely analytic expression. The diagrams are nevertheless

---

[1]Measure is formally defined later in Subsection 8.4.5.

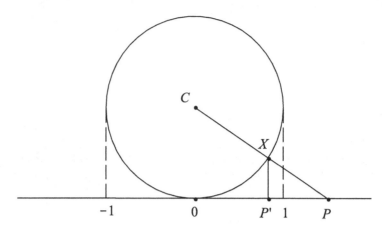

Figure 1.8 A geometric demonstration of the equipollence of the line $\mathbb{R}$ with the interval $(-1,1)$. A circle of radius 1 and centre $C = (0,1)$ sits on the real line at zero. An arbitrary point P in $\mathbb{R}$ is mapped to P' in the interval $(-1,1)$ as follows. Draw the line passing through C and P. This intersects the circle at X. The vertical line passing through X intersects the real line at P'. The correspondence $P \mapsto P'$ is a bijection.

invaluable. In the example of Figure 1.8 the bijection from $\mathbb{R}$ to $(-1,1)$ is given explicitly by $x \mapsto \frac{x}{\sqrt{x^2+1}}$. Of course, there are infinitely many bijections from $\mathbb{R}$ to $(-1,1)$, the latter, $x \mapsto \frac{x}{x^2+1}$ or $x \mapsto \frac{x}{|x|+1}$ being particularly simple examples.

3. The half line $[0,\infty)$ has the same cardinality as the interval $[0,1)$ (see Figure 1.9).

4. The closed interval $[0,1]$ has the same cardinality as the half-closed interval $[0,1)$. Let $\phi : [0,1] \to [0,1)$ be the map sending $\frac{1}{n}$ to $\frac{1}{n+1}$ for $n \in \{1,2,3,\ldots\}$ and fixing all other real numbers; ϕ is a bijection (see Figure 1.10). By a similar argument we see that $[0,1)$ has the same cardinality as $(0,1)$.

By contracting or expanding the intervals if necessary, the above bijections demonstrate that all intervals (excluding the degenerate case $[a,a] = \{a\}$) are equipollent.

REMARK

We should compare these results with corresponding statements for 'intervals' of rational numbers. For the purposes of this remark let us denote by $(a,b)_r$, $[a,b]_r$, etc. the sets of rational numbers in the intervals (a,b), $[a,b]$, and so on.

   (i) If q is a positive rational number then the map $x \mapsto qx$ is a bijection from $(a,b)_r$ to $(qa,qb)_r$.

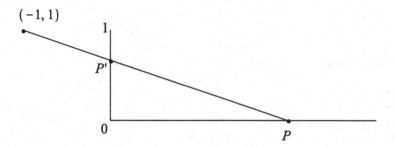

Figure 1.9 A geometric demonstration of the equipollence of $[0, \infty)$ and $[0, 1)$. The interval $[0, \infty)$ is drawn as the positive x-axis, and $[0, 1)$ the half-open segment on the y axis. An arbitrary point P on $[0, \infty)$ is mapped to P' in $[0, 1)$ by defining P' to be the intersection with $[0, 1)$ of the line passing through $(-1, 1)$ and P. The correspondence $P \mapsto P'$ is a bijection.

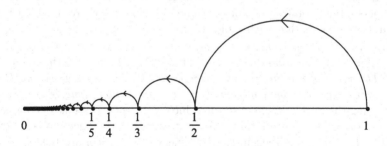

Figure 1.10 A picture illustrating the equipollence of $[0, 1]$ and $[0, 1)$. For each $n \in \{1, 2, 3, \ldots\}$ we map $\frac{1}{n}$ to $\frac{1}{n+1}$; all other points are fixed. This bijection removes the point 1 from $[0, 1]$. (Similar tricks show that $[0, 1]$ is equipollent to $[0, 1]$ minus any finite set of points.)

(ii) $\mathbb{Q}$ is equipollent to $(-1,1)_r$, but of course a bijection $\mathbb{R} \to (-1,1)$ need not map rational numbers to rational numbers. The map $x \mapsto \frac{x}{1+|x|}$ has inverse $x \mapsto \frac{x}{1-|x|}$, so we see it is a bijection $\mathbb{Q}$ to $(-1,1)_r$.

(iii) $[0, \infty)_r$ has the same cardinality as $[0,1)_r$, indeed the geometric demonstration also works for the rational case, and is given by the same map as (ii) above, restricted to the non-negative rationals.

(iv) $[0,1]_r$ has the same cardinality as $[0,1)_r$ and the bijection ϕ used in the real case, restricted to $[0,1]_r$, is clearly a bijection onto $[0,1)_r$.

All non-degenerate rational intervals are equipollent.

All intervals of real numbers, including the whole field of real numbers, are equipollent. One fast way of proving equipollence in general is to use the Cantor–Bernstein Theorem.

1.6.9   Introducing the Cantor–Bernstein Theorem

It is a triviality that for two finite sets A and B if each set has equal or fewer elements than the other then A and B must have the same number of elements. In other words, in the language of functions, if we can find an injection from A into B and an injection from B into A, then A and B are equipollent. One can prove that this result holds even if A and B are infinite. This result, an example of a phenomenon which carries over from the theory of finite magnitudes into the infinite, is called the Cantor–Bernstein Theorem.

In summary, the Cantor–Bernstein Theorem states that if for arbitrary sets A and B there exists an injection A into B and an injection B into A then there exists a bijection from A onto B. The bijection is not given explicitly, the theorem only tells us that such a bijection *exists*.

John Kelley in his *General Topology*[1] introduces an elegant proof of the Cantor–Bernstein Theorem due to Birkhoff and Mac Lane with the memorable words '*The proof of the theorem is accomplished by decomposing A and B into classes which are most easily described in terms of parthenogenesis*', very neatly summarizing the idea behind the proof (which we give for the sake of completion in Subsection 8.1.1). It is not quite correct to say that the bijection is not given explicitly. It is not described in the statement of the theorem, but examining the proof the set A is divided into two parts and the bijection is equal to the restriction of one injection to one part of A and the inverse of the other injection to the remaining part of A. Whether this can be regarded as an explicit description of the bijection or not depends on our ability to describe the partition, and this in turn depends on the complexity of the original injections. We shall comment on this again later, after giving the proof.

---

[1] Kelley [**121**].

REMARKS

1. Using the Cantor–Bernstein Theorem we can see instantly that $\mathbb{N}$ is equipollent to $\mathbb{N}^k$ for all $k \geq 1$. Clearly we can inject $\mathbb{N}$ into $\mathbb{N}^k$ simply by mapping n to $(n, 0, \ldots, 0)$. The challenge is to find an injection in the other direction. Prime factorization comes to the rescue. Suppose $p_1, p_2, \ldots, p_k$ are the first k primes (or indeed any k distinct primes), then the map $(n_1, \ldots, n_k) \mapsto p_1^{n_1} p_2^{n_2} \ldots p_k^{n_k}$ is an injection.

2. Prime factorization also suggests another non-trivial bijection. Let $\mathcal{S}_0$ be the set of sequences of natural numbers where all but finitely many of the terms are zero. Then the map $\mathcal{S}_0 \to \{1, 2, 3, \ldots\}$ sending $(n_1, n_2, n_3, \ldots)$ to $p_1^{n_1} p_2^{n_2} p_3^{n_3} \ldots$, where p_i is the ith prime, is a bijection. Thus $\mathcal{S}_0$ is countable.

If one can inject two sets into one another then they must be equipollent. This theorem can be used to establish some surprising bijections.

1.6.10 $\mathbb{R}$ versus $\mathbb{R}^n$

The Cantor–Bernstein Theorem could have been used to show, very briefly, some of the interval bijections described in Subsection 1.6.8 above, via a composition of translations and contractions, however it is often more illuminating to exhibit a direct bijection if possible.

The theorem is capable of producing surprises. For example, the square-bounded area $[0, 1) \times [0, 1)$ has the same cardinality as the interval $[0, 1)$. To avoid confusion let us write I in place of $[0, 1)$. It is clear that the function mapping x to the ordered pair $(x, 0)$ is an injection $I \to I \times I$. In the reverse direction let $(x, y) \in I \times I$ and write x, y in decimal form without recurring 9s: $x = 0.a_1 a_2 a_3 \ldots$, $y = 0.b_1 b_2 b_3 \ldots$. By interlacing these expressions we map (x, y) to $0.a_1 b_1 a_2 b_2 a_3 b_3 \ldots$, yielding the required injection.

The Cantor–Bernstein Theorem then tells us that there is a bijection $I \times I \to I$, and since $\mathbb{R}$ and I are equipollent, $\mathbb{R}^2$ has the same cardinality as $\mathbb{R}$. More generally $\mathbb{R}^n$ has the same cardinality as $\mathbb{R}$ for all natural numbers n (an inductive argument based on the obvious equipollence of $\mathbb{R}^{n+1}$ and $\mathbb{R} \times \mathbb{R}^n$ suffices).

This result so astonished Cantor when he first discovered it that he famously exclaimed in a letter to Dedekind (20 June 1887) 'I see it, but I don't believe it!'.

REMARKS

1. What happens if we replace $[0, 1)$ by the rational interval $[0, 1)_r$? We know already (since $[0, 1)_r$ is equipollent to $\mathbb{N}$) that $[0, 1)_r$ is equipollent to $[0, 1)_r^2$. The interlacing injection described above will also yield an injection from $[0, 1)_r^2$ to $[0, 1)_r$; our only worry is to confirm that the result is rational, but since rational numbers have eventually periodic expansions this is evident. We can even explicitly give the images of pairs of rational numbers under this mapping, for example $(\frac{1}{2}, \frac{1}{3})$ will map to $0.53\dot{0}\dot{3} = \frac{35}{66}$. If at least one of the pair is irrational there is generally no hope of giving an explicit closed form for the value of the interlaced number.

2. We could have used continued fractions to obtain the same result. This time we need to describe a way to combine $[0; a_1, a_2, \ldots]$ and $[0; b_1, b_2, \ldots]$. to give an injection $I \times I \to I$. We can't use exactly the same trick as before, i.e. mapping the pair to $[0; a_1, b_1, a_2, b_2, \ldots]$, since if at least one of the two numbers is rational we will run out of coefficients. One trick is to first apply the injection from $[0, 1)$ to itself which maps irrational $[0; a_1, a_2, \ldots]$ to $[0; a_1 + 1, a_2 + 1, \ldots]$ and rational $[0; a_1, a_2, \ldots, a_n]$ to $[0; a_1 + 1, a_2 + 1, \ldots, a_n, 1, 1, 1, 1, \ldots]$, extending it by a sequence of 1s. Having applied this injection first to each argument we can then apply the interlacing map. Notice that for this interlacing mapping we can give a closed expression not only for the image of pairs of rational numbers but also (unlike the decimal version) pairs of solutions to quadratic equations with rational coefficients. For example, to determine the image of the pair $(\frac{1}{2}, \sqrt{2} - 1)$ we first transform their respective continued fractions $[0; 2]$ and $[0; \dot{2}]$ to $[0; 3, \dot{1}]$ and $[0; \dot{3}]$. Interlacing these two yields $[0; 3, 3, \dot{2}, \dot{3}] = \frac{1}{7}(6 - \sqrt{15})$.

The set of real numbers is equipollent to the set of ordered pairs of real numbers comprising the plane, which in turn is equipollent to the set of ordered triples of real numbers comprising space, and so on. This is wildly counterintuitive at first glance. What we have not yet established is how the cardinality of the set of real numbers compares to that of the set of natural numbers.

1.6.11   The cardinality of $\mathbb{R}$ and beyond

We are now in a position to prove the most important result of this section. With this result we attain the goal set out earlier: *the set of real numbers has higher cardinality than the set of natural numbers.*

Suppose (a_n) is a sequence of elements in $[0, 1)$. We write each a_n in decimal form $0.b_{n,1}b_{n,2}b_{n,3}b_{n,4}\ldots$ without a tail of 9s. Next we consider the 'diagonal sequence' $b_{1,1}, b_{2,2}, b_{3,3}, b_{4,4}, \ldots$.

$$
\begin{array}{rcll}
a_1 & = & 0. & \mathbf{b_{1,1}} \quad b_{1,2} \quad b_{1,3} \quad b_{1,4} \quad b_{1,5} \quad b_{1,6} \quad \cdots \\
a_2 & = & 0. & b_{2,1} \quad \mathbf{b_{2,2}} \quad b_{2,3} \quad b_{2,4} \quad b_{2,5} \quad b_{2,6} \quad \cdots \\
a_3 & = & 0. & b_{3,1} \quad b_{3,2} \quad \mathbf{b_{3,3}} \quad b_{3,4} \quad b_{3,5} \quad b_{3,6} \quad \cdots \\
a_4 & = & 0. & b_{4,1} \quad b_{4,2} \quad b_{4,3} \quad \mathbf{b_{4,4}} \quad b_{4,5} \quad b_{4,6} \quad \cdots \\
a_5 & = & 0. & b_{5,1} \quad b_{5,2} \quad b_{5,3} \quad b_{5,4} \quad \mathbf{b_{5,5}} \quad b_{5,6} \quad \cdots \\
a_6 & = & 0. & b_{6,1} \quad b_{6,2} \quad b_{6,3} \quad b_{6,4} \quad b_{6,5} \quad \mathbf{b_{6,6}} \quad \cdots
\end{array}
$$

We form a new sequence (c_n) where c_n is chosen from $\{0, 1, 2, 3, 4, 5, 6, 7, 8\}$ (in fact $\{0, 1\}$ will do, we are avoiding 9s for the usual reason) subject to the single restriction that $c_n \neq b_{n,n}$. Then $0.c_1c_2c_3\ldots$ is the decimal representation of an element of $[0, 1)$ which differs from each a_n in at least one decimal position; it is, by construction, not in the original list. No matter which sequence of elements we begin with, we will always be able to construct an element of $[0, 1)$ which is not in the sequence. Hence, via the equipollence of $\mathbb{R}$ and $[0, 1)$, we see that $\mathbb{R}$ cannot be listed (it is *uncountable*). If one objects to the element of choice in this proof there are several easy fixes. We could let c_n be the smallest element in $\{0, 1\}$ not equal to $b_{n,n}$, for example.

The argument just presented is called Cantor's diagonal proof. This was not the first proof Cantor gave of the uncountability of $\mathbb{R}$. The first proof, employing a slightly more complex argument, was published three years earlier in 1874. Yet another proof is as follows. Let (a_n) be a sequence of real numbers and I_0 an interval. We define a nested sequence of intervals (I_n) as follows. For $n \geq 1$ partition I_{n-1} into three subintervals of equal length and define I_n to be the first of these subintervals (say in increasing order of initial endpoints) which does not contain a_n. The intersection of all of these intervals, $\cap I_n$, is non-empty, so there exists a real number in I which is not in the sequence (a_n).

The general method of diagonalization is a powerful technique which also features in Gödel's famous Incompleteness Theorems, in areas of theoretical computer science and elsewhere (see Appendix C – we shall be mentioning the Incompleteness Theorems throughout).

Now that we have established the existence of infinite sets which are not equipollent, $\mathbb{N}$ and $\mathbb{R}$, it is natural to ask if there are still higher sizes of infinity. It turns out that there do exist infinite sets which are neither countable nor equipollent to $\mathbb{R}$ (indeed, as we shall see later, there is a greater variety of infinities than can be assigned a cardinality). In classical mathematics the most familiar example is the set $\mathcal{F}$ of all functions $\mathbb{R} \to \mathbb{R}$. Suppose this set of functions were equipollent to $\mathbb{R}$, so that a bijection $\mathbb{R} \to \mathcal{F}$ exists mapping each real number x to a function f_x. Let $g(x) = f_x(x) + 1$. Since g is a function $\mathbb{R} \to \mathbb{R}$, it must coincide with an f_y for some y in $\mathbb{R}$; so in particular, $g(y) = f_y(y)$. But this implies $0 = 1$, a contradiction. This, again, is a diagonalization argument, only in this case we are 'diagonalizing' over the uncountable set $\mathbb{R}$ rather than the countable $\mathbb{N}$.

---

1. Let $\mathcal{S}$ be the set of all sequences of natural numbers. Diagonalization instantly tells us that $\mathcal{S}$ is uncountable: suppose $\mathcal{S} = \{S_1, S_2, S_3, \ldots\}$ is an enumeration of all sequences S_i of natural numbers, then define the nth term of a sequence D to be $S_{n,n} + 1$, where $S_{n,n}$ is the nth term of the sequence S_n. By design $D \in \mathcal{S}$ but differs from each sequence S_n in the nth term. Similarly we can see that for $m \geq 2$ the set $\mathcal{S}_m$ of all sequences with terms in $\{0, 1, 2, \ldots, m-1\}$ is uncountable: suppose $\mathcal{S}_m = \{S_1, S_2, S_3, \ldots\}$, then define the nth term of a sequence D to be $S_{n,n} + 1$ modulo m. Again, by design $D \in \mathcal{S}_m$ but differs from each sequence S_n in the nth term.

2. Consider the following construction of an infinite tree. Fix some natural number $m \geq 2$. We have a sequence of rows of vertices. In the first row there is one vertex. In the second row there are m vertices, in the third row there are m^2 vertices, in the fourth row there are m^3 vertices, and so on, so the $k + 1$th row has m^k vertices. These vertices are to be joined by edges as follows: the vertex in the first row is connected to each of the m vertices in the second row; the first vertex in the second row is joined to the first m vertices in the third row; the second vertex in the second row is joined to the second m vertices in the third row; and in general the ith vertex in the kth row is to be connected to the m vertices in the $k + 1$th row in positions $(i-1)m + 1, (i-1)m + 2, (i-1)m + 3, \ldots, (i-1)m + m$. A path in this infinite tree is an infinite sequence of vertices beginning with the vertex in the first row and such that the nth term in the sequence is in the nth row and is connected to the $n - 1$th vertex by an edge. Since each path uniquely determines an element of $\mathcal{S}_m$ (see the previous remark) and since any element of $\mathcal{S}_m$ uniquely describes such a path (i.e. the sequence $(s_1, s_2, \ldots)$ tells us to take the s_1th branch from the first vertex, the s_2th branch from the second vertex, and so on), the result of the previous remark tells us that, although the set of vertices in the tree is countable, the set of paths is uncountable.

3. Suppose we were to try Cantor's diagonalization on an enumeration of the rational numbers in $[0, 1)$. What would happen? The number formed from the diagonal is in $[0, 1)$, but is not in the original list, i.e. it will be irrational. So for example if we use the enumeration of the rationals in $[0, 1)$ given by: $0, \frac{1}{2}, \frac{1}{3}, \frac{2}{3}, \frac{1}{4}, \frac{3}{4}, \frac{1}{5}, \frac{2}{5}, \frac{3}{5}, \frac{4}{5}, \frac{1}{6}, \frac{5}{6}, \ldots$ then the corresponding diagonal number begins $0.003\,600\,000\,063\ldots$, so that using, say, the '$+1$ mod 9' rule for the new digits we form the irrational $0.114\,711\,111\,174\ldots$

4. In the diagonalization we could of course use any of the diagonals above the main diagonal to the same effect, for example $b_{1,2}, b_{2,3}, b_{3,4}, b_{4,5}, \ldots$, or indeed any subset of the $b_{i,j}$s provided that at least one term is chosen from each row.

5. Returning to our picture of the rational numbers as non-vertical lines passing through the origin and another point in the integer lattice $\mathbb{Z}^2$, this result tells us something quite startling. Not only do there exist lines through the origin that miss all other integer lattice points – *most* lines have this property!

The set of real numbers has strictly larger cardinality than the set of natural numbers. There are several different proofs of this, one of the most popular being Cantor's diagonalization argument. The set of real numbers models the continuum, a setting in which we can talk intelligibly about limits and continuity.

The order of the construction of the real numbers we have presented (from $\mathbb{N}$ to $\mathbb{R}$ via $\mathbb{Z}$ and $\mathbb{Q}$) is not the only possible route. One can, instead, from $\mathbb{N}$ construct non-negative rational numbers then positive real numbers, delaying the introduction of negative numbers until the end, or one might leap from $\mathbb{N}$ to positive rational numbers, then introduce negative rational numbers. Each method has its own benefits; the choice is wholly a matter of taste and the end result is the same.

The last quarter of the nineteenth century witnessed the great success of the so-called arithmetization of mathematics; that is, the embedding of real analysis – the formalization of calculus – in the theory of sets via Dedekind's construction of $\mathbb{R}$. Most mathematicians were content with this theory but had doubts about the further reduction of the natural numbers to more primitive entities.

1.7 Limits and continuity

Calculus required continuity, and continuity was supposed to require the infinitely little; but nobody could discover what the infinitely little might be... But at last Weierstrass discovered that the infinitesimal was not needed at all, and that everything could be accomplished without it.

– BERTRAND RUSSELL[1]

1.7.1 Geometric intuition

The pull of geometry is inevitably very strong owing to the dominance of our visual sense over all others. We can only speculate how mathematics would have developed in the hands of a blind species. What would have been the

---

[1] Russell [182]. Chapter V (Mathematics and the Metaphysicians).

starting point from which their mathematics evolved, and to what extent would it capture, in abstract form, geometric results that we can see at a glance? How easy would it be for us to share mathematical ideas?

Intuition concerning the continuum comes to us with surprising ease, something that is reflected in the emphasis, and pride of place, given to geometry in the early history of mathematics. It is impossible from our comfortable twenty-first century vantage point to make an accurate assessment as to why this was, outside the obvious argument that we are vision-biased creatures. Some of the difficulties would have arisen simply from a lack of suitable notation; centuries would pass before the powerful language of algebra arrived. Perhaps the noble desire to interpret everything in terms of geometry was ultimately a hindrance.

The algebraic notations we take for granted today are relatively recent inventions which only truly started to take hold in the sixteenth century. Before then, mathematical ideas were expressed in full, and often impenetrable sentences, or (by the Arabic mathematicians, for example, following Diophantus six centuries earlier) using a restricted notation.

Arithmetic and geometry used to be regarded as separate sciences, yet after Euclid there was some optimism that arithmetic could be fully absorbed into the realm of geometry. The work of nineteenth century mathematicians suggested the reverse, that geometry (and indeed all parts of mathematics) could be derived from a set theoretic formalization of arithmetic.

REMARKS

1. One must not underestimate the power of a good notation. With a suitably developed calculus a high school student can solve problems which would have troubled a master mathematician of antiquity. Indeed, when a notation reaches full maturity it is sometimes a little too easy to blur the distinction between the symbolism and the notion being symbolized.

2. There is a tendency in the literature to deny that the ancient Greeks did combinatorics or anything resembling it, however, there are counterexamples, such as Archimedes' *Ostomachion*, a dissection of a square into 14 pieces. It has been conjectured that it was Archimedes' intention to determine the number of distinct ways the pieces could be rearranged and still form a square. This might still be argued to be motivated by geometry, but there are further examples which seem to refer to an interest in enumerating possible combinations.

Geometry has been the bedrock of mathematics for millennia. Geometry and arithmetic, once regarded as separate sciences, are now unified, however, the nature of the unification would have surprised the ancient Greeks. They had thought that arithmetic might be describable entirely in terms of geometry, but late nineteenth century mathematicians instead formally embedded geometry and arithmetic in a theory of sets. The class of real numbers has properties that enable us to describe continuity and model motion.

1.7.2 Sequential convergence

The notion of limit is the central concept in differential and integral calculus, and, in various degrees of abstraction, in analysis in general. We are familiar with the notion of sequential convergence: the sequence $1, \frac{1}{2}, \frac{1}{3}, \frac{1}{4}, \ldots$ for example, gets closer and closer to 0 the further along it we travel. This intuitive idea is formally captured as follows. A sequence of real numbers (a_n) is said to converge to the real number a if for each real number $\varepsilon > 0$ there exists a natural number N such that for all $n \geq N$, $|a_n - a| < \varepsilon$. The number a is the *limit* of the sequence (a_n).

If (x_n) is a sequence of real numbers with limit x then we usually write

$$\lim_{n \to \infty} x_n = x$$

or we could say $x_n \to x$ as $x \to \infty$ ('x_n tends to x as n tends to infinity'). We must not make the error of thinking that the '∞' in this expression labels some tangible object approached by n, or even something capable of algebraic manipulation, the sentence '$x_n \to x$ as $x \to \infty$' is not to be split into its component parts with any meaning, it is simply an abbreviation for the formal definition given in the previous paragraph.

A sequence which does not converge is said to *diverge*. A sequence (a_n) is *bounded* if there is a $C > 0$ such that $|a_n| < C$ for all n. Divergent sequences may or may not be bounded, for example the divergent sequence of all natural numbers $0, 1, 2, 3, \ldots$ is unbounded but the alternating sequence $1, -1, 1, -1, \ldots$ is bounded.

Note the similarity of the formal definition of convergence with the definition of Cauchy sequences given earlier in Subsection 1.6.5. Cauchy sequences (and null sequences) of real numbers are defined in the same way as Cauchy sequences of rational numbers, but now that the real numbers have been defined we may permit ε to be real. (There is, however, no material gain in allowing such a real ε; the class of convergent sequences is not diminished by restricting ε to be rational.) It is easily shown that every convergent sequence is a Cauchy sequence. Conversely, every Cauchy sequence of real numbers converges (i.e. $\mathbb{R}$ is *Cauchy complete*). Cauchy completeness, which is slightly weaker than Dedekind completeness,[1] is used as the basis for a definition of completeness in more general (orderless) settings, for example in $\mathbb{R}^n$, $n > 1$, or in metric spaces in general. Let us expand on this.

A metric space is a set X together with a function $d : X^2 \to [0, \infty)$ satisfying:

(i) $d(x, x) = 0$;

---

[1]See the remarks to Subsection 1.7.11.

(ii) $d(x, y) = d(y, x)$; and

(iii) $d(x, z) \leq d(x, y) + d(y, z)$ for all $x, y, z \in X$ (the *triangle inequality*).

d gives the 'distance' between points x and y. Here a_n converges to a if for all $\varepsilon > 0$ there is an N such that for all $n \geq N$, $d(a_n, a) < \varepsilon$. A natural metric on $\mathbb{R}^n$ is given by $d(x, y) = \|x - y\|$, where the norm $\|\cdot\|$ is defined by $\|(x_1, \ldots, x_n)\| = (x_1^2 + \cdots + x_n^2)^{1/2}$, giving the Euclidean distance of the point $(x_1, \ldots, x_n)$ from the origin in $\mathbb{R}^n$.

When we say that the completion of $\mathbb{Q}$ is $\mathbb{R}$ we mean the completion induced by the usual measure of distance on $\mathbb{Q}$, $d(x, y) = |x - y|$, the completion being the set of all equivalence classes, modulo null sequences, of Cauchy sequences as described in Subsection 1.6.5. It is important that we don't take this convention for granted: there exist infinitely many alternative metrics on $\mathbb{Q}$ which give rise to (very useful) alternative completions, the fields of p-adic numbers.

The p-adic numbers were discovered at the beginning of the twentieth century by Kurt Hensel, inspired by analogies between the ring of integers and the ring of all polynomials with complex coefficients. For a fixed prime p we define on the field of rational numbers the p-adic absolute value $|\cdot|_p$ as follows. If we write $q \in \mathbb{Q}$ as $q = p^n \frac{a}{b}$ where a, b, n are integers and p is not a factor of a or b, then $|q|_p = \frac{1}{p^n}$ (we convene that $|0|_p = 0$). $|\cdot|_p$ satisfies all of the properties we usually associate with absolute values, i.e.

(i) $|a|_p = 0$ if and only if $a = 0$;

(ii) $|ab|_p = |a|_p |b|_p$; and

(iii) $|a + b|_p \leq |a|_p + |b|_p$ for all $a, b \in \mathbb{Q}$.

Most importantly, it turns out that any map $\mathbb{Q} \to \mathbb{Q}$ which satisfies these three properties is either equal to $|\cdot|^t$ (a power of the usual norm on $\mathbb{Q}$) or $|\cdot|_p^t$ for some prime p and positive real t, or it is the trivial absolute value defined by $|x| = 1$ if $x \neq 0$ and $|0| = 0$ (this is known as *Ostrowski's Theorem*). Note the novelty of the p-adic absolute value: with respect to $|\cdot|_p$, the sequence $1, p, p^2, p^3, \ldots$ is a null sequence! The completion of $\mathbb{Q}$ with respect to $|\cdot|_p$ is a field denoted by $\mathbb{Q}_p$, the field of p-adic numbers, and $|\cdot|_p$ extends in a natural way to all of $\mathbb{Q}_p$. p-adic numbers have many applications, especially in the solution of Diophantine equations. The intrigued reader may wish to seek a text on p-adic analysis for further information.

REMARKS

1. The isolated symbol ∞ does sometimes appear in measure theory in the context of the extended real line $\overline{\mathbb{R}} = \mathbb{R} \cup \{-\infty, \infty\}$ where it is merely a convenient way of avoiding an explosion of special cases. The new structure $\overline{\mathbb{R}}$ shares few algebraic properties with $\mathbb{R}$ and certain expressions ($\infty - \infty$, for example) are undefined.

2. Every element of the field of p-adic numbers $\mathbb{Q}_p$ can be written uniquely as a formal infinite series $a_n p^n + a_{n+1} p^{n+1} + a_{n+2} p^{n+2} + \cdots$, where n is some integer, possibly negative, $a_n \neq 0$ and $a_i \in \{0, 1, 2, \ldots, p-1\}$, which we can write in the usual way as a string of coefficients right to left, $\ldots a_{n+2} a_{n+1} a_n$, placing a point between coefficients with negative index and coefficients with non-negative index, as we might write a number in base p, except here there are finitely many terms to the right of the point and possibly infinitely many non-zero terms to the left of the point.[1] Although such a series diverges with respect to the usual absolute value on $\mathbb{Q}$ (see the next section), it converges with respect to $|\cdot|_p$.

In $\mathbb{Q}_2$ we have $\dot{1} + 1 = 0$, so $\dot{1}$ is the 2-adic '-1' (here the dot indicates recurrence to the left). Generally a positive integer is just given by its usual binary expansion and one can obtain its 2-adic negative by subtracting 1 and then changing all zeros to ones (including the infinite string of zeros at the beginning of the number) and all ones to zeros, for example, since 7 is 111, the 2-adic expansion of -7 is $\dot{1}001$. A p-adic number whose expansion in this form has no non-zero terms to the right of the point is a p-adic integer. The 2-adic expansion of $\frac{1}{2}$ is of course 0.1; $\frac{1}{3}$ is $\dot{0}\dot{1}011$ (so $\frac{1}{3}$ qualifies as a 2-adic *integer*); $\frac{1}{4}$ is 0.01; $\frac{1}{5}$ is $\dot{0}110\dot{1}$; and so on.

Since we know that all Cauchy complete ordered fields containing a dense copy of $\mathbb{Q}$ are isomorphic to $\mathbb{R}$ it follows that none of the (mutually non-isomorphic) p-adic fields $\mathbb{Q}_p$ have an ordering that is compatible with the field operations. While we are accustomed to thinking of $\mathbb{R}$ as a line, the geometry and topology of $\mathbb{Q}_p$ is much more elaborate, for example the p-adic integers form a compact subset of the locally compact space of p-adic numbers.

The intuitive idea of a sequence of real numbers getting closer and closer to a number can be formally defined without resorting to an ill-conceived notion of infinitesimals. With respect to the usual metric on the rational numbers the set of real numbers forms its completion (and alternative metrics lead to alternative completions). Using the notion of sequential convergence we define series convergence.

1.7.3   Series convergence

An infinite series

$$a_1 + a_2 + a_3 + \cdots = \sum_{n=1}^{\infty} a_n$$

---

[1] Some authors prefer to reverse the order of the coefficients, in which case one would 'carry to the right' when doing arithmetic, not to the left as usual.

is said to *converge* if the sequence of partial sums

$$a_1, \ a_1 + a_2, \ a_1 + a_2 + a_3, \ a_1 + a_2 + a_3 + a_4, \ldots$$

converges. Curiously this is quite an unstable property. There is a stronger form of convergence: the series $\Sigma_{n=1}^{\infty} a_n$ is *absolutely convergent* if $\Sigma_{n=1}^{\infty} |a_n|$ is convergent. All absolutely convergent series are convergent, but there are series which are convergent but not absolutely convergent; such series are said to be *conditionally convergent*. A series which does not converge is a *divergent series*.

A conditionally convergent series can be rearranged to sum to any real number one chooses, indeed it may be rearranged to diverge. This is not quite as bizarre as it may sound at first – perhaps the intuitive notion of series convergence is captured by absolute convergence but not by the weaker conditional convergence. To take a concrete example, the alternating series $\frac{1}{1} - \frac{1}{2} + \frac{1}{3} - \frac{1}{4} + \frac{1}{5} - \frac{1}{6} + \cdots$ converges to $\ln 2$,[1] however, we can permute the terms to make the series sum to, say, 12345, or to -4, or to π, or to any other number we choose, and we can even rearrange it so that the rearranged series diverges. The reason for this is that the series comprising just the positive terms $(\frac{1}{1} + \frac{1}{3} + \frac{1}{5} + \frac{1}{7} + \cdots)$ and the series comprising just the negative terms $(-\frac{1}{2} - \frac{1}{4} - \frac{1}{6} - \cdots)$ both diverge, but the terms of these divergent series tend to zero, so it is possible to set one's target, choose sufficiently many terms from the positive series until the target is overshot, then add terms from the negative series until the target is undershot, and so on. Repeating this oscillating process the amount by which the target is overshot or undershot diminishes owing to the decreasing magnitude of the terms being selected and so the sum gradually converges to the pre-set limit.

For example, if we were to rearrange the given series to sum to zero, according to this recipe the rearranged series would begin

$$\frac{1}{1} - \frac{1}{2} - \frac{1}{4} - \frac{1}{6} - \frac{1}{8} + \frac{1}{3} - \frac{1}{10} - \frac{1}{12} - \frac{1}{14} - \frac{1}{16} + \frac{1}{5} - \cdots$$

Rearrangements of the series which yield an unbounded sequence of partial sums or which oscillate without converging can be constructed similarly. To generate an unbounded sequence we choose enough positive terms to exceed 1, then add one of the negative terms, then add enough positive terms to exceed 2, add another negative term, add enough positive terms to exceed 3, and so on. A bounded divergent series is obtained by alternately adding sufficiently many positive or negative terms to exceed 1 and precede -1. An absolutely convergent series, on the other hand, can be rearranged as much as one likes, it will always sum to the same limit.

There is a natural way of attaching a sum to certain divergent series which is useful in Fourier analysis (in fact there are several different approaches to divergent series). If the partial sums of a series $\sum a_n$ are $s_1, s_2, s_3, \ldots$ then the series $\sum a_n$ is said to be *Cesàro 1-summable* if the sequence of Cesàro 1-means

$$s_1, \frac{s_1 + s_2}{2}, \frac{s_1 + s_2 + s_3}{3}, \frac{s_1 + s_2 + s_3 + s_4}{4}, \ldots$$

---

[1] The formal definition of the natural logarithm ln will be given in Subsection 1.7.8.

converges (the limit, the Cesàro 1-sum of $\sum a_n$, coincides with the usual sum of the series $\sum a_n$ if the latter is convergent). For example, the Cesàro 1-sum of the alternating divergent series $1 - 1 + 1 - 1 + 1 - \cdots$ is $\frac{1}{2}$. The Cesàro summation process can be iterated to accommodate ever wider classes of divergent series: if $t_1, t_2, t_3, \ldots$ are the Cesàro n-means of $\sum a_n$ then $\sum a_n$ is Cesàro $(n+1)$-summable if the sequence of Cesàro $(n+1)$-means $t_1, \frac{t_1+t_2}{2}, \frac{t_1+t_2+t_3}{3}, \ldots$ converges. The series $1 - 2 + 3 - 4 + 5 - \cdots$, for example, is Cesàro 2-summable with Cesàro 2-sum $\frac{1}{4}$ but is not Cesàro 1-summable. Divergent series with infinitely many positive terms and only finitely many negative terms, or vice versa, are not Cesàro n-summable for any n.

REMARKS

1. A permutation p which perturbs the limit of a conditionally convergent series has to be fairly wild. In particular, for any given natural number N, there must exist infinitely many n such that the difference between n and $p(n)$ exceeds N.

2. The general approach to divergent series is to present a class $\mathcal{C}$ of formal series, extending the class of all convergent series, together with a function on $\mathcal{C}$ whose value at any convergent series coincides with its usual sum and which respects certain algebraic properties of $\mathcal{C}$ (linearity, for instance, so that the limit of a sum is the sum of the limits). The class of Cesàro n-summable series is one such extension of the class of convergent series. A wider extension is given by the class of *Abel summable* series. The *Abel sum* of $\Sigma_{n=0}^{\infty} a_n$ is defined to be the limit as x increases to 1 of the function defined by $\Sigma_{n=0}^{\infty} a_n x^n$. This coincides with the usual sum if $\Sigma_{n=0}^{\infty} a_n$ converges, but one can find many divergent series which are Abel summable.

A series converges if its sequence of partial sums converges. One can strengthen or weaken the definition in several ways, for example a series is absolutely convergent if the sum of the absolute values of its terms converges. Series which are convergent but not absolutely convergent have some counterintuitive properties. Our next task is to formalize continuity.

1.7.4   Continuous functions

Our intuitive idea of a continuous function $[a, b] \to \mathbb{R}$ is of a function whose graph can be drawn without lifting the pen from the paper, something without disconnection. On the subject of graphs it should be remarked that, according to the set theoretic definition of a function as a set of ordered pairs, a function and its graph are the same object, so the terminology is a little redundant, but when speaking of the graph of a real-valued function the natural tendency is to think of it as a geometric entity, something one has drawn in order to visualize the associated function.[1]

Making the geometric notion of continuity precise, Weierstrass interpreted the expression $f(x)$ *tends to* $f(a)$ *as* x *tends to* a,

$$\lim_{x \to a} f(x) = f(a),$$

as meaning that for each real number $\varepsilon > 0$ there exists a real number $\delta > 0$ such that whenever $|x - a| < \delta$ we must have $|f(x) - f(a)| < \varepsilon$. Informally speaking this means: if x is close to a then $f(x)$ is close to $f(a)$. If this property holds then f is said to be *continuous* at a (otherwise it is *discontinuous* at a). A continuous function is a function which is continuous at all points of its domain.

Continuity is a heavy constraint, as is revealed by the fact that the set of continuous functions $\mathbb{R} \to \mathbb{R}$ has the same cardinality as $\mathbb{R}$, which as we saw in Subsection 1.6.11 is strictly smaller than the cardinality of the set of all functions $\mathbb{R} \to \mathbb{R}$. Note also that every monotonic function on an interval, that is, a function f such that $f(x) \le f(y)$ whenever $x \le y$ or a function f such that $f(x) \ge f(y)$ whenever $x \le y$, is discontinuous at at most countably many points.

The continuity of a function $f : \mathbb{R}^n \to \mathbb{R}^m$ is easily defined by generalizing the one-dimensional case using, in place of absolute value, the *norm* $\| \cdot \|$ (as defined in Subsection 1.7.2) on $\mathbb{R}^k$ ($k = n$ or m). The notion of continuity may be extended further into much deeper territory, to spaces which possess no notion of distance at all. The reader should consult a text on general topology for further details, but here is the basic idea. By an open subset of $\mathbb{R}$ we mean a union of open intervals. One can prove that a function $f : \mathbb{R} \to \mathbb{R}$ is continuous if and only if the counterimage $f^{-1}(U) = \{x \in \mathbb{R} : f(x) \in U\}$ of any open set U of $\mathbb{R}$ is open in $\mathbb{R}$. This removes the direct reference to the underlying metric in the definition of continuity. Hence we can use this new definition to describe continuous functions $f : X \to Y$ between arbitrary sets X and Y provided we can describe the 'open sets' of X and Y. Abstracting the basic set theoretic properties of open sets in $\mathbb{R}$, we call a class τ of subsets of X a *topology* (and the elements of τ *open sets* in X) if:

(i) $\varnothing \in \tau$ and $X \in \tau$;

(ii) the union of an arbitrary collection of sets in τ is again in τ; and

(iii) the intersection of any finite collection of sets in τ is in τ.

(For a given set X there will be many different topologies on X.) The set X together with the topology τ is a *topological space*. For arbitrary topological spaces (X, τ) and (Y, σ) a function $f : X \to Y$ is continuous (or (τ, σ)-continuous) if and only if $f^{-1}(U) \in \tau$ for all $U \in \sigma$.

---

[1]Paul Halmos makes a similar remark in Halmos [**91**], p. 31, problem 58.

The above definitions of metric convergence and of continuity only appeared in the nineteenth century in the hands of Weierstrass and others, and the topological generalization emerged in the first half of the twentieth century. Until then the notion of convergence had been rather fuzzy. Of course, great work in the subject had appeared before the invention/discovery of these precise definitions, but the new approach opened up another world, clarified the old achievements and was both powerful and precise enough to answer a host of unanswered questions.

Remarks

1. One can speak of left continuity and right continuity of a function at a point a. f is *left continuous* at a if, for all $\varepsilon > 0$, whenever $a - x < \delta$ we have $|f(x) - f(a)| < \varepsilon$ and right continuity is defined similarly, replacing $a - x$ by $x - a$. Clearly a function is continuous at a if and only if it is both left and right continuous at a.

2. For any continuous function $f : [a, b] \to \mathbb{R}$ and any $\varepsilon > 0$ there exists a polynomial $p : [a, b] \to \mathbb{R}$ with the property that $|f(x) - p(x)| < \varepsilon$ for all $x \in [a, b]$. This is the *Weierstrass Approximation Theorem*. It is fairly easy to see that one can choose the coefficients of the polynomial to be rational. We will see later that the class of all polynomials with rational coefficients is countable, and it is a short leap from here to show that every continuous function is the limit of a sequence from a countable set of functions. It follows that the set of all continuous functions $\mathbb{R} \to \mathbb{R}$ is equipollent to $\mathbb{R}$.

3. Although the idea of a topological space is motivated by the properties of metric spaces, it is an extremely wide generalization; general topological spaces can have very unfamiliar properties. A topological space need not be metrizable – i.e. it might be impossible to define a metric on the space which induces the topology, or worse, the space may contain pairs of distinct points which are inseparable in the sense that one cannot find two disjoint open sets each containing only one of the points. Some such spaces occur naturally in algebraic geometry and in analysis.

4. A curiously difficult enumeration problem is to determine the number of topologies on a finite set. For sets of cardinality $0, 1, 2, 3, 4, 5$ the number of topologies is $1, 1, 4, 29, 355, 6942$, respectively, and further values have been evaluated, however there is no known simple formula for the sequence.

Weierstrass' definition of continuity freed the calculus from the spectre of infinitesimals. A rigorous definition of the continuity of a function $\mathbb{R} \to \mathbb{R}$ is now available, and this extends naturally to general metric spaces and beyond.

This development was a turning point in analysis. Continuity is a fairly severe constraint on a function. Once strong continuity assumptions are imposed, counterintuitive properties are likely to evaporate. One such example concerns preservation of dimension.

1.7.5 Cardinality and dimension

Let us begin with a quick formal definition of (algebraic) dimension as an antidote to any pseudodefinitions that may have crept into the popular philosophical literature. The motivating example of the abstract structure we are about to describe is $\mathbb{R}^n$ with pointwise coordinate addition

$$(\alpha_1, \ldots, \alpha_n) + (\beta_1, \ldots, \beta_n) = (\alpha_1 + \beta_1, \cdots, \alpha_n + \beta_n)$$

and scalar multiplication

$$\lambda(\alpha_1, \ldots, \alpha_n) = (\lambda \alpha_1, \ldots, \lambda \alpha_n).$$

A set V together with a commutative and associative operation $+ : V \times V \to V$ and a scalar multiplication $\mathbb{R} \times V \to V$ $((\lambda, v) \mapsto \lambda v)$ is a *real vector space* if for $\alpha, \beta \in \mathbb{R}$ and $v_1, v_2 \in V$:

(i) there is an additive identity $\mathbf{0}$ in V;

(ii) every $v \in V$ has an additive inverse $-v$;

(iii) $\alpha(v_1 + v_2) = \alpha v_1 + \alpha v_2$;

(iv) $\alpha(\beta v) = (\alpha\beta)v$;

(v) $(\alpha + \beta)v = \alpha v + \beta v$ (from which it follows that $0v = \mathbf{0}$); and

(vi) $1v = v$.

In general an F-vector space for some field F is defined in the same way with F in place of $\mathbb{R}$.[1] A set of vectors (elements of V) $\{v_1, \ldots, v_n\}$ is *linearly independent* if the linear equation $\lambda_1 v_1 + \cdots + \lambda_n v_n = 0$ has only one solution $\lambda_1 = \cdots = \lambda_n = 0$. The set $\{v_1, \ldots, v_n\}$ *spans* V if every vector $v \in V$ is expressible in the form $v = \alpha_1 v_1 + \cdots + \alpha_n v_n$ for some scalars $\alpha_1, \ldots, \alpha_n$. A linearly independent set which spans V is a *basis* of V and it can be shown that every basis of a vector space has the same cardinality; this invariant is called the *dimension* of V. $\mathbb{R}^n$ has dimension n, as is easily proved.

Infinite-dimensional vector spaces (i.e. spaces which do not have a finite basis) are easily exhibited. For example the space of all real sequences (a_n) with addition $(a_n) + (b_n) = (a_n + b_n)$ and scalar multiplication $\lambda(a_n) = (\lambda a_n)$ is infinite-dimensional, as is its subspace l_2 of all sequences (a_n) such that $\sum a_n^2$

---

[1] The axiomatic definition of vector spaces was devised by Giuseppe Peano in the late 1880s.

converges. More generally, for $p > 0$, the infinite-dimensional space l_p is the vector space of all sequences (a_n) for which $\sum |a_n|^p$ converges. The space of all continuous functions $\mathbb{R} \to \mathbb{R}$ is also an infinite-dimensional vector space (with the obvious 'pointwise' addition and scalar multiplication).

This is the algebraic notion of dimension, but there are more general definitions. For example there is an extended notion of dimension (Hausdorff dimension) which, while assigning the usual dimension n to the space $\mathbb{R}^n$, assigns a non-integer dimension to certain sets (fractals). We shall not be using such generalized dimensions here.

The result that $\mathbb{R}^m$ and $\mathbb{R}^n$ have the same cardinality, regardless of m and n, seems to conflict with our intuitive perception of dimension. Dedekind responded to Cantor's alarming discovery suggesting that continuity needs to play a role if dimension is to be preserved. He conjectured that the existence of a continuous bijection $\mathbb{R}^n \to \mathbb{R}^m$ with continuous inverse (a *homeomorphism*)[1] must force $m = n$.

Dedekind's conjecture turned out to be true; however, it is non-trivial. L. E. J. Brouwer finally found a proof three decades after it was suggested.[2] It is notable that it is not too difficult to prove that no homeomorphism exists between $\mathbb{R}$ and $\mathbb{R}^2$; the real problems arise in higher dimensions.

Even if we slightly weaken the hypothesis of Brouwer's theorem the surprises return: difficult as it may be to imagine, there exist continuous *surjective* functions $[0, 1] \to [0, 1] \times [0, 1]$. Such strange space-filling mappings are called *Peano curves* or *space-filling curves* and can be described as the limit of recursively defined geometric constructions (see Figure 1.11).[3]

REMARKS

1. l_2 holds a distinguished position among the l_p spaces – it enjoys geometric properties (a natural notion of orthogonality, for example) similar to those of $\mathbb{R}^n$, in fact it is the simplest infinite-dimensional example of a whole class of spaces known as Hilbert spaces, which are natural geometric extensions of $\mathbb{R}^n$ and $\mathbb{C}^n$.

2. The assertion that every vector space has a basis is equivalent to the Axiom of Choice.[4]

---

[1]In informal terms, a homeomorphism between two topological objects describes a continuous deformation of one onto the other which neither tears holes in the object nor collapses it in on itself. When we mentioned the classification of 2-manifolds (surfaces) earlier we were referring to the most natural topological classification, where two surfaces are declared to be equivalent if they are homeomorphic.

[2]Brouwer, L., 'Beweis der Invarianz der Dimensionenzahl', *Math. Ann.*, Vol. **70**, 161–165, (1911).

[3]The existence of space-filling curves was first proved in Peano, G., 'Sur une courbe, qui remplit toute une aire plane', *Mathematische Annalen*, **36** (1), 157–160, (1890). The curve illustrated in Figure 1.11 appears in Hilbert, D., 'Ueber die stetige Abbildung einer Line auf ein Flächenstück', *Mathematische Annalen*, **38**, 459–460, (1891).

[4]Blass, Andreas, 'Existence of bases implies the axiom of choice'. *Contemporary Mathematics*, **31**, 1984.

Figure 1.11 The first six iterations of the construction which converges to a space-filling curve devised by David Hilbert in 1891. The iteration takes four smaller copies of the previous image, placing one in each square of a 2 × 2 grid, rotates the bottom left image clockwise by 90°, rotates the bottom right image anticlockwise by 90° and joins these four pieces together with three line segments.

3. Brouwer eventually rejected some of his early analytical/topological work on the basis of his strict Intuitionistic philosophy of mathematics. He had expressed his views on 'the untrustworthiness of the principles of logic' several years before proving Dedekind's conjecture, so the doubts he had were already present, but it was only after establishing his reputation among mathematicians that he became more outspoken about his new philosophy. Brouwer was able to reconstruct many of his earlier results in Intuitionistic terms.

4. With iterative constructions of space-filling curves one must emphasize that there is some non-trivial analysis that must be done in order to prove that the limiting curve is indeed space-filling (the diagrams are hopefully persuasive). Although Peano was the first to describe such a curve, his work was presented in a very formal analytical style with no accompanying diagrams. Hilbert was the first to provide an illustration of the process.

One of Cantor's most surprising discoveries, one that he found difficult to believe at first, was that all n-dimensional spaces $\mathbb{R}^n$ are equipollent. Dedekind realized that a bijection between spaces of different dimension would have to be fairly wild, and conjectured that no such bijection would be possible if we insist that the bijection is a homeomorphism, that is, if both the mapping and its inverse are continuous. Brouwer later proved this conjecture. The hypotheses cannot be weakened. Peano provided an example of a surjective continuous function $[0,1] \to [0,1] \times [0,1]$. The crucial property that sets the real field apart from the rational field is its metric completeness.

1.7.6 Completeness

It was mentioned earlier that a sequence of real numbers has a limit if and only if it is a Cauchy sequence (a result once called the 'General Convergence Principle', now generally summed up in the statement '$\mathbb{R}$ is complete'). This result exhibits the most important difference between $\mathbb{R}$ and $\mathbb{Q}$. It is easy to find Cauchy sequences of rational numbers which do not have a limit in $\mathbb{Q}$. One of the more interesting examples is given by the sequence (x_n) where

$$x_n = \left(1 + \frac{1}{n}\right)^n .$$

Its limit is Euler's number e ($= 2.718\,281\,828\ldots$), which is irrational (even better, it is transcendental – see Section 1.9).

   Proofs of the fact that every Cauchy sequence of real numbers (a_n) converges often include something of the form 'select an x_n such that $a_n < x_n < a_n + \frac{1}{n}$'. In other words we are asked to make an infinite number of arbitrary selections, a task which some view as 'uncomfortable' at best. In fact we can avoid making arbitrary choices; $\mathbb{Q}$ is countable, so we can fix some enumeration of it, perhaps the one given in Subsection 1.5.3. In this case we eliminate choice by saying 'let x_n be the first rational number (in the order of the enumeration) with the property $a_n < x_n < a_n + \frac{1}{n}$'. Many proofs in analysis may be freed from infinite choice in this fashion.

The completeness of $\mathbb{R}$ also tells us that $\mathbb{R}$ cannot be further extended by the same procedure that forged it from $\mathbb{Q}$, for if we were to form equivalence classes of Cauchy sequences of real numbers modulo null sequences we would find ourselves with a structure isomorphic to $\mathbb{R}$ itself, nothing larger or more exotic. Likewise, if we imitate the extension of $\mathbb{N}$ to $\mathbb{Z}$ and apply it to $\mathbb{Z}$ we obtain nothing new, we get an isomorphic copy of $\mathbb{Z}$. Similarly the construction of $\mathbb{Q}$ from $\mathbb{Z}$ applied to $\mathbb{Q}$ gives a copy of $\mathbb{Q}$, and the Dedekind extension of $\mathbb{Q}$ to $\mathbb{R}$, applied to $\mathbb{R}$, gives a copy of $\mathbb{R}$.

REMARKS

1. One can use $\lim_{n\to\infty}(1 + \frac{1}{n})^n$ as a *definition* of Euler's number. The fact that it converges can be proved in a variety of different ways, for instance one might prove that the sequence $((1 + \frac{1}{n})^n)$ is increasing and bounded above. Indeed, more generally, one can define the exponential function e^x by $\lim_{n\to\infty}(1 + \frac{x}{n})^n$.

2. The isomorphism that maps the set of equivalence classes of Cauchy sequences of real numbers modulo null sequences to $\mathbb{R}$ is precisely $\langle a_n \rangle \mapsto \lim_{n\to\infty} a_n$. If (A, B) is a 'Dedekind cut' of real numbers then the isomorphism from all such cuts to $\mathbb{R}$ is simply the map $(A, B) \mapsto \sup A$, where $\sup A$ denotes the least upper bound of A.

The field of real numbers is the metric completion of the field of rational numbers; it is equal to its own metric closure. Thanks to this dense limit-stable structure we can formally define the operations of calculus.

1.7.7   Calculus formalized

The familiar differentiation and integration of calculus are clearly geometric notions. The derivative of a function f at a point a is pictured as the gradient of the tangent to the curve of the graph of the function at $(a, f(a))$. The definite integral of a positive[1] function between two points a and b is the area bounded between the curve, the x-axis and the two vertical lines $x = a$ and $x = b$ (see Figure 1.12).

---

[1] The integral gives the total 'signed area' bounded between the curve and the axis: areas bounded *below* the real axis and the curve are taken to be negative (in calculating the Riemann sums one takes the product of the width of the partitioning intervals with the possibly negative 'height' of the function on those intervals).

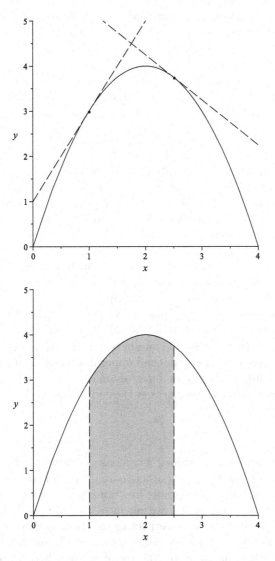

Figure 1.12 Geometrically motivated calculus. The two graphs above show the parabola $y = x(4 - x)$. At the top the two points on the curve at $x = 1$ and $x = \frac{5}{2}$ have been marked together with the dashed tangent lines passing through these points. The gradients, 2 and -1, of these lines represent the gradient of the curve at those points and are respectively the derivatives of the function $x \mapsto x(4 - x)$ at $x = 1$ and $x = \frac{5}{2}$. At the bottom, the shaded area bounded between the vertical lines $x = 1$ and $x = \frac{5}{2}$, the curve, and the x-axis is the definite integral $\int_1^{5/2} x(4 - x)\,\mathrm{d}x$. Both intuitively clear ideas are formally described by limiting processes which were only made precise in the nineteenth century.

Early work on the calculus was infested with references to 'infinitesimal quantities'. These ideal elements were assigned labels and were manipulated in symbolic form alongside regular superior sized magnitudes as if the two were perfectly compatible and satisfied similar algebraic properties. Such a treatment was bound to lead to problems, as was argued by Berkeley in *The Analyst*,[1] his 1734 criticism of this vague practice.

The epsilon-delta method, as the modern approach is sometimes called, freed the subject from the spectre of infinitesimals. The difficulties evaporate by interpreting the derivative $f'(a)$ of the function f at a as

$$\lim_{h \to 0} \frac{f(a+h) - f(a)}{h}.$$

The definite integral is also easily formally captured. To any finite partition – a disjoint union of intervals – of the interval $[a, b]$ (which we assume to be a subset of the domain of some function f) we associate upper and lower sums, as illustrated in Figure 1.13. The set of lower sums (indexed by the uncountable family of all finite partitions of $[a, b]$) is a set of real numbers bounded above and hence, by completeness, has a least upper bound, the *lower integral*. Similarly the set of upper sums (indexed by the uncountable family of all finite partitions of $[a, b]$) is a set of real numbers bounded below and hence, by completeness, has a greatest lower bound, the *upper integral*.[2]

It may be the case that the upper and lower integrals differ. For this to happen the function must be wildly discontinuous, for example the function which takes value 0 for all irrational numbers in $[0, 1]$ and value 1 for all rational numbers in $[0, 1]$ has lower integral 0 and upper integral 1. Functions $f : [a, b] \to \mathbb{R}$ for which the upper and lower integrals exist and coincide are said to be *(Riemann) integrable* (see the remarks at the end of this section) and the definite integral

$$\int_a^b f(x)\, dx$$

is defined to be the common value. The class of integrable functions includes all continuous functions and many other more exotic functions.

A limit of a sequence of integrable functions need not be integrable. There are other sophisticated types of integral, most notably the Lebesgue integral, which are much more stable with respect to their limiting behaviour.[3]

The famous Fundamental Theorem of Calculus tells us that integration and differentiation are in some sense the inverse of one another. More precisely, if f is a differentiable function with Riemann integrable derivative f', then

$$\int_a^b f'(x)\, dx = f(b) - f(a).$$

---

[1] Berkeley [**17**].

[2] If $A \subseteq \mathbb{R}$ has a lower bound then $\{-x : x \in A\}$ has an upper bound and consequently has a least upper bound a. It follows that $-a$ is the greatest lower bound of A.

[3] Jean Dieudonné goes a little further when demoting the Riemann integral in Dieudonné [**54**] (page 142), boldly claiming that the construction of the Riemann integral is only a mildly interesting exercise which, in the light of the Lebesgue integral, has far outlived its importance.

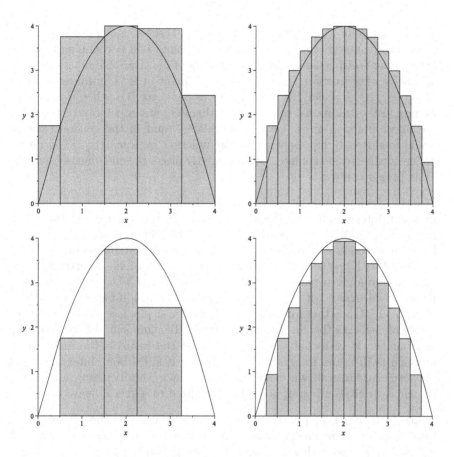

Figure 1.13 Two different partitions of the interval $[0, 4]$ into finitely many subintervals are depicted. For each partition the sum of the areas of the shaded rectangles is, for the top row, the upper sum, and for the bottom row, the lower sum of the illustrated function with respect to the given partition. These are two of infinitely many such partitions. Consider the set of all finite partitions of $[0, 4]$ and their respective upper and lower sums. The set of all lower sums is bounded above and the set of all upper sums is bounded below. By Dedekind completeness there is a least upper bound U and a greatest lower bound L of these sets respectively. If L and U coincide then the function is said to be (Riemann) integrable and the common value is the definite integral of the function on $[0, 4]$.

Following the precise epsilon-delta arguments (as may be found in any analysis text) removes all mystery from the processes of calculus.

REMARKS

1. The derivative of a function at some point is best thought of as the closest linear approximation to the function at that point. This becomes particularly transparent when generalizing to derivatives of multivariable functions. The derivative of a function $f : \mathbb{R}^n \to \mathbb{R}^m$ at a point in $\mathbb{R}^n$ is a linear *function* $\mathbb{R}^n \to \mathbb{R}^m$, not a scalar. In the one-dimensional case (i.e. when $n = m = 1$) we are, in the traditional way of presenting derivatives, simply representing the linear function $x \mapsto \lambda x$ by the scalar λ (and in the general case we would analogously represent the linear function by a matrix). The multi-dimensional extension of integration similarly places the one-dimensional case in its proper context.

2. Strictly speaking, the integral we have defined in this section is not the Riemann integral but the *Darboux integral*. In the usual definition of the Riemann integral for each partition of $[a, b]$ into n intervals $I_1, I_2, \ldots, I_n$ and for each selection of elements $x_i \in I_i$, the associated Riemann sum of f is defined as $\Sigma_{i=1}^n f(x_i)|I_i|$, where $|I_i|$ is the length of the interval I_i. The mesh size of a given partition $I_1, I_2, \ldots, I_n$ is $\max\{|I_n| : i = 1, \ldots, n\}$. f is Riemann integrable with Riemann integral s over $[a, b]$ if for any $\varepsilon > 0$ there exists a $\delta > 0$ such that for every partition of mesh size less than δ and every choice of elements therein the corresponding Riemann sum differs from s by less than ε. It turns out, however, that the end result is the same, i.e. a function is Riemann integrable if and only if it is Darboux integrable, and the resulting integrals will have the same value. The Riemann integral (or Riemann–Stieltjes integral) is easily generalized to higher dimensions.

3. Heuristically it is rather easy to see why the Fundamental Theorem of Calculus is true. For some fine partition of $[a, b]$ where $a = x_0 < x_1 < \cdots < x_n = b$ are equally spaced with $x_{i+1} - x_i = h$ we have $f(b) - f(a) = f(x_n) - f(x_0)$ which we write as

$$f(x_n) - f(x_{n-1}) + f(x_{n-1}) - f(x_{n-2}) + f(x_{n-2}) - f(x_{n-3}) + \cdots + f(x_1) - f(x_0)$$

and in turn as

$$h\left(\frac{f(x_n) - f(x_{n-1})}{h} + \frac{f(x_{n-1}) - f(x_{n-2})}{h} + \cdots + \frac{f(x_1) - f(x_0)}{h}\right)$$

and for small h the last expression is visibly close to $\int_a^b f'(x)\, \mathrm{d}x$.

4. The philosophy behind integration theory is somewhat analogous to that of divergent series. One would like to be able to integrate as many different types of function as possible while always assigning the same values to, say,

continuous functions. In addition we would like the integral to be fairly well-behaved with respect to algebraic operations (for example, $\int f+g = \int f + \int g$ and $\int \lambda f = \lambda \int f$ for functions f, g and $\lambda \in \mathbb{R}$). There is a rich theory of different types of integral.

Weierstrass' precise definition of limits and continuity puts calculus on a sure footing. Both differentiation and integration are limiting processes which can be described in 'epsilon-delta' terms, removing the uncomfortable use of infinitesimals that had been in use since the days of Leibniz and Newton. The integral can be used to generate a hierarchy of functions. One such function is the natural logarithm.

1.7.8   Exponentiation

The natural logarithm, ln, is often formally defined for any real number x in the interval $(0, \infty)$ by

$$\ln x = \int_1^x \frac{\mathrm{d}t}{t}.$$

Employing basic techniques of calculus it is an easy exercise to prove the familiar properties of ln from the definition, e.g. $\ln xy = \ln x + \ln y$; $\ln \frac{1}{x} = -\ln x$, etc. The exponential function exp is the inverse of ln (i.e. the unique function $f : \mathbb{R} \to \mathbb{R}$ satisfying $f(\ln x) = x$ for all $x \in (0, \infty)$). The power series

$$\sum_{n=0}^{\infty} \frac{x^n}{n!} = 1 + x + \frac{x^2}{2!} + \frac{x^3}{3!} + \cdots$$

converges to $\exp(x)$ for all real numbers x; in particular, Euler's number e is defined by the series $\exp(1) = 1 + 1 + \frac{1}{2!} + \frac{1}{3!} + \cdots = 2.718\,281\,828\ldots$. Alternatively we can reverse the order of this analysis, defining $\exp(x)$ by the power series and the logarithm $\ln : (0, \infty) \to \mathbb{R}$ as the inverse of exp and we can deduce the integral $\ln x = \int_1^x \frac{\mathrm{d}t}{t}$ as a theorem.

We define exponentiation for real numbers a, b with $a > 0$ by

$$a^b = \exp(b \ln a).$$

It is easy to prove that the familiar rules of powers hold and that for rational a and integer b the result coincides with that given previously. If $a = 0$ we make the convention that $a^b = 0$ unless $b = 0$, in which case the expression is undefined. (For many purposes, depending critically on whether one prefers a combinatorial or analytic interpretation of powers, the convention $0^0 = 1$ is sensible.) The definition of a^b for negative a will have to wait for the introduction of complex numbers (see Subsection 1.8.4).

1. Using the integral as a means of generating new functions of increasingly non-elementary character suggests the following hierarchical organization of real-valued functions. Let us define $\mathcal{E}_0$ to be the set containing just one function: the constant zero function $\mathbb{R} \to \mathbb{R}$. In general, we define $\mathcal{E}_{n+1}$ to be the class of all finite sums, finite products, quotients, composites and inverses of functions whose derivatives are in $\mathcal{E}_n$. This hierarchy grows quickly, and before long we encounter most of the functions that appear in general mathematics. $\mathcal{E}_1$ is the set of all constant functions. The set of functions with derivatives in $\mathcal{E}_1$ is simply the degree one polynomials, the linear functions $ax + b$. Forming the class $\mathcal{E}_2$ of all finite sums, finite products, quotients, composites and inverses of these yields all rational functions (quotients of polynomials), their inverses and all algebraic combinations thereof – already a substantial class. This means that $\mathcal{E}_3$ will contain, among many other functions, logarithms, exponentials and all trigonometric functions. Esoteric special functions soon emerge. And so on.

2. Later we will interpret n^m as the number of functions from a set of m elements to a set of n elements, so with this definition we have $0^0 = 1$, a definition which is also required to ensure that some series expressions such as $\sum_{n=0}^{\infty} x^n$ for $\frac{1}{1-x}$ (with $|x| < 1$) make sense. Nevertheless, from an analytical point of view, the function $f(x,y) = x^y$ is wildly discontinuous at $(0,0)$; it cannot be extended to a continuous function by defining $f(0,0)$ arbitrarily.

There are two approaches to exponentiation. Either we define the natural logarithm via the integral of the reciprocal function, then define the exponential function to be the inverse of the logarithm and derive its series expansion, or we define the exponential function as the series at the outset (deriving the integral as a theorem). Once the exponential function and logarithm are defined we can define real powers of positive real numbers. Calculus survived without a formal foundation for so long because any need for rigour seemed to be outshone by the extraordinary successes that the theory had won, especially in mechanics.

1.7.9   Mechanical origins and Fourier series

The techniques of calculus, from the geometer's point of view, are simply translations or modifications of ancient exhaustion methods.[1] The novelty in the modern approach is that differentiation and integration form a coherent purely algebraic approach to such geometric problems. The algebraic recasting of geometry was championed by Descartes in a development which ran against the

---

[1] See Thomas Heath's discussion in Chapter VII (Anticipations by Archimedes of the Integral Calculus) of Archimedes [4].

classical tradition of Euclid. Descartes' *La Géométrie*, an appendix to his *Discours de la méthode* (1637),[1] introduced cartesian coordinates and was the first published synthesis of algebra and geometry. Although Descartes cannot lay claim to all of the associated techniques of this fledgling theory, he was certainly the first to attempt to bring the various ideas together harmoniously in a single account. Symbols ('x' and 'y') were introduced to denote varying quantities regardless of their physical nature. Thus, in considering the graphs of functions, lengths were used to represent all types of continuous magnitude – length, area, volume or otherwise.

Berkeley's mockery of the lack of rigour to be found in the early calculus did not significantly influence mathematics at the time – mathematicians had become delirious with the explosion of activity in the subject, especially with regard to its hugely successful applications in the field that gave birth to it: mechanics.

In the context of mechanics differentiation is interpreted as giving the rate of change of one quantity (position) with respect to another (time); the first and second derivatives are then interpreted as velocity and acceleration, respectively. With this approach it is possible to speak of *instantaneous* positions, velocities and accelerations and, at least in simple cases, to derive explicit equations describing how these values change over time. This gave a clear response to ancient ideas regarding mechanics such as those provoked by discussions of Zeno's arrow paradox (see Subsection 1.13.3).

In one such application of calculus to mechanics in the eighteenth century Daniel Bernoulli studied the problem of how to give a mathematical description of the motion of a vibrating string. Based on a reasonable idealization of the properties of the string, he came to conclude that '*all sonorous bodies include an infinity of sounds with a corresponding infinity of regular vibrations...*'.[2] This problem led to developments which were to have a deep impact on mathematics both practical and pure.

Building on the work of Bernoulli and others, Jean-Baptiste Joseph Fourier formulated an interesting general and now familiar problem: which functions can be represented as a possibly infinite sum of waves?

If f is a periodic function with period 2π (i.e. $f(x+2\pi) = f(x)$ for all x) then the Fourier coefficients a_n and b_n of f (if the given integrands are integrable) are:

$$a_n = \frac{1}{\pi} \int_{-\pi}^{\pi} f(t) \cos nt \, \mathrm{d}t, \ n = 0, 1, 2, \ldots,$$

$$b_n = \frac{1}{\pi} \int_{-\pi}^{\pi} f(t) \sin nt \, \mathrm{d}t, \ n = 1, 2, \ldots,$$

and the series

$$\frac{a_0}{2} + \sum_{n=1}^{\infty} (a_n \cos nt + b_n \sin nt)$$

---

[1]Descartes [49].

[2]Bernoulli, D., 'Réflections sur les nouvelles Vibrations des Cordes exposés dans les Mémoires de D'Alembert et Euler', *Mém. Acad. Berl.* (1753).

is the *Fourier series* of f. By applying a simple change of variable the results of Fourier analysis apply to periodic functions of arbitrary period.

Mathematicians had to wait until the twentieth century before the deep questions concerning Fourier series were resolved. In 1966 Lennart Carleson[1] famously proved that if $|f|^2$ is Lebesgue integrable (f periodic with period 2π), and in particular if f is continuous, then its Fourier series converges to f at all points except possibly on a set of zero Lebesgue measure.[2] Jean-Pierre Kahane and Yitzhak Katznelson[3] later proved that this is the best possible result in the sense that for each set X of measure zero there exists a continuous function f for which the Fourier series of f fails to converge to f precisely at all points in X. So, in answer to Fourier's question, almost all periodic functions are expressible as a possibly infinite sum of waves (two examples are given in Figure 1.14). This result would have startled the early analysts.

The difficult questions concerning the convergence of general trigonometric series induced significant developments in analysis in the nineteenth century and ultimately led to Cantor's introduction of ordinal numbers (we shall see the connection in Subsection 1.11.7). All of these ideas were critical both in the development of the concept of function and in set theory in general.

Remarks

1. The most familiar example of Archimedean exhaustion is the method of estimating π by inscribing and enscribing a circle by regular polygons of ever higher numbers of edges, anticipating integration. Archimedes also produced the first known example of a summation of an infinite series in calculating the area of a segment of a parabola.

2. When Bernoulli claims that 'sonorous bodies include an infinity of sounds' he is referring to the fact that, say, the sound of a plucked string is a combination of its fundamental frequency f and its harmonics $2f$, $3f$, $4f$, etc. That is, the shape of the sound wave is an infinite linear sum of waves having these frequencies, the coefficients sculpting the final sound. In practice the very high harmonics are sufficiently inaudible to be ignored.

3. Fourier series generate a treasure trove of infinite sums. As already mentioned in Figure 1.14 the square and sawtooth waves tell us 'for free' that $1 - \frac{1}{3} + \frac{1}{5} - \frac{1}{7} + \cdots = \frac{\pi}{4}$ and $\frac{1}{1^2} + \frac{1}{3^2} + \frac{1}{5^2} + \frac{1}{7^2} + \cdots = \frac{\pi^2}{8}$. From this second series, since $\frac{1}{2^2} + \frac{1}{4^2} + \frac{1}{6^2} + \frac{1}{8^2} + \cdots = \frac{1}{4}(1 + \frac{1}{2^2} + \frac{1}{3^2} + \frac{1}{4^2} + \cdots)$, it is easy to find the solution to the famous *Basel problem* on the sum of reciprocals of squares: $1 + \frac{1}{2^2} + \frac{1}{3^2} + \frac{1}{4^2} + \cdots = \frac{\pi^2}{6}$. (Euler's original solution to this problem involved factorizing $\frac{\sin x}{x}$ as an infinite product of linear terms.)

---

[1] Carleson, L., 'On convergence and growth of partial sums of Fourier series', *Acta Mathematica* **116**, 135–157, (1966).

[2] Lebesgue measure will be defined in Subsection 8.4.5.

[3] Kahane, J.-P. and Katznelson, Y., 'Sur les ensembles de divergence des séries trigonométriques', *Studia Math.*, **26** , 305–306, (1966).

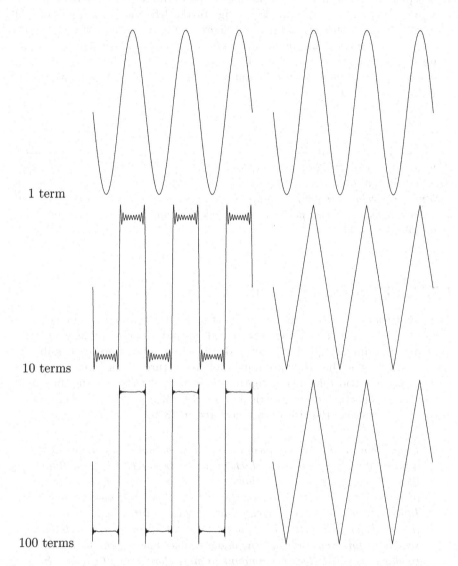

1 term

10 terms

100 terms

Figure 1.14 A familiar example which is usually given early on in the study of Fourier series is that of the square wave with period 2π and amplitude 1 (see the left column above). The Fourier series of this wave is $\frac{4}{\pi}\left(\frac{\sin t}{1} + \frac{\sin 3t}{3} + \frac{\sin 5t}{5} + \cdots\right)$. This series converges to the wave at all points t at which the wave is continuous (i.e. everywhere except at the set of points $D = \{n\pi : \ n \in \mathbb{Z}\}$) and converges to 0 for $t \in D$. If we substitute $t = \frac{\pi}{2}$ we obtain the series $1 - \frac{1}{3} + \frac{1}{5} - \frac{1}{7} + \cdots$ for $\frac{\pi}{4}$. The sawtooth wave of period 2π and amplitude 1 (see the right column above) has Fourier series $\frac{8}{\pi^2}\left(\frac{\sin t}{1^2} - \frac{\sin 3t}{3^2} + \frac{\sin 5t}{5^2} - \frac{\sin 7t}{7^2} + \cdots\right)$; substituting $t = \frac{\pi}{2}$ we obtain the series $\frac{1}{1^2} + \frac{1}{3^2} + \frac{1}{5^2} + \frac{1}{7^2} + \cdots$ for $\frac{\pi^2}{8}$.

4. If a piecewise continuous function has a discontinuity at a then the Fourier series of f will converge to the average of the left and right limits of f at a. Looking at the left column in Figure 1.14 we can see that the wave seems to overshoot the square wave at each point of discontinuity, leaping up or dipping a little too far before returning to the line. This behaviour of Fourier series at points of discontinuity, well understood, is known as the *Gibbs phenomenon*.

Almost all periodic functions are expressible as a possibly infinite sum of waves. As analysis developed it was able to describe more and more complex functions with progressively more peculiar properties. This took the subject further from its geometric origins. It was possible to describe in analytic form functions which were not visualizable.

1.7.10   Pathological functions

The belief that a function should be identified with its graphical depiction, a point of view favoured by Euler, was tested beyond breaking point when some 'pathological functions' were discovered. These functions are expressible as, say, an infinite sum, but they are impossible to picture. Some highly respected mathematicians had difficulty accepting these new objects as 'true' functions, blaming the departure from 'practicality' on a trend toward overgeneralization or overformalization. Henri Poincaré declared in 1899:

> *Logic sometimes makes monsters. For half a century we have seen a mass of bizarre functions which appear to be forced to resemble as little as possible honest functions which serve some purpose. More of continuity, or less of continuity, more derivatives, and so forth. Indeed, from the point of view of logic, these strange functions are the most general; on the other hand those which one meets without searching for them, and which follow simple laws appear as a particular case which does not amount to more than a small corner. In former times when one invented a new function it was for a practical purpose; today one invents them purposely to show up defects in the reasoning of our fathers and one will deduce from them only that. If logic were the sole guide of the teacher, it would be necessary to begin with the most general functions, that is to say with the most bizarre. It is the beginner that would have to be set grappling with this teratologic museum.[1]*

---

[1] As quoted in Kline [126]. As evidenced in this quote, Poincaré was not a fan of the 'reduction of mathematics to logic'. See Gray [85] for more details.

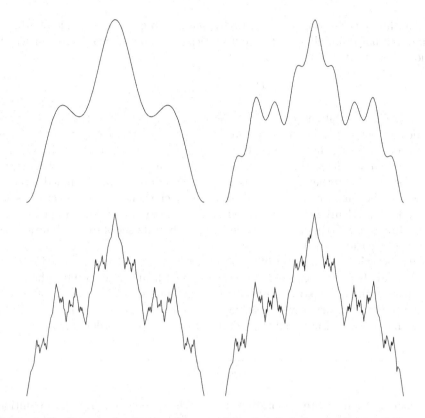

Figure 1.15   Four graphs showing the sum of the first 2, 3, 5 and 9 terms of the series $\sum_{n=0}^{\infty} \frac{\cos(3^n x)}{2^n}$ on the interval $[-\pi, \pi]$, a continuous but nowhere differentiable function. The axes have been omitted for clarity.

Weierstrass' everywhere continuous but nowhere differentiable function, the graph of which might be described as being 'made entirely of corners', is perhaps the most famous example. A family of such functions is given by

$$f(x) = \sum_{n=0}^{\infty} b^n \cos(a^n x)$$

where a is an odd integer greater than 1, $0 < b < 1$ and $ab > 1$ (see Figure 1.15).

The first example of a continuous nowhere differentiable function seems to have been discovered by Bernhard Bolzano in the 1830s, but his example went unnoticed, only to be discovered and published much later, and the phenomenon did not enter into the realm of common mathematical knowledge until Weierstrass revealed his example.[1]

---

[1]Presented by Weierstrass to the *Königliche Akademie der Wissenschaften* on July 18, 1872.

Another surprise was provided by Riemann who gave an example of a function which has infinitely many discontinuities between any two points but which is nonetheless integrable:

$$f(x) = \sum_{n=1}^{\infty} \frac{(nx)}{n^2},$$

where (x) is the difference $x - k$, k the nearest integer to x and $(x) = 0$ if x is midway between two integers (see Figure 1.16).[1] The mapping $x \mapsto (x)$ is discontinuous at the half-integers $0, \pm\frac{1}{2}, \pm 1, \pm\frac{3}{2}, \ldots$ and the effect of Riemann's function is to condense these discontinuities, forming a dense subset of $\mathbb{R}$. However, for any $\varepsilon > 0$ there are only finitely many jumps in any bounded interval for which the function f leaps a distance greater than ε, and this ensures its integrability. (Henri Lebesgue proved half a century later[2] the stronger result that a function is Riemann integrable if and only if its set of discontinuities has Lebesgue measure zero.)

Here we witness the power of algebra to overthrow faulty geometric intuition (implicit in this statement is the wholesale identification of the line, the plane and space with $\mathbb{R}$, $\mathbb{R}^2$ and $\mathbb{R}^3$ respectively). By the end of the period of reform in the nineteenth century it was possible to prove results in the calculus *rigorously* from what we would now call the axioms of the complete ordered field $\mathbb{R}$.

REMARKS

1. If a definition creates 'monsters' then it is time to modify the definition (perhaps add some stricter conditions). Today we are accustomed to vast zoos of functions (differentiable, bounded, n-times differentiable, analytic, Riemann integrable, Lebesgue integrable, Lipschitz, C_0, etcetera) and we choose to restrict ourselves to such subclasses as we see appropriate. The question of 'what a function should be' is a peculiar one.

2. The class of continuous but nowhere differentiable functions given here is G. H. Hardy's neater version of Weierstrass' original family.[3] Weierstrass' original was

$$f(x) = \sum_{n=0}^{\infty} b^n \cos(a^n \pi x),$$

where $0 < b < 1$, a is a positive odd integer and $ab > 1 + \frac{3}{2}\pi$.

3. Topologically speaking *most* continuous functions are nowhere differentiable in the sense that the complement of the set of nowhere differentiable continuous functions in the class of continuous functions is a countable union of nowhere dense sets (density being intended in the uniform sense). This is a

---

[1] This is an example given by Riemann in his *Habilitationsschrift* (1854).
[2] Lebesgue [136].
[3] See Hardy, G. H., 'Weierstrass's nondifferentiable function', *Trans. Amer. Math. Soc.*, **17**, 301–325, (1916).

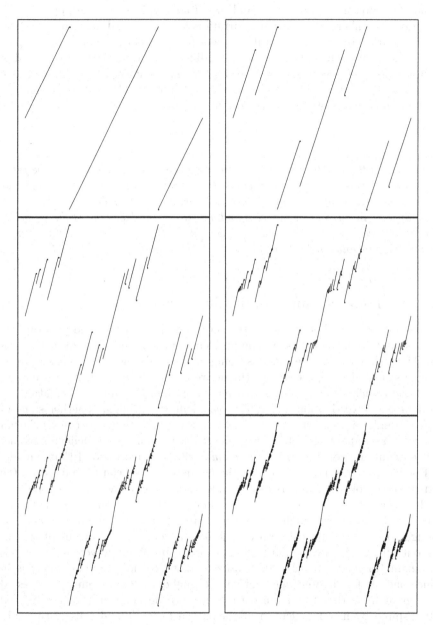

Figure 1.16 Six graphs showing the sum of the first 1, 2, 4, 8, 16 and 32 terms of the series $\sum_{n=1}^{\infty} \frac{(nx)}{n^2}$ converging to Riemann's everywhere discontinuous but integrable function on the interval $[-1, 1]$. Note that most of the discontinuities are too small to see at this scale. (See the text for the definition of (nx) – as before the axes have been omitted for clarity.)

familiar theme. The most intuitively graspable objects in mathematics need not be 'typical' in any sense, and are often atypical, despite appearances: rational numbers are dwarfed by irrational numbers and algebraic numbers are dwarfed by transcendental numbers (as we shall see later); the apparently pathological continuous nowhere differentiable functions are typical of functions as a whole. In an amusing turn of events nowhere differentiable functions have turned out to be of relevance in physical models too; consider the sample paths of Brownian motion, for example.

At the end of the nineteenth century mathematicians were able to describe functions which would not have qualified as 'true' functions in the earlier geometric sense. The graphs of these new functions appeared to be made entirely of corners, or were infinitely disconnected and yet still integrable. One of the surprises of twentieth century logic was that it is possible, after all, to define infinitesimals in a consistent and workable way.

1.7.11   Foundations; infinitesimals anew

In the nineteenth and early twentieth century mathematicians and philosophers became quite anxious to resolve the problems in the foundations of mathematics. The task of freeing the calculus from infinitesimals and other quasi-intuitive artefacts started this process and the desire to understand infinity, to uncover the basic principles of mathematical reasoning, and to unify mathematics, fuelled further investigation. In 1899 David Hilbert[1] brought together some of these strands, ancient and new, clarifying the connection between Euclidean geometry and analysis. Hilbert provided twenty axioms for Euclidean geometry (patching up some gaps in *The Elements*) and demonstrated that a model of Euclidean geometry, essentially uniquely determined, can be built from the formal theory of real numbers as embedded in Cantor's new theory of sets.

It was clear by this time that one could do analytical geometry without any need for the 'infinitely small'; after Weierstrass the idea of 'infinitesimals' was thought of as an historical monster, at best a non-rigorous means of sketching out an idea before demonstrating a formal proof. However, in the 1960s Abraham Robinson[2] developed a subject known as non-standard analysis in which one *can* legitimately speak of infinitesimal quantities. It must be stressed with great force that this is an entirely modern development, an unexpected gift inspired by new frontiers in mathematical logic. The details of this subject, which still has a relatively small following, are set out in Robinson's book *Non-Standard Analysis*.[3]

---

[1] As described in Hilbert [**100**].
[2] Robinson, A., 'Non-standard analysis', *Nederl. Akad. Wetensch. Proc.*, Ser. A, **64**, 432–40, (1961).
[3] Robinson [**176**].

Robinson, inspired by Thoralf Skolem's non-standard models of first-order Peano Arithmetic,[1] proved the existence of field extensions *$\mathbb{R}$ of $\mathbb{R}$ which contain both infinitesimal elements (elements ι with $0 < \iota < r$ for all positive real r) and infinite elements (elements m with $0 < r < m$ for all positive real r), and yet which do not differ significantly from $\mathbb{R}$ in terms of logical properties. Indeed they are so logically similar that the construction provides a new method of proof: given a statement about real numbers and real functions one transforms it in a natural way into an associated statement about the extended field and its functions. In this new setting one can make use of the novel features of *$\mathbb{R}$. One then appeals to the result that the original statement is true if and only if its transformed statement is true. The operations of calculus are then expressed in terms of infinitesimals mirroring Leibniz's 260 year old 'incomparably small quantities'[2] and so the extended field is arguably an interpretation in precise modern terms of what Leibniz and his contemporaries sought to describe.

The existence of non-standard models of the real numbers is a consequence of the Compactness Theorem for first-order logic (see Subsection 10.2.2). The novelty in Robinson's construction is that the field is 'close enough' to $\mathbb{R}$ to be of some use. There are various approaches to these new field extensions of $\mathbb{R}$: one can introduce *$\mathbb{R}$ axiomatically, $\mathbb{R}$ appearing as a distinguished subfield, or one can start with $\mathbb{R}$ and describe an extension to *$\mathbb{R}$. Here we will focus on an algebraic construction of such an extension *$\mathbb{R}$, the field of *hyperreals*, which is as concrete as can reasonably be expected. In order to do this we need a few elementary results from ring theory.

A *ring* is a set A together with two binary operations $A^2 \to A$, addition and multiplication, such that addition is commutative and associative (and A possesses an additive identity element 0), multiplication is associative, and multiplication distributes over addition. Of the structures we have met so far, $\mathbb{Z}$, $\mathbb{Z}_n$, $\mathbb{Q}$ and $\mathbb{R}$ are all examples of rings.

Consider the ring $\mathcal{R}$ of all real sequences. This is simply the set of all sequences (a_n) of real numbers on which we define operations of addition and multiplication by: $(a_n) + (b_n) = (a_n + b_n)$ and $(a_n)(b_n) = (a_n b_n)$. The subring $\mathcal{N}$ of $\mathcal{R}$ comprising those sequences which only have finitely many non-zero terms is an *ideal* of $\mathcal{R}$, meaning that not only is it a ring in its own right, but for any element $r \in \mathcal{R}$ and any element $n \in \mathcal{N}$, we have $rn \in \mathcal{N}$. Given any ideal $\mathcal{I}$ of $\mathcal{R}$ the relation $r_1 \sim r_2$ on $\mathcal{R}$ defined by $r_1 - r_2 \in \mathcal{I}$ is an equivalence relation, and the class of all equivalence classes under this relation, denoted $\mathcal{R}/\mathcal{I}$, is a ring. If we denote the equivalence class containing r by $r + \mathcal{I}$, then addition and multiplication are defined by $(r + \mathcal{I}) + (s + \mathcal{I}) = r + s + \mathcal{I}$ and $(r + \mathcal{I})(s + \mathcal{I}) = rs + \mathcal{I}$.

Among all ideals of $\mathcal{R}$ which have $\mathcal{N}$ as a subring (and hence as an ideal) there are maximal ideals, i.e. ideals $\mathcal{M}$ with the property that if $\mathcal{I}$ is an ideal

---

[1] Skolem, T., 'Über die Nicht-charakterisierbarkeit der Zahlenreihe mittels endlich oder unendlich vieler Aussagen mit ausschliesslich Zahlenvariablen', *Fundamenta Mathematicae*, **23**, 150–161, (1934).

[2] A description used by Leibniz in a 1702 letter to Pierre Varignon. Robinson pays tribute to Leibniz by calling infinitesimal neighbourhoods of standard real numbers *monads*.

of $\mathcal{R}$ with $\mathcal{M}$ as a proper ideal, then $\mathcal{I} = \mathcal{R}$. We know that such ideals exists by Zorn's Lemma (see Section 9.2). Let us fix one such maximal ideal $\mathcal{M}$. It is a general result of ring theory that the ring ${}^*\mathbb{R} = \mathcal{R}/\mathcal{M}$ is a field. $\mathbb{R}$ embeds in ${}^*\mathbb{R}$ via $r \mapsto (r) + \mathcal{M}$ where (r) is the constant sequence $(r, r, \ldots)$.

The order $<$ on $\mathbb{R}$ extends naturally to an order on ${}^*\mathbb{R}$. For $m = (m_n) \in \mathcal{M}$ let $Z(m) = \{n \in \mathbb{N} : m_n = 0\}$. Then we say $(a_n) + \mathcal{M} < (b_n) + \mathcal{M}$ if there is an $m \in \mathcal{M}$ with $\{n : a_n < b_n\} = Z(m)$. An example of an infinitesimal element in ${}^*\mathbb{R}$ is $(1, \frac{1}{2}, \frac{1}{3}, \frac{1}{4}, \ldots) + \mathcal{M}$ and an example of an infinite element is $(1, 2, 3, 4, \ldots) + \mathcal{M}$.

A problem with this construction is that there is no canonical way of forming $\mathcal{M}$; and there are many candidates to choose from. However, if we further assume the Continuum Hypothesis (see Section 9.3) we know that for any two such maximal ideals $\mathcal{M}_1$ and $\mathcal{M}_2$, $\mathcal{R}/\mathcal{M}_1$ and $\mathcal{R}/\mathcal{M}_2$ are isomorphic.

To any function $f : \mathbb{R} \to \mathbb{R}$ we associate a unique function ${}^*f : {}^*\mathbb{R} \to {}^*\mathbb{R}$ by defining

$$ {}^*f(((a_k), (b_k), (c_k), \ldots) + \mathcal{M}) = ((f(a_k)), (f(b_k)), (f(c_k)), \ldots) + \mathcal{M}. $$

(The function ${}^*f$ is well-defined by this construction, although this is perhaps not immediately obvious.) One similarly defines ${}^*f$ for functions f on $\mathbb{R}^n$. Any statement ϕ about $\mathbb{R}$ and functions on $\mathbb{R}^n$ then translates, via the mapping $f \mapsto {}^*f$, to a statement ${}^*\phi$ about ${}^*\mathbb{R}$ and functions on ${}^*\mathbb{R}^n$. One can prove for a wide range of statements ϕ (essentially the class of first-order statements about real functions and real numbers) that ϕ is true if and only if ${}^*\phi$ is true. We are getting a little ahead of ourselves here; more logical machinery is needed to make this slightly vague statement satisfactory, and there are some subtleties that can easily be overlooked at first, but I hope the main thrust of the idea is clear. It is this transference principle which is of interest; it may be that the proof of ${}^*\phi$, in the new setting of ${}^*\mathbb{R}^n$, is easier, or more intuitive, than the proof of ϕ.

We must not make the mistake of assuming that *any* field extension of $\mathbb{R}$ which admits infinitesimals enjoys such transference. Consider, for example, the field $\mathbb{R}(x)$ of rational functions, this being the set of all expressions of the form

$$ \frac{a_n x^n + a_{n-1} x^{n-1} + \cdots + a_1 x + a_0}{x^m + b_{m-1} x^{m-1} + \cdots + b_1 x + b_0} $$

where $n, m \in \mathbb{N}$, $a_i, b_i \in \mathbb{R}$. The above element is deemed positive if $a_n > 0$ and we order $f \prec g$ if $g - f$ is positive. This is a field extension of $\mathbb{R}$ (the order $\prec$ extending the usual order on $\mathbb{R}$) which contains infinitesimals ($\frac{1}{x}$ for example) and infinite elements (x for example). Let $a > 0$. Then $a - \frac{1}{x} = \frac{ax-1}{x} \succ 0$, so $0 \prec \frac{1}{x} \prec a$ for all $a > 0$. Similarly, $x - a \succ 0$ for all $a > 0$. But, in a sense, $\mathbb{R}(x)$ is too different and too large in relation to $\mathbb{R}$ to cast any light on the analysis of $\mathbb{R}$; there is no transference principle from $\mathbb{R}$ to $\mathbb{R}(x)$.

Non-standard analysis did not replace the Weierstrass theory and continues to be a minority sport, but it nevertheless gives an interesting new perspective on the difficulties of pre-Weierstrassian analysis and has given rise to some diverting philosophical debates and criticisms.

Another more exotic extension of $\mathbb{R}$ appeared in the 1970s when John H. Conway, inspired by game theory, demonstrated how the idea of Dedekind cuts can be generalized and, starting from a very sparse foundation, generated a huge algebraic structure (a proper class) which not only includes a copy of Robinson's non-standard field but also includes *all* ordinal numbers (which we shall meet in Section 1.11). This extension, coined the class of *surreal numbers* by Donald Knuth, nevertheless satisfies the field axioms.[1]

REMARKS

1. A rejection of infinitesimals can be traced back to Archimedes and beyond (interestingly, and perhaps surprisingly, Cantor also seemed to be opposed to the notion of infinitesimals, although this was admittedly the common view following Weierstrass' demonstration that one can do analysis without them). In his Physics (Book III (6)) Aristotle says '*every finite magnitude is exhausted by means of any determinate quantity however small*'. Archimedes puts it this way: '*any magnitude when added to itself enough times will exceed any given magnitude*'. Nevertheless, like his successors, Archimedes used the infinitesimal notion heuristically, only eliminating it in formal solutions.

   In honour of this notion, an ordered field F with an absolute value $|\cdot|$ is said to be *Archimedean* if for all non-zero x there exists a natural number n such that $|x| + \cdots + |x|$ (with n summands) is greater than 1. Both $\mathbb{Q}$ and $\mathbb{R}$ are Archimedean. Indeed, when working through the fine details of the Dedekind cut construction of $\mathbb{R}$ one makes critical use of the fact that $\mathbb{Q}$ is Archimedean, and the Archimedean property of $\mathbb{R}$ is evident. Moving away from particular models of $\mathbb{R}$, the Archimedean property follows from Dedekind completeness alone: suppose ι is a positive infinitesimal in $\mathbb{R}$, i.e. $0 < n\iota < 1$ for $n \in \mathbb{N}$. Since the set $\{n\iota : n \in \mathbb{N}\}$ has an upper bound (namely 1), it has by completeness a least upper bound u. Then $\frac{u}{2}$ is not an upper bound of $\{n\iota : n \in \mathbb{N}\}$, which means that for some $n \in \mathbb{N}$, we have $\frac{u}{2} < n\iota$. But then $u < 2n\iota$, a contradiction. Alternatively we can show by a very similar argument that $\mathbb{R}$ has no infinite elements. Suppose I is an infinite positive element of $\mathbb{R}$, so that $I > n$ for all $n \in \mathbb{N}$. Then since $\mathbb{N}$ has an upper bound in $\mathbb{R}$, namely I, it has a least upper bound m. But then $m - 1$ is not an upper bound of $\mathbb{N}$, so there is an $n \in \mathbb{N}$ with $m - 1 < n$, so $m < n + 1$, a contradiction.

2. It is the Archimedean property which provides the important link between Dedekind completeness (every set with an upper bound has a least upper bound) and Cauchy completeness (every Cauchy sequence converges). An ordered field is Dedekind complete if and only if it is Cauchy complete and

---

[1]The full details of Conway's construction are too lengthy to include here. For a full exposition, see Conway [**37**]. See also Knuth's popularization [**128**].

Archimedean, and of course in that case it is isomorphic to $\mathbb{R}$. One can find examples of Cauchy complete ordered fields which are not Dedekind complete (simply take the Cauchy completion of your favourite non-Archimedean field).

3. Every Archimedean field is isomorphic to a subfield of $\mathbb{R}$. In other words, $\mathbb{R}$ can be characterized as the maximal Archimedean field. This leads to a rather slick way of proving that all complete ordered fields are isomorphic.[1]

4. Provided we are allowed to call a proper class a field then the class of surreal numbers is the largest ordered field in the sense that it contains an isomorphic copy of every ordered field. This also highlights how strong the Archimedean property is: assume it, then the maximal Archimedean ordered field is $\mathbb{R}$, remove it and the maximal field is the enormous field of surreal numbers. Various different ways of representing surreal numbers have been developed.

Abraham Robinson discovered in the early 1960s that one can extend the real field to accommodate infinitesimal elements in such a way that the logical properties of the extended field mirror those of the real field. This provided a new way of doing analysis that was celebrated by Kurt Gödel as 'the analysis of the future'. Although the real field seems to be last word as far as metric completeness is concerned, it still suffers a certain algebraic incompleteness: the simple equation $x^2 + 1 = 0$ has no real solutions. One can remedy this by extending to a new field.

1.8   Complex numbers

Between two truths of the real domain, the easiest and shortest path quite often passes through the complex domain.

– PAUL PAINLEVÉ[2]

Out of context this seems like a very mysterious statement, especially when the jargon words 'real' and 'complex' are mistaken for their colloquial counterparts – giving the wrong impression entirely! The point is that the theory of complex functions, a beautiful synthesis of geometry and analysis, often provides remarkably clear short cuts and insights into problems of real analysis.

---

[1]See, for example, Chapter 8 of Potter [**170**].

[2]Painlevé [**162**]. This is sometimes mis-attributed to Jacques Hadamard, who alludes to it in *An Essay on the Psychology of Invention in the Mathematical Field* (1945) [**87**]: 'It has been written that...'. Several different loose translations of the quote have appeared in print.

1.8.1   The construction of $\mathbb{C}$

Having made the formal definitions we tend to gradually forget the fine details and embeddings and return, with some peace of mind, to the naive picture of an increasing family of number systems $\mathbb{N} \subset \mathbb{Z} \subset \mathbb{Q} \subset \mathbb{R}$. Appended to the end of this chain one usually finds the set of complex numbers $\mathbb{C}$.

The complex numbers may be defined as the set of all ordered pairs of real numbers with addition

$$(a, b) + (c, d) = (a + c, b + d)$$

and multiplication

$$(a, b)(c, d) = (ac - bd, ad + bc).$$

$\mathbb{C}$ is a field and has $\mathbb{R}$ as a subfield via the embedding $x \mapsto (x, 0)$. The important and motivating property of the given multiplication is that $(0, 1)^2 = (-1, 0)$; $\mathbb{C}$ has a 'square root of minus one', indeed it has exactly two such square roots, the other one being $(0, -1)$.

This may seem like a humble extension of $\mathbb{R}$. Certainly from the point of view of cardinality the extension is not profound – $\mathbb{C}$ is equipollent to $\mathbb{R}$ – however, the incorporation of $\sqrt{-1}$ and all that goes with it unlocks an extraordinary amount of algebraic and analytic structure which simply doesn't appear in real analysis. It is invariably the complex numbers, not the real numbers, which supply a natural setting for the study of problems in analysis, physics and elsewhere.[1] One can fill a terrifying number of volumes on the subject of complex analysis alone.[2] It is with the greatest effort of will that I resist discussing the topic in any detail here – to open the floodgates just a little would result in a torrent of information distracting us from our principal theme of set theory. Instead, we focus on a few core properties.

The complex number $(0, 1)$ is traditionally denoted by i, for *imaginary*, and one writes $a + bi$ for (a, b), adding and multiplying as if this were a real polynomial in i subject to the single rule $i^2 = -1$.[3]

What is remarkable is that this augmentation of $\mathbb{R}$ by a single new element, which allows us to factor $z^2 + 1$ as $(z + i)(z - i)$, suffices to split into linear factors *all* polynomials with coefficients in $\mathbb{C}$. This is the content of the Fundamental Theorem of Algebra – one of the strongest purely algebraic hints that the thin field $\mathbb{R}$ is, at least in terms of elegance, subordinate to $\mathbb{C}$.

> If $p(z)$ is a non-constant polynomial then there is a complex number a with $p(a) = 0$ (a is a *zero* of p).
>
> Corollary: if $p(z) = c_d z^d + c_{d-1} z^{d-1} + \cdots + c_1 z + c_0$ ($c_d \neq 0$) has zeros $a_1, \ldots, a_n \in \mathbb{C}$, then, for some constant c and natural $\alpha_1, \ldots, \alpha_n$, $p(z) = c(z - a_1)^{\alpha_1} \cdots (z - a_n)^{\alpha_n}$ where $\alpha_1 + \cdots + \alpha_n = d$.

---

[1] Roger Penrose enthuses about complex analysis, with an eye on physics, in Penrose [164].

[2] Conway [36] provides a nice introduction to the basics. See also the geometrically motivated Needham [160].

[3] A formal definition of the algebra of polynomials over a field, i.e. the set of expressions of the form $a_n v^n + a_{n-1} v^{n-1} + \cdots + a_1 v + a_0$, v an indeterminate and the a_is members of the field in question, is given in Section 1.9.

A host of results in real analysis which need to be split into awkward special cases translate to much simpler, easier to state and more powerful results in complex analysis. Complex analysis invariably gives us the big picture.

The formal definition of complex numbers as ordered pairs may seem like a very modern idea, but it can be traced back to the 1830s, long before the comparably formal definitions of the other number systems. William Hamilton defined $\mathbb{C}$ in this way in an 1835 paper, and Gauss says in a letter to Wolfgang Bolyai (1837) that he had been familiar with the representation since the beginning of the decade.

The algebraic construction of $\mathbb{C}$ is simple: it is just the cartesian product of $\mathbb{R}$ with itself with certain operations of addition and multiplication. What is surprising is that analysis in this extended field far surpasses real analysis in the simplicity, elegance and scope of its results. The complex numbers had a difficult birth – their troubling but highly useful existence was once a closely guarded secret.

1.8.2   Origins

Complex numbers were used with a sense of bewildered embarrassment and suspicion in the sixteenth century when Niccolo Tartaglia, Scipione del Ferro, Gioralmo Cardano (learning of the method from Tartaglia), Lodovico Ferraria and Rafael Bombelli worked on the general cubic and quartic equations:

$$ax^3 + bx^2 + cx + d = 0$$
$$ax^4 + bx^3 + cx^2 + dx + e = 0.$$

It was discovered that the real solutions to some such equations could be found, but not without first journeying through the apparently mysterious realm of complex numbers. The solvers were then placed in the difficult position of having found demonstrably correct solutions without fully understanding how they got there.

This is just one of many occasions where $\mathbb{C}$ is seen to play a critical role in *real* analysis. The general solution $x = \frac{-b \pm \sqrt{b^2 - 4ac}}{2a}$ to the quadratic equation $ax^2 + bx + c = 0$ is easily derived. The cubic and quartic equations also have (more complicated) explicit solutions in terms of their coefficients, rational functions (i.e. a quotient of two polynomials) and radicals (i.e. kth roots) – a so-called 'solution by radicals'. These solutions are found as follows.

(a) The general cubic can be reduced by a simple change of variables to the form $x^3 + px + q = 0$. Setting $x = A - B$, with $3AB = p$, the equation reduces to $A^3 - B^3 + q = 0$. Multiplying by $3^3 A^3$ eliminates B, leaving us with a quadratic in A^3: $27A^6 + 27qA^3 - p^3 = 0$, and so on.

(b) The general quartic can be reduced by a simple change of variables to the form $x^4 = -px^2 - qx - r$. Introducing a new variable A we have $(x^2 + A)^2 = (-p + 2A)x^2 - qx + (-r + A^2)$. The quadratic expression on the right-hand side we design to have a single root, so that it is of the form $C(x + D)^2$; this will be the case if and only if the discriminant $q^2 - 4(-p + 2A)(-r + A^2)$ is zero. The latter is a cubic equation in A which is solvable by the method of (a) above, and so on.

The search for a general solution by radicals to the quintic equation and beyond came to a surprising end when it was proved by Évariste Galois and Niels Henrik Abel, independently in the 1820s, that no such solution exists. This result is a major landmark in modern mathematics. The proof can be found in most Galois Theory texts. In a jargon-filled nutshell: to each polynomial equation is affiliated a natural algebraic object associated with permutations of its roots, its Galois group. If an equation is solvable by radicals then its Galois group has a nested finite sequence of subgroups each a normal subgroup of prime index in its predecessor, the final subgroup in this descending sequence being the single element group comprising the identity element alone. Any group possessing such a sequence is said to be *solvable*. One can prove the converse: if the Galois group of an equation is solvable then the equation is solvable by radicals. It is then a matter of proving that the Galois groups of the general equations of degree ≥ 5 are not solvable (in fact it suffices to consider the degree 5 case alone).[1]

It should be stressed that some particular quintic (and higher degree) equations *can* be solved by radicals (at least in theory – the number of symbols required to write out the solutions can be enormous); the Galois/Abel result simply tells us that there is no general radical solution which will work for all quintics. Techniques for solving higher degree equations that go beyond radicals exist; these express the solutions in terms of generalized hypergeometric functions.

The sixteenth century achievements were made long before the geometry of complex analysis was revealed in all its glory. $\sqrt{-1}$ was regarded as a mysterious object which was used guiltily alongside 'normal' numbers as if it was no more peculiar than a common integer. The description 'imaginary' reflects a sense of unease with the new concept, but it is a deeply unfortunate label that has stuck – imaginary numbers are no more imaginary than any other type of number.[2] Even Leonhard Euler in his 1768 elementary algebra text *The Elements*

---

[1] Galois' work *Mémoire sur les conditions de résolubilité des équations par radicaux* (1826) is translated in Appendix 1 to Edwards [61]. For Abel's results, see [1].

[2] I admit, however, that as a young teenager with a love of the surreal the term may have helped to pull me in to the subject in the first place. In answer to Euler's quote above, *all* numbers exist only in our fancy or imagination.

of Algebra struggled over the question of the existence of complex numbers:

> *This circumstance leads us to the concept of numbers, which by their very nature are impossible, and which are commonly called* imaginary numbers *or* fancied numbers *because they exist only in our fancy or imagination.*

Gauss (who mastered the geometric interpretation) laments the unsuitable terminology in the introductory review to his *Theoria residuorum Biquadraticum (Commentatio Secunda)* (1831) suggesting that $+1$, -1, $\sqrt{-1}$ might have been called direct, inverse and lateral unity instead, removing the taste of obscurity.

> *If this subject has hitherto been considered from the wrong viewpoint and thus enveloped in mystery and surrounded by darkness, it is largely an unsuitable terminology which should be blamed.*

It is a great shame that Gauss' terminology, or something similarly sober, has not been adopted.

REMARKS

1. Heron of Alexandria, working in the first century AD, came dangerously close to making use of an imaginary number when trying to calculate the volume of a truncated pyramid, but gave up trying to interpret the meaning of the 'impossible value' that his calculations had unearthed.

2. One of the simplest quintic equations that cannot be solved by radicals is $x^5 - x + 1 = 0$. An example of an irreducible quintic that *can* be solved by radicals is $x^5 + 5x^2 + 3 = 0$, which has one real solution

$$
\frac{1}{5}\left(-\frac{3125}{4} - \frac{625}{4}\sqrt{5} - \frac{125}{4}\sqrt{15 - 6\sqrt{5}} - \frac{2625}{4}\frac{\sqrt{5}}{\sqrt{15 - 6\sqrt{5}}} \right)^{1/5}
$$

$$
+ \frac{1}{5}\frac{\frac{375}{4} + \frac{125}{4}\sqrt{5} + \frac{125}{4}\sqrt{15 - 6\sqrt{5}} + \frac{375}{4}\frac{\sqrt{5}}{\sqrt{15 - 6\sqrt{5}}}}{\left(-\frac{3125}{4} - \frac{625}{4}\sqrt{5} - \frac{125}{4}\sqrt{15 - 6\sqrt{5}} - \frac{2625}{4}\frac{\sqrt{5}}{\sqrt{15 - 6\sqrt{5}}} \right)^{3/5}}
$$

$$
+ \frac{1}{5}\frac{-\frac{125}{4} - \frac{25}{4}\sqrt{5} + \frac{25}{4}\sqrt{15 - 6\sqrt{5}} - \frac{75}{4}\frac{\sqrt{5}}{\sqrt{15 - 6\sqrt{5}}}}{\left(-\frac{3125}{4} - \frac{625}{4}\sqrt{5} - \frac{125}{4}\sqrt{15 - 6\sqrt{5}} - \frac{2625}{4}\frac{\sqrt{5}}{\sqrt{15 - 6\sqrt{5}}} \right)^{2/5}}
$$

$$
- \frac{\sqrt{5}}{\left(-\frac{3125}{4} - \frac{625}{4}\sqrt{5} - \frac{125}{4}\sqrt{15 - 6\sqrt{5}} - \frac{2625}{4}\frac{\sqrt{5}}{\sqrt{15 - 6\sqrt{5}}} \right)^{1/5}}.
$$

3. In 1877 Felix Klein published his *Lectures on the Icosahedron and the Solution
   of Equations of the Fifth Degree*[1] in which he described a complete solution to
   the general quintic equation, relating it to the symmetries of an icosahedron
   and presenting the solutions in terms of hypergeometric functions.

---

*Complex numbers were first used in the sixteenth century to find solutions of
the general cubic and quartic equations. These real solutions were demonstrably
correct, but the route to discovery led the mathematicians through a thoroughly
mystifying new realm of numbers that they were unable to understand. It took
centuries to clarify the nature of this new class of objects. As the new discoveries
continued to flood in the strange new field seemed to become more and more
fundamental to mathematical analysis and eventually to physics.*

1.8.3   A few basic features of complex analysis

The geometric representation of a complex number $a + bi$ as the point (a, b)
in the plane is much more than a convenient illustration; as complex analysis
develops it reveals itself to be deeply tied to geometry.

The modulus $|z|$ of a complex number $z = a + bi$ is the Euclidean distance
$\sqrt{a^2 + b^2}$ of z from the origin 0. The real numbers a and b are respectively the
real and *imaginary* parts of z. A subset G of $\mathbb{C}$ is *open* if it is the (possibly
infinite) union of open discs $\{z : |z - a| < r\}$ ($a \in \mathbb{C}$, $r > 0$) and is *connected*
if between any two points of G we can find a continuous path which does not
stray outside G. We will focus on functions defined on *regions* of $\mathbb{C}$, by which
we mean open connected subsets of $\mathbb{C}$.

Limits and continuity for complex functions are defined just as in the real
case except we use modulus (the appropriate measure of distance from the
origin) in place of absolute value. A function $f : G \to \mathbb{C}$, G a region, is *complex
differentiable* at $a \in G$ if the limit

$$\lim_{h \to 0} \frac{f(a + h) - f(a)}{h}$$

exists and is independent of the path in $\mathbb{C}$ that h traverses to approach zero.
The map $x + yi \mapsto x - yi$ (*complex conjugation*), for example, is *not* complex
differentiable at any point because if h approaches zero along the real line we
obtain the limit 1 but if h approaches zero along the imaginary axis (i.e. h is of
the form λi, for real λ) we obtain the limit $-i$.

If $f : G \to \mathbb{C}$ is complex differentiable at every point of G, it is *holomorphic*.
Holomorphic functions are automatically differentiable arbitrarily many times,
in contrast to real differentiable functions which can have non-differentiable

---

[1] Available online. The spelling 'Ikosohedron' is used in the 1888 translation.

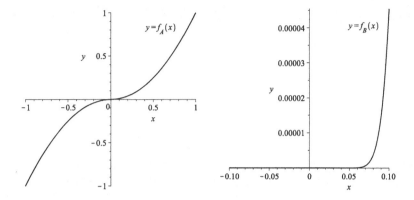

Figure 1.17 The real function illustrated on the left (called f_A in the text) is everywhere differentiable but has no second derivative at the origin. The real function illustrated on the right (called f_B in the text) is differentiable arbitrarily many times everywhere, however, it is not analytic at the origin. Both are in contrast to the good behaviour of complex functions: if a complex function is differentiable at a point then it is differentiable there arbitrarily many times and furthermore it is analytic at that point.

derivatives. For example, the function $f_A : \mathbb{R} \to \mathbb{R}$

$$f_A(x) = \begin{cases} x^2, & x \geq 0, \\ -x^2, & x < 0, \end{cases}$$

is differentiable everywhere, but $f_A'(x) = 2|x|$ which is not differentiable at 0. Furthermore, every holomorphic function has a power series expansion in a neighbourhood of every point of its domain (it is *analytic*), again in contrast to the real case. A well-known example that illustrates this is

$$f_B(x) = \begin{cases} 0, & x \leq 0, \\ e^{-\frac{1}{x}}, & x > 0, \end{cases}$$

which is differentiable arbitrarily many times, yet for no $r > 0$ is there a power series $\sum_{n=0}^{\infty} a_n x^n$ which coincides with f_B in the interval $(-r, r)$ (see Figure 1.17). So real analytic functions are not the same functions as infinitely differentiable real functions. But in the complex case every holomorphic function is analytic and vice versa, so the terms are interchangeable.

We define

$$\exp(z) = 1 + z + \frac{z^2}{2!} + \frac{z^3}{3!} + \frac{z^4}{4!} + \cdots$$

which converges for all $z \in \mathbb{C}$. (In power series such as this we are defining z^n recursively, for natural n, as 1 if $n = 0$ and $z^{n-1}z$ if $n \geq 1$.)

Defining $\cos \theta$ and $\sin \theta$ to be the real and imaginary parts of $\exp(i\theta)$ we have *Euler's formula*:

$$\exp(i\theta) = \cos \theta + i \sin \theta.$$

This yields the familiar series

$$\sin\theta = \theta - \frac{\theta^3}{3!} + \frac{\theta^5}{5!} - \cdots \,;$$
$$\cos\theta = 1 - \frac{\theta^2}{2!} + \frac{\theta^4}{4!} - \cdots$$

One can deduce from the series expansion of $\cos\theta$ that there is a smallest positive real number τ with $\cos\tau = 0$. We can then *define* $\pi = 2\tau$ and, from some fairly simple analysis, we recover the trigonometric origins of $\sin\theta$ and $\cos\theta$ from their analytical definitions. From Euler's formula alone it is possible to quickly derive a large number of trigonometric identities. In particular *de Moivre's formula*

$$(\cos\theta + \mathrm{i}\sin\theta)^n = \cos n\theta + \mathrm{i}\sin n\theta$$

is immediate, so the task of expressing $\sin n\theta$ and $\cos n\theta$ in terms of $\sin\theta$ and $\cos\theta$ ($n \in \mathbb{N}$, say) becomes an exercise in elementary algebra.

Key to this retrieval of the geometry from the analysis is the observation that exp is a periodic function with period $2\pi\mathrm{i}$, i.e. for all $n \in \mathbb{Z}$, $\exp(z + 2n\pi\mathrm{i}) = \exp(z)$, and the mapping $t \mapsto \exp(\mathrm{i}t)$ is a surjection of $\mathbb{R}$ onto the unit circle $\{z : |z| = 1\}$ of $\mathbb{C}$. It follows that for each non-zero $z \in \mathbb{C}$ there exists a θ (unique modulo the addition of integer multiples of 2π) such that

$$\frac{z}{|z|} = \exp(\mathrm{i}\theta).$$

Any such value of θ is an *argument*, $\arg z$, of z. The *principal argument* of z is the unique argument in the interval $(-\pi, \pi]$. The geometric interpretation of $\arg z$ is clear, it is the angle (measured in radians) in anticlockwise orientation between the positive real axis and the line passing through the origin and z (see Figure 1.18).

REMARK

Since a function $f : \mathbb{C} \to \mathbb{C}$ can be regarded as a function $\mathbb{R}^2 \to \mathbb{R}^2$ it is of some interest to see how the statement that f is holomorphic translates to a statement about real differentiability. Suppose $f(x + y\mathrm{i}) = u(x, y) + v(x, y)\mathrm{i}$ is holomorphic where $u, v : \mathbb{R}^2 \to \mathbb{R}$. Then it is easy to show that

$$\frac{\partial u}{\partial x} = \frac{\partial v}{\partial y} \quad \text{and} \quad \frac{\partial u}{\partial y} = -\frac{\partial v}{\partial x}.$$

These are known as the *Cauchy–Riemann equations*. Conversely, if $u, v : \mathbb{R}^2 \to \mathbb{R}$ are real differentiable and the partial derivatives satisfy the Cauchy–Riemann equations then the function $f(x + y\mathrm{i}) = u(x, y) + v(x, y)\mathrm{i}$ is holomorphic.

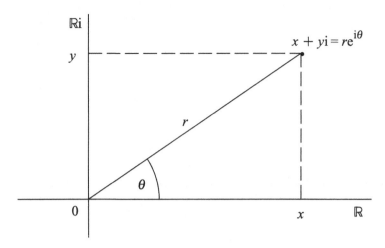

Figure 1.18 A complex number $z = x + yi$ represented as a point in the plane, the polar representation $re^{i\theta}$ is indicated, where $r = |z|$ and $\theta = \arg z$.

The results of complex analysis are generally far more elegant than those of real analysis. As soon as a complex function is differentiable it is differentiable arbitrarily many times. A complex function which is differentiable is automatically analytic. Contrast this to real functions which can be differentiable n times but not n + 1 times, or which are differentiable but not analytic. Trigonometry also seems to effortlessly fall out of complex analysis alongside a host of miraculous geometric properties.

1.8.4   The geometry of complex analysis

Fixing some complex number α of modulus 1 and argument θ, the map $z \mapsto \alpha z$ corresponds to a rotation of the plane through an anticlockwise angle of θ radians about the origin. For real positive r the map $z \mapsto rz$ is a radial dilation (a contraction if $r < 1$) by a factor r, centred at the origin. Thus, for a general complex number α of modulus r and argument θ, the map $z \mapsto \alpha z$ is the composite of a radial dilation by a factor r and an anticlockwise rotation by the angle θ. Of course, for arbitrary complex α the map $z \mapsto z + \alpha$ is a linear translation of the plane by α units in the direction of α. So even the most elementary arithmetic operations of complex analysis are geometric transformations, being composites of dilations, rotations and translations. It should come as no surprise, then, that once we delve a little deeper into the theory the subject soon reveals itself to be deeply tied to geometry. For example, complex differentiable functions preserve angles at those points where the derivative does not vanish, and the integral of a function over a simple closed curve is determined by the residues of the singularities of the function in the interior of the curve (this powerful technique can yield, with great ease, results in real analysis which are otherwise very difficult to obtain).

These statements perhaps deserve some further explanation. The reader who does not want to get caught up in this slightly technical excursion (or who knows this material already) may wish to skip to the next section.

Conformal mapping

The angle between two paths $\gamma_1 : [a, b] \to \mathbb{C}$ and $\gamma_2 : [a, b] \to \mathbb{C}$ passing through some point $z_0 = \gamma_1(s) = \gamma_2(t)$ is the difference $\arg \gamma_2'(t) - \arg \gamma_1'(s)$ where the derivative $\gamma_k'(t)$ is defined as expected:

$$\lim_{h \to 0} \frac{\gamma_k(t + h) - \gamma_k(t)}{h}.$$

(The paths must be suitably differentiable at the point of intersection and the angle is determined only up to the addition of an integer multiple of 2π.)

A continuous function $f : \mathbb{C} \to \mathbb{C}$ which is complex differentiable at z_0 maps paths γ_1 and γ_2 intersecting at z_0 to new paths δ_1 and δ_2 intersecting at $f(z_0)$ and if the angle between δ_1 and δ_2 equals that between γ_1 and γ_2 (modulo addition of an integer multiple of 2π) for any such paths γ_1 and γ_2 we say f 'preserves angles' at z_0 (here both the magnitude and orientation of the angle is preserved).

If a function $f : G \to \mathbb{C}$ (G a region of $\mathbb{C}$) is complex differentiable at $a \in G$ and $f'(a) \neq 0$, then f preserves angles at a. Conversely if $f : G \to \mathbb{C}$ preserves angles at a and

$$\lim_{z \to a} \frac{|f(z) - f(a)|}{|z - a|}$$

exists, then f is complex differentiable at a and $f'(a) \neq 0$.

A holomorphic function $f : G \to \mathbb{C}$ for which $f'(a) \neq 0$ for all $a \in G$ (and hence preserves angles at all points) is a *conformal mapping*. We say a pair of regions G_1 and G_2 of $\mathbb{C}$ are *conformally equivalent* if there is a bijective conformal mapping of G_1 onto G_2 (this is an equivalence relation on the class of all regions; the inverse mapping is automatically conformal). (A conformal mapping is illustrated in Figure 1.19.) A region is *simply connected* if it has no holes. This can be captured in more formal terms – 'all closed curves are homotopic to zero' – which less formally simply means we can continuously retract any loop in the region to a point.

The remarkable *Riemann Mapping Theorem* tells us that every simply connected proper subregion G of $\mathbb{C}$ is conformally equivalent to the open disc $D = \{z : |z| < 1\}$ (and consequently all such regions are conformally equivalent to one another). Indeed, for each $a \in G$ there is a *unique* conformal mapping f of G onto D with $f(a) = 0$ and $f'(a) > 0$. This might be a little counterintuitive at first, one might expect global angle preservation to be quite a restrictive condition, however, the fact is that among the simply connected regions of $\mathbb{C}$ there are only two conformal equivalence classes, one containing all proper simply connected subregions of $\mathbb{C}$ and the other containing only $\mathbb{C}$.

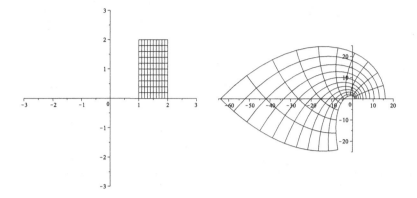

Figure 1.19 The graph on the right shows the effect of the transformation $z \mapsto z^4$ on the grid depicted on the left. Notice the preserved orthogonality of the deformed grid lines. Except at the origin, all angles are preserved by this map. More generally, any holomorphic function preserves angles at all points at which its derivative is non-zero.

Singularities

For $r > 0$ and $a \in \mathbb{C}$ let $D_r(a)$ be the open disc of radius r centred at a, $\{z : |z - a| < r\}$, and $D_r(a)_0$ the disc $D_r(a)$ minus the point a. A function f has an *isolated singularity* at a if there exists an $r > 0$ such that the restriction of f to $D_r(a)_0$ is defined and analytic but the restriction of f to $D_r(a)$ is not.

There are three flavours of isolated singularity.

(i) a is *removable* if there is an analytic function on $D_r(a)$ which restricts to f on $D_r(a)_0$ (this is the case if and only if $\lim_{z \to a}(z - a)f(z) = 0$). For example $\frac{1}{z}\sin z$ has a removable singularity at 0.

(ii) a is a *pole* of f if for any $N > 0$ there exists an $r > 0$ such that $|f(z)| > N$ whenever $z \in D_r(a)_0$ (this is the case if and only if there is a positive integer n, a positive real r and an analytic function $g : D_r(a) \to \mathbb{C}$ such that $f(z) = g(z)/(z - a)^n$ on $D_r(a)_0$). For example $\frac{1}{(z-a)^n}$ has a pole at a, where n is a positive integer.

(iii) All other isolated singularities are *essential*. For example 0 is an essential singularity of $\exp(\frac{1}{z})$ (for all z on the imaginary axis minus the origin we have $|\exp(\frac{1}{z})| = 1$).

A striking result concerning essential singularities (the *Great Picard Theorem*) is that in *every* neighbourhood of an essential singularity of f, f assumes each complex number (with one possible exception) infinitely many times; for example, for any complex number $\alpha \neq 0$ and every $r > 0$ there are infinitely many z in $D_r(0)_0$ such that $\exp(\frac{1}{z}) = \alpha$.

An example of a non-isolated singularity is given by the function $f(z) = \frac{1}{\sin\frac{1}{z}}$ at 0 (f has infinitely many singularities in every open neighbourhood of 0 – or put another way, $\frac{1}{\sin\frac{1}{z}}$ is not holomorphic on $D_r(0)_0$ for any $r > 0$).

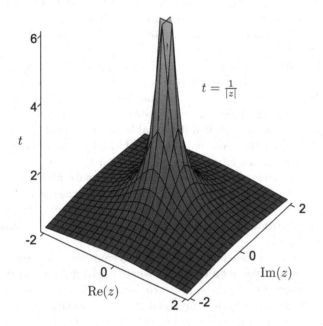

Figure 1.20 One can visualize a singularity of a function $f : \mathbb{C} \to \mathbb{C}$ by considering the surface given by the modulus mapping $z \mapsto |f(z)|$. The singularity illustrated above is the simple pole at 0 of the function $z \mapsto \frac{1}{z}$; the 'peak' continues to taper indefinitely – here we have truncated it at an appropriate height. Such singularities might cluster around a non-isolated singularity, or (more difficult to illustrate) the singularity might be essential.

Curves and integrals

A *path* (or a *curve*) in $\mathbb{C}$ is a continuous map $\gamma : [a, b] \to \mathbb{C}$. If $\gamma(a) = \gamma(b)$ then γ is a *closed curve*. If a closed curve does not intersect itself, i.e. if $x, y \in [a, b]$ and $\gamma(x) = \gamma(y)$ implies $x = y$ or $x, y \in \{a, b\}$, then the curve is said to be a *Jordan curve*. The proof of the apparently intuitively clear fact that some points are inside the region enclosed by a Jordan curve calls for a surprising amount of mathematical machinery. It is worth detailing the salient points here.

Suppose $\gamma : [a, b] \to \mathbb{C}$ is a Jordan curve. Let us denote by Γ the subset $\{\gamma(x) : x \in [a, b]\}$ of $\mathbb{C}$, so when we refer to the curve we mean this set of points Γ, not the function γ. The continuous image Γ of the compact set $[a, b]$ is itself compact. This means we can find a disc D such that $\Gamma \subset D$. Choose a point $x \in \mathbb{C}$ not in this disc D. We can now define the *outside* of Γ, $\mathrm{Out}(\Gamma)$, to be the set of $y \in \mathbb{C}$ such that there exists a continuous path joining y and x which does not intersect Γ. It would then be natural to define the *inside* of Γ, $\mathrm{Ins}(\Gamma)$, to be the complement of $\Gamma \cup \mathrm{Out}(\Gamma)$ in $\mathbb{C}$, in other words the set of points $y \in \mathbb{C}$ for which every continuous path joining y and x intersects Γ. The crucial question is: *is* $\mathrm{Ins}(\Gamma)$ *non-empty*? Intuitively it seems obvious that it is non-empty, but this may be because we are deceived by simple diagrams. Certainly if we add further restrictions to the function γ, for instance making it piecewise linear, so that Γ is a polygon, or making it differentiable at some point so we can define a normal to the curve, then one can prove that $\mathrm{Ins}(\Gamma)$ is non-empty. But even these special cases are not trivial. We might begin to doubt that it is always non-empty by considering possible modifications of space-filling curves such as those discovered by Peano and Hilbert. The general case is actually a very deep and difficult theorem in topology called the *Jordan Curve Theorem*: Every Jordan curve Γ partitions the plane into three non-empty path-connected sets $\mathrm{Out}(\Gamma)$, Γ and $\mathrm{Ins}(\Gamma)$.

In complex analysis the curves one generally considers are *rectifiable*. This means that they are curves of finite length, where the length is defined as follows. Choose a finite sequence of points $a_0 = \gamma(a), a_1, a_2, \ldots, a_n = \gamma(b)$ on the curve $\gamma : [a, b] \to \mathbb{C}$ and consider the polygonal path comprising the line segments joining a_0 to a_1, a_1 to a_2, a_2 to a_3 and so on up to a_n. Define the length of the polygonal path to be the sum of the lengths of the component line segments. Consider the set of all lengths of all such polygonal paths. In some pathological cases this set will have no upper bound. If the curve is reasonably well-behaved, however, this set of lengths will have an upper bound and its least upper bound is defined to be the length of the curve. Thus the rectifiable curves are those which can be closely approximated by polygonal paths and includes all piecewise smooth curves, i.e. curves formed by joining a finite collection of smooth curves end to end.

If $\gamma : [a, b] \to \mathbb{C}$ is a rectifiable path in $\mathbb{C}$ we define the complex integral of a function $f : G \to \mathbb{C}$ (where G contains the path γ) by:

$$\int_\gamma f(z)\, \mathrm{d}z = \int_a^b f(\gamma(t))\gamma'(t)\, \mathrm{d}t.$$

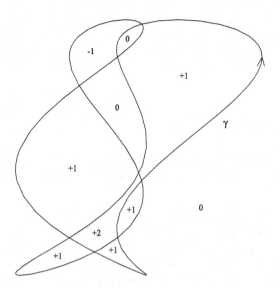

Figure 1.21 A self-intersecting closed smooth curve γ oriented as indicated by the arrow. The curve divides the complex plane into ten regions including the unbounded region outside the curve. The restriction of the index function $a \mapsto \operatorname{ind}(\gamma, a) = \frac{1}{2\pi i} \int_\gamma \frac{dz}{z-a}$ to any one of these regions is constant and is equal to an integer. This integer counts the number of times the curve winds around the point a in the anticlockwise orientation. These integer values are indicated for each region.

Fix $a \in \mathbb{C}$ and suppose γ is a closed rectifiable curve in $\mathbb{C}$ which does not pass through a. The *index* of γ with respect to a, or the γ-*index of* a, is

$$\operatorname{ind}(\gamma, a) = \frac{1}{2\pi i} \int_\gamma \frac{dz}{z-a}.$$

The index is an integer, it counts the number of times the curve γ winds around the point a in the anticlockwise orientation (so a result of -2, for example, means the path winds twice around a in the clockwise orientation). The path γ naturally splits the complex plane into path-connected regions. If the curve does not cross itself then one of these connected regions is unbounded and is identified precisely as the set of points in $\mathbb{C}$ with γ-index zero, these points are 'outside' the region enclosed by the curve, and from the Jordan Curve Theorem we know that there is one interior region (all points therein having index 1 or -1 depending on the orientation of the path). If the curve crosses itself there will still be one unbounded region, all points of which have γ-index zero, but there may also be bounded regions of index zero and several 'interior' regions of various non-zero indices (see Figure 1.21).

It is easily shown that

$$\int_\gamma \frac{dz}{z} = 2\pi i,$$

where γ is any closed rectifiable path in $\mathbb{C}$ which winds anticlockwise around 0 once. Building on this one can prove that if f has isolated singularities $a_1, a_2, \ldots$

then $\int_\gamma f(z)\,\mathrm{d}z$, for a closed path γ which does not pass through any of the singularities, is equal to

$$2\pi\mathrm{i}\sum_i \mathrm{res}(f,a_i)\mathrm{ind}(\gamma,a_i),$$

where $\mathrm{res}(f,a_i)$, the *residue* of f at a_i, is the coefficient c_{-1} in the Laurent expansion $f(z) = \sum_{n\in\mathbb{Z}} c_n(z-a_i)^n$ of f at a_i.

Multi-valued mappings and Riemann surfaces
We define the complex logarithm for non-zero z by[1]

$$\ln z = \ln|z| + \mathrm{i}\arg z,$$

where the logarithm appearing on the right-hand side is the real logarithm introduced in Subsection 1.7.8. This is clearly not an arbitrary definition. We want the equality $\exp(\ln z) = z$ to hold for all $z \neq 0$. If $\ln z = a + b\mathrm{i}$ then we must have $\exp(a)\exp(b\mathrm{i}) = z = |z|\exp(\mathrm{i}\arg z)$, and by the uniqueness of polar decomposition $\exp(a) = |z|$ and $\exp(b\mathrm{i})=\exp(\mathrm{i}\arg z)$. We conclude that $a = \ln|z|$ and $b = \arg z$. As the argument of z is only determined modulo 2π, $\ln z$ is determined modulo the addition of some integer multiple of $2\pi\mathrm{i}$, so $\ln$ is a *multi-valued mapping* $\mathbb{C} - \{0\} \to \mathbb{C}$, in other words it is a function which maps each non-zero complex number to a *set* of complex numbers (in this case a countably infinite set). For example, $\ln(1+\mathrm{i}) = \{\ln\sqrt{2} + \frac{\pi}{4}\mathrm{i} + 2n\pi\mathrm{i} : n \in \mathbb{Z}\}$.

The multi-valued mapping $\ln z$ is 'locally holomorphic' in the following sense. For any non-zero $z_0 \in \mathbb{C}$ we can find a neighbourhood U of z_0, fixing some argument for z_0, this fixed argument determining a continuous single-valued argument function $\theta : U \to \mathbb{R}$, so that the argument-restricted complex logarithm $z \mapsto \ln|z| + \mathrm{i}\theta(z)$ is a single-valued *holomorphic* function $U \to \mathbb{C}$. We can take U to be any region in $\mathbb{C}$ which does not contain a loop which winds around 0 (arg has no continuous single-valued restriction on such a loop).

For non-zero $z \in \mathbb{C}$ and arbitrary $a \in \mathbb{C}$ we define

$$z^a = \exp(a\ln z).$$

Multi-valuedness generally persists. We find, for example, that $\mathrm{i}^{\mathrm{i}} = \{\mathrm{e}^{-\frac{\pi}{2}+2n\pi} : n \in \mathbb{Z}\}$.

The somewhat unsatisfactory appearance of multi-valued mappings in complex analysis has a beautiful resolution due to Riemann. A suitably 'locally holomorphic' multi-valued mapping f on $\mathbb{C}$ determines a surface $\mathcal{R}_f$ (an example of a *Riemann surface*) in such a way that f can be naturally recast as a single-valued function $\mathcal{R}_f \to \mathbb{C}$. These surfaces are generally described by making certain identifications on multiple (possibly infinitely many) sheets, each one a copy of $\mathbb{C}$ (minus the points at which f has a singularity); the original multi-valued map retrieved, in a sense, by collapsing all of the sheets on top of

---

[1] It is quite standard to use the notation log here, rather than ln, however I shall stick to ln throughout.

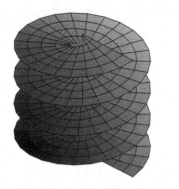 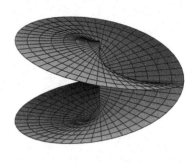

Figure 1.22 Two Riemann surfaces depicted in $\mathbb{R}^3$. On the left the Riemann surface of $\ln(z)$ is an infinite corkscrew. On the right the Riemann surface of $z^{1/2}$ is obtained by gluing two copies of $\mathbb{C}$ together as shown. The self-intersection is unavoidable unless we pass to a higher-dimensional space. Replacing the domain of these multi-valued functions by their respective Riemann surfaces, single-valuedness is restored.

one another. The Riemann surface of $\ln z$, for example, looks like an infinite corkscrew winding around an axis (at zero); the Riemann surface of $z^{1/2}$ is obtained by gluing two copies of $\mathbb{C}$ together in such a way that from some point $p \neq 0$ on one sheet one full rotation around the origin will lead us to a point p' on the second sheet and repeating the full rotation (in the same direction) will lead us back to the original point p (the surface $\mathcal{R}_{z^{1/2}}$ cannot be embedded in $\mathbb{R}^3$ without self-intersection). (See Figure 1.22.)

Riemann's name is also attached to a simple construction which 'compactifies' the complex plane, the *Riemann sphere*. If we imagine a sphere sitting on the complex plane at the origin then the stereographic projection from the 'north pole' P of the sphere defines a bijection between $\mathbb{C}$ and the sphere minus P. Via this projection, and by including P, we make sense of a 'point at infinity' of $\mathbb{C}$, so that the map $z \mapsto \frac{1}{z}$, for example, is viewed as a map from the Riemann sphere to itself where 0 maps to P and vice versa, and we can also make sense of 'singularities at infinity'. The projection is usually presented so that the complex plane passes through the 'equator' of the sphere rather than simply touching the south pole, identifying the southern hemisphere with the unit disc of $\mathbb{C}$ (see Figure 1.23).

The Mandelbrot set

Passing to another type of geometry, the complex plane is also the canvas for the infinite decorations popularly known as (planar) fractals. The famous Mandelbrot set, $\mathcal{M}$, for example, is the set comprising all $c \in \mathbb{C}$ for which the sequence

$$f_c(0), \ f_c(f_c(0)), \ f_c(f_c(f_c(0))), \ldots$$

of composed iterates of the function $f_c(z) = z^2 + c$ is bounded – the points of $\mathcal{M}$ are typically coloured black and all else white; colours are introduced to indicate

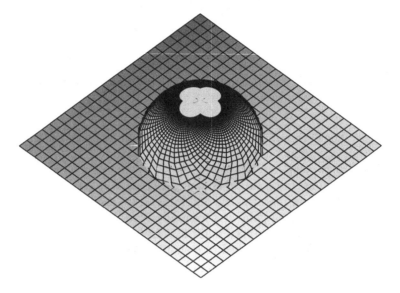

Figure 1.23 The complex plane depicted passing through the equator of a unit sphere. By projecting the plane onto the sphere via a ray passing through the north pole we map the entire plane onto the sphere leaving just the north pole empty, identifying the unit disc with the southern hemisphere. Here we illustrate the projection of a large square grid centred at the origin. The entire sphere, including the north pole, represents a compactification of the plane, and via this device we can speak intelligibly about the point at infinity. This realization of $\mathbb{C}$ is known as the Riemann sphere.

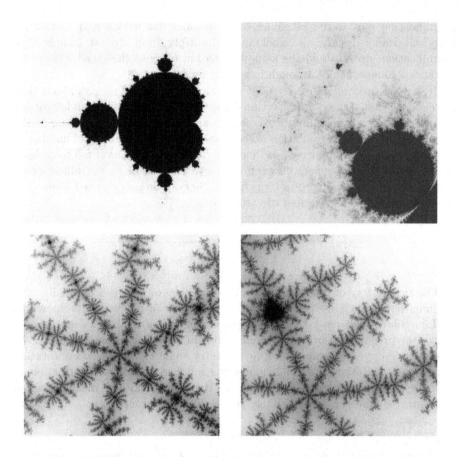

Figure 1.24 The Mandelbrot set shown in full at the top left and three progressively closer zooms. Reading from top left to bottom right the widths and bottom left coordinates are: $[width = 2.6, coordinates = (-2, -1.3)]$; $[width = 0.1, coordinates = (-0.7, 0.4)]$; $[width = 0.02, coordinates = (-0.68, 0.45)]$; $[width = 0.002, coordinates = (-0.675, 0.464)]$.

speed of divergence and the results are, as is well documented, awe inspiring (see Figure 1.24).

Glossy colour images of the Mandelbrot set and related objects are undeniably impressive, and are accessible to all, however, there are even more beautiful abstract structures that dwell in the mathematics behind the images. The theory of complex dynamics is a deep field filled with profoundly difficult questions which have attracted many exceptional mathematicians over the last century.

We have all enjoyed zooming in on quasi-self-similar regions of such sets and we know that, although we are limited by a finite array of pixels, a finite amount of time and a finite amount of computing power the set is, in a strong sense, completely determined to an arbitrary level of precision. It is the 'almost' self-similarity that makes the Mandelbrot set such an interesting object, miniature versions of the whole image appearing initially as crumb-sized clones, but on magnification most often revealing some peculiar distortions and mutations. Fully self-similar designs, the whole image multiply duplicated at some level of magnification, do not hold the same fascination because the steady 'zoom in' becomes a monotonous (although hypnotic) repetition.

It is worth making an obvious but sometimes overlooked remark about this, anticipating our discussion in Section 1.13. By displaying the Mandelbrot set or related images on a computer screen we are discretizing it according to the screen resolution. This resolution determines a very large finite total number of possible images so, no matter how fine it is, it will at some level fail to pick up on the filamental differences between an image and one of its magnifications – it will only be able to present a large finite gallery of images selected from what is genuinely an infinite number of variations.

REMARKS

1. The rich variety of conformal mappings in evidence here is a decidedly two-dimensional phenomenon. Liouville proved (one of many results known as 'Liouville's Theorem') that angle preserving maps in dimension three or greater are much more sober than those in two dimensions. Indeed, every such map is a composite of an isometry (reflection/translation/rotation and their combinations), a homothety (radial expansion or contraction of space from a point) and an inversion. By an inversion we mean inversion in an n-sphere, i.e. if the sphere has radius r we map $(x_1, x_2, \ldots, x_n)$ to $\left(\frac{r^2 x_1}{\sum_{j=1}^{n} x_j^2}, \frac{r^2 x_2}{\sum_{j=1}^{n} x_j^2}, \ldots, \frac{r^2 x_n}{\sum_{j=1}^{n} x_j^2} \right)$.

2. This is as good a place as any to mention one of the most important functions in mathematics. First consider the function $\zeta(s) = \sum_{n=1}^{\infty} \frac{1}{n^s}$. This is defined for all $s \in \mathbb{C}$ with real part greater than 1 (and, marking an early connection with prime numbers, it is equal to the product $\Pi_p \frac{1}{1-p^{-s}}$ where p ranges over all primes). One can extend $\zeta(s)$ uniquely to a function (we continue to call it ζ) which is holomorphic everywhere in $\mathbb{C}$ except at 1, where it has a simple pole. This extended function is called the *Riemann zeta function*. ζ has zeros at the negative even numbers (called trivial zeros), but it also has other zeros (called non-trivial zeros). The most famous open problem in pure mathematics is the *Riemann Hypothesis*, which states that all non-trivial zeros of the Riemann zeta function have real part $\frac{1}{2}$. If the Riemann hypothesis is true (and the evidence is overwhelmingly in its favour) then there are many consequences in the rest of mathematics, in particular it tells us some critical information about the distribution of the prime numbers.

3. A general Riemann surface is a manifold which locally looks like $\mathbb{C}$, on which one can define holomorphic functions. Riemann surfaces are necessarily orientable.

Every aspect of complex analysis has a beautiful geometric component: the basic algebraic operations describe simple geometric transformations of the plane; holomorphic functions describe deformations of the plane which preserve angles at most points; integrals around curves are determined by values associated with singularities and their relative position with respect to the curve; multivalued functions determine surfaces on which they become single-valued; and iterations of analytic functions give rise to impossibly complicated objects such as the Mandelbrot set. In the bigger picture, the field of complex numbers is just one of infinitely many extensions of the real numbers, each having its own novel algebraic and geometric properties.

1.8.5   Quaternions and octonions

Having barely touched on the beautiful subject of complex analysis we leave it behind, but before we leave it completely, there are some closely related higher algebraic structures which deserve a mention.

The *quaternions*, a four-dimensional extension of $\mathbb{R}$ denoted by the letter $\mathbb{H}$ in honour of its discoverer William Hamilton, is the associative algebra comprising the elements $a + b\mathbf{i} + c\mathbf{j} + d\mathbf{k}$, $a, b, c, d \in \mathbb{R}$, where $\mathbf{i}^2 = \mathbf{j}^2 = \mathbf{k}^2 = \mathbf{ijk} = -1$. The new elements $\mathbf{i}$, $\mathbf{j}$, $\mathbf{k}$ commute with any real number, but $\mathbf{ij} = -\mathbf{ji}$, $\mathbf{ik} = -\mathbf{ki}$ and $\mathbf{jk} = -\mathbf{kj}$, thus we lose commutativity in general. (Hamilton discovered the quaternions in 1843, on an impulse *'unphilosophical as it may have been'* carving the identity

$$\mathbf{i}^2 = \mathbf{j}^2 = \mathbf{k}^2 = \mathbf{ijk} = -1$$

into a stone on Brougham Bridge in Dublin.)

The algebraic construction that takes us from $\mathbb{R}$ to $\mathbb{C}$ applied to $\mathbb{C}$ leads to $\mathbb{H}$, and the same technique (multiplication suitably twisted – the order being critical now due to lack of commutativity) leads from $\mathbb{H}$ to the less well-known algebra of *octonions* $\mathbb{O}$. $\mathbb{O}$, an eight-dimensional extension of $\mathbb{R}$, is neither commutative nor associative.[1]

This so-called Cayley–Dickson process can be continued indefinitely (the next stop is the 16-dimensional sedenions). Each new Cayley–Dickson algebra has twice the dimension of its predecessor but has fewer pleasant algebraic properties; the sedenions have zero divisors for example. Further algebraic structures (infinite in variety and all describable in purely set theoretic terms) are beyond the scope of this book.

---

[1] Conway and Smith **[38]** investigates the geometry of $\mathbb{H}$ and $\mathbb{O}$.

Remarks

1. It seems that Carl Friedrich Gauss was the discoverer of quaternions more than 20 years before Hamilton. Gauss' short note 'Mutationen des Raumes' (1819) includes the quaternionic formulae, but it was only published in 1900, in his collected works.[1]

2. One can give a reasonable algebraic characterization of the classes of real numbers, complex numbers, quaternions and octonions among all real algebras: a *real algebra* is a real vector space A with a multiplication operation (where $(\lambda a)(\mu b) = \lambda\mu ab$ for all scalars λ, μ and $a, b \in A$, and where multiplication distributes over addition). Such an algebra is *quadratic* if it has a multiplicative identity 1 and, for all $x \in A$, there exist $a, b \in \mathbb{R}$, depending on x, such that $x^2 = a1 + bx$; it is *alternative* if $x(xy) = x^2 y$ and $(xy)y = xy^2$ for all $x, y \in A$. If $A \neq \{0\}$ is an alternative quadratic real algebra with no zero divisors then A is isomorphic to one of $\mathbb{R}$, $\mathbb{C}$, $\mathbb{H}$ or $\mathbb{O}$.

3. A theory of quaternionic functions analogous to complex function theory was initiated by R. Fueter in the mid-1930s. One of the early hurdles is to identify the appropriate class of functions which will play the role of the holomorphic functions. Definitions of holomorphy which are equivalent over $\mathbb{C}$ give rise to different classes of functions over $\mathbb{H}$. The usual limit definition, for example, is far too restrictive (only linear functions satisfy it). It turns out that an appropriate generalization of the Cauchy–Riemann equations is the right approach, and from this one can deduce various quaternionic versions of the theorems of complex analysis.

4. Quaternions are used as an effective way of describing rotations in three-dimensional space. Less mundane applications to physics have also been made, for example Maxwell's equations, usually presented as four partial differential equations, can be expressed as a single quaternionic differential equation. The octonions too have found physical relevance. But let us not pretend that a structure is to be judged by its applications. $\mathbb{H}$ and $\mathbb{O}$ are interesting simply because they exist.

The quaternions and octonions are respectively four- and eight-dimensional extensions of $\mathbb{R}$. Unlike $\mathbb{C}$, they are not commutative, and the octonions also fail to be associative. $\mathbb{R}$, $\mathbb{C}$, $\mathbb{H}$ and $\mathbb{O}$ are the initial terms in an infinite sequence of algebraic structures which in turn form a subset of an even wider class of objects. Having defined the complex numbers we are now able to describe the last of the classical number systems as a certain subset of $\mathbb{C}$.

---

[1] See Carl Friedrich Gauss, *Werke*, volume 8, Cambridge University Press (2011), pages 357–362.

1.9   Algebraic numbers

> *Mathematical discoveries – small or great, and whatever their con-*
> *tent (new subjects of research, divination of methods or of lines to*
> *follow, presentiments of truths and of solutions not yet demonstrated,*
> *etc.) – are never born by spontaneous generation. They always pre-*
> *suppose a ground sown with preliminary knowledge and well prepared*
> *by work both conscious and subconscious.*
>
> – Jules Henri Poincaré[1]

1.9.1   The definition of $\mathbb{A}$

In our growing progression of number systems we have skipped over an important class, the set of algebraic numbers, which we shall denote by $\mathbb{A}$. The set of
algebraic numbers occupies a position strictly between $\mathbb{Q}$ and $\mathbb{C}$; every rational
number is algebraic but there are algebraic numbers which are not rational and
complex numbers which are not algebraic. So far we have carefully ascended
through a hierarchy of ever more complicated algebraic structures, the definition of one based on the properties of its smaller predecessor. It is, however,
more usual in mathematics, and perfectly natural, to define something as a certain distinguished substructure of a larger object. If we were to build $\mathbb{A}$ from
$\mathbb{Q}$ directly without prior knowledge of $\mathbb{C}$ (or $\mathbb{R}$), then either $\mathbb{C}$ would be constructed 'by accident' in the process, or it would soon become apparent that
the combinatorial complexity of the construction would be greatly reduced by
introducing such a superstructure.

A complex number α is said to be algebraic (over $\mathbb{Q}$) if there exist rational
numbers $a_0, a_1, \ldots, a_n$, not all zero, such that

$$a_n \alpha^n + a_{n-1} \alpha^{n-1} + \cdots + a_1 \alpha + a_0 = 0.$$

In fact, it suffices to find *integers* $a_0, \ldots, a_n$ since we can multiply the above
equation through by, say, the product D of the denominators of the coefficients
$a_0, a_1, \ldots, a_n$, each Da_i then being an integer. A complex number β is said to
be *transcendental* if it is not algebraic.

These notions are normally expressed in terms of polynomials. Very formally
the class of polynomials (with coefficients in some number system F) is the set
of 'eventually null' sequences (a_n) in F (a sequence (a_n) is eventually null if
there exists an $N \in \mathbb{N}$ such that for all $n \geq N$, $a_n = 0$) with addition and

---

[1]A truncated version of this statement (deleting 'and whatever their content' and the par-
enthetical text that follows it) is very widely quoted but almost never adequately sourced. The
text is an extract from an answer Poincaré gave to part of a questionnaire for *L'Enseignement
Mathèmatique* published in parts between 1905 and 1908 and collected in 1909 (Poincaré was
one of more than a hundred mathematicians to provide responses). A short review by R. C.
Archibald, including a fuller version of the quote, appears in the *Bulletin of the American
Mathematical Society*, Volume **22**, Number 3 (1915), 125–136.

multiplication defined by:

$$(a_n) + (b_n) = (a_n + b_n); \text{ and}$$
$$(a_n)(b_n) = (c_n),$$

where $c_n = \sum_{i+j=n} a_i b_j$. The point of giving such a formal definition will become clear later. It is important that all of our defined structures (geometric, analytic, algebraic or otherwise) can be fully captured in a formalized logical language of aggregates.

Choosing some neutral symbol, x, we usually denote the 'formal' polynomial $(a_0, a_1, \ldots, a_n, 0, 0, \ldots)$ by the expression $a_0 + a_1 x + a_2 x^2 + \cdots + a_n x^n$, giving an explicit algebraic description of how to evaluate the polynomial at x. One can then think of a polynomial as a function (of x). If $a_n \neq 0$, n is the *degree* of the polynomial $a_n x^n + \cdots + a_1 x + a_0$. The view of a polynomial as a function is the one that is most familiar to the man on the street who learns about polynomials only in the context of calculus.

An element α is a *root* or *zero* of the polynomial p if $p(\alpha) = 0$. Hence a complex number α is algebraic (over $\mathbb{Q}$) if it is a root of some non-zero polynomial with rational coefficients. The number $\sqrt{2}$, for example, is algebraic since it is a root of the polynomial $x^2 - 2$.

The qualification 'over $\mathbb{Q}$' appears here because the idea naturally extends to the more general setting where for any subfield F of a field G we say $x \in G$ is algebraic over F if there is a non-zero polynomial f with coefficients in F such that $f(x) = 0$. The study of fields, their extensions and the relation to groups of automorphisms forms a fundamental part of the wonderful subject of Galois theory.

Remark

It was Galois theory that was to provide an insight into the old geometric problem of what is constructible with a straight edge and compass. In particular one can prove that it is impossible to trisect a general angle by such means. The angle θ can be trisected if and only if the polynomial $4x^3 - 3x - \cos\theta$ is reducible over the field $\mathbb{Q}(\cos\theta)$, so for instance a 60° angle cannot be trisected.

The set of constructible points in $\mathbb{C}$ is a subfield K of $\mathbb{A}$ whose elements are characterized as follows: $x \in K$ if and only if there is a finite tower of fields $\mathbb{Q} = K_0 \subseteq K_1 \subset K_2 \subseteq \ldots \subseteq K_n$ with $x \in K_n$ and such that K_{i+1} is a quadratic extension of K_i, i.e. K_{i+1} is obtained from K_i by appending a root of a degree 2 polynomial which is irreducible over K_i. The constructible field K is the smallest field extension of $\mathbb{Q}$ which has the two properties: (i) if $z \in K$ then $\sqrt{z} \in K$; and (ii) if $z \in K$ then $\bar{z} \in K$.

An algebraic number is a complex number which is a root of a polynomial with rational coefficients. One of the achievements of nineteenth century mathematics was to prove that certain famous constants are not algebraic and to further describe the nature of the set of algebraic numbers.

1.9.2 Properties of $\mathbb{A}$

If α and β are arbitrary algebraic numbers, say with $\alpha \neq 0$, then by studying the rings $\mathbb{Q}[\alpha]$, $\mathbb{Q}[\beta]$ and $\mathbb{Q}[\alpha, \beta]$ generated by α and β one can show without too much difficulty that $\alpha + \beta$, $\alpha\beta$ and α^{-1} are all algebraic. In other words, $\mathbb{A}$ is a field.

Since $\mathbb{A}$ is a field, any composite of sums, products and quotients of rational powers of rational numbers will again yield an algebraic number. For instance we can see immediately that the messy expression

$$\frac{(2^{7/11} + 1)^{1/19} + 3^{1/23}}{1 + 5^{1/7}}$$

is an algebraic number without troubling ourselves to find a polynomial with rational coefficients having this as a root.

$\mathbb{A}$ is the *algebraic closure* of $\mathbb{Q}$, i.e. it is the smallest field extension F of $\mathbb{Q}$ for which every non-constant polynomial with coefficients in F has a root in F ($\mathbb{A}$ is *algebraically closed*). Neither $\mathbb{Q}$ nor $\mathbb{R}$ are algebraically closed since neither contains a root of $x^2 + 1$, for example. The field $\mathbb{C}$ is algebraically closed, indeed it is the algebraic closure of $\mathbb{R}$.

It was shown by Charles Hermite in 1873 that the constant e is transcendental.[1] Modifying Hermite's technique, Carl Lindemann proved in 1882 that the constant π is transcendental,[2] a fact which was undiscovered when Cantor was working on his theory of cardinal numbers. The transcendence, indeed even the irrationality, of some commonly occurring constants remains undecided, a famous example being the Euler–Mascheroni constant γ defined as the limit of the sequence $(1 + \frac{1}{2} + \frac{1}{3} + \cdots + \frac{1}{n} - \ln n)$ (its decimal expansion begins $0.577\,215\,664\ldots$). The harmonic series

$$1 + \frac{1}{2} + \frac{1}{3} + \frac{1}{4} + \cdots$$

diverges (very slowly) – simply observe that the 2^nth partial sum exceeds $\frac{n}{2}$ – and the result that $1 + \frac{1}{2} + \frac{1}{3} + \frac{1}{4} + \cdots + \frac{1}{n} - \ln n$ converges tells us precisely how slow this divergence is. (The sum of the first million terms of the harmonic series does not exceed 14.) The overwhelming feeling amongst mathematicians is that γ *must* be transcendental, but no proof nor even a hint of a successful approach to the problem has been forthcoming.

REMARKS

1. In proving that π is transcendental Lindemann also settled the long unsolved problem of squaring the circle, that is, given a circle, the problem of constructing with straight edge and compass a square having the same area as

---

[1] Hermite, C., 'Sur la fonction exponentielle', *C. R. Acad. Sci. Paris*, **77**, 18–24, 74–79, and 226–233, (1873).

[2] Lindemann, C. L. F. von, 'Über die Zahl π', *Math. Ann.*, **20**, 213–225, (1882).

that enclosed by the circle. Since π is transcendental, so is $\sqrt{\pi}$, therefore the construction is impossible.

2. The Gelfond–Schneider Theorem (proved independently by Aleksandr Gelfond and Theodor Schneider in 1934) tells us that if $\alpha, \beta \in \mathbb{A}$, $\alpha \notin \{0, 1\}$ and $\beta \notin \mathbb{Q}$ then (any of the possibly multiple values of) α^β is transcendental. If we consider the number e^π, for example, then the theorem would at first sight seem to be wholly inappropriate since neither e nor π is algebraic. However, $e^\pi = (e^{i\pi})^{\frac{1}{i}} = (e^{i\pi})^{-i} = (-1)^{-i}$, so in fact we can use the theorem: e^π is transcendental. It is not known whether π^e is transcendental, but I think it would be a huge shock to discover that it is not.

The set of algebraic numbers forms a field, indeed it is the smallest field extension F of $\mathbb{Q}$ for which all non-constant polynomials with coefficients in F have roots in F. It wasn't until the end of the nineteenth century that mathematics' most famous constants e and π were proved to be transcendental. The existence of transcendental numbers can be proved by comparing the cardinality of the set of algebraic numbers with that of the complex numbers.

1.9.3   The cardinality of $\mathbb{A}$

Before the appearance of the theory of infinite sets, Joseph Liouville had found a number of results via which it was possible to explicitly exhibit certain transcendental numbers. Liouville's proof of the existence of transcendental numbers requires only a little analysis. A real number x is a *Liouville number* if for each natural n there is a rational number $\frac{p}{q}$ with $q > 1$ such that

$$\left| x - \frac{p}{q} \right| < \frac{1}{q^n}.$$

The series $\sum_{n=0}^{\infty} r^{n!}$, for example, yields a Liouville number for all rational r in the open interval $(0, 1)$. The definition of Liouville numbers is motivated by the following result: if a real algebraic number z is the root of a polynomial of degree $n > 1$ then there exists a positive integer M such that

$$\left| z - \frac{p}{q} \right| > \frac{1}{Mq^n}$$

for all integers p and q, $q > 0$.[1] The fact that all Liouville numbers are transcendental is an easy corollary.

We can bypass this, proving the existence of transcendental numbers by considering cardinality alone, for it turns out that there are only countably

---

[1] The only result from analysis needed to prove this is the Mean Value Theorem.

many algebraic numbers. To prove this we need the following two facts (for the rest of this section by 'polynomial' we mean a polynomial with rational coefficients).

(i) Any polynomial of degree n has at most n distinct roots.

(ii) The set of all polynomials is countable.

(i) follows from the Fundamental Theorem of Algebra and induction and (ii) was proved (in lightly disguised form) in the remarks to Subsection 1.6.9.

Fix some enumeration of the set of all non-constant polynomials p_1, p_2, p_3, p_4, ... We then list all the distinct roots of p_1, followed by the distinct roots of p_2 (excluding any shared roots with p_1), then the distinct roots of p_3 minus any that have already been listed, and so on.

So, although the algebraic numbers are the numbers we might encounter most frequently in casual mathematical doodles (excepting the transcendental π and e), the set $\mathbb{A}$ is, in the sense of cardinality, tiny in comparison with the less familiar set of transcendental numbers (if the set of transcendental numbers were countable then its disjoint union with the algebraic numbers, namely $\mathbb{C}$, would also be countable, a contradiction).

This simple cardinality proof of the existence of transcendental numbers, due to Cantor, was a very successful advertisement for his new theory. It was partly this that led to the theory being fairly swiftly embraced by some of Cantor's contemporaries. The method is exquisitely simple: to show that there are elements of a set X which are not elements of a set Y, one simply shows that X is in some sense 'bigger' than Y. The difficulty, of course, is to find a suitable notion of 'bigger' that has the right properties. Besides cardinality, there are other fruitful notions of relative size that can be used to prove existence theorems, for example measure theoretic differences (nullsets versus sets of positive measure) or sets that differ in terms of density properties (Baire first category sets versus Baire second category sets).[1]

REMARK

Cantor's proof of the existence of transcendental numbers is not just an existence proof. It can, at least in principle, be used to construct an explicit transcendental number. We first need to fix an explicit enumeration of the non-constant polynomials with integer coefficients. Define the *height* of a polynomial to be the sum of the absolute values of its coefficients. For fixed positive integers d and h there are finitely many polynomials with degree d and height h. These can be 'lexicographically' ordered by coefficients. For example, the ten degree 2 polynomials with height 2 are ordered in the sequence: $-2x^2$, $-x^2 - x$, $-x^2 - 1$, $-x^2 + 1$, $-x^2 + x$, $x^2 - x$, $x^2 - 1$, $x^2 + 1$, $x^2 + x$, $2x^2$. The *size* of a polynomial will be defined to be the sum of its height and degree. For a fixed s there are finitely many polynomials of size s. These can be ordered first by listing the

---

[1]For a discussion of these notions and their relationship to one another, see Oxtoby [161].

degree 1, height $s-1$ polynomials (as described above), then the degree 2 height $s-2$ polynomials and so on. Thus by listing the size 2 polynomials followed by the size 3 polynomials and so on we list in a very explicit way all polynomials with integer coefficients. Next, as described earlier, we can list the roots of each polynomial (ignoring repetitions)[1] ordered first by magnitude of their real parts then by imaginary parts. We can then perform a diagonalization on the generated list of all algebraic numbers, choosing the new number to differ from the nth algebraic number in the nth decimal place of, say, its real part in some agreed way, for example by setting it equal to 0 unless it is already 0, in which case we can make it 1.

The set of algebraic numbers is countable and the set of complex numbers is uncountable, therefore there exist (uncountably many) transcendental numbers. This simple application of Cantor's notion of cardinality was an elegant alternative to the analytic construction described earlier by Liouville and also serves as a prototype of other existence proofs that refer to measure theoretic notions of size. We have met countably infinite sets such as $\mathbb{N}$, $\mathbb{Z}$, $\mathbb{Q}$, $\mathbb{A}$ and their cartesian products and uncountable sets such as $\mathbb{R}$ and $\mathbb{C}$ and their cartesian products. We have also seen that these cardinalities are exceeded by sets of functions on $\mathbb{R}$. How high can we reach, and how might we get there?

1.10   Higher infinities

No one shall be able to drive us from the paradise that Cantor created for us.

                                            – DAVID HILBERT[2]

1.10.1   Power sets

There are sets, $\mathbb{R}$ being our first example, with cardinality greater than that of the set of natural numbers. As we have seen, there is a still higher infinity, that of the set of all functions $\mathbb{R} \to \mathbb{R}$. Can we construct sets of an even larger cardinal magnitude? The answer is yes, and there is a simple construction which we can use to generate from any set a new set of strictly greater cardinality.

The construction suggested by the function example on $\mathbb{R}$ works in the general case too (but is not the one we shall be pursuing below): if X is a set with at least two elements then the set $\mathcal{F}(X)$ of all functions $X \to X$ has cardinality strictly larger than that of X. The proof, yet another application of the general diagonalization technique, is a direct lift of the proof for the special case $X = \mathbb{R}$: Suppose there is a bijection $x \mapsto f_x$ of X onto $\mathcal{F}(X)$. Let us define a function

---

[1] For the purpose of generating a transcendental number we needn't worry about repetitions, the diagonalization will still work.

[2] 'Über das Unendliche', *Mathematische Annalen*, **95**, (1926).

$g \in \mathcal{F}(X)$ as follows: for each $x \in X$, choose $y \in X$ not equal to $f_x(x)$ and set $g(x) = y$. As $g \in \mathcal{F}(X)$, by the assumed bijection there must exist a $t \in X$ such that $g = f_t$, but $f_t(t) \neq g(t) = f_t(t)$, a contradiction.

The *power set* $P(X)$ of a set X is the collection of all subsets of X (including itself and the empty set). The power set of the set $\{0, 1\}$, for example, is the set $\{\varnothing, \{0\}, \{1\}, \{0, 1\}\}$. If X is a finite set with n elements then $P(X)$ has 2^n elements since for any given element $x \in X$ and each subset A of X there are two possibilities; either x is a member of A or it is not.

If X is an infinite set then $P(X)$ is equipollent to the set $\mathcal{F}(X)$ (this follows from the equipollence of $X \times X$ with X, however, the statement that $X \times X$ is equipollent with X for all infinite sets X is equivalent to the Axiom of Choice – see Subsection 9.1.3).

The power set of $\mathbb{N}$, it turns out, has the same cardinality as $\mathbb{R}$. To see why this is we proceed as follows. As we know that the set of real numbers is equipollent to the interval $[0, 1)$ and that equipollence is an equivalence relation (in particular it is transitive) it suffices to show that $P(\mathbb{N})$ is equipollent to $[0, 1)$. The Cantor–Bernstein Theorem will be invoked, so we need to exhibit two injections, one from $P(\mathbb{N})$ into $[0, 1)$ and the other from $[0, 1)$ into $P(\mathbb{N})$. The latter is quite straightforward. We first represent a given element x of $[0, 1)$ in binary form $0.x_0 x_1 x_2 \ldots$ (without a tail of 1s). This binary representation of x determines a subset A of $\mathbb{N}$ via the condition $i \in A$ if and only if $x_i = 1$. The function mapping x to A is clearly an injection. It is not a bijection; we need only observe that $\mathbb{N}$ itself is not the image of any x (nor is any subset of $\mathbb{N}$ which has a subset of the form $\{n : n \geq a\}$ for some $a \in \mathbb{N}$). An injection $f : P(\mathbb{N}) \to [0, 1)$ is obtained by defining $f(A)$, for $A \in P(\mathbb{N})$, to be the binary number $0.a_0 0 a_1 0 a_2 0 a_3 0 a_4 0 \ldots$ where a_i is 1 if and only if $i \in A$. The alternate splicing of 0s is here in order to prevent a tail of 1s. By the Cantor–Bernstein Theorem we infer the existence of a bijection $P(\mathbb{N}) \to [0, 1)$ and hence a bijection $P(\mathbb{N}) \to \mathbb{R}$.

REMARK

Taking the power set of the set of natural numbers (or indeed any infinite set) is a wildly non-constructive operation. Later, when we begin to introduce some axioms, we will use as an axiom: *if A is a set then $P(A)$ is a set.* Objecting to the non-constructive nature of the power set operation, some constructivists have replaced this axiom with more sober alternatives.

The power set of a set is the set comprising all of its subsets. If the set is finite and has n elements then its power set has 2^n elements. The power set of the set of natural numbers is equipollent to the cardinally larger set of real numbers. Cantor's Theorem tells us that this cardinal leap occurs no matter which set we begin with.

1.10.2   Cantor's Theorem

The power set of any set X has cardinality strictly greater than X. This important result, known as Cantor's Theorem, has the following simple proof. Suppose $f : X \to P(X)$. Let $B = \{x \in X : x \notin f(x)\}$, then $B \in P(X)$. If f is surjective then there exists a $b \in X$ with $f(b) = B$. From the definition of B we have $b \in B$ if and only if $b \notin f(b) = B$, a contradiction. Therefore there is no surjection $X \to P(X)$.

In particular $P(\mathbb{R})$ has cardinality greater than that of $\mathbb{R}$, $P(P(\mathbb{R}))$ has cardinality greater than that of $P(\mathbb{R})$, and so on; an infinite hierarchy of infinities is revealed. Whether one can find a set with cardinality strictly between that of an infinite set X and that of $P(X)$ is the substance of the Generalized Continuum Hypothesis (which states that no such set exists).

The relative abundance of elements in $P(\mathbb{N})$, as compared to $\mathbb{N}$, is ensured by our ability to extract infinite subsets of $\mathbb{N}$, for if only finite subsets of $\mathbb{N}$ were allowed we would not be able to escape countability: the set Fin of finite subsets of $\mathbb{N}$ is countable – the function mapping A to $\sum_{n \in A} 2^n$, where we make the convention that $\varnothing$ maps to 0, is a bijection Fin $\to \mathbb{N}$. It follows that the set of subsets X of $\mathbb{N}$ with a finite complement $\mathbb{N} - X$ (so-called cofinite sets) is also countable, so the bulk of $P(\mathbb{N})$ is made up of infinite subsets with infinite complements. One can continue to peel off countable subsets of $P(\mathbb{N})$ of increasing complexity. This is a useful exercise, if only to develop an appreciation of the size of $P(\mathbb{N})$. More generally, the set of all finite subsets of any infinite set X has the same cardinality as X.

REMARK

Consider the set $\mathcal{G}$ of infinite subsets of $\mathbb{N}$ with infinite complements. We can describe an explicit bijection f from $\mathcal{G}$ to the set of all infinite subsets of $\mathbb{N}$ as follows. Suppose $A \in \mathcal{G}$. Let us denote by $[n, m]$ the set of all integers k such that $n \leq k \leq m$. The set A is uniquely expressible as a union $\cup_{i=0}^{\infty} [n_i, m_i]$ with $n_i \leq m_i$ and $m_i + 1 < n_{i+1}$ for all i. We now count the lengths of these component blocks and the gaps between them, starting with the possibly empty initial gap. The sequence begins $n_0, m_0 - n_0 + 1, n_1 - m_0 - 1, m_1 - n_1 + 1, n_2 - m_1 - 1, m_2 - n_2 + 1, \ldots$. From this sequence (call it (a_i)) we generate the set of cumulative sums $f(A) = \{a_0, a_0 + a_1, a_0 + a_1 + a_2, \ldots\}$. So, for instance, if we were to start with the set of even numbers then the sequence of gap and block lengths begins $0, 1, 1, 1, 1, \ldots$ and taking the cumulative sums of this sequence the generated set is $\mathbb{N}$. If we start with the set of odd numbers then the generated set is the set of positive natural numbers. f is a bijection.

Using this we can describe an explicit bijection g from the set of infinite subsets of $\mathbb{N}$ to $P(\mathbb{N})$. If A is an infinite subset of $\mathbb{N}$ then either it has an infinite complement, in which case we define $g(A) = f(A)$ as described above, or it has a finite complement F in which case we define $g(A) = F$. The composite $g \circ f$ yields an explicit bijection from $\mathcal{G}$ to $P(\mathbb{N})$.

Cantor proved that the cardinality of an infinite set is always strictly less than the cardinality of its power set. This means we can repeatedly take the power set of, say, $\mathbb{N}$ and produce a sequence $\mathbb{N}, P(\mathbb{N}), P(P(\mathbb{N})), \ldots$ of infinite sets of ever larger cardinality. We might then take the union of this collection and repeat the process. In labelling these further iterations we seem to have run out of natural numbers. A new larger class of numbers extending the natural numbers is required.

1.11 From order types to ordinal numbers

> Abstractness, sometimes hurled as a reproach at mathematics, is its chief glory and its surest title to practical usefulness.
>
> – ERIC TEMPLE BELL[1]

1.11.1 Order types

We shall meet ordinal numbers in finer detail later, nevertheless a rough introduction including an outline of the origins of the idea is needed. The central property we are interested in when considering ordinal numbers from a classical point of view is that of well-ordering, but first we look at ordered sets in general. By an ordered set we shall mean a set X together with a transitive antisymmetric relation $<_X$ on X having the property that for distinct $x, y \in X$ we have $x <_X y$ or $y <_X x$. We also assume that $<_X$ is irreflexive, meaning that there are no $x \in X$ with $x <_X x$. This extra condition is included to avoid unnecessary technical difficulties later – the order on well-ordered sets will be modelled using the irreflexive relation $\in$.

Two ordered sets $(A, <_A)$ and $(B, <_B)$ are *order equivalent* if there is a bijection $\phi : A \to B$ such that $\phi(x) <_B \phi(y)$ if and only if $x <_A y$. Such a ϕ is called an *order isomorphism*. If a set X is equipollent to an ordered set $(M, <_M)$, then X may be given a natural order $<_X$ in such a way that it is order isomorphic to M simply by fixing some bijection $\phi : X \to M$ and defining $a <_X b$ if and only if $\phi(a) <_M \phi(b)$.

The relation of order isomorphism partitions the class of all ordered sets into equivalence classes, each of which we call an *order type*. In practice, however, we tend to picture an order type as a particular representative ordered set rather than the class to which it belongs. When we say 'the order type of X' we mean the order equivalence class containing the ordered set X.

The order type of $\mathbb{N}$, that is, the class of ordered sets order isomorphic to $\mathbb{N}$ with the usual ordering $0 < 1 < 2 < 3 < \cdots$, is denoted by ω. The reverse of an ordered set $(X, <_X)$ is the ordered set $(X, >_X)$ where $>_X$ is defined, as we might expect, by the condition $x >_X y$ if and only if $y <_X x$. The reverse

---
[1]Bell [14]. Chapter 1.

of an order type α, denoted α^*, is the order type of the reverse of any of its representatives so, for example, ω^* is represented by $\mathbb{N}$ with the ordering $>$.

If $(X, <_X)$ is an ordered set then every subset of X inherits the order (but not necessarily the order type) of X in the natural way, becoming an ordered set in its own right. If α and β are order types we say $\alpha \leq \beta$ if α is the order type of a subset of some representative of β. Order types are not generally comparable; neither $\omega \leq \omega^*$ nor $\omega^* \leq \omega$, for example.

REMARKS

1. Some authors use the notation $^*\alpha$ for the reverse order type of α.

2. It is easy to see that for any two orders $<$ and $<'$ on some fixed finite set F the ordered set $(F, <)$ is order isomorphic to $(F, <')$, so each finite set determines a unique order type. This changes drastically as soon as we move into infinite sets; on any infinite set one can define many different, i.e. non-isomorphic, orders.

The class of ordered sets is partitioned into equivalence classes under the relation of order isomorphism. Each equivalence class is an order type. One can define natural operations of addition and multiplication on order types, creating a novel algebraic structure.

1.11.2   Addition of order types

The ordered sum $\alpha + \beta$ of order types α and β is defined as follows. First choose a representative $(M, <_M)$ of the order type α and a representative $(N, <_N)$ of the order type β such that M and N are disjoint (such representatives can always be found, for we are free to relabel elements if necessary). We insist on disjoint representatives so that $M \cup N$ can be ordered in such a way that all elements of M precede all elements of N. The sum of ordered sets $M + N$ is the union $M \cup N$ ordered so that $x < y$ if:

(i) $x, y \in M$ and $x <_M y$; or

(ii) $x, y \in N$ and $x <_N y$; or

(iii) $x \in M$ and $y \in N$.

The sum $\alpha + \beta$ is then defined to be the order type of $(M + N, <)$. This sum is, as one would hope, independent of the choice of representatives M and N.

For example, we form the sum $\omega + \omega$ by selecting two disjoint copies of $\mathbb{N}$, say $0 < 1 < 2 < 3 < \cdots$ and $0' < 1' < 2' < 3' < \cdots$, and ordering the union of the two copies by:

$$0 \prec 1 \prec 2 \prec 3 \prec \cdots \prec 0' \prec 1' \prec 2' \prec 3' \prec \cdots$$

It is clear that $\omega + \omega$, of which the set $S = \{0, 1, 2, 3, \ldots, 0', 1', 2', 3', \ldots\}$ with the above ordering $\prec$ is a representative, is not equal to ω, for otherwise we would be able to find an order isomorphism between $(\mathbb{N}, <)$ and $(S, \prec)$, entailing the existence of an $n \in \mathbb{N}$ (corresponding to $0'$ in S) with the property that there exists an $m < n$ and for every such m, $m + 1 < n$. No such element exists, as follows by induction: suppose $n \in \mathbb{N}$ has the stated property and let A be the set of all m with $m < n$. Then $0 \in A$ and if $n \in A$, $n + 1 \in A$, so induction tells us that $A = \mathbb{N}$, yet $n \notin A$.

If we denote by the natural number n the order type of the set

$$\{0, 1, 2, 3, \cdots, n - 1\}$$

with its usual ordering (or indeed any of its other orderings – all finite ordered sets of the same cardinality being order isomorphic) we see that $n + \omega = \omega$ and $\omega^* + n = \omega^*$ for all n, the order types $\omega, \omega + 1, \omega + 2, \ldots$ are all distinct and the order types $\omega^*, 1 + \omega^*, 2 + \omega^*, \ldots$ are all distinct.

Observing, for example, that $1 + \omega \neq \omega + 1$, we see that addition of order types is not commutative. One can show, however, that the associative law of addition holds, i.e. $\alpha + (\beta + \gamma) = (\alpha + \beta) + \gamma$ for all order types α, β, γ.

REMARKS

1. For arbitrary order types α and β it is clear that $(\alpha + \beta)^* = \beta^* + \alpha^*$.

2. The set of order types that can be formed from finite sums of ω, ω^* and n (n being the order type of a set of n elements) forms an easily comprehendible proper subset of the class of all order types of countable sets. It does not include, for example, the order type of the rational numbers. Algebraically it is a fairly trivial system comprising all such finite sums subject only to the identifications described above. Indeed, if one was so inclined, one could set up this algebraic system in disguise, entirely abstractly, describing it as the set $\mathcal{A}$ of all finite sums of elements n^o, n^* and n', where n is a natural number, subject to the rules:

   (i) $n^o + m^o = (n + m)^o$ for all n, m;

   (ii) $n^* + m^* = (n + m)^*$ for all n, m;

   (iii) $n' + m' = (n + m)'$ for all n, m;

   (iv) $0^o = 0^* = 0'$;

   (v) $n' + m^o = m^o$ for all n, m;

(vi) $m^* + n' = m^*$ for all n, m;

(vii) $(a + b) + c = a + (b + c)$ for all $a, b, c \in \mathcal{A}$.

The natural interpretation of this algebraic system is, of course, that n^o is the sum of n copies of the order type ω, n^* is the sum of n copies of the order type ω^* and n' is the order type of a set of n elements. There may, however, be alternative interpretations. We will build on this system later.

Two order types are added together by taking disjoint representatives of each order type, ordering their union in such a way that all elements of the first set precede all elements of the second set, and then defining the sum to be the order type of this new ordered set. Restricted to the (order types of the) natural numbers this definition coincides with the usual operation of addition, but in general addition is not commutative.

1.11.3   Multiplication of order types

The ordered product $\alpha\beta$ of order types α, β is defined as follows. First, as before, choose representatives $(M, <_M)$ and $(N, <_N)$ of α and β respectively. These representatives need not be disjoint. The product of ordered sets MN is then the cartesian product $M \times N$ together with the antilexicographic ordering:

$(m_1, n_1) < (m_2, n_2)$ if:

(i) $n_1 = n_2$ and $m_1 <_M m_2$; or

(ii) $n_1 <_N n_2$,

and $\alpha\beta$ is the order type of MN. Again, this product is independent of the choice of representatives M and N.

Multiplication of order types is not commutative as we can see by considering 2ω and $\omega2$. The former may be represented by

$$(0,0) < (1,0) < (0,1) < (1,1) < (0,2) < (1,2) < (0,3) < (1,3) < \cdots ,$$

which is order isomorphic to $\mathbb{N}$ in its natural ordering, i.e. $2\omega = \omega$. The product $\omega2$, on the other hand, may be represented by

$$(0,0) < (1,0) < (2,0) < (3,0) < \cdots < (0,1) < (1,1) < (2,1) < (3,1) \cdots ,$$

i.e. $\omega2 = \omega + \omega$, which as we saw earlier is not equal to ω. One can prove that the associative law of multiplication holds, i.e. $(\alpha\beta)\gamma = \alpha(\beta\gamma)$ for all order types α, β, γ.

The order type ω^2, to give a slightly more exotic example, may be represented by the set $\mathbb{N} \times \mathbb{N}$ ordered by

$$(0,0) < (1,0) < (2,0) < \cdots$$

$$< (0,1) < (1,1) < (2,1) < \cdots$$

$$< (0,2) < (1,2) < (2,2) < \cdots$$

$$< (0,3) < (1,3) < (2,3) < \cdots$$

The antilexicographic (rather than lexicographic) ordering of products of order types gives rise to left distributivity, i.e. $\alpha(\beta + \gamma) = \alpha\beta + \alpha\gamma$ for all order types α, β, γ. However, right distributivity fails as is demonstrated by considering $(\omega + 1)2$ and $\omega 2 + 2$, which are equal to $\omega + \omega + 1$ and $\omega + \omega + 2$, respectively.

REMARKS

1. Consider $\omega\omega^*$. We would picture this as a sequence of copies of ω running right to left: $\cdots + \omega + \omega + \omega$. Reversing the order of multiplication, $\omega^*\omega$ is pictured as a sequence of copies of ω^* running left to right: $\omega^* + \omega^* + \cdots$

2. We can extend our abstract algebraic system $\mathcal{A}$ to accommodate multiplication by defining it to be the set generated by all finite sums and products of n^o, n^* and n' (n natural) subject to the earlier rules and the following:

(viii) $n^o m' = (nm)^o$ for all n, m;

(ix) $n^* m' = (nm)^*$ for all n, m;

(ix) $n'm' = (nm)'$ for all n, m;

(x) $(ab)c = a(bc)$ for all $a, b, c \in \mathcal{A}$;

(xi) $a(b + c) = ab + ac$ for all $a, b, c \in \mathcal{A}$.

Multiplication of two order types is defined via a natural antilexicographic ordering of the cartesian product of any of their respective representatives. Like addition, multiplication coincides with the usual operation of multiplication on the (order types of the) natural numbers, but fails to be commutative in general. By imposing cardinal and order restrictions on certain sets we are able to classify their possible order types.

1.11.4   Dense ordered sets

The order type of $(\mathbb{Z}, <)$ is $\omega^* + \omega$. Denote by η the order type of $(\mathbb{Q}, <)$ and λ the order type of $(\mathbb{R}, <)$. The intervals $[0, 1)$, $(0, 1]$ and $[0, 1]$ have order types $1 + \lambda$, $\lambda + 1$ and $1 + \lambda + 1$, respectively and the sets of rational numbers lying in these intervals have order types $1 + \eta$, $\eta + 1$ and $1 + \eta + 1$, respectively. An ordered set X is said to be *dense* if it contains at least two elements and has the property that for any pair $a < b$ in X there exists an $x \in X$ such that $a < x < b$. The property of density is clearly preserved by order isomorphism. One can describe the order type of all *countable* dense sets quite easily: Cantor proved that all dense, countable sets which neither possess a first nor last element are order isomorphic to $\mathbb{Q}$; furthermore, an arbitrary dense countable ordered set has one of the four order types $\eta, 1 + \eta, \eta + 1$ or $1 + \eta + 1$.

In particular Cantor's result tells us that $\eta 2 = \eta + \eta = \eta$. It is also true that $\lambda + \lambda = \lambda$. There also exist non-dense sets whose order type α satisfies $\alpha + \alpha = \alpha$, perhaps the simplest being 2η (by left distributivity $2\eta + 2\eta = 2(\eta + \eta) = 2\eta$). We picture 2η as a hybrid copy of $\mathbb{Q}$ where each point has been replaced by an adjacent pair of points separated by a small gap.

Remarks

1. We sketch a proof of Cantor's result on dense countable ordered sets (this proof employs the so-called 'back-and-forth' method). Suppose A and B are two dense countable ordered sets without first or last elements. We fix an enumeration (a_n) and (b_n) of A and B respectively. We need to pair each a_n with a b_n in such a way that the correspondence respects the orders on A and B. The pairing is recursive. Suppose we have paired off some set of a_ns with b_ns. We choose the first a_n that hasn't yet been paired with an element from B. We pair a_n with the first b_k in the list that hasn't been used already and which is consistent with the ordering (thanks to the density and the absence of a first and last element such an element will always exist). Then we choose the first b_n on the list which hasn't yet been paired with an element of A, choose the first a_k in the list compatible with such a pairing, then go back to B, and so on.

2. Note that $\eta = \eta^*$. For arbitrary order types α and β, it is *not* the case in general that $(\alpha\beta)^* = \beta^*\alpha^*$ (consider 2η and $\eta 2$, for example).

3. Consider $\eta\omega$. This is a dense countable order type with no first or last element, so it is equal to η. The order type $\omega\eta$, on the other hand, is countable and has no first or last element, but it is not dense. Similarly we see that $\eta\omega^* = \eta$ and $\omega^*\eta$ is the reverse of $\omega\eta$. $\eta\eta$ is dense and countable with no first or last element, so is again equal to η.

4. Incorporating the new order type η into our abstract algebra is straightforward. We introduce a new element, e say, and declare the algebra to be

generated by all finite sums and products of n^o, n^*, e and n' (n natural) subject to the earlier rules and the following:

(xi) $e^2 = e$;

(xii) $en' = en^o = en^* = e$ for all n.

5. Every countable order type can be realized as a subset of $\mathbb{Q}$ with the usual induced order. For finite order types and for ω and ω^* this is obvious. Since the set $\{1, \frac{1}{2}, \frac{1}{3}, \frac{1}{4}, \ldots\}$ of order type ω^* is contained in the interval $(0, 1]$, and since the set of rational numbers in $(0, 1)$ is order isomorphic to $\mathbb{Q}$ we begin to see how we can string ordered sets together to generate more elaborate order types. For more complicated sets we can use back-and-forth style embeddings.

6. There are uncountably many countable order types. Let

$$A = \{a_0, a_1, a_2, \ldots\}$$

be a subset of $\mathbb{N}$. We associate with A the countable order type

$$a_0 + (\omega^* + \omega) + a_1 + (\omega^* + \omega) + a_2 + (\omega^* + \omega) + a_3 + \cdots$$

This clearly describes an injection from the uncountable set $P(\mathbb{N})$ into the set of countable order types.

Since, furthermore, there is an injection from the set of countable order types into $P(\mathbb{Q})$ (see the previous remark) we see that the number of countable order types is equipollent to $P(\mathbb{N})$ and hence to $\mathbb{R}$. As the number of order types describable in terms of finite combinations of sums and products of $\omega, \omega^*, n, \eta$ is countable, we are far from describing all countable order types – indeed since the latter class is uncountable there is no such finitistic algebraic description of all countable order types. Some more exotic examples will be given later.

7. For fixed $p \geq 2$ the set Q_p of rational numbers of the form $\frac{n}{p^m}$, $n \in \mathbb{Z}$, $m \in \mathbb{N}$, is a dense countable ordered set with no first and last element and hence is order isomorphic to $\mathbb{Q}$. Let us use Q_2 to label a disjoint set of closed intervals of $\mathbb{R}$ with rational end-points. We map an integer n to the interval of width $\frac{1}{2}$ with mid-point n. The set of all such intervals is pairwise disjoint. We want to associate with each gap between these intervals a new interval centred at the mid-point of the gap and of width $\frac{1}{4}$, so the interval between the intervals labelled n and $n+1$ will be the interval labelled $\frac{2n+1}{2}$. Next we construct intervals of width $\frac{1}{8}$ centred at the mid-points of the gaps left, so that in general the interval centred at the mid-point in the gap between the intervals of width $\frac{1}{2^{m+1}}$ labelled $\frac{n}{2^m}$ and $\frac{n+1}{2^m}$ will be the interval of width $\frac{1}{2^{m+2}}$ labelled $\frac{2n+1}{2^{m+1}}$. The set of all such Q_2-labelled intervals is pairwise disjoint by construction. The set of all end-points of these intervals has order type 2η. If we add to this set the mid-points of each interval we have a set of order type 3η and in general if within each interval we place a set of order type α then the resulting subset of $\mathbb{Q}$ will have order type $\alpha\eta$.

*Every countable dense set without a first or last element is order isomorphic to
the rational numbers with respect to the order $<$. So, denoting the corresponding
order type by η, we have $\eta + \eta = \eta$. Likewise, if λ is the order type of the set
of real numbers with respect to $<$, then $\lambda + \lambda = \lambda$. If we restrict to the class
of order types of well-ordered sets the algebraic operations, although still non-
commutative, resemble more closely those of the natural numbers.*

1.11.5   Well-ordering: introducing ordinal numbers

An ordered set contains at most one first element and at most one last element,
possibly none (all possibilities are exhibited in the four examples $\mathbb{R}$, $(-\infty, 1]$,
$[0, \infty)$, $[0, 1]$, where the ordering is the natural ordering inherited from $\mathbb{R}$). An
order on a set S is a *well-ordering* if every non-empty subset of S has a smallest
element. The prototypical example of a well-ordered set is $(\mathbb{N}, <)$.

Note that we can give $\mathbb{Q}$ (or indeed any countable set) an *alternative* ordering
which turns it into a well-ordered set: simply choose a bijection $\phi : \mathbb{N} \to \mathbb{Q}$ and
say $\phi(n) \prec \phi(m)$ if and only if $n < m$, then $(\mathbb{Q}, \prec)$ is a well-ordered set, even
though $(\mathbb{Q}, <)$ is not. Perhaps the most natural alternative ordering of $\mathbb{Q}$ which
is a well-ordering is that given in Section 1.5: $0 \prec 1 \prec -1 \prec \frac{1}{2} \prec -\frac{1}{2} \prec 2 \prec$
$-2 \prec \frac{1}{3} \prec -\frac{1}{3} \prec 3 \prec -3 \prec \frac{1}{4} \prec -\frac{1}{4} \prec \frac{2}{3} \prec \cdots$. Whether it is possible to
well-order uncountable sets is an issue of great importance which we come to
later.

The ordinal numbers are classically regarded as the equivalence classes of
well-ordered sets under the relation of order isomorphism, that is, an ordinal
number is the order type of a well-ordered set. There are other approaches
to ordinal numbers where one focuses instead on canonical representatives of
each equivalence class (we will see this later). The class of ordinal numbers
is a proper class which is central to nearly all questions pertaining to the set
theoretic universe.

Ordinal numbers remain so under addition and multiplication. In other
words, the ordered sum and the ordered product of two well-ordered sets is
again well-ordered. Ordinal numbers are well-behaved in a number of other
respects too, for example the class $\mathcal{O}n$ of all ordinal numbers is itself well-
ordered (restricted to ordinals the order type ordering $\alpha \leq \beta$ is equivalent to
the existence of an ordinal number γ such that $\alpha + \gamma = \beta$). So, despite its
much richer structure, the class $\mathcal{O}n$ shares with the humble natural numbers
the property that every strictly decreasing family of its elements is *finite*. From
here on we shall refer to ordinal numbers simply as *ordinals*.

REMARKS

1. Every well-ordered subset of $\mathbb{R}$ (with respect to the usual order on $\mathbb{R}$) must
   be countable.

2. Although considerably more sober than countable order types in general, one shouldn't be fooled into assuming that the countable ordinals are very much easier to grasp – it continues to be an uncountable set, a fact that we shall come to shortly.

3. In a well-ordered set S every element x, with one possible exception, has a well-defined successor, namely the smallest element in the subset $\{y \in S : x < y\}$. The one exception, if it exists, is the largest element of S.

4. Since $\omega 2$ is the infinite sum $\omega + \omega + \cdots$ we have $\omega 2 = \omega + \omega 2 = \omega 2 + \omega 2 = \cdots = \omega n + \omega 2$ for all finite ordinals (i.e. natural numbers) n. Similarly, smaller powers of ω are absorbed into higher powers simply by left distributivity: suppose $n < m$, then for all natural k we have $\omega^n k + \omega^m = \omega^n(k + \omega^{m-n}) = \omega^n(\omega^{m-n}) = \omega^m$.

An ordinal number is the order type of a well-ordered set. The class of ordinal numbers shares some familiar properties with the class of natural numbers, for example it is linearly ordered, the sum and product of two ordinal numbers is again an ordinal number and every non-empty set of ordinal numbers has a least element. Cantor's original description of ordinal numbers was not as concrete as this order theoretic definition.

1.11.6 Historical interlude: ordinals via principles of generation

The purely set theoretic presentation of ordinals as order types of well-ordered sets resolves some philosophical worries that were present in Cantor's original presentation where new 'next' ordinals were apparently plucked from nowhere. Cantor described the structure and arithmetic of his new numbers in terms of principles of generation and limitation.[1] He called the class of finite ordinals $\{0, 1, 2, 3, 4, ...\}$ the First Number Class and introduced further Number Classes as follows.

The First Principle of Generation For any ordinal number α there exists an ordinal β which is greater than α and such that all ordinals less than β are equal to or less than α, i.e. β is the 'next' ordinal number '$\alpha + 1$' after α. Ordinals generated in this way are called *successor ordinals*. It is convenient to include 0 in the class of successor ordinals, despite the absence of a predecessor.

The Second Principle of Generation For any set of ordinals X there exists an ordinal α which is greater than all elements of X and such that no ordinal less than α exceeds all elements of X. An ordinal so generated which is not a

---

[1] See Cantor [**29**].

successor is called a *limit ordinal*. The smallest limit ordinal, which we obtain
by applying the Second Principle of Generation to the First Number Class, is
ω.

Each ordinal formed after ω using the two principles of generation has the
property that the set of ordinals preceding it is countable. The class of all such
ordinals is called the Second Number Class and is uncountable. Furthermore,
the cardinality of the Second Number Class is the next highest after the cardinality of the First Number Class. (The Continuum Hypothesis, more of which
later, states that this is equal to the cardinality of $\mathbb{R}$.)

Applying the Second Principle of Generation to the Second Number Class
gives us an ordinal ω_1. This launches us into the Third Number Class, the class
of all ordinals α having the property that the set of all ordinals preceding α
has the same cardinality as the Second Number Class. The smallest ordinal
exceeding all ordinals in the Third Number Class is ω_2.

More generally ω_n is the smallest ordinal exceeding all ordinals in the $(n +
1)$th Number Class.

The *General Principle of Limitation* tells us that all ordinal numbers formed
after ω_τ, using the two Principles of Generation, exceed a set of ordinals with
the cardinality of the $(\tau+1)$th Number Class and collectively form the $(\tau+2)$th
Number Class.

The $(\tau+1)$th Number Class has the same cardinality as the union of all the
Number Classes preceding it. For a limit ordinal β the βth Number Class is the
class of all ordinals which exceed a set of ordinals with cardinality equal to that
of the union of the αth Number Classes with $\alpha < \beta$, and so on.

REMARK

The proof that ω_1 is uncountable is a direct consequence of its definition. If ω_1
were countable then it would be greater than itself. Similarly the fact that the
cardinality of the Second Number Class is the next highest after the cardinality
of the First Number Class, and so on, is also immediate from the definition.

---

*Cantor introduced the ordinal numbers in terms of principles of generation,
postulating the existence of successors and limit ordinals following any given
class of ordinals. Despite its abstract nature the idea was grounded in a very
concrete problem in analysis.*

1.11.7   Analytical origins

Cantor's work on trigonometric series led him to consider more and more intricate sets of real numbers. In studying the problems of the uniqueness of the

expansion of a function as a trigonometric series, he considered the operation of forming the derived set of a set of real numbers. Let us expand on this.

Each subset P of $\mathbb{R}$ has a well-defined set $P^{(1)}$ of limit points, the *derived set* of P. $P^{(1)}$ is the set of points x in $\mathbb{R}$ having the property that every punctured neighbourhood of x, that is, every set of the form $U - \{x\}$ where U is an open set with $x \in U$, contains at least one element of P. The derived set of a set can be much larger than the original, for example $\mathbb{Q}^{(1)} = \mathbb{R}$, and it can be much smaller, e.g. $\mathbb{Z}^{(1)} = \varnothing$. The operation can be repeated: $P^{(2)}$ is the derived set of $P^{(1)}$, $P^{(3)}$ is the derived set of $P^{(2)}$, and so on. It may be that there is some n such that $P^{(n)}$ is empty, or alternatively $P^{(n)}$ could be non-empty for all $n \in \mathbb{N}$. In a certain sense the number of iterations of the derivation operator needed to achieve emptiness (if this is ever attained) is a type of density measure of the original set P.

But one need not stop here, for if $P^{(n)}$ is not empty for any finite n and we define the set $P^{(\omega)}$ to be the intersection of all $P^{(n)}$, $n \in \mathbb{N}$, then $P^{(\omega)}$ need not be empty, and we can apply the derivation operation to it, yielding a new set $P^{(\omega+1)}$, and so on, $P^{(\alpha+1)}$ being the derived set of $P^{(\alpha)}$ and for limit ordinals β, $P^{(\beta)} = \cap_{\alpha < \beta} P^{(\alpha)}$.

Examples where $P^{(\omega)} \neq \varnothing$ are easy enough to find, but what about examples where $P^{(\omega)} = \varnothing$ and yet $P^{(n)} \neq \varnothing$ for all $n < \omega$? Consider the following construction. We define a sequence of sets (A_n) by setting $A_0 = \varnothing$; $A_1 = \{0\}$; $A_2 = \{\frac{1}{n} : n \in \{1,2,3,\ldots\}\}$; and further A_n defined recursively as follows: for each x in A_n there exists an $\varepsilon_n(x) > 0$ such that $(x, x + \varepsilon_n(x)) \cap A_n = \varnothing$, let B_x be the infinite set $\{x + \frac{1}{n} : n \in \{1,2,3,\ldots\}$ and $\frac{1}{n} < \varepsilon_n(x)\}$ and set $A_{n+1} = \cup\{B_x : x \in A_n\}$. Then $A_{n+1}^{(1)} = A_n$ for all n, and in particular $A_n^{(n)} = \varnothing$. Each A_n, $n \in \{1,2,3,\ldots\}$, is a bounded subset of $\mathbb{R}$ so we can glue translated copies of the sets $A_1, A_2, A_3, \ldots$ end to end forming a subset P of $\mathbb{R}$ with the property that $P^{(n)} \neq \varnothing$ for all $n < \omega$ but $P^{(\omega)} = \varnothing$. Similar constructions (perhaps scaling the concatenated copies of the A_ns so that their union occupies a bounded region) can be applied to obtain comparable results for higher limit ordinals.

Cantor was interested in trigonometric series of the form

$$s(x) = \sum a_n \cos nx + b_n \sin nx,$$

where the a_n, b_n are arbitrary, i.e. not necessarily Fourier coefficients. In an 1872 paper[1] Cantor proved that if $s(x)$ converges to zero on $G = \{x \in [a,b] : x \notin A\}$ and $A^{(k)}$ is empty for some $k \in \mathbb{N}$, then the coefficients a_n, b_n must all be zero. In other words, if two series $s(x)$ and $t(x)$ coincide on such a G then the series are identical. Incidentally this is the same paper in which Cantor introduces the Cauchy sequence construction of the real numbers.

Thus the new infinite ordinal numbers made their first appearance as indices of the iterations of an operation designed to unravel the structure of elaborate

---

[1]Cantor, G., 'Über die Ausdehnung eines Satzes aus der Theorie der trigonometrischen Reihen', *Math. Annalen*, **5**, 123–132, (1872).

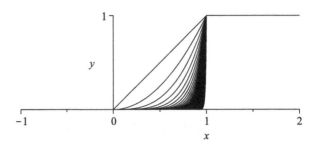

Figure 1.25 The sequence of continuous functions (f_n) defined by $f_n(x) = x^n$ on the interval $[0, 1]$ and as the constant functions 0 and 1 for $x < 0$ and $x > 1$, respectively, has a pointwise limit which is discontinuous at $x = 1$. The figure above shows the graph of $y = f_n(x)$ for $n \in \{1, \ldots, 100\}$.

subsets of $\mathbb{R}$ and to demystify the representation of functions on $\mathbb{R}$ by trigonometric series. Mathematics is often spawned in this indirect way; attempts to solve problems in one subject, even if unsuccessful, often give rise to interesting new avenues of research in another. There are few true dead ends in mathematics, and one cannot predict which parts of mathematics will influence one another in the future.

Another classical example of the indicial use of ordinals is the following. For any collection F of functions $\mathbb{R} \to \mathbb{R}$ we let $L(F)$ be the set of functions which are pointwise limits of sequences of functions in F. That is, $L(F)$ is the set of functions $g : \mathbb{R} \to \mathbb{R}$ having the property that there exists a sequence (f_n) in F such that for any given $x \in \mathbb{R}$, $f_n(x) \to g(x)$ as $n \to \infty$. The set C of continuous functions $\mathbb{R} \to \mathbb{R}$ is not stable under the formation of sequential pointwise limits, i.e. $L(C) \neq C$. For example, if we define $f_n(x)$ by

$$f_n(x) = \begin{cases} 0, & x < 0, \\ x^n, & 0 \le x \le 1, \\ 1, & 1 < x, \end{cases}$$

then all functions f_n are continuous, but the pointwise limit of f_n is the *discontinuous* function

$$g(x) = \begin{cases} 0, & x < 1, \\ 1, & 1 \le x. \end{cases}$$

So $g \in L(C)$ but $g \notin C$ (see Figure 1.25).

It is a natural problem, first studied by René-Louis Baire in 1898,[1] to determine the smallest set of functions containing C that is stable under pointwise limits. If we define $L^{n+1}(C) = L(L^n(C))$ and $L^0(C) = C$, then we might be

---

[1] Baire, R.-L., 'Sur les fonctions des variables reelles', *Ann. Mat. Pura Appl.*, **3**, 16–30, (1899).

tempted to believe that this smallest set is

$$L^\omega(C) = \bigcup_{n \in \mathbb{N}} L^n(C),$$

but it is not. The sequence continues $L^{\omega+1}(C), L^{\omega+2}(C), \ldots$ and so on.

Baire classified functions as follows. A function $\mathbb{R} \to \mathbb{R}$ is *Baire Class zero* if it is continuous and, for any ordinal α, is of *Baire Class $\alpha + 1$* if it is the pointwise limit of a sequence of functions of Baire Class α. For limit ordinals β the βth Baire Class (what we have called $L^\beta(C)$) is the union of all the Baire Classes of order less than β. This process does eventually stabilize. The ω_1th Baire Class $\mathcal{B}$, i.e. the union of the αth Baire Classes where α is either finite or exceeds a countable number of ordinals, called the *Baire Class*, is the class we are looking for; the pointwise limit of any sequence in $\mathcal{B}$ is again in $\mathcal{B}$. This result generalizes to functions defined on arbitrary subsets of metric spaces and marks the starting point of much deep work in measure theory.

It is clear, then, that ordinal numbers naturally index many processes that occur in mathematics. Importantly, we use them within set theory itself to label the levels of the cumulative set hierarchy, that is, the class of all sets pictured as a recursively generated tower of classes. In this sense the ordinals form the 'spine' of set theory. Two of the basic techniques of set theory, transfinite induction and transfinite recursion (see Sections 6.3 and 6.4), which are extensions of the familiar induction and recursion on the natural numbers, are based on this generalized iterative concept.

Remarks

1. The example P of a subset of $\mathbb{R}$ we constructed with $P^{(n)} \neq \emptyset$ for all $n < \omega$ but $P^{(\omega)} = \emptyset$ has order type $\omega + \omega^2 + \omega^3 + \cdots$, which we define to be the smallest ordinal which is larger than each of the ordinal numbers $\omega, \omega + \omega^2 (= \omega^2), \omega + \omega^2 + \omega^3 (= \omega^3), \ldots$. This ordinal is ω^ω according to the definition of ordinal exponentiation and ordinal limits given in Chapter 7.

2. *There are few true dead ends in mathematics.* Major breakthroughs in mathematics often bring with them a sense of loss, the most famous recent examples being Andrew Wiles' proof of Fermat's Last Theorem[1] and Grigori Perelman's proof of the Poincaré conjecture[2]. Such major problems induce much research and give birth to novel ideas and techniques, often in subject areas seemingly far removed from the area in which the problem first arose. By solving such a problem the momentum of the subject that formed in its wake may be temporarily weakened, however, owing to the wealth of new ideas often introduced in the proof, it is not usually long before another distant goal presents itself.

---

[1] Wiles, A., 'Modular elliptic curves and Fermat's Last Theorem', *Annals of Mathematics*, **141** (3), 443–551, (1995).

[2] Submitted to arXiv:math in a series of three papers (2002–2003).

3. Continuous limits of sequences of continuous functions are guaranteed if we take convergence to be in the *uniform* sense rather than the much weaker pointwise sense, and there are various other modes of convergence too. While pointwise convergence of a sequence of functions $\mathbb{R} \to \mathbb{R}$ focuses on the properties of the sequence of real numbers $(f_n(x))$ for each $x \in \mathbb{R}$, uniform convergence looks at the function as a whole. Formally $f_n \to f$ uniformly if for all $\varepsilon > 0$ there exists an $N \in \mathbb{N}$ such that if $n \geq N$ then for all $x \in \mathbb{R}$, $|f_n(x) - f(x)| < \varepsilon$. The sequence (f_n) in the example given above does not converge to g uniformly because for any $\varepsilon > 0$ if $\sqrt[n]{\varepsilon} < x < 1$ then $|f_n(x) - g(x)| > \varepsilon$.

The need for ordinal numbers first arose in analysis as a means of indexing the iterations of natural operations which fail to stabilize at the sequential limit. Cantor's principles of generation tried to capture this extension of the natural numbers, but the over-generous scope of the Second Principle of Generation led to a new paradox.

1.11.8   The Burali-Forti paradox appears

Cantor's theory had a mixed reception at first. Dedekind celebrated it, Kronecker criticized it, and most philosophers either ignored it or were indifferent.[1] Despite this turbulent start, by the early 1890s set theory was widely applied in analysis and geometry. It was during this period of relative calm and optimism for the future that Cantor came face to face with the first of the paradoxes in 1895: the Burali-Forti paradox (named after Cesare Burali-Forti who independently discovered it).[2] This difficulty arises when we ask what happens if we apply the Second Principle of Generation to the class of all ordinals. It would seem that there exists an ordinal which is strictly greater than all ordinals, *including itself*.

There was an optimism that the Burali-Forti paradox could be resolved, however, things were to take a turn for the worse when, at the beginning of the twentieth century, Russell announced a new paradox which now bears his name. While the details of the Burali-Forti paradox may have seemed obscured by the complications of a new theory, Russell's paradox appeared at an elementary level, involving only very basic notions of logic. This fresh nightmare was clearly going to be more difficult to banish.

---

[1] For an account of the intellectual climate in which Cantor's ideas emerged, see Dauben [41].

[2] Burali-Forti, C., 'Una questione sui numeri transfiniti', *Rend. Circ. Mat. Palermo*, **11**, 154–64, (1895) and 'Sulle classi ben ordinante', *Rend. Circ. Mat. Palermo*, **11**, 260, (1895) (the second correcting an error in the first). There is still some confusion in the second paper which was to be clarified by Bertrand Russell, who seems to be responsible for the name 'Burali-Forti paradox'.

REMARK

As Cantor stated it, the Second Principle of Generation applies to arbitrary classes of ordinals. The Burali-Forti paradox signals that the principle needs to be restricted so that it can only apply to certain relatively sober collections of ordinals which must necessarily exclude the class of all ordinals. We will see later that in the modern realization of ordinals the principle applies to any *set* of ordinals and the class of all ordinals is not a set.

Applying the Second Principle of Generation to the class of all ordinals, we create an ordinal number which is greater than all ordinals, including itself. This is the Burali-Forti paradox. The discovery of a more fundamental problem in the foundations, Russell's paradox (which makes no reference to ordinal numbers) together with several other paradoxes revealed that set theory was still in an unstable infancy. While ordinal numbers label the order isomorphism equivalence classes of well-ordered sets, the cardinal numbers label equipollence equivalence classes of arbitrary sets.

1.12   Cardinal numbers

> *Aristotle maintained that women have fewer teeth than men; although he was twice married, it never occurred to him to verify this statement by examining his wives' mouths.*
>
> – BERTRAND RUSSELL[1]

1.12.1   Classical cardinality

We have already met one approach to cardinal numbers in our earlier interpretation of 'the number three'. The cardinal number of a set X is, classically, the class of all sets equipollent to X. This describes a much coarser division of the class of ordered sets than that given by the ordinal numbers. The coarsening is illustrated quite dramatically when one considers the extraordinary uncountable variety of countable well-ordered sets, all of which are accounted for in a single cardinal number. The basic idea behind cardinality has a somewhat older history than that of order types; however, although Bolzano came close to defining cardinality in his *Paradoxien des Unendlichen* (1851),[2] Cantor was the first to attempt a precise distillation of the theory.

It can be argued that, for the purposes of mathematics, putting aside philosophical issues, it doesn't matter which objects we choose cardinal numbers to be (perhaps equivalence classes or representatives thereof), all we ask is that two sets have the same cardinal number if and only if the sets are equipollent.

---

[1]Russell [**188**].
[2]Bolzano [**21**].

The same may be said of the representation of ordinal numbers; it doesn't matter which objects we take to be ordinal numbers provided two well-ordered sets have the same ordinal number if and only if they are order isomorphic. Indeed, the remark applies to the rigorous presentation of any mathematical structure whatsoever; the lines and points of Euclidean geometry, for example, may be represented by daydreams and clouds so long as the axioms all hold (each pair of distinct clouds determines a unique daydream, etc.).[1]

When we come to the subject of cardinal numbers and their arithmetic in more detail we will view it from a different perspective which is intimately tied to the theory of ordinal numbers: we shall define the cardinal number of a set X to be the smallest ordinal number α with the property that X is equipollent to the set $\{\beta : \beta < \alpha\}$ (and the definition of ordinals we shall use will equate this set with the ordinal α). Assuming that every set can be well-ordered, this gives a theory equivalent to the one sketched in this section. If a set X has no well-ordering then X has a 'classical' cardinal number (the class of sets equipollent to X) but no 'modern' cardinal number (there being no ordinal α for which X is equipollent to the well-ordered set $\{\beta : \beta < \alpha\}$).

REMARKS

1. In answer to the question of whether infinite sets exist objectively, Bolzano proposed the following: Let T_0 be the statement 'there is a true statement'. T_0 is a statement of absolute truth – suppose to the contrary that there are no true statements, then the statement 'there are no true statements' is true, a contradiction. Let T_1 be the statement 'T_0 is true', T_2 the statement 'T_1 is true', and so on. T_0, T_1, T_2, and so on are all distinct statements. We see that the set of all truths is infinite.

2. Mirroring Bolzano's example, Richard Dedekind presents Theorem 66 in his *The Nature and Meaning of Numbers*: 'There exist infinite systems'. He argues that his own realm of thoughts is such a system, i.e. the totality S of all things which can be objects of thought. If s is an element of S then the thought s' that s is an object of my thought is itself an element of S, and so on.

3. Although one uses the intuitive origins of an abstractly presented system as a source of conjectures and insights, it can often be beneficial to forget about the original motivation and find an alternative model of the same axioms, or of a modified set of axioms. The most famous example of this is in the construction of models of non-Euclidean geometry, which we shall come to later.

---

[1]For Hilbert, 'point', 'line' and 'plane' could be replaced by 'table', 'chair' and 'beer mug', respectively.

Classically the cardinal number of a set is the class of all sets equipollent to that set. In the modern theory, ordinal numbers are certain concrete well-ordered sets and the cardinal number of a set A is defined to be the smallest ordinal number equipollent to A. One can define natural operations of addition, multiplication and exponentiation on cardinal numbers.

1.12.2   Operations on cardinal numbers

The definitions of cardinal addition and multiplication are similar to those we gave for addition and multiplication of order types earlier, indeed they are much simpler since we do not have to worry about defining an order on the given sums and products. The cardinal operations and relations defined in this subsection apply regardless of which objects we take to be cardinal numbers.

Let m and n be the cardinal numbers of disjoint sets M and N respectively. We then define the sum $m + n$ to be the cardinal number of the union $M \cup N$ and the product mn the cardinal number of the cartesian product $M \times N$. Both cardinal addition and multiplication are commutative and associative.

The reason for insisting on disjoint representatives in order to define addition should be transparent; this is cardinality viewed at its most primitive intuitively immediate level – we picture addition in terms of the augmentation of one collection of objects by further elements not in the original collection. Disjointness is not needed for the definition of multiplication.

We define $m \leq n$ if there exist sets M and N with cardinal numbers m and n, respectively, and an injection $M \to N$, and we write $m < n$ if $m \leq n$ but $m \neq n$, i.e. if there is an injection, but no bijection, $M \to N$.

The Axiom of Choice plays such a critical role in questions relating to cardinality it is most often tacitly assumed; without it we are generally unable to compare two arbitrary cardinal numbers. Even assuming Choice, we cannot conclude from the two inequalities $m_1 < n_1$ and $m_2 \leq n_2$ that the strict inequality $m_1 + m_2 < n_1 + n_2$ holds because, for example, for every finite cardinal number n and every infinite cardinal number a (i.e. the cardinal number of an infinite set) we have $n < a$ and $a \leq a$ yet $a + n = a + a = a$.

Cardinal addition and multiplication where at least one of the summands is infinite is trivial: if a is an infinite cardinal number and b is any cardinal number with $b \leq a$, then $a + b = a$; if a is infinite and b is any non-zero cardinal number with $b \leq a$, then $ab = a$ ($0 \cdot \alpha = 0$ for all cardinal numbers α as follows from the set equality $\emptyset \times N = \emptyset$). I use the term 'trivial' only in the sense that the correct answer can be determined instantaneously. The results themselves are non-trivial set theoretic theorems. The addition result says if A and B are disjoint sets, A is infinite and there exists an injection $B \to A$, then there is a bijection $B \cup A \to A$. The product result says that if A is infinite, B is non-empty and there is an injection $B \to A$ then there is a bijection $A \times B \to A$.

These results block any reasonable attempt to define subtraction and division on the class of all cardinal numbers in imitation of the same operations on the natural numbers. If we were to attempt to copy the construction of $\mathbb{Z}$ from $\mathbb{N}$, using the set of all cardinal numbers in place of $\mathbb{N}$, then proceedings would come to an abrupt halt immediately, for the relation $(a,b) \sim (c,d)$ defined by $a+d = b+c$ is not transitive on the class of ordered pairs of cardinal numbers: $(0,0) \sim (a,a)$ and $(a,a) \sim (1,0)$ for all infinite cardinals a but $(0,0)$ is not related to $(1,0)$ (indeed every pair (b,c) with $b \leq a$ and $c \leq a$ is related to (a,a)).

Let a and m be arbitrary cardinal numbers, then m^a is defined to be the cardinal number of the set of functions $A \to M$ where the sets A and M have cardinal numbers a and m, respectively ($m^0 = 1$ since a function $\varnothing \to M$ is a subset of $\varnothing \times M = \varnothing$, so the set of all such functions is the singleton $\{\varnothing\}$). If the cardinal number of M is m then the cardinal number of $P(M)$ is 2^m, the cardinal number of the set of all functions $M \to \{0,1\}$. The reasoning behind this is a direct extension of the case for finite M: A bijection $f \mapsto A_f$ between the set of all functions $M \to \{0,1\}$ and $P(M)$ is determined by the condition $x \in A_f$ if and only if $f(x) = 1$.

Cardinal exponentiation enjoys some familiar properties:

(i) $m^{a_1} m^{a_2} = m^{a_1+a_2}$;

(ii) $m_1^a m_2^a = (m_1 m_2)^a$;

(iii) $(m^a)^b = m^{ab}$;

(iv) $m_1 \leq m_2$ implies $m_1^a \leq m_2^a$;

(v) if $a_1 \leq a_2$ then $m^{a_1} \leq m^{a_2}$,

where $m_1, m_2, m_3, a_1, a_2, a_3$ are arbitrary cardinal numbers. The definition of exponentiation is compatible with the definition of multiplication.

REMARKS

1. The statement that all infinite sets are equipollent to their squares (i.e. the cartesian product of the set with itself), in other words that $a^2 = a$ for all infinite cardinal numbers a, is equivalent to the Axiom of Choice. The result $a + a = a$ for infinite cardinals follows from Choice but is not equivalent to it (see Section 9.1.3).

2. If we were to try to define an integer analogue from the class of all ordinal numbers (which, after all, does have a *one-sided* subtraction), again imitating the construction of $\mathbb{Z}$ from $\mathbb{N}$, then bearing in mind that ordinal addition is not commutative we would have to define $(a,b) \sim (c,d)$ if and only if $a + d = c + b$, reversing the sum $b + c$, to ensure reflexivity of $\sim$. This relation is also symmetric but it too fails to be transitive: $(1,1) \sim (0,0)$ and $(0,0) \sim (\omega,\omega)$ but $(1,1)$ is not related to (ω,ω).

The sum of two cardinal numbers a and b is the cardinal number of the disjoint union of two sets X_a and X_b with cardinal numbers a and b, respectively, and the product ab is the cardinal number of the cartesian product $X_a \times X_b$. The cardinal power a^b is defined as the cardinal number of the set of functions $X_b \to X_a$. Both cardinal addition and multiplication are commutative and all cardinal operations share some familiar algebraic properties with those of the natural numbers. The class of cardinal numbers is, assuming the Axiom of Choice, linearly ordered and to each cardinal number there is a next highest cardinal number. We introduce a notation to help us express this.

1.12.3   Aleph: $\aleph$

Cantor introduced the Hebrew letter aleph, $\aleph$, to the theory of cardinal numbers ($\aleph$ being the first letter of the Hebrew *ein sof* – which we will roughly translate as 'unending'). The First Number Class has cardinal number $\aleph_0$. The next highest cardinal number, the cardinality of the Second Number Class, is $\aleph_1$, and so on. More generally $\aleph_\alpha$, for some ordinal α, is the cardinal number of the $(\alpha+1)$th Number Class and $\aleph_{\alpha+1}$ is the next highest cardinal number after $\aleph_\alpha$.

In the modern ordinal based definition of cardinal numbers we have $\aleph_0 = \omega$, $\aleph_1 = \omega_1$, $\aleph_2 = \omega_2$,... Cantor's Theorem uncovers a strictly increasing family of distinct infinite cardinal numbers: $\aleph_0$, $2^{\aleph_0}$, $2^{2^{\aleph_0}}$, $2^{2^{2^{\aleph_0}}}$,.... The question of where we should position $\aleph_1$, or any other aleph greater than $\aleph_0$ for that matter, within this power hierarchy has a very surprising answer which we come to later.

The collection of all cardinal numbers, like the collection of all ordinal numbers, is a proper class and so we cannot assign a cardinal number to it: it is larger than any of the infinities that its members describe. This settles (perhaps unsatisfactorily) a question asked by all beginners upon discovering that there is an infinity of infinities. Given our earlier comment that cardinality offers a much coarser gradation of the class of well-ordered sets than ordinality, one could be fooled into picturing the class of cardinal numbers as, in some naive sense, being 'smaller' than the class of ordinal numbers. The existence of the ordinal indexed class of distinct cardinal numbers $\aleph_\alpha$, $\alpha \in \mathcal{O}n$, soon shatters this illusion.

Without any axioms to tell us how sets may be formed from other sets, the boundary between sets and proper classes will inevitably seem vague (even arbitrary). Such axioms will be introduced in subsequent chapters. A critical reader will therefore be perfectly entitled to be dissatisfied with the distinction for the time being.

REMARK

When we say the collection of all cardinal numbers is a proper class we are leaping ahead of ourselves a little. The point is that we quickly arrive at a contradiction if we try to assign a cardinal number to the class of all cardinal numbers, so we need some modification of the theory.

The smallest infinite cardinal is $\aleph_0$, the next smallest infinite cardinal is $\aleph_1$, then $\aleph_2$ and so on, so that $\aleph_\alpha$ is the αth infinite cardinal for arbitrary ordinal number α. Another strictly increasing collection of cardinal numbers is given by the exponential tower $\aleph_0, 2^{\aleph_0}, 2^{2^{\aleph_0}}, 2^{2^{2^{\aleph_0}}},...$ Comparing these two collections of cardinal numbers poses some interesting questions with some surprising answers. Do we need this hierarchy of infinities in the real physical world? From an information theoretic perspective, aren't large finite numbers 'enough'?

1.13   A finite Universe

> *If Hilbert's illness did not lend a tragic aspect, this ink war would for me be one of the most funny and successful farces performed by that sort of people who take themselves deadly seriously... I do not intend to plunge as a champion into this frog-mice battle with another paper lance.*
>
> – ALBERT EINSTEIN[1]

Einstein's quote is a reference to the Intuitionist versus Formalist debate between Brouwer and Hilbert. The feud between Hilbert and Brouwer had become quite public, Brouwer eventually being expelled from the editorial board of *Mathematische Annalen* in 1928.[2]

1.13.1   The total library

Some of Cantor's critics based their argument on a conviction that logic can only possibly apply to finite sets. There are no physically infinite sets in the Universe and, said the critics, accordingly it is unreasonable to expect the methods of logic to apply to the infinite without running into serious difficulties. One can try to counter such doubts but the argument will go back and forth without resolution, opposing philosophies unable to agree on the ontological status of infinity.

A great deal of muddled thinking results from a careless blurring of purely mathematical notions and physical phenomena. The critics, to be fair, probably

---

[1] Letter to Max Born (27 November 1928). Reprinted by permission of the Hebrew University of Jerusalem.
[2] See van Dalen [218] for details.

weren't being so crude as to suggest that the absence of a physically infinite aggregate of matter would have any bearing on the results of logic, but were instead imposing a fanciful limit on the scope of logic based on nothing more than intuition. This peculiar kind of censorship is a matter for philosophers to ponder. *All* mathematical objects are Platonic abstractions (as are most of the ideas of common parlance). Some have close ties to physical objects or physical phenomena, others are multiply abstracted and have no physical counterparts at all. Do we draw an arbitrary boundary permitting us to talk about circles but not Dedekind cuts, natural numbers but not Hilbert spaces, planes but not holomorphic functions? All of these notions, and the notion of an infinite set, can be defined in a formal language in finite terms – this point cannot be stressed enough – it is only the interpretation of these finite strings of symbols which causes difficulty. The only boundary that a mathematician will refuse to cross is that between a consistent theory and an inconsistent one, otherwise he is free to introduce any number of new notions provided they can be described in the underlying language.

If we are to reject infinite sets on the grounds that they have no physical counterparts, then we ought to reject *all* such Platonic notions. Once we explore this route we head towards a strict form of finitism which places a finite bound on the cardinality of objects we can speak of intelligibly. (But we cannot possibly define 'intelligibility' in this naive way, via cardinality alone, for there will be a natural number N for which all sets of cardinality N are intelligible but all sets of cardinality $N + 1$ are unintelligible, so a more subtle distinction is called for; this begins to resemble Eubulides' sorites paradox: exactly when does a heap of sand cease to become a heap if we remove grains one by one?) If we choose to accept some Platonic notions without physical counterparts but not others such as infinite sets, this begs the question of why we are being so prejudiced against the latter; how is this boundary to be drawn?

Most mathematicians, who are quite accustomed to operating in a purely Platonic universe, ignore all of this and simply get on with the business of producing mathematics.

The use of finitistic language and techniques to deduce facts about infinite aggregates was predicted to end in disaster and, when it was discovered at the end of the nineteenth century that there were some serious problems in the foundations of mathematics, the critics felt that they had won the argument. These problems arose from working unquestioningly with arbitrary collections of the form 'the set of all x such that ϕ' where ϕ is some statement involving x. Indeed, the problems were so great that several major projects were abandoned or substantially revised, Frege's work being a famous example.

It is an enduring puzzle that we are able to talk about infinite sets at all. Humans, spear-throwing primates, seem to have a thin intuitive grasp of all but the smallest of finite sets. The following popular example, inspired by Jorge Luis Borges,[1] highlights our inability to cope with large finite sets. Versions of this

---

[1] See Borges' characteristically short piece *The Library of Babel* and, for a compact history of the idea, Borges' *The Total Library* [**25, 27**].

idea appear throughout the literature; Borges credits the nineteenth century experimental psychologist Gustav Theodor Fechner as the inventor of the Total Library but traces glimpses of it back to Aristotle's *Metaphysics*.

Let us suppose that an alphabet has 100 characters, including upper and lower case letters, punctuation, spaces and various other symbols one might encounter in print. To permit languages which use different scripts we generously assume that we have 100 such alphabets and that a gargantuan library stands before us, each book therein with, say, 400 pages, a page holding 5000 characters. Let us further assume that the library is complete: every possible combination of letters, in all possible orders, can be found somewhere on its shelves. Then, in total, there are $10^{8\,000\,000}$ different books. Reducing the pages of this library to a pulp we could form a tight papery ball of such terrifyingly large diameter that it is difficult to a find a physical comparison that can do it justice; making reasonable assumptions about page width and the dimensions of the book, if we were to repeatedly double the diameter of the observable Universe somewhere in the region of 8 or 9 million times then the ball would just be able to fit inside this unimaginably vast expanse.

Included in this incomparable library will be the text of all books ever written and ever to be written using the characters of our 100 alphabets (it will also contain painstaking calligraphic instructions for all alphabets not included in our original 100). The vast majority of these books will contain nothing but meaningless strings of symbols. Some will contain an occasional meaningful word or sentence, but these will be exceedingly rare gems amidst the confusion. We might wish to employ a legion of librarians to sift out the nonsense. If one of our librarians discovers any typographical error whatsoever he will be entitled to throw the spoiled book on to the scrap heap, safe in the knowledge that there is a perfect version of the book elsewhere in the library. For any book in the library there will exist a myriad of copies (destined for eventual destruction) differing from the original only in some trivial respect, perhaps an ill-placed comma or some unfortunate meaningless permutation of letters or words. Indeed, the number of 'close neighbours' of a given book is vast. If by an n-neighbour of a book we mean one that differs from it only in n characters (taken from any alphabet) then each book has $19\,998\,000\,000$ 1-neighbours and approximately 2×10^{20}, 1.3×10^{30}, 1.6×10^{41} and 3.2×10^{51} 2-, 3- ,4- and 5-neighbours, respectively. Add to this the similarly vast number of volumes which have a few missing letters ('`mssing lettrs`'). Since the number of tolerable mistakes far exceed these small numbers we see that the number of volumes which are recognizable minor corruptions of a given book is enormous.

Having removed from the library all such repetitions modulo minor typographical errors and all the books which contain unintelligible streams of symbols (some of which will be retained on suspicion that they are experimental literature or works of genius written in a language of the future) we are left with the ultimate collection. Every text one could dream of may be found somewhere in this library. Books which would normally exceed 400 pages may be split across two or more volumes, our trusty librarians doing a fine job of collecting together such works and cataloguing all of the alternative endings (of course, they don't really need to catalogue these as the many-volumed catalogue itself exists somewhere in the library, alongside its multiple incorrect versions).

The librarians will respond to the work of new authors pointing out that their supposedly new book already exists in the library, as do all earlier drafts and all translations of it into every known language. These mighty book keepers would no doubt get above themselves, feeling that they are the privileged guardians of a concrete realization of a Platonic realm plagiarized by writers on a daily basis.

But where do we draw the line with this example? A much smaller library could be stocked with leaflets of one page rather than 400 pages which, assuming more hard work by the librarians, could be regarded as single page serializations of books. All books would still be represented, but in a more fragmented form. What then of a library of paragraphs, or even of sentences? As we downgrade to ever smaller libraries our librarians are transformed from mere cataloguists at one extreme into fully fledged writers at the other, and we begin to see the deceit of the whole fantasy. W. V. O. Quine takes the idea to its simplest extreme, reducing the library to two symbols, with the understanding that a suitable coding is used to translate binary strings into characters.[1]

This is all laughably artificial, of course, but anyone with heavily Platonic leanings must think of themselves, when indulging in their chosen art, as plucking something from a lofty body of potential ideas. Certainly when researching mathematics there is a strong sense (or illusion) that one is making fresh footprints through an unexplored but seemingly very real territory. This is reinforced when different researchers working in different areas independently discover the same mathematical results. Far from being a surprising coincidence, this synchronicity is exactly what one would expect. Each new advance in mathematics shines light on future directions. It is inevitable that the shortcuts through the logical web that are partially uncovered by these advances will be thoroughly explored and that several researchers will find their way to the notable regions of interest. We should be no more surprised to find that two mathematicians have discovered the same theorem independently than to find that two scientists have, say, independently discovered bacteriophages, electrons, the anthrax bacillus, the structure of RNA, or the principle of photography. There is no 'collective unconscious' at work here. In any case, the most improbable universe is the one that admits no coincidences.

Human curiosity being what it is, even the ultimate collection of all texts will eventually seem insufficient and we will begin to crave images. Let us assume that our books (still 400 pages in length) accommodate ten million pixels on each page, each pixel having, say, 500 different possible colours.[2] Then there will be $500^{4\,000\,000\,000}$ different books, including all of the books in the original text library (rendered in all fonts), plus reproductions of all paintings,

---

[1]See Quine [175].

[2]I have chosen completely arbitrary round numbers for convenience – computer scientists might prefer a power of 2 here, but it makes no difference – later I abandon specific figures completely. Five hundred different colours for each pixel is an overkill given the resolution, a much smaller number would be adequate!

photographs, doodles and other images ever produced and capable of being produced. Large images may be spread over hundreds of pages in varying degrees of magnification. Most of the books in the image library will consist of page after page of muddy uniform washes (the visual equivalent of white noise).

It is worth remarking that a single image can be encoded in text simply as a tedious list of instructions specifying pixel coordinates together with the code of the pixel's colour, for example something along the lines of 'colour the pixel at position $(0,0)$ colour 162; colour the pixel at position $(0,1)$ colour 19;...', or more efficiently just a long sequence of pixel codes preceded by instructions on how to decipher the sequence. Unfortunately no matter how clearly we try to encode the image in an intelligible way, even if we exploit self-similarity to a significant degree (using the cleverest means of data compression) the chances are that any one of our 400 page text books will only be able to list sufficient instructions for a small patch of a given image. Any given picture is a collage of such patches so it is certainly possible to gather a collection of instructional texts providing enough information to assemble the whole image. Although the text library is considerably smaller that the image library there is no paradox here, for the set of all ordered *subsets* of books in the text library is more than adequate to describe all possible images in this pixel-by-pixel fashion.

Again, we crave more. We demand moving images (let us call these 'animations'). Assuming a screen with 100 million pixels each with 500 possible colours, a screen refresh rate of 100 times per second and a given animation lasting 50 000 seconds, then there are $500^{500\,000\,000\,000\,000\,000}$ different possible animations. This incorporates the previous collection; a book may be depicted as a 'slide show', proceeding page by page at a suitable speed. Also included is every conceivable (and inconceivable) two-dimensional moving image.

Bored with this, we have a desire to add sound to our animations. All sound parameters (say, within some large fixed span including the human hearing range) will be discretized to a very high resolution, much smaller that is perceivable. We can fix some length of time for each sound clip. We then have an enormous, but finite, collection of all possible sounds including all conversations ever conducted *or that ever will be conducted*, all music, and lots and lots of white noise. This sound library can be combined with the animation library to create a collection of all possible moving two-dimensional images with sound.

Feeling constrained by the flat screen, we pine for three-dimensional images. Leaping up one dimension is no obstacle, our three-dimensional 'pixels', or 'voxels', now occupy a cube of some appropriately large size.[1] Here we may witness, with sound if desired, all possible and impossible events in three dimensions, including the image of me writing this sentence, a supernova in a tea cup, a hydrogen atom spontaneously exploding into a choir of talking hummingbirds, each one reciting the complete works of Shakespeare backwards.

In this last example we picture the observer as a passive outsider, unable to interact with the simulation they find themselves immersed in. We could, if

---

[1] A voxel – volumetric pixel – is a small cubic unit used in computer rendering of three-dimensional objects. Strictly speaking a voxel is then a virtual three-dimensional cube that is only ever realized in two dimensions, but let us not quibble over obscure terminology.

we were so inclined, continue to add discretized elements of other perceivable physical attributes or, for a more consistent experience, simulate subatomic phenomena upwards (according to an ad hoc physical theory) within a space of size, say, the observable Universe. With sufficient resolution, and assuming the physical theory models our Universe accurately, the effect of one such simulation from the point of view of a central observer will be indistinguishable from the 'reality' of our common experience except for the giveaway passiveness of it all – the observer is not part of the system. But such a complex system should be able to harbour its own life forms, and to such native simulacra the simulation *is* the Universe.[1]

One can play with this endlessly – it is a fruitful setting for some philosophical speculations. Science (fiction and fact) writers are much enamoured with the idea, ultimately suggesting that we are nothing more than flickering aggregates of data in a simulated universe. The thorny subjects of consciousness and free will provide an incredibly rich source of discussion in this cellular automaton context, but, at the risk of sounding like a coward, I'm afraid such sagacious excesses, fascinating as they are, would lead us too far afield here.[2]

REMARK

Borges' glimpses of the Total Library in Aristotle refer to Aristotle's discussion of '*Lucippus and his associate Democritus*'. In *On Generation and Corruption*, Book I (2), on rearrangements of atoms he says '*Hence – owing to the changes of the compound – the same thing seems different to different people: it is transposed by a small additional ingredient, and appears utterly other by the transposition of a single constituent. For Tragedy and Comedy are both composed of the same letters.*' In *Metaphysics*, Book I (4), '*And as those who make the underlying substance one generate all other things by its modifications, supposing the rare and the dense to be the sources of the modifications, in the same way these philosophers say the differences in the elements are the causes of all other qualities. These differences, they say, are three – shape and order and position.*'

As soon as we quantize information into finite collections of packets the totality of all possible worlds becomes a huge but nevertheless still finite set. These gargantuan sets are difficult to imagine. By contrast many infinite sets such as the set of natural numbers are easy to grasp because of their simple defining features (as characterized, for example, by Peano's Postulates). It is not the large cardinality of a set that makes it difficult to imagine but the complexity of its

---

[1]There is an assumption here that a finite model of physics is possible, or at least that the sort of universes that are able to generate and sustain life can be finite.

[2]This is a much discussed subject. Something of this 'game of life' flavour can be found in the writings of Daniel Dennett, in particular see **[48]** (and its 'pilot' **[45]**), **[46]** and **[47]** and the references therein for thought-food in this direction.

definition. If finite sets are enough to describe an information theoretic universe then, some argue, the same ought to be true of the mathematical universe.

1.13.2 Finitism

This indulgence in fantasy is not really my purpose here. I have laboured the point enough: all of the sets just described are *finite*. They are absurdly huge, but not infinite. It is just as difficult to comprehend the magnitude of these sets as it is to comprehend any infinite set. In fact, paradoxically, it seems in many ways more difficult to fathom these finite sets than it is to picture the infinite set $\mathbb{N}$. There is a relative simplicity in $\mathbb{N}$; we can 'almost' picture it at a glance, perhaps as a perspective arrangement of points vanishing on the horizon or, what amounts to the same thing, as a collection of points crammed into a finite region with a single limit point ($\{\frac{1}{n} : n \in \{1, 2, 3, \ldots\}\}$ for example) – the structure suggested by the notation '$1, 2, 3, \ldots$' does not exercise our imagination too much. It is also the homogeneity of $\mathbb{N}$ that aids its visualization (likewise for $\mathbb{Z}, \mathbb{Q}, \mathbb{R}, \mathbb{C}$). The more symmetry an object has the easier it is to reproduce, only a small part of it together with the symmetric means of duplication are needed to recreate the whole.

The large finite sets described in our generalized total library, although 'indexable' by a set of natural numbers, are burdened with such weighty conceptual baggage that they seem impossible to grasp. This suggests that our imaginings of infinite sets are not so much an encapsulation of an aggregate of objects but an extrapolation of easily pictured small finite phenomena. The presentation of a set either as a list of objects or as $\{x : \phi\}$ both involve only a finite amount of information – as the property 'ϕ' is a finite collection of symbols the conceptual difficulty of grasping the potentially infinite set $\{x : \phi\}$ is no more demanding than comprehending the finite content of the criterion ϕ. The finitely expressed property 'n is a natural number' is easy to grasp and hence so is the set $\mathbb{N}$; one does not have to count all of the members in order to declare that our understanding of the set is complete.

Finitists reject any talk of infinite sets, and any theorems about them are deemed meaningless. There are several different species of Finitism. The strict Finitist (or Ultrafinitist) rejects infinite sets completely and might further partition finite sets into 'small' and 'large' in some sense, believing there to be a bound on the numbers with which we can deal intelligibly. How the boundary between small and large might be formally defined seems to be a difficult issue, as we mentioned earlier; models of this small universe of discourse may be found in computers (real or theoretical) which undoubtedly explains why computer scientists are sometimes more sympathetic to the Finitist position than most mathematicians and physicists. If we adhere strictly to this argument we must also insist that only a finite number of theorems are valid in mathematics, for there will be theorems (or perhaps I should say conjectures, if they are without proof) which cannot be proved using a 'small' number of symbols (and one can only form chains of a 'small' number of corollaries). Among many other things,

the strict Finitist will object to nearly every ellipsis and etcetera appearing in
this book. The classical Finitist does not impose such cognitive bounds on the
finite sets we are allowed to work with, but nevertheless rejects infinite sets.

The mathematics employed in modern physics makes constant and critical
use of infinite structures. Attempts have been made to avoid this, but the fact
remains that the underlying differential geometric foundation of much of physics
in its present form is far from a finite combinatorial model of the Universe. The
Finitist must come to terms with the fact that these fictions make predictions
about observable phenomena that are verifiable to a very high level of precision.
A physicist might happily concede that these smooth infinite structures are
convenient simplifications of computationally intractable interactions of large
grainy aggregates, but the strict Finitist denies the validity of these models at
the outset. Perhaps it doesn't matter – it could be that a large finite combinatorial physical Universe is so well-modelled by continua that the observable
phenomena of each are indistinguishable.[1]

A more pressing challenge to the strict finitist can be found within mathematics itself. One can give examples of propositions concerning sequences
of natural numbers which are unprovable in first-order Peano arithmetic, but
which can be solved with the aid of infinite ordinal numbers (for an example,
see the account of Goodstein sequences in Subsection 7.2.5). We should not
be surprised by this – repeatedly throughout mathematics we see solutions to
problems which are achievable only by appealing to higher structures quite remote from the landscape in which the problem originally arose. At a purely
formalist level this is entirely expected; by admitting new concepts (new valid
sentences) the theory is clearly going to be able to prove more that its unaugmented predecessor. All of the philosophical difficulties and arguments are the
result of attaching certain *interpretations* to finite strings of symbols. Should
we impose any special importance on the distinction between the class of finite
sets and the class of infinite sets just because the perceivable physical crumbs of
reality about us do not fall into the latter class? Should the fleshy limitations
of biology and the restrictions of the physical Universe influence the scope of
mathematics? There is sufficient diversity in the philosophy of mathematics to
find advocates of both a strong positive and a strong negative answer to this
question.

REMARKS

1. Let us imitate Bolzano's example of an infinite set, replacing truth by existence. Suppose that something exists (that there is, so to speak, something
   rather than nothing). Call it E_0. Then the set $E_1 = \{E_0\}$ exists (doesn't
   it?), and hence so does the set $E_2 = \{E_1\}$, and so on, so that $E_{n+1} = \{E_n\}$,
   and we can presumably carry on in this way, forming a distinct collection of
   objects until any proposed ultrafinitistic upper limit on the size of a collection is breached. (If $E_0 = \emptyset$ this forms the Zermelo model of the natural

---

[1]A semi-popular account of these physical issues may be found in Penrose [164].

numbers, which we will meet later.) To deny that this is a well-defined collection one must either argue against the validity of such a recursive definition (so that 'and so on' is ill-defined, or that in some subtle way the growing set cannot reach the finite upper limit), or one must simply deny the existence of one of $E_1, E_2, E_3, \ldots$, but which one? Anything other than E_1 would seem arbitrary. But denying E_1 means denying the ability to form a subset of the Universe with just one object – quite a heavy restriction.

There have been some interesting attempts to put Ultrafinitism on a sound logical foundation; we can see that it is an apparently very difficult problem which most probably requires some radical departures from classical set theoretic notions.[1] The general idea is, in some interpretations, partly motivated by the hypothesis that the Universe only has finitely many states, but there seems to be more to it that this. It also aims to draw attention to abstract operations such as the notion of collecting arbitrary elements together to form subsets. Questioning and examining such things can only be a healthy exercise.

2. The notion of a finite model of the Universe being physically indistinguishable from a continuum-based model in the sense that no physical observation can favour one model over the other is an amusing one. There are many examples of relatively macroscopic physical phenomena where we know that the underlying process is finitistic (the typical behaviour of astronomically large numbers of molecules colliding with one another in an enclosed space being the most familiar example). And yet continuum based statistical models can describe these processes with extraordinary accuracy; replacing continuum models with appropriately large discrete systems one tends to find the same phenomena occurring – often nothing is lost by replacing the large scale combinations by idealized purely imaginary smooth functions (expressed in terms of a very surprisingly small number of abstract symbols).

The strictest Finitists say that there are no sets of cardinality N where N is some very large finite number. Other Finitists say there are sets of all finite cardinalities but none of infinite cardinality. For a Formalist such restrictions are simply alternative axioms whose consequences we can study. In the theory we are describing here one thing is certainly true: all definitions and theorems are finite strings of symbols, even definitions of infinite sets and theorems about real numbers.

1.13.3 Zeno's paradoxes

For Plato there was no controversy in the infinite, but his most famous student Aristotle forcefully disagreed, arguing that an infinite object or process exists

---

[1] For further reading, see Lavine [135], which includes a discussion of a finitary Zermelo–Fraenkel set theory with Choice.

only as a potential that can never actually be attained. Aristotle's interpretations of Zeno's paradoxes in his *Physics*, which we describe below in a slightly modified form, fuelled the Finitist position for centuries.

Little is known of Zeno himself, and most of the biographical information that is known is drawn from the beginning of Plato's *Parmenides*, where he appears in fictional form as an interlocutor. Perhaps all we can confidently claim is that Zeno was born in the early years of the fifth century BC. His book of paradoxes is also explicitly referenced in Simplicius' commentary on Aristotle's *Physics*, but is now lost. All known information concerning Zeno's book is taken from these two sources, Aristotle and Simplicius, and it is difficult to judge the extent to which Zeno's original statements have been distorted. The origins of the paradoxes themselves are equally shrouded in darkness. Some claim they were formulated in order to support his teacher Parmenides' philosophy that the perception of motion is an illusion.[1]

There were apparently forty paradoxes in Zeno's book. Many have been forgotten and, judging by those that survived, it seems likely that the forty could be subdivided into a small number of equivalent statements.

Even in Zeno's lifetime the paradoxes were generally regarded as easy to resolve absurdities, however there is still some discussion today among philosophers who argue that the paradoxes have not been properly resolved and that the mathematical treatments are invalid. Bertrand Russell gives Zeno due respect in *The Principles of Mathematics*:[2]

> *One of the most notable victims of posterity's lack of judgment is the Eleatic Zeno. Having invented four arguments, all immeasurably subtle and profound, the grossness of subsequent philosophers pronounced him to be a mere ingenious juggler, and his arguments to be one and all sophisms. After two thousand years of continual refutation, these sophisms were reinstated, and made the foundation of a mathematical renaissance, by a German professor [Weierstrass], who probably never dreamed of any connection between himself and Zeno.*

As mathematicians we are free to construct alternative models of space and motion which, although logically consistent, bear little or no resemblance to the phenomena witnessed in our physical Universe. We ought to turn to raw observation (or even turn to the limits of observation itself) in order to make a sensible judgement about the paradoxes. The peculiarities of quantum mechanics, for example, paint a picture of motion which is so far removed from the smooth transitions we find in continuum-based models of spacetime that the paradoxes might deserve to be re-examined in that context. However, we

---

[1] In what seems to be either a spectacular misunderstanding of Parmenides' argument or – what is more likely given his taste for provocation – a genuine point, Diogenes the Cynic sought to demonstrate the absurdity of Parmenides' theory simply by standing up and walking away – the original case of *solvitur ambulando*, solution by walking.

[2] Russell [181] (Section 327).

shouldn't get too carried away with this popular idea; quantum mechanics is intimately tied to $\mathbb{C}$ and is far from a 'discrete' theory.[1]

The dichotomy paradox

Suppose that there are no indivisible units of space or time, so that any given length and any given interval of time can be split into parts indefinitely. A model of spacetime based on $\mathbb{R}$ (regarded as a fusion of intervals) has this property, but we must avoid thinking of it as a collection of points for the purpose of this example – our pieces will be intervals, always divisible into other intervals. $\mathbb{Q}$ is also infinitely divisible, but its metric incompleteness arguably makes it a less satisfactory candidate.

A runner sprinting to his destination must first reach the halfway point. Having reached this point, before reaching the end he must reach the halfway point of the remaining course, and so on (see Figure 1.26). There will always be a small part of the course to complete before finishing and so, says Zeno, he will never finish. As Aristotle beautifully puts it:

> That which is in locomotion must arrive at the half-way stage before it arrives at the goal.
>
> (*Physics* VI: 9)

By the same reasoning our runner cannot start either, for in order to reach the endpoint he must first reach the halfway point, and before he reaches the halfway point he must reach the first quarter, and so on. If the track has length 1 then we would interpret the paradox in terms of the limit of a series: $1 = \frac{1}{2} + \frac{1}{4} + \frac{1}{8} + \cdots$, likewise for the divisions of time.

The paradox of Achilles and the tortoise

Again assuming that space and time are divisible without limit, suppose a race is to take place between Achilles and a tortoise. Achilles confidently lets the tortoise start some distance down the track. Achilles, of course, runs at a much faster rate than the tortoise and, in a short time, he has reached the position where the tortoise started. However, in the time it has taken Achilles to reach this point the tortoise has made some progress. Whenever Achilles reaches the

---

[1]Grünbaum [**86**] stresses the same point. Although the answer should have no restrictive influence on the development of mathematics, the question of whether spacetime is discrete or not is nevertheless fascinating and has given rise to a cascade of deep thinking in theoretical physics. The naive approach of taking Planck scale objects as 'spacetime voxels' isn't too satisfactory and doesn't seem to work out very well as a serious model, but more sophisticated alternatives have been proposed. All one can say with any confidence is that below this tiny scale the usual notions of distance cease to make sense, so we can't get away with modelling spacetime by familiar metric structures 'all the way down'. My favourite answer to the question of the local topological nature of spacetime is that it is *neither* continuous nor discrete and/or it is *both* continuous and discrete. This is founded only on the fact that Nature has a habit of biting back with conceptual conundrums when we think we have finally understood it. The duality might be compared to the question of whether light is a wave or a particle. What is needed is a novel new mathematics, or an interpretation of existing mathematics, to describe this unusual state of affairs.

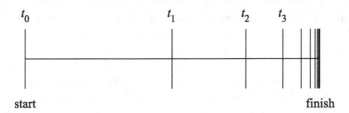

Figure 1.26 A runner sets off down a track, reaching the halfway point t_1. In order to reach the finish he must reach the halfway point of the remaining part of the track, t_2, then the halfway point of the remaining track, t_3, and so on. Zeno claimed a limitless division of space implies that the runner cannot reach his destination.

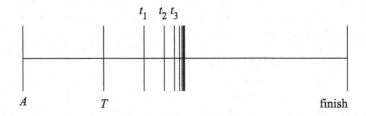

Figure 1.27 Achilles, starting at A, gives a tortoise a head start in a race. By the time Achilles has reached the tortoise's starting point T the tortoise has reached t_1, but once Achilles reaches t_1, the tortoise has reached t_2, and so on. Zeno held that a limitless division of space implies that Achilles can never overtake the tortoise.

tortoise's previous position, no matter how little time it takes him, the tortoise will have moved some distance ahead (see Figure 1.27). Consequently, claimed Zeno, Achilles can never catch the tortoise.

> *In a race, the quickest runner can never overtake the slowest, since the pursuer must first reach the point whence the pursued started, so that the slower must always hold a lead.*
>
> (*Physics* VI:9)

Let us assume that Achilles runs 100 times faster than the tortoise and that the tortoise starts halfway down the track (we will simply call this point $\frac{1}{2}$). By the time Achilles reaches $\frac{1}{2}$ the tortoise will have reached $\frac{1}{2}+\frac{1}{200}$, and by the time Achilles reaches $\frac{1}{2}+\frac{1}{200}$ the tortoise will have reached $\frac{1}{2}+\frac{1}{200}+\frac{1}{20000}$. We see that the two will cross at the limit of the series $\frac{1}{2}(1+\frac{1}{100}+\frac{1}{100^2}+\frac{1}{100^3}+\cdots)=\frac{50}{99}$.

The (modern) stadium paradox

Let us assume that space and time have limits of division with magnitude, that is, we assume that there are tangible irreducible units of time and distance so that at some scale objects in motion judder from one state to the next as in a film slowed down to slide-show speed. We must build into the statement of the paradox an assumption (rarely explicitly stated) that an object in motion cannot skip space units – to move from α to β it must at some time occupy all space units on a path between α and β. This in turn imposes a global maximum speed where an object advances one space unit per time unit. Here, of course, in the classical world, time is an all encompassing regular universal pulse beating at the same rate for all objects, regardless of whether they are moving or not. Anything moving slower than this discrete world 'speed of light' will appear to linger in certain space units for more than one time unit. This continuity of motion is essential if the paradox is to remain valid.

Suppose two objects A and B are travelling at top speed (i.e. one space unit per time unit) in opposite directions within a stadium. Then after one unit of time A and B have moved one space unit relative to the stadium but two space units relative to one another.[1] But if B has moved two space units relative to A there must have been a time when it had moved just one space unit relative to A and at this time will have moved 'half' a space unit relative to the stadium, so our minimal spacetime division seems to be divisible after all (see Figure 1.28). (The same argument applies as soon as A and B move simultaneously, regardless of how fast they are moving.) As we arrive at a contradiction we must reject at least one of the assumptions: the existence of irreducible space units, the existence of irreducible time units or the (discrete) 'continuity' of motion. (The ancient Greeks did not, of course, contemplate the possibility that each object has its own individual 'clock'.)

This is a modern 'quantized' recasting of the stadium paradox based on that given in Tiles' *The Philosophy of Set Theory*.[2] The original paradox seems to be a confusion over relative velocity. Aristotle describes the stadium paradox in terms of three bodies in the following (less than transparent) words:

> ...equal bodies which move alongside equal bodies in the stadium from opposite directions – the ones from the end of the stadium, the others from the middle – at equal speeds, in which he [Zeno] thinks it follows that half the time is equal to the double. The fallacy consists in requiring that a body travelling at an equal speed travels for an equal time past a moving body and a body of the same size at rest... let the stationary equal bodies be AA; let BB be those starting from the middle of the As..., and let CC be those starting from the end... Now it follows that the first B and the first C are at the end at the same time, as they are moving past one another. And it follows that the C has passed all the As and the B half; so that the time is half, for each of the two is alongside each for an equal time. And at the same time it follows that the first B has passed all the Cs. For at the same time the first B and the first C will be at opposite ends... because both are an equal time alongside the As. That is the argument, and

---

[1] This already contradicts the maximum speed assumption, but let us view the contradiction from the point of view of the irreducible units of space and time.

[2] Tiles [212]. This in turn is modified from G. E. L. Owen's 'Zeno and the mathematicians', *Proceedings of the Aristotelian Society*, 1957–1958.

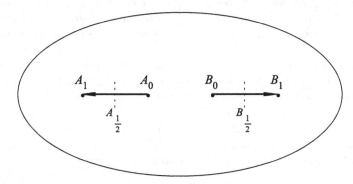

Figure 1.28 A and B travel at maximum speed in opposite directions in a stadium starting at A_0 and B_0 respectively. After one irreducible unit of time each has moved one irreducible unit of distance to the new positions A_1 and B_1. B has moved one distance unit relative to the stadium but two distance units relative to A, so (assuming a reasonable continuity of motion) there must have been a time when B had moved one unit relative to A, but such time $\frac{1}{2}$ cannot exist, nor can the corresponding distance moved by B (and A) relative to the stadium. As Zeno might have concluded, neither space nor time have limits of division with magnitude.

> *it rests on the stated falsity.*
>
> (*Physics*, VI:9)

The arrow paradox

Finally, suppose that the limits of division of space and time have no magnitude. This is the assumption that space and time are pluralities of points, not to be confused with the assumption made in the dichotomy and Achilles/tortoise paradoxes, where it is assumed that spacetime is not a plurality but simply a divisible whole.

Zeno asks us to picture an archer firing an arrow. At any fixed instant in time the released arrow navigates a locus with the same volume as the arrow at rest, hence it does not move. The flight of the arrow is composed of moments, at each of which it is motionless. Zeno concludes that the arrow never moves.

> *If everything when it occupies an equal space is at rest, and if that which is in locomotion is always in a now, the flying arrow is there-fore motionless.*
>
> (*Physics* VI:9)

Bertrand Russell has the following to say on the matter in *The Principles of Mathematics*:[1]

> *Weierstrass, by strictly banishing all infinitesimals, has at last shown that we live in an unchanging world, and that the arrow, at every*

---

[1] Russell [**181**] (Section 327).

*moment of its flight, is truly at rest. The only point where Zeno
probably erred was in inferring (if he did infer) that, because there
is no change, therefore the world must be in the same state at one
time as at another. This consequence by no means follows, and
in this point the German professor is more constructive than the
ingenious Greek. Weierstrass, being able to embody his opinions
in mathematics, where familiarity with truth eliminates the vulgar
prejudices of common sense, has been able to give his propositions
the respectable air of platitudes; and if the result is less delightful to
the lover of reason than Zeno's bold defiance, it is at any rate more
calculated to appease the mass of academic mankind.*

Zeno's unfortunate analysis, as recorded by Aristotle, is difficult to take
seriously these days, but, despite the (presumably illusory) contrary evidence
provided by observation, it helped to obstruct the development of dynamics for
nearly two thousand years. From our comfortable vantage point we see that the
error is the insistence that the arrow is 'at rest' at any given moment of time.
Sure enough motion in zero time is impossible but the velocity of an object at a
given moment can be described (nowadays via the limiting processes of calculus,
which assumes a continuum model of motion via $\mathbb{R}$).[1]

Aristotle concluded from Zeno's paradoxes that neither space nor time *nor
any continuous magnitude* may be supposed to be an aggregate of indivisible
parts. Any such continuum is obviously divisible into subsections, but these
sections are not the building bricks of the whole. Such pieces exist only as a
result of division of a pre-existing entity, in other words: *the whole is more than
the sum of its parts*.

By modelling spacetime on $\mathbb{R}$-structures we conflate the Zeno spacetime of
the arrow paradox with that of the dichotomy and Achilles/tortoise paradoxes.
$\mathbb{R}$ is divisible without limit, in the sense that each interval may be partitioned
into any number of subintervals, but it is also a plurality of points. This model
of the continuum, as we have seen, was only clarified at the end of the nineteenth
century.

REMARK

The philosophical objection to (continuous) motion is that it can be viewed as
a 'supertask', that is, the completion of infinitely many tasks in a finite amount
of time. If space is infinitely divisible then one can partition any length into
infinitely many sublengths. So a runner running from one point to another
must complete the supertask of running each of these subintervals. This applies
not only to a continuum based model of space but also to a $\mathbb{Q}$-based model.
Discretizing space frees us from supertasks but presents us with the (modern)
stadium paradox. The message is that space, time and motion are not quite

---

[1]There is much more that can be said of Zeno's paradoxes. The collection of essays in
Salmon **[191]** provides a readable introduction which might be of interest. See also Sainsbury
[190].

the simple things we would have liked them to be, and that some conceptual wrestling is needed to come to terms with these difficult notions. Even in a continuum model the stadium paradox as we have presented it, with its global maximum speed, shows that something is missing in the classical view, and that some new idea is needed. (Not surprisingly this version of the stadium paradox has been used as a pedagogical means of introducing relativity.)

Zeno considered the paradoxical consequences of a space and time which are either infinitely divisible or which are made up of indivisible atoms. The modern model of the continuum has both features: given any interval we can split it into any finite number of subintervals indefinitely, but each interval is also an aggregate of indivisible points. The continuum model of physical space and time is wildly successful. Is this simply because the combinatorics of huge finite sets so closely resembles the structures modelled by real manifolds that the difference is undetectable?

1.13.4   Later developments

Once Cantor's theory and the rigorous redevelopment of the calculus had been assimilated, Zeno's original paradoxes (for those who were satisfied with the mathematical resolutions) and the apparent 'counterintuitive' behaviour of infinite sets generally came to be regarded as historical curiosities and confusions.

The question of the existence of infinite sets does not perturb science at the immediate empirical level, which is necessarily confined to interpretations of finite error-prone measurements. One must instead turn to the mathematical theory that lurks behind the measurements – both the probabilistic theory used for the interpretation of the measurements and the models that inspired the measurements in the first place. These extraordinarily successful models are often infinitistic in some respect (a motto might be that large scale combinatorial phenomena are sufficiently smooth in a statistical limit to be faithfully modelled by continuous structures). It may seem paradoxical at first, but it is often far more tractable to model an idea using idealized infinite sets than to model it using a large finite set. Indeed, in order to describe and understand the physical world it often seems necessary to appeal to purely mathematical concepts which have no physical counterparts.

It is unlikely that any mathematical or observable physical considerations will persuasively decide the case either way on whether it is in any sense valid to consider infinite sets, one side convinced of the correctness of its Platonic censorship (perhaps inspired by properties of the physical world) the other side pointing out the absence of a contradiction in assuming certain infinite aggregates in mathematics and the overwhelming success of the ensuing theory not only as a subject within itself but also as a model for Nature. Then there is the argument that the theory of infinite sets is nevertheless 'semantically finite', i.e. it is expressed entirely in finite logical sentences, so that the Platonic abstractions these finite descriptions are intended to represent are just convenient fictions which inform some of the formal axioms and rules of inference.

Growing partly out of a concern that so many proofs in mathematics were non-constructive, Brouwer's school of Intuitionism, a twentieth century development of which there are now several variations, imposed a more restrictive notion of truth in mathematics.[1] Some of the principles of Aristotelian logic were compatible with Brouwer's philosophy but others were not, for example the principle of the excluded middle (that 'ϕ or not ϕ' always holds) was rejected, meaning that we cannot argue by contradiction. A set, for the Intuitionist, has been constructed using some definite collection of rules from other sets without making any purely existential appeals. The imposition of such finitistic combinatorial methods in analysis, which is classically a highly 'non-constructive' branch of mathematics, gives the subject a very different flavour. Indeed, mathematics itself assumes a peculiarly distorted shape under the alternative viewpoint of Intuitionism (not a 'wrong' shape by any means – simply one that differs from the familiar picture of modern mathematics as modelled by set theories based on classical logic), and consequently only a tiny minority of mathematicians have been willing to pursue it.

The Intuitionist approach to mathematics leads to some very interesting problems. From a Formalist point of view Intuitionism is just an alternative set of rules which we must adhere to in order to play the same game (as are the various flavours of Finitism). This limitation, if we can call it that, sometimes results in some ingenious strategies which might have been overlooked had the Intuitionistic restrictions not been imposed – one can be forced by Haiku-like constraints into unexpectedly beautiful new proofs. On reflection, it is odd to restrict to one set of rules. Why not examine as many possible logical universes as possible and treat each as an individual object of study as one might study different groups or manifolds?

REMARKS

1. David Hilbert once said '*Taking the principle of the excluded middle from the mathematician would be the same, say, as proscribing the telescope to the astronomer or to the boxer the use of his fists.*'[2] This turned out to be rather unfounded, as Errett Bishop and others discovered.[3] It is possible to do a surprising amount of analysis constructively. Nevertheless, I have to confess that I find the rejection of the excluded middle profoundly more counter-intuitive than any of the purely existential appeals that the constructivists

---

[1]Intuitionism is too often misrepresented as a type of Finitism (it would perhaps be more accurate to describe Finitism as an extreme species of Intuitionism). To help correct this misunderstanding I recommend to the reader the biography by Dirk van Dalen [**218**], which describes how Brouwer's ideas were misunderstood in his own lifetime.

[2]Hilbert, David, 1928, 'Die Grundlagen der Mathematik', *Hamburger Mathematische Einzelschriften*, **5**, Teubner, Leipzig. Translated in van Heijenoort [**219**].

[3]See, for example, Bishop [**19**].

were originally resisting. The initial idea of insisting that everything is constructible is understandable up to a point, although it does seem to be an extraordinarily conservative attitude, but the subsequent logical contortions one has to go through in order to make the system work, such as rejecting the excluded middle, would surely serve as an indication that the idea isn't so intuitive after all. The physicist David Deutsch adopts this position, describing Intuitionism as a form of mathematical solipsism, an inward looking over-reaction to the uncertainty of our knowledge of the wider world.[1]

2. An obvious contender for a physically infinite notion, one that I have shied away from mentioning, is the physical Universe. Recent results seem to support that it is 'flat', which is to say that it is (apart from local perturbations) Euclidean. But flat spaces can be bounded, like a torus, so this doesn't really tell us anything about its extent, only reducing the possible shapes it may assume. I fear that this is one of those questions which will be forever out of reach.

Several alternative ways of looking at mathematics have evolved, motivated either by finitistic ideas or by physical realism. Sometimes these restrictions can shed light on the unrestricted theory, revealing novel new proofs. By studying all such modifications one gets a healthy overview of the subject. Occasionally a principle with far-reaching consequences will reveal itself to be apparently unprovable. Three such postulates appeared early in the life of set theory.

1.14   Three curious axioms

> *I have deeply regretted that I did not proceed far enough at least to understand something of the great leading principles of mathematics, for men thus endowed seem to have an extra sense.*
>
> – CHARLES DARWIN[2]

As set theory developed, three seemingly unprovable principles emerged from the fog that appeared to have deep significance. These were the Axiom of Choice, the Continuum Hypothesis and the Generalized Continuum Hypothesis.

1.14.1   The Continuum Hypothesis and its generalization

With the aid of his new language of sets Cantor was able to give sense to a question which had previously been so imprecise as to be meaningless: *how many points are there in a line?* We must admit that if a line can be thought of as a set of points (modelled by $\mathbb{R}$, for example), then this set is infinite and, in this case,

---

[1]Deutsch [**50**], Chapter 10.
[2]Darwin [**40**]. Chapter II (Autobiography).

we are compelled to ask which of Cantor's newly discovered cardinal numbers labels this infinity. Cantor's conjectured solution, the Continuum Hypothesis, is that there does not exist a set with cardinality greater than that of $\mathbb{N}$ and smaller than that of $\mathbb{R}$, equivalently there is no set with cardinality greater than that of $\mathbb{N}$ and smaller than that of $P(\mathbb{N})$. Translating to the language of alephs, the hypothesis is that $2^{\aleph_0} = \aleph_1$.

The Generalized Continuum Hypothesis makes a similar statement about S and $P(S)$ where S is any *infinite* set. Note that this hypothesis concerns itself strictly with infinite sets; if S is finite and has at least two elements then we can always find a set with cardinality greater than that of S and smaller than that of $P(S)$. We will dwell on the Continuum Hypothesis and its generalization in more detail later.

REMARKS

1. Prior to the modelling of the line by the set $\mathbb{R}$, before the knowledge of uncountable sets and before the discovery of infinite ordinals, the notion of exhausting the line by points was inconceivable: between any two points one can find another point, so no process of construction indexed, say, by the natural numbers can possibly end. But might we be able to reach it via an ordinal indexed construction? Or, if not by an explicit construction, can we nevertheless prove the existence of a labelling of the points of $\mathbb{R}$ by some ordinal? If this were possible, then it would not only tell us that it is possible to well order $\mathbb{R}$, but it would also reveal how large $\mathbb{R}$ is. We come to that question in the next subsection. Let us suppose that it *is* possible to well-order the real line (or, if you prefer, $P(\mathbb{N})$), then we can ask which ordinal is the smallest ordinal number that can label this collection. The answer may come as a surprise.

2. There is an amusing topological object called the *long line*. The idea is this. The real line is a countable chain of intervals $\dots [-1,0) \cup [0,1) \cup [1,2) \dots$, i.e. a countable union of copies of $[0,1)$ joined end to end. Suppose we replace the countable chain by an uncountable chain. If ω_1 is the first uncountable ordinal then we form a long ray which we think of as a collection of joined copies of $[0,1)$ indexed by ω_1 in the natural way. To make the ray extend in both directions we can use the order type $\omega_1^* + \omega_1$. So formally we can define the long line as the cartesian product $(\omega_1^* + \omega_1) \times [0,1)$ with the natural lexicographic order (I leave the reader to fill in the details). This space looks locally like $\mathbb{R}$, but it is not compact, nor is it metrizable, so it doesn't meet the usual conditions for being a manifold. It has a number of other properties which make it an interesting counterexample to some naive conjectures in topology. Of course one can use larger ordinals to construct even 'longer' lines.

The Continuum Hypothesis states that the cardinality of the continuum (equivalently the cardinality of $P(\mathbb{N})$) is the next highest after the cardinality of the natural numbers. The Generalized Continuum Hypothesis states that for any infinite set a the cardinality of $P(a)$ is the next highest after the cardinality of a. The early set theorists struggled to prove or disprove these statements for decades without success. Another principle which resisted proof was the Axiom of Choice.

1.14.2   The Axiom of Choice

The Axiom of Choice, which has a huge number of interesting equivalent formulations, asserts that for any set A of non-empty sets, there exists a set that has exactly one element in common with each set in A. Some texts insist that the collection A is pairwise disjoint, i.e. every pair of distinct elements of A have empty intersection. It is easy to show that this condition is not necessary (the trick used to prove this is that for an arbitrary class $\{A_i\}_{i \in \Lambda}$ of distinct sets the class $\{A_i \times \{A_i\}\}_{i \in \Lambda}$ is pairwise disjoint).

Choice had been used tacitly for many years before set theorists distilled it into a recognizable isolated principle. In 1904 when Zermelo gave a proof of the Well-Ordering Theorem[1] (the assertion that every set has a well-ordering) E. Borel observed that Zermelo's proof used a property of sets, precisely the Axiom of Choice, which is itself a consequence of the principle he was trying to prove. Consequently the Axiom of Choice and the Well-Ordering Theorem are equivalent. Ever since those pioneering days the Axiom of Choice has been at the centre of much heated discussion between philosophically minded mathematicians.

Some reject Choice, believing that it is impossible to make infinitely many selections (even in an abstract setting) or because the axiom does not give an explicit rule for making the choice. Others accept the axiom in the countable case and reject it in the uncountable case.

The Axiom of Choice has important applications in every branch of mathematics which deals with infinite sets. For example one of the easiest to grasp and seemingly self-evident equivalent formulations of Choice within set theory itself is the principle of trichotomy: if a and b are sets then there either exists an injection $a \to b$ or an injection $b \to a$. It is so often used in analysis that its use is taken for granted and rarely explicitly mentioned. This abundance of utility is sometimes cited in support for the acceptance of the axiom. On the other hand the axiom also has some consequences which some regard, perhaps naively, as paradoxical.

The fact that something can be shown to be equivalent to both an intuitively plausible and an intuitively implausible statement (e.g. trichotomy versus

---

[1]Zermelo, E., 'Beweis, dass jede Menge wohlgeordnet werden kann', *Mathematische Annalen*, **59**, 1904. The paper is based on part of a letter Zermelo sent to Hilbert, who thought the result deserved publication. A translation appears in van Heijenoort [**219**].

the Well-Ordering Theorem) underlines how embarrassingly weak, indeed non-existent, our intuition is when it comes to infinite sets. 'Implausible' may be too strong a word for the Well-Ordering Theorem, nevertheless a well-ordering of, say, $\mathbb{R}$ is certainly impossible to visualize. Indeed this is the whole point – if such an ordering could be visualized then that visualization could be formalized and incorporated without question into whatever alternative set theory that could spawn from (and seem natural to) a species able to imagine such a thing; but the fact is the axioms of ZF, which are strong enough to model many of our intuitive ideas in mathematics, can tell us nothing about the existence of a well-ordering of $\mathbb{R}$.

Remarks

1. The case of forming a choice set from one set is trivial: if X is non-empty this means precisely that there exists an element $x \in X$, so we can form our set $C = \{x\}$. Then by induction we can prove that if $X_1, X_2, \ldots, X_n$ are non-empty sets then there exists a set C with exactly one element in common with each X_i (note that C needn't have n elements: the X_i might overlap or even be subsets of one another, in which case we can choose one element from their respective intersections). So the Axiom of Choice is really only an issue when we are choosing from infinitely many sets. We should observe that this proof is non-constructive – we have not chosen a specific element from X, we simply know that one exists and from that we infer the existence of the choice set C. If one wanted to unambiguously choose a particular element from X then there has be a prior well-ordering on X, so that we could let x be the smallest element of X with respect to that ordering.

2. Let us naively attempt to set up a well-ordering of $P(\mathbb{N})$. Two approaches immediately spring to mind. Firstly, we have the set of all ordinals ready to do the indexing so we can slice off a few infinite sets one at a time, and if we agree that $P(\mathbb{N})$ is a well-orderable set, then there are more ordinals than there are elements of $P(\mathbb{N})$, so we certainly can't run out of indices. The well-ordering ought to start with the empty set. Next we can list the single element sets in the natural way $\{0\} < \{1\} < \{2\} < \cdots$. So at this point we have $1 + \omega(= \omega)$ labelled sets. Next come sets of two elements ordered by $\{0,1\} < \{0,2\} < \{0,3\} < \cdots < \{1,2\} < \{1,3\} < \cdots\{2,3\} < \cdots$, a collection with order type ω^2. Then we have the set of three element sets, similarly ordered lexicographically, which has order type ω^3, and so on, so that eventually we have all finite sets laid out in an ordering with order type $1 + \omega + \omega^2 + \omega^3 + \cdots = \omega^\omega$, a countable ordinal. We still have infinite subsets of $P(\mathbb{N})$ to deal with. We have met this problem before. We could go on to list the cofinite sets, so that cofinite sets E and F satisfy $E < F$ if and only if $E' < F'$, where E' and F' are the (finite) complements of E and F. Having done this we would have arranged the finite and cofinite subsets of $\mathbb{N}$ in a list of order type $\omega^\omega + \omega^\omega$. But what next? No matter how cleverly

we try to describe an explicit well-ordering of $P(\mathbb{N})$ we seem to be doomed to failure, only being able to break off ever more elaborate initial segments.

One can argue that this first attempt fails simply because such constructive descriptions are necessarily going to produce only countably many sets, and hence can never exhaust all of $P(\mathbb{N})$. What one needs is some algorithm which tells us how to determine which is the 'smaller' of any given pair of subsets of $\mathbb{N}$ (in much the same way that the linear ordering of two distinct real numbers can be determined from a simple lexicographic inspection of their decimal expansions), and then to prove that such a rule determines a well-ordering on $P(\mathbb{N})$. This second approach appears to be more promising since the uncountability of $P(\mathbb{N})$ is no longer the principal obstacle, nevertheless the reader will find that any proposed explicit rule for such an ordering (something that does not rely on some form of Choice) fails to be a well-ordering.

The first experiment in this direction would be a lexicographic ordering. If A and B are two distinct subsets of $\mathbb{N}$ (the smallest element of a non-empty set X being denoted by X_1) then we define $A \prec B$ recursively by: (i) $\varnothing \prec B$ for all non-empty B; (ii) $A \prec B$ if A and B are both non-empty and $A_1 < B_1$; and (iii) $A \prec B$ if A and B are both non-empty, $A_1 = B_1$ and $A\backslash\{A_1\} \prec B\backslash\{B_1\}$. This ordering is well-defined, and is linear, but it is not a well-ordering since it is easy to exhibit an infinite decreasing sequence of elements: if n' denotes the complement of $\{n\}$ in $\mathbb{N}$, then $0' \succ 1' \succ 2' \succ 3' \succ 4' \cdots$. This ordering shows in general that if X is linearly ordered then $P(X)$ can be linearly ordered without appealing to any type of Choice.

3. Since $P(\mathbb{N})$ is equipollent to the interval $[0, 1)$, any attempt to explicitly define a well-ordering of the interval $[0, 1)$ will, of course, also fail. One can imagine beginning with an enumeration of the finite decimals, i.e. rational numbers in $[0, 1)$ in reduced form $\frac{a}{b}$ where $b = 2^n 5^m$ for some natural n, m. What then? Unfortunately, although we can begin to make a few ad hoc orderings of the remaining numbers, the desired well-ordering seems beyond reach. Similarly any ordering explicitly defined in terms of two general decimal expansions will fail to be a well-ordering.

Given any collection of infinite sets the Axiom of Choice tells us that there exists a set which has one element in common with each of the sets in the collection. Choice, which seems to be an intuitively sound principle, is equivalent to the much less plausible statement that every set has a well-ordering. Although many tried to prove Choice, they only seemed to be able to find equivalent statements which were just as difficult to prove. It wasn't until the early 1960s that the reason for their failure was fully understood.

1.14.3   Independence

The question of whether the Axiom of Choice should be regarded as an axiom at all, and the status of the Continuum Hypothesis, was not clarified until Kurt Gödel, in 1938,[1] proved them both to be consistent with ZF, and Paul Cohen, in 1963,[2] proved that they are each *independent* of ZF. That is, assuming the axioms of ZF are consistent, the further assumption of either the Axiom of Choice or its negation will not lead to contradiction, and likewise for the Continuum Hypothesis. (In fact the theory Gödel worked in differed a little from ZF – see the next section – nevertheless the result is still stands.)

It is a matter of philosophical prejudice whether one chooses to accept or reject these principles; Gödel's consistency result shows that there are no purely logical reasons to reject Choice and Cohen's result shows that we are free to assume its negation. Most mathematicians, but not all, assume, implicitly or otherwise, that the Axiom of Choice (or some restricted form of it) is true. On the other hand, proofs which use Choice are not always as informative as they could be, and if a more constructive proof is possible then it is often given instead. We witness, then, a certain reluctance in some quarters to use Choice despite the general consensus that it is 'true'. In contrast, the current feeling among Platonically minded researchers in logic and set theory is that the Continuum Hypothesis is *false* (they seek alternative axioms of set theory which are just as plausible as those of ZF, but which decide the Continuum Hypothesis). Such difficulties in finding answers to fundamental questions seem to be an intrinsic feature of mathematics that we must learn to live with. Of course, the Finitists regard both Choice (from infinitely many sets) and the Continuum Hypothesis as meaningless.

This wasn't the first independence result in mathematics. The most famous predecessor is in geometry: the independence of the parallel postulate from the other axioms of Euclidean geometry, which we will come to in more detail later. The basic method of establishing independence is essentially the same, i.e. one constructs models of the other axioms in which the parallel postulate is either true or false, but of course the technical details are wildly different.

In 1930 Kurt Gödel proved the relative consistency of the Axiom of Choice and the Continuum Hypothesis with the other axioms of his set theory. Assuming the other axioms are consistent among themselves there exists a model of set theory in which both Choice and the Continuum Hypothesis are true. In 1963 Paul

---

[1]See Gödel, K., *The Consistency of the Axiom of Choice and of the Generalized Continuum Hypothesis with the Axioms of Set Theory.* Princeton University Press (1940).

[2]Cohen, P. J., 'The Independence of the Continuum Hypothesis', *Proc. Nat. Acad. Sci. U.S.A.*, **50** (6), 1143–1148, (1963) and Part II in volume **51** (1), 105–110, (1964).

Cohen described a technique which proved that there are models of set theory in which Choice and the Continuum Hypothesis are false. So the reason the early set theorists were unable to prove these statements is that they are in a very concrete sense unprovable. By assuming or rejecting these principles one describes alternative set theoretic universes. Many different set theories have been developed in the last century.

1.15   A theory of sets

...philosophers subsequent to Kant, in writing on mathematics, have thought it unnecessary to become acquainted with the subject they were discussing, and have therefore left to the painful and often crude efforts of mathematicians every genuine advance in mathematical philosophy.

– BERTRAND RUSSELL[1]

1.15.1   Russell's bombshell

For Cantor a set was any conceivable collection of objects. This seemingly harmless idea leads us head first into the classical paradoxes, the most notorious of which is Russell's paradox, which arises in the following way. Let **Ru** be the collection of all sets A having the property that A is not a member of A. There is nothing troublesome about this at first sight; **Ru** is a very large class, indeed if we are asked to think of an arbitrary collection it will most likely be an element of **Ru**: the set of natural numbers is not itself a natural number, the set of all functions $\mathbb{R} \to \mathbb{R}$ is not a function, the set of all finite sets is not a finite set, and so on. It is more difficult to think of a set that *is* a member of itself. One such example, if we are to stick to Cantor's broad all-encompassing definition, is the collection of all sets. The 'set of all ideas', which is itself an idea, is another rather vague alternative. Let us assume that **Ru** is a set and ask whether **Ru** is an element of itself. According to its definition **Ru** is an element of itself if and only if it is not an element of itself, a highly paradoxical state of affairs.

If we are to develop a satisfactory theory of sets from simple foundations we need to make sure that the Russell class is strictly out of reach. The most obvious way to do this, a very severe restriction, is to abandon arbitrary classes entirely. But the technique of gathering together all objects satisfying a given property is practically indispensable in mathematics, so an alternative solution is desired. A less drastic course of action is to carve out in some manner different 'types' of collections including a class comprising well-behaved objects that we call 'sets'. Several ways of developing such a theory evolved in the twentieth century.

---

[1]The closing words of a book review in *Mind: A Quarterly Review of Psychology and Philosophy*. Volume IX. The Aberdeen University Press Limited (1900).

REMARKS

1. If a set S is a member of itself then we have $\cdots \in S \in S \in S \in S \in \cdots$, an infinite regress of membership. We can see, then, why we might want to isolate the Russell class as a first step towards pruning the class of sets, removing its more pathological members. To avoid Russell's paradox we somehow need to obstruct $\mathbf{Ru} \in \mathbf{Ru}$.

2. Russell illustrated his paradox with the following puzzle, known as the *barber paradox*. A barber is a man who shaves those (and only those) who do not shave themselves. The puzzle is: who shaves the barber? The answer cannot be the barber himself, since then he would be a person who shaves himself. But if it is not the barber then, being someone who does not shave himself, he must be shaved by ...

Originally a set was simply an unrestricted collection of elements. Paradoxes such as the Burali-Forti paradox indicated that some restriction was necessary; the set of all sets, for example, should not be allowed to be a set. Since self-membership is not an intuitively attractive property it is natural to consider the class of sets which are not members of themselves. This is the Russell class. The assumption that the Russell class is a set leads immediately to a contradiction. To remedy this, and other problems in the foundations, various alternative set theories have been explored.

1.15.2   Some set theories

Russell and Whitehead in their famous work *Principia Mathematica* (1910)[1] resolved the difficulties with a so-called theory of types. The central idea first appears in Russell's *Principles of Mathematics* (1903):[2]

> *The doctrine of types is here put forward tentatively, as affording a possible solution of the contradiction; but it requires, in all probability, to be transformed into some subtler shape before it can answer all difficulties.*

Russell and Whitehead described and elaborated on a hierarchy of sets in which a set x can be a member of a set y only if y is one level higher in the hierarchy than x. One unsatisfactory feature of this construction is that each level has its own primitive notions, so that there are infinitely many primitive notions in the theory at large. This early theory of types was not entirely successful.

---

[1] Whitehead and Russell [**225**].
[2] Russell [**181**] (Appendix B).

Axioms for a general set theory were given by Zermelo in 1908.[1] The later development of this axiomatization by Fraenkel (and significant others) is now known as Zermelo–Fraenkel set theory and has been the most successful and influential approach to the foundations of mathematics to date. ZF and the closely related Gödel–Bernays set theory (also called Von Neumann–Bernays–Gödel set theory – this theory was initiated by John von Neumann in 1925 and later modified by Paul Bernays and Kurt Gödel) were to play a significant role in the development of twentieth century set theory. Both theories have many features which are more attractive than those of Russell and Whitehead's type hierarchy; in particular they each have a small finite number of primitive notions.

Gödel–Bernays set theory avoids the paradoxes by dividing classes into two types: sets and proper classes. Sets are permitted to be members of other classes; proper classes have sets as elements but cannot themselves be elements of other classes. There are three primitive notions: set, class and membership. Together with the logical operations the formal language uses set variables, class variables and the membership symbol '$\in$'. Gödel's work on the relative consistency of the Axiom of Choice and the Continuum Hypothesis was done in Gödel–Bernays set theory (or at least something very close to it). The features of this theory had been developed with these problems of consistency in mind.

In ZF there are only two primitive notions: set and membership. There *is* a notion of class, but it is introduced as a defined term, that is to say a class is a construction from more primitive notions, and any statement about classes can be translated to an equivalent (but usually longer) statement about sets. In addition to the logical operations, only set variables and the membership symbol '$\in$' are used in the formal language. In ZF we can say 'there exists a set...' and 'for all sets...' but statements of the form 'there exists a class such that...' or 'for all classes...' (or, what amounts to the same thing, 'there exists a property...' or 'for all properties...') are prohibited, while in Gödel–Bernays set theory the latter are allowed. As a result of its ability to quantify over classes there are theorems in Gödel–Bernays set theory that are not theorems in ZF. However, Gödel–Bernays set theory is not so much grander than ZF; any statement *about sets* provable in Gödel–Bernays set theory is provable in ZF. We say Gödel–Bernays set theory is a *conservative extension* of ZF. Generally a theory $E(T)$ is a conservative extension of a theory T if every theorem of T is a theorem of $E(T)$ and if every theorem of $E(T)$ which is expressible in the language of T is a theorem of T. Theories that *are* stronger than ZF when it comes to sets, non-conservative extensions of ZF, also exist, for example Morse–Kelley set theory (based on the theory outlined in Morse's *A Theory of Sets*[2] and in Kelley's *General Topology*[3]).

It should be mentioned that Gödel–Bernays set theory has only finitely many axioms while one of the axioms of ZF, as we shall see, is an axiom schema – a

---

[1]Zermelo, E., 'Untersuchungen über die Grundlagen der Mengenlehre', *Math. Annalen*, **65**, 261–281, (1908). Translated in van Heijenoort [**219**].
[2]Morse [**155**].
[3]Kelley [**121**] (Appendix).

collection of axioms indexed by certain 'well-formed formulas' of the theory –
so strictly speaking ZF has infinitely many axioms. This unavoidable plenitude
of axioms is not as unsatisfactory as it may sound at first. Set theories, such
as Gödel–Bernays set theory, which are able to express such axiom schemas
as a single axiom (so-called *predicative extensions* of set theory) can be useful
when it comes to stating certain ideas economically, but as mentioned above
the extension does not necessarily result in a materially stronger theory.

It is not universally held that Gödel–Bernays set theory is less intuitive than
ZF, but some view it, and other predicative extensions of set theory, as more
artificial or abstract than ZF, and the accommodation of class quantification
can involve some difficult tricks. In the final analysis, however, the theory one
chooses to use is simply a matter of taste – the interpretation of any given
theoretical framework, for the logician, is of no importance whatsoever so long
as it seems to be free of contradiction. Philosophers tend to have opposite
concerns, focusing on interpretation and regarding the fine details of theorem
building to be comparatively unimportant.

Of all the theories of sets that have appeared, ZF is by far the most widely
quoted and used. Cohen's independence results, for example, brief details of
which are discussed in Chapter 11, were proved in ZF. I will be more than a
little biased towards ZF in this book, not solely because of its popularity (which
is a poor reason to favour anything) but simply to avoid the confusion of mixing
languages.

REMARK

The hierarchical idea behind the basic theory of types, only allowing $x \in y$ if y
is at a higher level than x, instantly blocks the Russell paradox since it prevents
any set from being an element of itself, so that the argument for the paradox
cannot even get started. The Zermelo–Fraenkel approach, making a distinction
between sets and proper classes, might seem more artificial at the moment, but
we will see that this first impression is misleading. Ideally, we aim to identify
those properties ϕ for which $\{x : \phi(x)\}$ is intuitively what we want to be a set
(for example, if we take $\phi(x)$ to be 'x is a natural number', then we would like
the collection of all natural numbers $\{x : \phi(x)\}$ to be a set) but for which this
assumption does not lead to contradiction (for example, if $\phi(x)$ is 'x is not a
member of x', then the assumption that $\{x : \phi(x)\}$ is a set quickly leads to a
contradiction, so we must exclude $\phi(x)$ from our list of admissible predicates).

Among the many approaches to the foundations of mathematics that were developed in the twentieth century, by far the most successful was the theory initiated by Zermelo and which is now known as Zermelo–Fraenkel set theory (ZF). Alongside other set theories, approaches to the foundations that are not set theoretic have also developed in parallel.

1.15.3   Axiomatics and other approaches

There are many reasons to examine the foundations of arithmetic, not all of them governed by anxieties about internal contradiction. Even if the paradoxes had remained hidden, a foundational theory would still have appeared, for the wheels of such a paradox-naive theory were already in motion in the late nineteenth century.

Dedekind and Frege, for example, were interested in distilling the principles which underlie arithmetic; they asked if a small set of assumptions exists which is capable of generating all of arithmetic, or even all of mathematics. Frege, in his *Begriffsschrift*,[1] proposed that:

> ...we divide all truths that require justification into two kinds, those
> for which the proof can be carried out purely by means of logic and
> those for which it must be supported by facts of experience... Now,
> when I came to consider the question to which of these two kinds the
> judgement of arithmetic belong, I first had to ascertain how far one
> could proceed in arithmetic by means of inference alone,...

He believed that all of arithmetic could be so reduced to logic, but geometry was forced, at least in part, to rely on matters of experience.

The axiomatic method is at least as old as Euclid (*circa* 300BC). It is an extremely powerful technique, allowing its users to describe with penetrating clarity the features of the theory it encompasses. After Euclid 'axiomatics' seemed to fade into the background for centuries, excepting commentaries, interpretations and imitations of *The Elements*, only to be revived in relatively recent times. This is perhaps not too surprising given that an axiomatization of a subject is a sign of its maturity, only being feasible after much lengthy experimentation. It was to be the clarification of geometry once again which motivated an axiomatic approach to mathematics in the nineteenth century. Hilbert's formalization of geometry[2] gathered together these ideas making critical use of the set theoretic ideas of the time, referring to a model of the continuum as an aggregate of points in the style of Cantor and Dedekind.

Matters of 'human experience', as Frege might have put it, still play a vital role in choosing the axioms of set theory, so we can't pretend that we have replaced intuitive mathematical ideas by some sort of detached Platonic wondermachine unentangled from sensory prejudices. However, it would be very pessimistic indeed to claim that the observed faults of visual intuition had simply been replaced by just as many as yet undiscovered contradictions in set theory. The basic motivation behind set theory is to tame the difficulties associated with the infinite and at the same time to provide a language capable

---

[1]Frege, G., *Begriffsschrift: eine der arithmetischen nachgebildete Formelsprache des reinen Denkens*. Halle, 1879. A translation by Stefan Bauer-Mengelberg appears in van Heijenoort [219]. *Begriffsschrift*, which is the name of the system being described in Frege's book, is translated as 'Concept Script'.

[2]Hilbert [100].

of accurately describing the reasoning employed in mathematics in general. In both respects set theory has been stunningly successful.

Alternative routes have been explored. Category theory, for example, might be used in place of set theory as a foundation of mathematics (more accurately, topos theory – a topos is a type of category, the category of sets being an example). Category theory, affectionately known as *abstract nonsense*, is, as its nickname suggests, a highly abstract body of work the roots of which are based on the metadescription of mathematical systems in a unified way as a collection of morphisms joining objects: in topology, for example, topological spaces are joined by continuous maps, in group theory groups are joined by group homomorphisms, and so on. While set theory emphasizes membership, category theory emphasizes function.[1] Other views of the possibility of a foundational theory of mathematics are a little more pessimistic.[2]

Present day research in set theory, for the most part, either focuses on the consequences of new axioms or studies the metamathematics of the theory itself, perhaps viewing all the alternative set theories as forming a coherent 'multi-verse'. Set theory also serves a practical purpose; it often happens in research that theorems (or at least the underlying ideas) from one branch of mathematics are used in another. The existence of a unified framework for mathematics guarantees that we can indulge in such cross-pollination without worrying about foundational issues. This is an invisible supporting role; mathematicians do not think about, and certainly do not worry about, such foundational conflicts in the general course of research.

REMARKS

1. We should stress that the axiomatic method is not a method of *producing* mathematics. That is to say, when trying to gain insight into a topic or when attempting to prove a conjecture the approach is as it has always been: intuitive, built on analogies and creative leaps (and failures) based on previous work on the subject, exploring one dead end after another until eventually hitting upon the right idea. Mathematics, like any other creative pursuit, is about having good ideas, original insights, free exploration, and the underlying system of axioms is largely almost an irrelevance, unless of course the research is about the foundations themselves. Of course, once the dust has settled one can forensically examine the ideas and see in principle what minimal assumptions are needed for a proof to be valid, but the proofs

---

[1] See Mac Lane [**142**] for one of the first accounts of category theory. Category theorists probably view the set theoretical foundation of mathematics as terribly old fashioned, and they certainly have a point! Philosophers of mathematics ought to learn category theory. Topos theory in particular is an extraordinary unification of ideas from topology, algebraic geometry, logic and set theory and has also been proposed as a promising way of thinking about physics. A topos is a sufficiently broad notion that it can be designed to cater to the user's philosophical needs, whether they are finitist, constructivist, physicist or otherwise.

[2] See, for example, Lakatos [**133**].

themselves generally come from stream of consciousness leaps of imagination, not from any systematic exhaustion of cold logical implications.

2. Here ends our cultural introduction to the subject. Now we begin again, focusing on the mathematics – what logical machinery might we need to construct a theory of sets and how might we use that machinery to prove such deep results as the independence of the Continuum Hypothesis from the axioms of ZF?

The modern view of set theory is increasingly multiversial, so that ZF is considered as just one of infinitely many set theories that one can construct, perhaps by modifying or augmenting ZF itself. Now that we see why a set theory, or some such foundation, is needed, we can take a closer look at how such a theory might be developed.

2

Logical foundations

2.1 Language

The universe cannot be read until we have learnt the language and become familiar with the characters in which it is written.

– GALILEO GALILEI[1]

2.1.1 Building a theory of sets

The axiomatic development of set theory is among the most impressive accomplishments of modern logic. It can be used to give precise meaning to concepts which were beyond the grasp of its vague predecessors. A successful set theory describes clearly the logical and extra-logical principles of mathematics.

We want a theory of sets to be at least powerful enough to cope with the concepts of classical mathematics, in particular we need to be able to speak about the number systems discussed in the introduction. We have seen that the systems of integers, rational numbers, real numbers, complex numbers and algebraic numbers (and beyond) can be built from the natural numbers using a handful of logical constructions. Thus our theory needs to be capable of describing a model of the natural numbers, that is, a collection of sets with a successor operator satisfying Peano's Postulates, together with such notions as ordered pairs, functions and other relations of various kinds. At the same time, and this is where the creative tension comes into play, we don't want the theory to be so loose and overconfident with its assignment of sets to admit such horrors as Russell's paradox.

The basic symbols of the language must be stated. We need the important notion of membership, which we denote by the symbol $\in$, a symbol which has evolved from its earlier incarnation as the letter epsilon, abbreviating the Greek 'is'.[2] We would like to say $x \in y$, the principal interpretation being 'x is an

---

[1]Galilei [75].

[2]This, and a significant proportion of set theoretic notation in general, is due to Giuseppe Peano.

element of the collection y', where x and y are intended to be substitutes for sets. In order to do this we need a supply of variables, which we will habitually denote by lower case roman letters (subscripted if necessary). For the time being let us be content that we have made available an assembly of variables for manipulation and not distract ourselves too much with interpretations.[1]

REMARK

In his Autobiography,[2] Bertrand Russell describes a turning point in his intellectual life, meeting Peano at the International Congress of Philosophy in Paris, 1900: '*In discussions at the Congress I observed that he was more precise than anyone else, and that he invariably got the better of any argument upon which he embarked. As the days went by, I decided that this must be owing to his mathematical logic. [...] It became clear to me that his notation afforded an instrument of logical analysis such as I had been seeking for years, and that by studying him I was acquiring a powerful technique for the work that I had long wanted to do.*'

Intuitively speaking, the beginnings of the theory are simple. We need only the notion of set membership as a primitive, together with variables representing sets. To this we need to add axioms restricting what our sets will look like and to draw deductions from these axioms we need an underlying logic.

2.1.2   Logical operations

A glance at even the most elementary definitions, theorems and proofs in mathematics reveal in abundance the simple logical notions: 'not'; 'or'; 'and'; 'implies'; 'equivalent'; 'for all'; 'there exists'. These must be formally incorporated into the theory at the outset. The usual notation for the operators and quantifiers in the order just listed is $\neg$, $\vee$, $\wedge$, $\rightarrow$, $\leftrightarrow$, $\forall$ and $\exists$. Together with these are added auxiliary symbols (various parentheses, used as a form of punctuation and sometimes used cosmetically to ease readability). A page of such symbols can be an impressive spectacle, and conveys something of the hard precision of the subject, however I don't think it would benefit us to use these symbols without lapsing into natural language. It is the concepts that we are interested in, and these, with some exceptions, can be adequately expressed in English

---

[1] The symbol $\in$ is so loaded with prior interpretation ('membership') that one can convincingly advance a case for abandoning it in favour of an entirely neutral symbol, say $\square$. However, later we will see that one can still do rather a lot by restricting to standard models of set theory, that is, models in which our abstract primitive binary relation $\square$ is indeed modelled by the relation of set membership. So we shall continue to use $\in$, as is traditional. We should perhaps pause to appreciate the remarkable fact that such a spare language, having just one primitive binary relation, no constants and no primitive operators, can be used to describe all of classical mathematics.

[2] Russell [**189**], Volume I.

without too much loss of information. We use the symbols in those cases where natural language fails. Often we make an aesthetic decision and use a sensible mixture of the two.

We have a sound *intuitive* grasp of truth; the truth value of a logical combination of statements is easily determined from the truth value of its component parts according to the principles described below. In particular the negation of a true statement is false and the negation of a false statement is true. The axioms we begin with form our initial stock of true statements and, assuming the theory is sound, we confidently assign the value 'true' to all statements provable from these axioms using the rules of inference.

For some privileged theories, the negation complete theories, of which Alfred Tarski's first-order version of Euclidean geometry is an example, this class of provable statements describes all truths expressible in the underlying language, however for a large class of theories, and in particular for any theory to which Gödel's First Incompleteness Theorem applies,[1] there will be statements ϕ such that neither ϕ nor its negation is provable. If all statements are to be assigned a truth value (where we take 'true' to mean 'satisfied in some prior fixed model of the theory' – see Chapter 10), then one of ϕ or its negation must be true, so we conclude that in such theories there is a true statement which is not provable. Thus we must not confuse truth with provability – the class of provable statements in a theory is, with respect to a given model of the theory, often much smaller than the class of true statements.

Assigning a truth value to an undecidable statement is either a philosophical decision or, depending on the nature of the statement and the theory, possibly an arbitrary one, made simply to study a larger theory (and hence restrict its models). Strict Platonists have a grander notion of truth used, say, when expressing the opinion that the Axiom of Choice is true, even though it is known to be undecidable in ZF – so that ZF is viewed as a first approximation to the Platonist's ideal mathematical universe.

The logical operators and quantifiers are as follows.

NEGATION

If a statement ϕ is true then NOT ϕ is false, and vice versa. The formal notation for NOT ϕ is $\neg\phi$ ($\sim\phi$ is also found in the literature).

DISJUNCTION

The statement ϕ OR ψ is true if at least one of the statements ϕ or ψ is true and is false only when both ϕ and ψ are false. The formal notation for ϕ OR ψ is $\phi \vee \psi$. In English there is an ambiguity in the use of 'or' – it is sometimes confused with 'exclusive or'; ϕ XOR ψ is a true statement only when one of its

---

[1]There are plenty of familiar theories to which Gödel's Incompleteness Theorems do not apply but which are nevertheless still negation incomplete. The theory of groups is one example, the simple statement $(\forall a)(\forall b)[ab = ba]$ being an example of an undecidable sentence. The strength of Gödel's result is that it proves that some theories, such as Peano Arithmetic, are effectively *incompletable*. See the remarks to Subsection 2.4.1.

arguments is true and the other is false. In terms of our logical symbols ϕ XOR ψ is $(\phi \wedge \neg\psi) \vee (\neg\phi \wedge \psi)$.

CONJUNCTION

The statement ϕ AND ψ is true if both ϕ and ψ are true and is false if at least one of ϕ or ψ is false. The formal notation for ϕ AND ψ is $\phi \wedge \psi$.

IMPLICATION

ϕ implies ψ, written $\phi \to \psi$, means: if ϕ is true then ψ is true. We will either write this symbolically, or say 'ϕ implies ψ', or 'if ϕ then ψ', whichever is most harmonious in context. In semi-formal mathematical shorthand one finds $\phi \Rightarrow \psi$ in place of $\phi \to \psi$ (and in some very formal contexts a distinction is made between $\Rightarrow$ and $\to$, a distinction which we shall not make here). Logicians often write $\phi \supset \psi$ rather than $\phi \to \psi$, however we will avoid this as it clashes with our use of $\supset$ as the superset relation.

We have to be careful when talking informally about implication. When a mathematician says 'sentence A implies sentence B' the intended meaning is: 'assuming all of the axioms of the underlying theory are true and the sentence A is true, then the sentence B is true'. Put another way, in all models of the theory in which A is true, B is also true. It is possible that the implication may fail to hold in a weaker theory.

EQUIVALENCE

This is implication 'both ways'; ϕ and ψ are equivalent if each implies the other. The notation for the equivalence of ϕ and ψ is $\phi \leftrightarrow \psi$ ($\phi \equiv \psi$ is sometimes found in logic texts or $\phi \Leftrightarrow \psi$ in less formal presentations). We will also say 'ϕ is equivalent to ψ', or 'ϕ if and only if ψ'. To show ϕ is equivalent to ψ one must prove both implications $\phi \to \psi$ and $\psi \to \phi$.

It is not uncommon to find in mathematical texts a list of proposed equivalent statements $\phi_1, \phi_2, \ldots, \phi_n$. In this case the soundest strategy is to arrange the statements in a cycle in such a way that the proofs of the implications $\phi_1 \to \phi_2$, $\phi_2 \to \phi_3$,..., $\phi_{n-1} \to \phi_n$, $\phi_n \to \phi_1$ are as easy as possible. The mutual equivalence of $\phi_1, \ldots, \phi_n$ then follows by the transitivity of the implication relation (see Subsection 2.1.3 below).

Again we should stress the dangers of the casual use of the word 'equivalent'. The informal statement 'sentence A is equivalent to sentence B' means that if one assumes A *and all of the axioms of the underlying theory* then B is true and if one assumes B *and all of the axioms of the underlying theory* then A is true. Put another way, given a model of the underlying theory, A and B are either both true or both false. It may be that in a weaker theory only one of the implications is true.

One mustn't get the impression that equivalences are uninteresting trivialities. For example, assuming a very sparse arithmetical foundation, one can prove that if n is a natural number the following two statements are equivalent:

(i) n is expressible as the sum of two squares.

(ii) All of the prime factors of n which are congruent to 3 modulo 4 have even exponent.

It is far from obvious that these statements are equivalent. In this example the equivalent statements both refer to the same object, but this need not be the case in general – there are some surprising equivalences to be found spanning different parts of mathematics. A statement in one corner of the subject can turn out to be equivalent to another statement in an apparently distant realm. Further scrutiny often reveals deep similarities between two bodies of work. The ideal is to describe a meta-isomorphism where two independently studied collections of theorems, previously strangers, are united as one; apart from differences in notation and terminology, and in interpretation, the two theories are the same, only exhibited differently by an accident of history. The union of two such theoretical twins can give rise to some very powerful results.

More humbly, the notion of equivalence also gives us access to alternative definitions of concepts. If a definition ϕ is equivalent to a condition ψ, then we can use ψ as a definition in place of ϕ without perturbing the theory. This is very useful when abstracting concepts from one setting to another. For example, consider the characterization of a continuous function in terms of counterimages of open sets, which leads to general topology, freeing the notion of continuity from the confines of metric spaces (as was described briefly in Subsection 1.7.4).

In many cases the proof of the equivalence of two statements may be easy in one direction but the proof in the reverse direction fiendishly difficult, indeed many celebrated theorems and unsolved conjectures are converses (or partial converses) of relatively easy results.

THE UNIVERSAL QUANTIFIER
We often need to make a statement of the type *'for all x the statement ϕ is true'* where ϕ is some statement involving x. This is formally denoted $(\forall x)\phi$. The symbol $\forall$ is the universal quantifier.

THE EXISTENTIAL QUANTIFIER
Existential quantification appears in the statement *'there exists an x such that ϕ'*, which is given in logical notation by $(\exists x)\phi$. The symbol $\exists$ is the existential quantifier. 'Such that' is a natural language joiner which is in danger of becoming incredibly clumsy when used repeatedly. To avoid confusion in the case of multiple quantification we will make use of the symbols $\exists$ and $\forall$ instead. Natural language often cannot cope very well with quantifier-rich mathematical and logical statements without appearing bloated and ambiguous.

REMARKS

1. We have already mentioned the equivalence of the Axiom of Choice and the Well-Ordering Principle in ZF. Just to re-emphasize what this means: if we assume all of the axioms of ZF and the Axiom of Choice then we can prove

that every set has a well-ordering. Similarly, if we assume all of the axioms of ZF and that every set has a well-ordering then we can prove the Axiom of Choice. If we replace well-ordering by linear ordering then the equivalence breaks down into an implication. ZF plus Choice (henceforth abbreviated ZFC) implies every set has a linear ordering, but ZF plus the assumption that every set has a linear ordering does not imply the Axiom of Choice.

2. Definitions can be just as beautiful as theorems and finding the right definition of a concept can be as much of a challenge as proving a difficult conjecture. The correct path often only reveals itself after much difficult work.

3. Perhaps the most famous recent connection between previously distant subjects is that of so-called *Monstrous Moonshine* which revealed an unexpected link between the Monster Group (a huge finite group, with

$$808\,017\,424\,794\,512\,875\,886\,459\,904\,961\,710\,757\,005\,754\,368\,000\,000\,000$$

elements – the largest sporadic group – think of it as the set of symmetries of an exceptionally large multi-dimensional object of particular theoretical interest) and modular forms (certain types of analytic functions). Richard Borcherds proved the conjectured connection in 1992 using ideas from string theory. The proof has revealed a deep relationship between different parts of mathematics and theoretical physics.

There are seven basic logical symbols representing the intuitive notions of 'not'; 'or'; 'and'; 'implies'; 'equivalent'; 'for all'; and 'there exists'. We need to describe how these symbols can be combined, what logical axioms govern their properties, and what basic logical properties follow from these axioms.

2.1.3   Properties of the logical operations

The logical operations enjoy some familiar properties:

De Morgan's Laws
$$\neg(\phi \wedge \psi) \leftrightarrow \neg\phi \vee \neg\psi$$
$$\neg(\phi \vee \psi) \leftrightarrow \neg\phi \wedge \neg\psi$$

Associative Laws
$$(\phi \wedge \psi) \wedge \eta \leftrightarrow \phi \wedge (\psi \wedge \eta)$$
$$(\phi \vee \psi) \vee \eta \leftrightarrow \phi \vee (\psi \vee \eta)$$

Distributive Laws

$\phi \wedge (\psi \vee \eta) \leftrightarrow (\phi \wedge \psi) \vee (\phi \wedge \eta)$

$\phi \vee (\psi \wedge \eta) \leftrightarrow (\phi \vee \psi) \wedge (\phi \vee \eta)$

Transitive Law

$[(\phi \to \psi) \wedge (\psi \to \eta)] \to (\phi \to \eta).$

If a statement is true regardless of the truth values of its component parts then it is said to be a *tautology*, for example the following are all tautologies:

$$\phi \;\to\; \phi$$
$$\phi \wedge \psi \;\to\; \psi \wedge \phi$$
$$\phi \;\to\; \phi \wedge \phi$$
$$\phi \wedge \psi \;\to\; \psi$$

If a statement is false regardless of the truth values of its component parts then it is a *contradiction*, for example $\phi \wedge \neg\phi$. If τ is a tautology then $\neg\tau$ is a contradiction.

There is already some redundancy to be observed here. We could make do with the symbols $\neg, \to$ and $\forall$ alone, or certain other small collections of symbols ($\neg$ together with one quantifier and one of $\to, \vee$ or $\wedge$ will suffice), since:

$\phi \vee \psi$	is equivalent to	$\neg\phi \to \psi$;
$\phi \wedge \psi$	is equivalent to	$\neg(\phi \to \neg\psi)$;
$\phi \leftrightarrow \psi$	is equivalent to	$(\phi \to \psi) \wedge (\psi \to \phi)$;
	which in turn is equivalent to	$\neg[(\phi \to \psi) \to \neg(\psi \to \phi)]$;
$(\exists x)\phi$	is equivalent to	$\neg(\forall x)\neg\phi.$

We will make use of such a small collection of symbols when the occasion demands it, but in general confining ourselves to such small alphabets would hardly make for a clear exposition. In fact each of the symbols $\neg, \wedge, \vee, \to, \leftrightarrow$ can be expressed entirely in terms of parentheses and a single symbol, the *Sheffer stroke* $|$, which one might alternatively call NOR, defined by the condition $\phi|\psi$ *is true if and only if at least one of ϕ or ψ is false.*[1] The basic logical operations are re-expressed in terms of the Sheffer stroke as follows:

$\phi \wedge \psi$	$\equiv$	$(\phi|\psi)|(\phi|\psi)$
$\neg\phi$	$\equiv$	$\phi|\phi$
$\phi \vee \psi$	$\equiv$	$(\phi|\phi)|(\psi|\psi)$
$\phi \to \psi$	$\equiv$	$\phi|(\phi|\psi)$ or alternatively $\phi|(\psi|\psi)$
$\phi \leftrightarrow \psi$	$\equiv$	$(\phi|\psi)|((\phi|\phi)|(\psi|\psi)).$

---

[1] Named after Henry Sheffer who published the result in Sheffer, H. M., 'A set of five independent postulates for Boolean algebras, with application to logical constants', *Transactions of the American Mathematical Society*, **14**, 481–488, (1913). Charles Peirce discovered, but did not publish, the same result in 1880.

Similarly the five basic logical symbols can be expressed entirely in terms of the one symbol $\downarrow$ (NAND) where $\phi \downarrow \psi$ is true if and only if both ϕ and ψ are false. $\mid$ and $\downarrow$ are related via $\phi \downarrow \psi \leftrightarrow \neg(\neg\phi \mid \neg\psi)$. It can be shown that $\mid$ and $\downarrow$ are the only binary operators which are capable of singularly defining *all* binary logical operations, and indeed all n-ary logical operations.

We can go one step further. Parentheses can be avoided entirely by using the notation of Jan Łukasiewicz (so-called 'Polish notation', invented in the mid-1920s). Operators precede operands, so if A is a binary operator then we would write Axy rather than xAy. Repeating this principle we free complex expressions of all parentheses, for example if A and B are binary operators then the expression $xB(xA((yBz)Az))$ is, in Polish notation, $BxAxAByzz$.

So by using Polish notation and the Sheffer stroke together the equivalences we listed above become:

$$
\begin{aligned}
\phi \wedge \psi &\equiv \ \mid\mid\phi\psi\mid\phi\psi \\
\neg\phi &\equiv \ \mid\phi\phi \\
\phi \vee \psi &\equiv \ \mid\mid\phi\phi\mid\psi\psi \\
\phi \to \psi &\equiv \ \mid\phi\mid\phi\psi \text{ or alternatively } \mid\phi\mid\psi\psi \\
\phi \leftrightarrow \psi &\equiv \ \mid\mid\phi\psi\mid\mid\phi\phi\mid\psi\psi.
\end{aligned}
$$

These facts have practical applications in circuit building; one need only manufacture one type of logic gate (the NAND gate or alternatively the NOR gate) from which all other gates can be fabricated. For now, we will continue to favour clarity over economy and use all seven logical symbols.

REMARKS

1. One of many approaches to studying logical identities, although not necessarily the most elegant, is to work in the ring $\mathbb{Z}_2$ of two elements $\{0, 1\}$ (interpreting 0 and 1 as false and true respectively). The logical operations can then be recast as arithmetical operations as follows where $\langle s \rangle$ is the truth value of s:

$$
\begin{aligned}
\langle \phi \wedge \psi \rangle &= \langle \phi \rangle \langle \psi \rangle \\
\langle \phi \vee \psi \rangle &= \langle \phi \rangle + \langle \psi \rangle + \langle \phi \rangle \langle \psi \rangle \\
\langle \neg\phi \rangle &= \langle \phi \rangle + 1 \\
\langle \phi \to \psi \rangle &= \langle \phi \rangle + \langle \phi \rangle \langle \psi \rangle + 1 \\
\langle \phi \leftrightarrow \psi \rangle &= \langle \phi \rangle + \langle \psi \rangle + 1
\end{aligned}
$$

and we have $\langle \phi \mid \psi \rangle = \langle \phi \rangle \langle \psi \rangle + 1$ and $\langle \phi \downarrow \psi \rangle = \langle \phi \rangle + \langle \psi \rangle + \langle \phi \rangle \langle \psi \rangle + 1$.

2. The (naive) set theoretic interpretation of the logical operations is this. Assuming a universe V of all objects, we associate with the statement $\phi(x)$ the class $[\phi]$ of all objects x satisfying ϕ. Then we obtain the familiar Boolean algebra of describable subclasses of V:

(i) $[\phi \vee \psi] = [\phi] \cup [\psi]$.

(ii) $[\phi \wedge \psi] = [\phi] \cap [\psi]$.

(iii) $[\neg\phi] = V \backslash [\phi]$ (the complement of $[\phi]$ in V).

(iv) $[\phi \rightarrow \psi] = [\neg\phi \vee \psi] = [\neg\phi] \cup [\psi] = (V \backslash [\phi]) \cup [\psi]$.

(v) $[\phi \leftrightarrow \psi] = ((V \backslash [\phi]) \cup [\psi]) \cap ((V \backslash [\psi]) \cup [\phi])$.

(vi) $(\forall x)\phi(x)$ if and only if $[\phi] = V$.

(vii) $(\exists x)\phi(x)$ if and only if $[\phi] \neq \emptyset$.

In particular from (iv) and (vi) we have $(\forall x)(\phi(x) \rightarrow \psi(x))$ if and only if $[\phi] \subseteq [\psi]$.

The logical operations satisfy a number of familiar rules which match our intuitive understanding of classical logic. One can considerably reduce the number of logical symbols necessary for the theory by adopting so-called Polish notation (which eliminates the need for parentheses) and using only certain small subsets of symbols which still suffice to express any logical sentence. This is of theoretical interest, but, as one realizes when trying to write even mildly complicated notions entirely in logical symbols, it is usually better to stick to a mixture of natural language, jargon and symbols.

2.1.4 An historical aside

In the late nineteenth century Peano published three pioneering logical works, each a testing ground for his new ideas, all of them written entirely in logical symbols save for brief explanatory material: *Arithmetices Principia nova methodo exposita* (1889),[1] a treatise on formal arithmetic and part of set theory; *I Principii di Geometrica logicamente esposti* (1889),[2] an analysis of descriptive geometry; and *Demonstration de l'intégrabilité des équations différentielles ordinaires* (1890),[3] a demonstration of the integrability of a system of ordinary real differential equations. Peano's grand ambition, to which these three works formed a prelude, was a *Formulaire de Mathématiques*, which was to be a collaborative attempt to compile a purely symbolic collection of all mathematical truths. The last edition, called *Formulario Mathematico*,[4] appeared in 1908 but it was quite different from the book Peano had originally envisaged (and not only because it was now written in an artificial language Peano had invented called *Latino sine flexione* – Latin without inflections – a stripped down version of Latin with some borrowed English, French and German words); only a

---

[1] 'The principles of arithmetic, presented by a new method'. See van Heijenoort [219].
[2] Torino: Fratelli Bocca, (1889).
[3] *Mathematische Annalen*, **37**, 182–228, (1890).
[4] Torino: Fratelli Bocca, (1908).

tiny proportion was devoted to logic, more traditionally presented mathematics occupying over 400 pages.

Peano's shift from pure logical symbols to natural language is understandable. In practice, although formal logical proofs may be visually and philosophically attractive, they are rather tedious and difficult to read. It is far better to indicate that a proof *could* be formalized if necessary and present it in a more efficient metalanguage, usually a jargon-rich version of natural language decorated with a sprinkling of logical symbols when required.

REMARK

We will see, when stating our axioms, why it is necessary to use natural language when explaining mathematics to humans. For now let us be masochistic and try to express a number theoretic theorem in the language of Peano Arithmetic (see Appendix A): *There are infinitely many primes.* First of all we need to rephrase it such a way that it is expressible in the underlying language: *There exists a prime number and if p is a prime then there exists a prime q with $p < q$.* We need to define divisibility, strict inequality and the notion of primeness. Putting all of this together we have:

$a < b$ is an abbreviation for $\neg[a = b] \wedge (\exists c)[a + c = b]$.

$a|b$ is an abbreviation for $(\exists c)[a \cdot c = b]$.

$\pi(p)$ ('p is prime') is an abbreviation for[1]

$$s0 < p \wedge (\forall a)[a|p \rightarrow [a = s0 \vee a = p]].$$

Finally, 'there are infinitely many primes' is expressed as

$$(\exists a)[\pi(a)] \wedge (\forall a)[\pi(a) \rightarrow (\exists b)[a < b \wedge \pi(b)]].$$

The latter expression does not look too bad, but consider that it is made up of several abbreviations. If we were to replace each abbreviated expression with its full form (being careful not to mix up the various variable names) the statement would be transformed into something considerably larger and considerably less readable. The determined reader is invited to do this.

Peano's pioneering work in logic led him to attempt some expositions entirely in terms of logical symbols. It soon became apparent that this was wholly impractical and instead a natural language had to be employed, explaining how

---

[1] An alternative way to express 'p is prime' is $p > s0 \wedge (\forall a)(\forall b)\neg[p = (ssa) \cdot (ssb)]$, which has the advantage of avoiding divisibility in its definition. The resulting expression for 'there are infinitely many primes' using this definition will be noticeably shorter than the one produced using the other definition. My choice of predicate is simply chosen to match more literally the statement '$p > 1$ and the only divisors of p are itself and 1' that one tends to hear when prime numbers are defined.

everything could, in principle, be translated into the primitive symbols. These days we are more acutely aware of this through the hierarchy of languages we see in use in computer science. Even a simple definition, when unravelled, can result in an enormous explosion of logical symbols.

2.1.5   Definitions

So far we have only introduced bare symbols with no rules for their meaningful combination. We could simply have given a list of symbols and variables and declared that the sentences of our language will be assembled from strings of these characters according to some fixed principles. This sort of cold formalism is fine, if not a little exasperating, and it would be unreasonably perverse to introduce ∧, for example, without stating that the model we have in mind for its behaviour is the familiar 'and' of intuitive logic.

The idea is that all notions, all theorems and definitions, can in principle, given sufficient patience, be decoded back into a string of primitive symbols. As in any language it is often convenient to assign a new symbol to a given cluster of symbols (a recurring theme deserving its own signature) thereby increasing clarity and shortening proofs. Even the simplest theorem of calculus, say, when broken down into its primitive logical parts would be of an unreadable and unwieldy length. The effect of such a reduction to primitive terms is comparable in its comprehension-hindering magnitude to that of a novel transformed so that each instance of a word is replaced by its full dictionary definition; the intended meaning of each sentence lost in a jumble of information.

Having made such fresh definitions and assignments of new symbols, no effort is made to highlight the fact that the new molecules rather than their component atoms are being used – we won't fuss over whether we are using primitive or defined symbols.

Any definition must be eliminable in the sense that each occurrence of the concept defined can be replaced by its definition without perturbing the meaning of the statement in which it appears. Since such reduction can only be understood if it terminates in a finite number of steps, any logically sound exposition must include concepts which are not defined in terms of others: these are the primitive elements of the theory. To avoid an infinite regress this applies both to the defined terms and to the principles of proof used to describe their properties, the primitive elements of the theory being its axioms, assumed at the outset and on which the theorems of the theory are built using the underlying logical principles.

The axiomatic method has been an undeniable source of strength for mathematics. The spectacular explosion in mathematical activity over the last century, which is sometimes wildly miscredited to influences from technological advances, owes a lot of its existence to the modern axiomatic approach. However, let us not overstate its influence; deeply creative mathematical insights are somewhat independent of our knowledge of the axiomatic foundations of a subject.

EMARK

A formal system will have many different possible interpretation besides the
natural one that motivated it. Consider, for example, the associative, non-
commutative, left distributive algebra generated by the symbols n^o, n^*, n', e (n
natural) described in the remarks to Section 1.11. If we had been presented
with this system out of the blue, not knowing anything about order types, how
might we have interpreted it?

Here is an example of a simpler system, presented without motivation. There
are two symbols a and b which we can multiply together subject to only three
rules: (i) $a^5 = a$; (ii) $b^2 = b$; (iii) $ba = a^3b$.

After a moment's thought we see that any combination of as and bs can
be uniquely reduced to one of the eight elements a, a^2, a^3, a^4, ab, a^2b, a^3b,
a^4b. Perhaps after a while it might occur to us that this might be viewed in
terms of permutations of a row of four symbols $[ABCD]$ where a corresponds to
the cyclic permutation mapping $[ABCD]$ to $[DABC]$ and b corresponds to the
permutation mapping $[ABCD]$ to its reverse $[DCBA]$, where xy means 'apply
permutation y first followed by permutation x'. Either directly, or via this
representation, we might realize that there is a simple geometric interpretation
of our abstract system. a is a ninety degree clockwise rotation of a square and b
is the reflection of the square in the axis passing horizontally through its centre.
The same system can have many different models. Conversely, it is possible to
give many different axiomatizations of natural structures.

*Our primitive symbols and primitive assumptions (axioms) are necessary to
avoid an infinite regress. Although we could treat these as meaningless starting points, their behaviour and properties are clearly inspired by intuitive ideas.
We are at the outset trying to describe a notion of 'set', wary that the unrestricted classes of old lead to contradiction. All of our axioms are seemingly
sound, but we have no a priori way of telling that they cannot lead to contradiction. It is inevitable that many different alternatives and variations of set
theory would have developed, including variations in the underlying logic.*

2.1.6   A menagerie of set theories and logics

The means of inference in use before the new clarifications brought to light in
the nineteenth and twentieth centuries was Aristotelian logic, which is centred
around the theory of syllogisms. A syllogism infers a relation between two terms
from their relation to a third, for example 'all A are B, all B are C, therefore
all A are C'. Traditionally the relations which are permitted to appear in a
syllogism are: 'all A are B'; 'some A are B'; 'no A are B'; and 'some A are
not B'. Aristotle described all of the syllogistic forms in his *Prior Analytics*,
regarding as invalid some forms which were accepted by later scholars.

The difficulty in considering negations when using syllogisms is that one is, in modern terms, essentially considering the complement of a set – the collection of objects not having some property – and in the absence of some fixed universe of objects in which to take the complement the concept leads to difficulties. Augustus De Morgan,[1] who with George Boole[2] helped to pioneer a mathematical approach to logic in the nineteenth century, recognized this problem of an ill-defined universe. This mathematical recasting of logic reformed the subject, freeing it from its Aristotelian limitations, and in the process uncovered some deeper problems.

Zermelo developed his original set theory partly to resolve the set theoretic paradoxes but largely in response to criticism of the use of the Axiom of Choice in his proof of the Well-Ordering Theorem. Russell and Whitehead's *Principia Mathematica* was a more direct attack on the set theoretic paradoxes, but a less successful one. In reaction to the awkwardness of Russell and Whitehead's early form of type theory W. V. O. Quine developed another system (including contributions of B. Rosser, H. Wang and others) which had peculiarities and variations of its own. Many other experiments in set theory were carried out in the first half of the twentieth century.

Most recent developments assume from the outset the so-called principle of purity, that there are no atoms, which is to say that all sets are built via set theoretic constructions from the empty set (which itself is first defined as an abstract class and is then deduced to be a set) – no other objects are needed. Even if one wants to use the theory to model a physical collection of objects (viruses or clusters of galaxies, for example), it is not necessary to introduce these objects as atoms (i.e. as set theoretic constants). Instead one can model physical objects by certain well-chosen pure sets, as is done implicitly by all physicists whenever they appeal to ZF-describable mathematics in their simulations of real worldly phenomena.

It is from this perspective that one can confidently make the claim that applied mathematics *is* pure mathematics. No sensible criterion distinguishes between the two practices, only the intent and level of rigour of its practitioners, and possibly mutual misunderstanding, giving rise to the illusion of separation. Granted, applied mathematics bases its primitive assumptions on conjectured laws governing observable phenomena, these in turn inspired by many empirical measurements (assumptions which are vindicated if the consequent mathematical theory makes testable predictions confirmed by further experimentation). But the objects of applied mathematics (for example, partial differential equations on structures modelled by regions of $\mathbb{R}^4$) are abstractions (set theoretic models). Pure mathematicians are also experimenters in exactly the same fashion, the matter of the experiments being only a few levels of abstraction removed from the 'real world' stuff of physical observation. Pure mathematicians might occasionally start with perverse abstractions, but these often turn out to have widespread applications (indeed more so than the parts of mathematics spe-

---

[1]De Morgan [44].

[2]Boole [22], following his *Mathematical Analysis of Logic* (1847).

cially designed to solve a specific physical problem); so we are simply looking at two parts of the same elephant.

Even in logic itself there are alternative approaches. For example, Brouwer's Intuitionism rejected the law of the excluded middle and other similar principles.

What exactly does Intuitionistic logic allow? The mathematical assertions of Intuitionism are the results of finitistic constructions. This imposes restrictions on the rules of negation. We can only assert $\neg\phi$ if there is a finitistic construction which leads from ϕ to a contradiction. For this reason we cannot assume the law of the excluded middle $(\phi \vee \neg\phi)$ or that $\neg\neg\phi \rightarrow \phi$. In order to illustrate why this is we need to rely on presently unsolved problems (this reliance on what is not known at some arbitrary point in time seems a little unsatisfactory, but let us persevere with it). The discussion below is based on an account in Monk's *Mathematical Logic*,[1] however, since Monk uses two examples which are now resolved[2] we shall instead use the notoriously unsolved Goldbach conjecture (1742): *every even number ≥ 4 is expressible as the sum of two primes.*[3]

Let ϕ be the statement 'there is an even $n \geq 4$ which is not expressible as the sum of two primes'. The intuitionistic interpretation of ϕ is that such an even number has explicitly been exhibited while the intuitionistic interpretation of $\neg\phi$ is that starting from the assumed existence of an even $n \geq 4$ which is not expressible as the sum of two primes a contradiction has been deduced. We cannot assert ϕ, for no such even number has been found to date, and since no-one has proved that the existence of such an n leads to a contradiction we cannot assert $\neg\phi$ either, so $\phi \vee \neg\phi$, which is a tautology in classical logic, cannot be asserted.

Staying with the Goldbach example, let $G(n)$ be the number of ways n can be written as a sum of two primes, so that, for example, $G(100) = 6$ since 100 can be written as a sum of two primes only in the following six ways: $3 + 97$, $11 + 89$, $17 + 83$, $29 + 71$, $41 + 59$, $47 + 53$. We construct a real number $\alpha \in (0, 1)$ with the following decimal expansion defined by G: the nth decimal place of α is equal to 3 unless there is an $m \leq n$ with $G(2m + 2) = 0$ in which case it is 0. Let ϕ be the statement 'α is rational'. The intuitionistic interpretation of ϕ is 'integers m and n have been constructed so that $\alpha = \frac{m}{n}$'. Classically we happily assert that α is rational because we know that it is either equal to $\frac{1}{3}$ or a terminating decimal. However the intuitionistic interpretation of ϕ claims that a specific rational number has been constructed, and in order to do this one must know whether the Goldbach conjecture is true $(\alpha = \frac{1}{3})$ or not $(\alpha$ is a terminating decimal). So intuitionistically we cannot assert ϕ. But we *can* inuitionistically assert the double negative $\neg\neg\phi$ since if we assume $\neg\phi$ then α

---

[1] Monk [**149**] (Chapter 8).

[2] Fermat's Last Theorem and Brouwer's example of the occurrence of the string 0123456789 in the decimal expansion of π (which is now known to appear beginning at the 17 387 594 880th digit, and has also been found in five other later positions).

[3] As the final draft of this book was nearing completion Harald Helfgott of the École Normale Supérieure in Paris announced a proof of the *ternary Goldbach conjecture*: every odd number greater than 5 is the sum of 3 primes. The binary Goldbach conjecture we are discussing here implies the ternary conjecture (now theorem) but still remains stubbornly open.

cannot be of the form $0.3\ldots3\dot{0}$, so $G(2n) > 0$ for all $n \geq 2$, hence $\alpha = 0.\dot{3} = \frac{1}{3}$, so we conclude from $\neg\phi$ that ϕ is true, which is a contradiction in both classical and intuitionistic logic. Putting these two arguments together we see that we cannot intuitionistically assert the classical implication $\neg\neg\phi \to \phi$.

Perhaps even more esoteric are the many-valued logics which incorporate values other than those interpreted as 'true' or 'false'. The natural temptation in many-valued logic is to treat the values 'true' and 'false' as extremes, the other truth values indicating interpolated levels of indeterminacy. One then makes a decision how to evaluate the truth values of $\neg\phi$ and (one of) $\phi \wedge \psi$, $\phi \vee \psi$ or $\phi \to \psi$ for general ϕ and ψ, perhaps inspired by probabilistic ideas.

Mathematics, at the foundational level just as much as elsewhere, is a highly creative discipline, its innovators are driven forward by a desire for elegance and are repelled only by the possibility of internal contradiction. All of the alternative set theories, each being apparently contradiction-free, are equally 'valid'.

REMARKS

1. The first statement of a syllogism, called the *major premise*, is of one of four forms: all A are B; some A are B; no A are B; some A are not B (traditionally one uses 'is' rather than 'are', but I find this a bit jarring). The second statement, the *minor premise*, needs to relate one of A or B to C – since we can swap the major and minor premises there is some symmetry, so we can assume without any loss that the minor premise relates B and C. There are eight possibilities: all B are C; some B are C; no B are C; some B are not C; all C are B; some C are B; no C are B; some C are not B. The conclusion likewise has eight possible forms relating A and C. So in total there are 256 syllogisms. Of these only 24 are valid, in the sense that the conclusion follows logically from the major and minor premises. The letters A, E, I, O were used in medieval times to stand for each of the basic forms 'all', 'no', 'some' and 'some not', so that each syllogistic form could be given a three letter code. For a given combination of three letters there may be up to four valid syllogisms. For example there are four valid syllogisms of the type EIO:

   (i) No A are B and some C are B, therefore some C are not B.

   (ii) No A are B and some C are B, therefore some C are not A.

   (iii) No A are B and some A are C, therefore some C are not B.

   (iv) No A are B and some B are C, therefore some C are not A.

   Each of the 24 valid syllogisms was given a mnemonic: *Modus Barbara* for the only valid AAA syllogism, *Modus Celarent* and *Modus Celare* for the two valid EAE syllogisms, *Modus Darii* and *Modus Datisi* for the two valid AII syllogisms, etcetera. Nine of these 24 syllogism are valid only if one of the categories in question is non-empty, which would have caused some doubt in medieval times. Each syllogism is typically illustrated using a Venn diagram.

2. Had the Goldbach conjecture been a conjecture in the empirical sciences it would long have been declared to be true. Not only has it been verified up to astronomically high numbers, but there is also some suggestive structure in the number of representations of an even number as a sum of two primes. If we plot n against the number of distinct representations of $2n$ as a sum of two primes then the result, illustrated in Figure 2.1, known as *Goldbach's comet*, strongly indicates that on average the number of representations increases. A rogue value of zero seems incredibly unlikely. (The various dense bands in the comet correspond to certain modularly equivalent values of n.)

For centuries the only logic was that described by Aristotle's syllogisms. When logic was mathematicized in the nineteenth century the much richer language of classical logic emerged. Later some alternative logics were proposed which differed from classical logic in a fundamental way. ZF takes as its logic (classical) first-order logic. As this theory developed, Platonism seemed to be under attack, and a new philosophy began to emerge.

2.1.7   Elucidation – competing schools of thought

Mathematical realists consider primitive elements not to be 'meaningless' arbitrary starting positions in a sterile game but as being potentially interpretable, that is it may be possible to provide a natural interpretation of the primitive elements of a theory in terms of ordinary language, by reference to some observable phenomena (what Frege called *elucidations*). Such points of reference are not needed in the symbolic development of a subject but are critical in forming an early appreciation, and in choosing the basic principles, of an abstract theory.

Realists in the above sense fall into at least two camps: the Platonist takes ideal objects to have an existence independent of human scientific and artistic activity, occasionally touched upon or modelled in various degrees of accuracy; a Constructivist takes an object to exist only if it can be formally constructed by some definite finitistic process. One might think of the Platonist's ideal universe as the perfect limit of all that may be approached, but possibly not attained, through Constructivist means.

The primitives 'point', 'line' and 'plane' in the formal axiomatic development of Euclidean geometry, for example, are heavily bound to their real-world approximations and the axioms governing their behaviour come straight from visual experience. Euclidean geometry is the most obvious, and is the first, example of a formal theory whose primitives are tightly tied to our interpretation of idealized aspects of the physical world. It was this seemingly unbreakable coupling of primitive element and its 'real world' interpretation that drove the

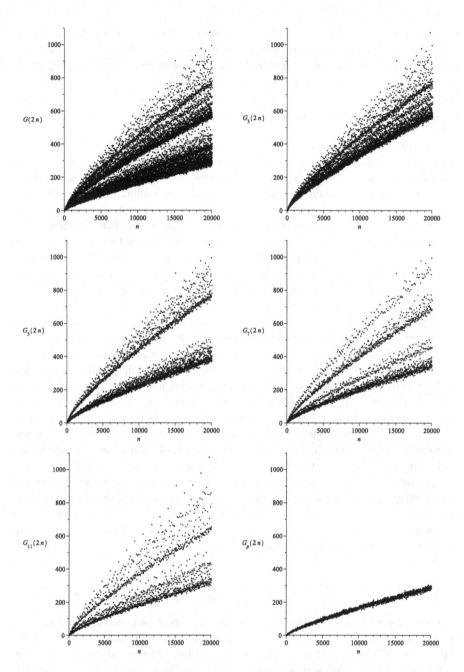

Figure 2.1 If we plot n against $G(2n)$, the number of distinct ways of representing $2n$ as a sum of two primes, then we obtain what is called Goldbach's comet (top left), giving a neat pictorial illustration of why most people are convinced that the Goldbach conjecture is true. By considering certain subsets of integers, for example, n a multiple of 3 (top right), 5 (middle left), 7 (middle right), 11 (bottom left), or where n is prime (bottom right), and so on, we get an idea of some of the fine band structure of the comet.

initial development of the subject (we can literally 'see' what the definitions are and what the theorems are telling us – conjectures can be formed and proofs can be sketched by drawing a few diagrams), however the same coupling was to obstruct the resolution of the status of the parallel postulate (see Section 2.3). Only by releasing a primitive from its accompanying elucidation was it possible to perceive enough freedom to construct models in which the parallel postulate fails.

The gradual clarification of Euclidean geometry and the realization that it is merely one among many geometries perfectly illustrates the strength of the axiomatic method. The interplay between axiom selection and model building also serves as an effective mathematical model of understanding. The first stage usually begins with the raw phenomenon, physical or mathematical. From this we extract some recognizable properties, forming our list of axioms. We study the logical consequences of these axioms and assess the extent to which the consequent theory faithfully describes what we observe. We experiment with the axioms, negating some of them, adding or removing others, giving us further insight into the inner workings of the theory. The set of models of these perturbations of the original theory suggest further developments. We end up not only with an effective model of the subject we originally sought to understand, but also a deeper understanding of its position in a wider context.

During the nineteenth century, triggered in part by the appearance of different axiomatizations of geometry, another family of philosophies all falling under the umbrella of 'Formalist' came to prominence. Formalists rejected the Platonic notion of absolute truth. Instead, although inspired by existing physical or mathematical phenomena, axioms were to be thought of as merely a list of consistent statements which we assume only to demonstrate consequent properties. The axioms had been demoted from their status as Platonic truths and a freedom was felt to replace a given axiom with another that contradicts it, opening a window to a new universe. The Formalist's truth is conditional: *if* any structure satisfies the axioms, *then* it satisfies the theorem. The more liberating Formalist philosophy had been adopted by the majority of, but by no means all, mathematicians by the 1920s and was inevitably the underlying force in the development of set theory.

REMARK

This categorizing of mathematicians (or any group of thinkers, for that matter) into clumpy schools of thought is oversimplistic and often unnecessary and so I have been deliberately sloppy with the definitions. Survey all mathematicians as to their private philosophical position 'I'm an Xist, subject to qualifications Y and Z', and at the survey's end we shouldn't be too surprised to find the same number of philosophical species as mathematicians. The notion of Platonism seems too vague and quasi-mystical to admit a satisfactory analysis. Many mathematicians confess to a high degree of apathy or even antipathy when it comes to such matters (Paul Cohen includes himself in this category in a very interesting personal talk, captured on film, given at the Gödel Centennial, Vienna, 2006 – Cohen's work is a model of how one can cut through philosophical fluff with mathematics). The fact that all of these individuals can produce work in the same subject and agree that one another's arguments are *logically* sound (ignoring the prohibitions of philosophical bias) shows that ontological matters are unlikely to divide the community to any great extent. Mathematical conferences are not filled with people agonizing over the meaning of the 'existence' asserted by the axioms and theorems of whichever theory is being discussed. In fact the axioms themselves are usually invisible.

Beginning with the discovery of non-Euclidean geometry and growing with the new developments in logic and set theory, the stricter forms of Platonism seemed to fade. The Formalist philosophy tended to be increasingly adopted by mathematicians working in these fields. Irrespective of our philosophical standpoint, whether we regard our sentences as meaningless strings of symbols or as labelling Platonic notions, we need to know how to form them. What are our basic rules of formation?

2.2   Well-formed formulas

> *I concluded that I might take as a general rule the principle that all things which we very clearly and obviously conceive are true: only observing, however, that there is some difficulty in rightly determining the objects which we distinctly conceive.*
>
> – RENÉ DESCARTES[1]

2.2.1   Construction of well-formed formulas

Well-formed formulas, the sentences of our language, are constructed from variables and the logical symbols according to the following rules.

(i) If x and y are variables then $x \in y$ is a well-formed formula.

(ii) If ϕ and ψ are well-formed formulas then so are $\neg\phi$, $\phi \vee \psi$, $\phi \wedge \psi$, $\phi \rightarrow \psi$ and $\phi \leftrightarrow \psi$.

(iii) If ϕ is a well-formed formula and x is a variable then $(\forall x)\phi$ and $(\exists x)\phi$ are well-formed formulas.

(iv) A string of symbols is a well-formed formula if and only if it can be formed using the three rules above.

---

[1]Descartes [49].

Formulas of the form $x \in y$ are said to be *irreducible* or *atomic* because they cannot be formed from smaller well-formed formulas using rules (ii) and (iii) above.

We will generally use ϕ, ψ and η to denote arbitrary well-formed formulas.

It is conceivable that set theory could expose itself to misplaced criticism at this stage by referring, in the metalanguage, to a 'countable set of variables' or 'a class of variables: $x_0, x_1, x_2, \ldots$', invoking a sense of cardinality long before cardinality has been formally defined in the object language. This confusion between the principles of the object language (the language of the theory itself) and the metalanguage (the language used to talk about the object language) was the cause of much debate in the early days of set theory; it is clear why the distinction is necessary – a language cannot build itself. Had this been a genuine issue, the only cardinal-tainted property of the collection of variables that we need is something similar to the meta-schema *if ϕ is a well-formed formula then there exists a variable which does not appear in ϕ.*

REMARK

We should observe that there are countably many well-formed formulas and that the class of well-formed formulas can be recursively generated in a fairly natural way. For example, suppose at generation n we have formed a finite list $\mathcal{V}$ of variables and a finite list $\mathcal{F}$ of well-formed formulas. Then for the $n + 1$th generation we systematically do the following: add the variable x_{n+1} to $\mathcal{V}$; running through all $\phi, \psi \in \mathcal{F}$ and all $x_i, x_j \in \mathcal{V}$ form the formulas $x_i \in x_j$, $\neg\phi$, $\phi \vee \psi$, $\phi \wedge \psi$, $\phi \rightarrow \psi$, $\phi \leftrightarrow \psi$, $(\forall x_i)\phi$ and $(\exists x_i)\phi$; delete from this list those formulas which already appear in $\mathcal{F}$; add the new formulas to the end of the list $\mathcal{F}$, forming the new $\mathcal{F}$. Starting at the '-1st generation' with empty $\mathcal{V}$ and $\mathcal{F}$, at the 0th generation we have $\mathcal{V} = \{x_0\}$ and $\mathcal{F} = \{x_0 \in x_0\}$, and so on.

The class of well-formed formulas is built up recursively from the set of variables using the membership symbol and the logical symbols via simple rules of construction. One notion that we shall often need is that of free and bound variables.

2.2.2   Free and bound variables

The notions of bound and free variables of a well-formed formula will appear in the logical axioms and rules of inference to come. They are defined as follows.

FREE VARIABLES

(i) If x occurs in ϕ and ϕ is irreducible, then x is free.

(ii) If x is free in ϕ or ψ then the same occurrence of x is free in: $\neg\phi$, $\phi \to \psi$, $\phi \lor \psi$, $\phi \land \psi$, $\phi \leftrightarrow \psi$ and, for y distinct from x, in $(\forall y)\phi$ and $(\exists y)\phi$.

BOUND VARIABLES

(i) The occurrence of x immediately following the quantifier in $(\forall x)\phi$ and in $(\exists x)\phi$ is bound.

(ii) All occurrences of x in ϕ are bound in $(\forall x)\phi$ and in $(\exists x)\phi$.

(iii) If x is bound in ϕ or ψ then the same occurrence of x is bound in: $\neg\phi$, $\phi \to \psi$, $\phi \lor \psi$, $\phi \land \psi$, $\phi \leftrightarrow \psi$ and, for any y, in $(\forall y)\phi$ and $(\exists y)\phi$.

A well-formed formula with at least one free variable is a *predicate*. We shall call a well-formed formula with exactly n free variables an *n-predicate*. A well-formed formula with no free variables is a *sentence*. If we want to emphasize that $x_1, \ldots, x_n$ are among the free variables of a predicate ϕ we shall write $\phi(x_1, \ldots, x_n)$.

The collection of sentences comprising the non-logical axioms of ZF specify how we want sets to behave by simultaneously controlling which types of predicates should be allowed to describe sets (avoiding paradoxes) and generating new sets from old based on the intuition of a set as an arbitrary collection of objects.

REMARK

A 1-predicate $\phi(x)$ describes a property that a set x may or may not have, and the class of sets associated with ϕ is precisely the class of sets with property ϕ. We can bind the variable in one of two ways, forming one of the two sentences $(\forall x)\phi(x)$ or $(\exists x)\phi(x)$, which respectively say that the class associated with ϕ is the class of all sets or is a non-empty class. An n-predicate describes a property that an ordered n-tuple of sets may or may not have. By preceding an n-predicate by n quantifiers we can form a number of different sentences, but note that, for example, $(\forall x)(\forall y)\phi(x,y)$ will be equivalent to $(\forall y)(\forall x)\phi(x,y)$ and $(\exists x)(\exists y)\phi(x,y)$ will be equivalent to $(\exists y)(\exists x)\phi(x,y)$, so many of the $n!2^n$ sentences formed by binding each of the variables in an n-predicate will be equivalent.

A variable in a well-formed formula is bound if it appears in the scope of a quantifier and is free otherwise. Well-formed formulas without any free variables are sentences. A selection of sentences, based on properties we intuitively want sets to have, will form our basis of non-logical axioms. We need some machinery that will allow us to make logical deductions from these axioms.

2.3   Axioms and rules of inference

*A scientist can hardly meet with anything more undesirable than to
have the foundations give way just as the work is finished. I was put
in this position by a letter from Mr. Bertrand Russell when the work
was nearly through the press.*

– GOTTLOB FREGE[1]

Frege goes on to identify the source of the problem as one of his axioms (Axiom
(V) – the Axiom of Abstraction), which he admits he had always been suspicious
of.

2.3.1   Logical principles

At this stage we essentially only have a mechanism to create sequences of symbols, and a list of symbol clusters certainly does not make a language. Guided
by experience we create logical axioms, rules of inference and non-logical (not
to be confused with illogical) axioms. The details of the non-logical axioms
are governed, to a greater extent, by the 'needs of mathematics'. The task of
choosing the 'correct' set of axioms is a direct descendant of Frege's project to
separate the logical assumptions needed for arithmetic (the logical axioms) from
those that appeal to matters of experience (the non-logical axioms). Each logical axiom is an axiom schema, i.e. an infinite collection of axioms. For example
the axiom $\phi \to (\psi \to \phi)$ is a doubly indexed schema, meaning for any fixed
well-formed formulas ϕ and ψ we have $\phi \to (\psi \to \phi)$.[2]

Logical Axioms

1. $\phi \to (\psi \to \phi)$.

2. $[\phi \to (\psi \to \eta)] \to [(\phi \to \psi) \to (\phi \to \eta)]$.

3. $(\neg\phi \to \neg\psi) \to (\psi \to \phi)$.

4. If x does not have a free occurrence in ϕ then $(\forall x)(\phi \to \psi) \to [\phi \to (\forall x)\psi]$.

5. If x has no free occurrence in a well-formed subformula of ϕ of the form
   $(\forall a)\psi$ then $(\forall x)\phi(x) \to \phi(a)$.

Here $\phi(a)$ is the formula obtained by replacing, in a delicate sense to be described
in Subsection 2.5.1, every free occurrence of x in $\phi(x)$ by a.

---

[1]Part of a note appearing in the appendix to the work in question, Frege's *Grundgesetze
der Arithmetik*. This translation is taken from Geach and Black [**79**].

[2]There are many alternative (but equivalent) ways of stating the logical axioms, some more
intuitive than others.

Rules of inference

1. (modus ponens)[1] From ϕ and $\phi \to \psi$ we infer ψ.

2. (generalization) From ϕ we infer $(\forall x)\phi$.

REMARK

We should give a proper definition of 'subformula' since it appears in Logical Axiom 5. By the formation rules for well-formed formulas a well-formed formula η is either irreducible (in which case it has one subformula, η itself) or it is of one of the forms $\neg\phi$, $\phi \vee \psi$, $\phi \wedge \psi$, $\phi \to \psi$, $\phi \leftrightarrow \psi$, $(\forall x)\psi$, $(\exists x)\psi$ for some well-formed formula(s) ϕ and ψ and possibly some variable x. Each of ϕ and ψ are subformulas of η. The definition is completed by declaring the relation of being a subformula to be transitive: subformulas of subformulas of η are also subformulas of η.

Five logical axioms and two rules of inference describe how we infer one logical statement from another. Any theory with a collection of variables and well-defined well-formed formulas together with these logical axioms and rules of inference is a first-order theory. Several different ways of working in first-order logic have been developed.

2.3.2   Some basic tautologies

It is sometimes useful to distinguish between sentences that are theorems of ZF and those sentences that have the stronger property of being obtainable using the logical axioms and the rules of inference alone. Formulas ϕ and ψ are *logically equivalent* if the well-formed formula $\phi \leftrightarrow \psi$ can be proved without appealing to any non-logical axioms. Some of the logical results we might obtain, most of which are taken for granted and used without explicit reference, include the laws (De Morgan's, Associative, Distributive and Transitive) and abbreviations listed in Subsection 2.1.3, and such results as:

$$\neg\neg\phi \leftrightarrow \phi \qquad [\phi \wedge \psi] \leftrightarrow [\psi \wedge \phi] \qquad [\phi \vee \psi] \leftrightarrow [\psi \vee \phi]$$
$$[\phi \vee \phi] \leftrightarrow \phi \qquad [\phi \wedge \phi] \leftrightarrow \phi \qquad \phi \to \phi$$

The tautologies above are easily verified using the familiar 'truth table' approach, where all combinations of the truth values of the component well-formed formulas are tabulated and the validity of the whole formula is confirmed in each case. For example, if we wish to verify the De Morgan Law $[\neg(\phi\vee\psi)] \leftrightarrow [\neg\phi\wedge\neg\psi]$ we list all possible combinations of truth values for ϕ and ψ, using 0 for 'false'

---

[1] Also called *detachment*.

and 1 for 'true', and for each case observe that the corresponding truth values of each side of the proposed equivalence coincide (i.e. the fourth and seventh columns in the truth table below are identical).

ϕ	ψ	$\phi \vee \psi$	$\neg(\phi \vee \psi)$	$\neg\phi$	$\neg\psi$	$\neg\phi \wedge \neg\psi$
0	0	0	1	1	1	1
0	1	1	0	1	0	0
1	0	1	0	0	1	0
1	1	1	0	0	0	0

There are more efficient and elegant ways to establish the truth of such logical statements, for example the method of tableaux.[1] An analytic tableau for a formula ϕ is an ordered tree rooted at ϕ which, by a well-defined and efficient algorithm, branches out with formulas at each junction. A proof of ϕ corresponds to a tableau rooted at $\neg\phi$ every branch of which terminates. Of course it needs to be proved first that these procedures are valid. The purist might prefer a first principles approach, demonstrating such results from the logical axioms and rules of inference alone. The reader who wishes to attempt this may need to refer to Subsection 2.1.3 for the expression of all of the logical operations in terms of the two symbols $\neg$ and $\to$ (with this in mind, we note also that $\phi \vee \psi$ can be expressed entirely in terms of $\to$: it is equivalent to $(\phi \to \psi) \to \psi$).

Hiding in the list of tautologies at the end of the first paragraph of this subsection is the tell-tale $\neg\neg\phi \to \phi$ which reveals that our logic is a classical (not intuitionistic) one. We can also easily derive the principle of the excluded middle (i.e. $\phi \vee \neg\phi$; equivalently $\phi \to \phi$). It is worth detailing exactly how we obtain these two results from the axioms and rules of inference.

As a first step let us prove $\phi \to \phi$. By Logical Axioms 1 and 2 we have

$$\phi \to [(\phi \to \phi) \to \phi] \text{ and}$$

$$[\phi \to [(\phi \to \phi) \to \phi]] \to [[\phi \to (\phi \to \phi)] \to (\phi \to \phi)]$$

respectively. Combining these, via modus ponens, we have

$$[\phi \to (\phi \to \phi)] \to (\phi \to \phi).$$

Via modus ponens again (since $\phi \to (\phi \to \phi)$ follows from Logical Axiom 1) we have $\phi \to \phi$.

In particular, for any ϕ we have the implication $\neg\phi \to \neg\phi$ (or in other words $\neg\neg\phi \vee \neg\phi$). To prove $\neg\neg\phi \to \phi$ we assume $\neg\neg\phi$, then by Logical Axiom 1 $\neg\neg\neg\neg\phi \to \neg\neg\phi$, and using Logical Axiom 3 twice we have $\neg\neg\phi \to \phi$. We infer ϕ by modus ponens.

[1]See Smullyan [199] for precise details and the origins of the idea.

REMARKS

1. Each tautology generates infinitely many other tautologies (simply replace each subformula in the tautology by another tautology).

2. Let us make a few obvious remarks to rule out the possibility of a pathological case.

   Consider the tautology $x \in y \leftrightarrow x \in y$. By generalization we have $(\forall y)[x \in y \leftrightarrow x \in y]$. Call the latter predicate $\phi(x)$. By generalization again we have $(\forall x)\phi(x)$.

   Now consider the sentence $(\forall x)\phi(x) \wedge (\forall x)\neg\phi(x)$. If this were true then each of its components $(\forall x)\phi$ and $(\forall x)\neg\phi$ would be true, but then by Logical Axiom 5 each of $\neg\phi(a)$ and $\phi(a)$ is true, implying $\neg\phi(a) \wedge \phi(a)$ is true, a contradiction. Therefore (via Logical Axiom 3) the original sentence $(\forall x)\phi(x) \wedge (\forall x)\neg\phi(x)$ is false, so its negation $\neg(\forall x)\phi(x) \vee \neg(\forall x)\neg\phi(x)$ is true, but this is simply the implication $(\forall x)\phi(x) \rightarrow (\exists x)\phi(x)$. By modus ponens we conclude that $(\exists x)\phi(x)$.

   This shows that the universe of discourse is non-empty. There exists a set. This will be important later.

Proving logical statements from first principles can be difficult. More mechanical procedures are available to help unburden the proofs. It took some time to fully axiomatize set theory, the early experiments sometimes alluding to general principles but rarely explicitly stating them.

2.3.3 An aside: implicit axioms in Cantor's work

Cantor did not explicitly mention axioms in his work, however, he tacitly appeals to three principles which are immediately recognizable today.

(i) The Axiom of Extensionality for sets: two sets are identical if they have the same members. We will give a more precise statement of this axiom in Section 2.5.

(ii) The Axiom of Choice: given an infinite collection of sets there exists a set which has exactly one element in common with each set in the collection.

(iii) The source of the paradoxes: the Axiom of Abstraction.

The first explicit formulation of the Axiom of Abstraction seems to be in Frege's *Grundgesetze der Arithmetic*[1] (1893) (as 'Axiom V'). When Russell pointed out

---

[1] 'Basic Laws of Arithmetic'.

to Frege that this axiom leads to internal contradiction, Frege was understand-ably devastated, as evidenced in the quote at the beginning of this section. Formally the Axiom of Abstraction asserts an infinite number of statements, each predicate $\phi(x)$ yielding the axiom $(\exists y)(\forall x)[x \in y \leftrightarrow \phi(x)]$.

REMARK

Cantor did explicitly propose the Well-Ordering Theorem but wavered as to whether it should be taken as an axiom or if it was provable from the other principles of the theory (this, of course, predates Zermelo's proof of its equiv-alence with the Axiom of Choice). He also recognized the need to restrict sets in some way to avoid the paradoxes associated with the 'set of all sets', but the exact nature of this distinction was rather vaguely stated.

Cantor's initial experiments in set theory implicitly assumed the three Axioms Choice, Extensionality and Abstraction. The unrestricted Axiom of Abstraction, which intuitively says that every describable collection is a set, is the cause of the early set theoretic paradoxes. Decades would pass before the underlying logic of set theory was clarified. Zermelo–Fraenkel set theory is a first-order theory, i.e. the underlying logic is first-order. Higher-order languages can be described, each of which is able to express with more power or with more economy the ideas of its predecessors.

2.3.4   Orders of logic

In its most general form a first-order theory comprises the following elements.

1. A collection (possibly empty) of primitive n-ary operator symbols ($n \in \{0, 1, 2, \ldots\}$) where 0-ary operators are interpreted as constants.

2. A collection of primitive n-ary relation symbols ($n \in \{1, 2, 3, \ldots\}$).

3. A countable collection of variables x_0, x_1, x_2, $\ldots$

4. By a *term* we mean (i) a variable; or (ii) any expression of the form $O(t_1, \ldots, t_n)$ where O is a primitive n-ary operator and $t_1, \ldots, t_n$ are terms (this includes all constants, i.e. 0-ary operators).

5. Atomic formulas, these being expressions of the form $R(t_1, \ldots, t_n)$, where R is a primitive n-ary operator and $t_1, \ldots, t_n$ are terms.

6. Well-formed formulas, which comprise all atomic formulas together with all formulas of the form $\neg\phi$, $\phi \wedge \psi$, $\phi \vee \psi$, $\phi \to \psi$, $\phi \leftrightarrow \psi$, $(\forall x)\phi$ and $(\exists x)\phi$, where ϕ and ψ are well-formed formulas and x is a variable. A well-formed formula with no free variables is a *sentence*, otherwise it is a *predicate*.

7. Logical axioms (the axiom schemas described in Subsection 2.3.1), rules of inference (modus ponens and generalization) and a collection of sentences comprising the non-logical axioms.

It is possible to dispense with operator symbols; every n-ary operator is a particular type of $n + 1$-ary relation. More precisely, an $n + 1$-ary relation R is an operator if for each n-tuple of constants $(a_1, \ldots, a_n)$ there exists a unique x such that $(a_1, \ldots, a_n, x) \in R$. Furthermore, one can regard variables as 0-ary relations, so the whole theory can be reduced to one of relations. We shall not pursue this here. In future I will generally use the more euphonious terms *constant*, *unary* and *binary* in place of 0-ary, 1-ary and 2-ary, respectively.

The theory ZF, for example, has no primitive operator symbols; variables $x_0, x_1, x_2, \ldots$; one primitive binary relation symbol $\in$; the notion of a term is redundant (these being just the variables); the atomic formulas are formulas of the form $x_i \in x_j$ for variables x_i, x_j; and the well-formed formulas are built from these as described in Subsection 2.2.1. We have given the schemas of sentences comprising the logical axioms and rules of inference for first-order logic in Subsection 2.3.1 above, and the non-logical axioms of ZF are to come.

The theory of Peano Arithmetic has four primitive operator symbols 0 (constant), s (unary), $+$ (binary), $\cdot$ (binary); one primitive binary relation symbol $=$; variables $x_0, x_1, x_2, \ldots$; the terms are all variables together with the constant 0 and all expressions of the form $s\tau$, $\tau + \sigma$ and $\tau \cdot \sigma$, where τ and σ are terms; the atomic formulas are formulas of the form $\tau = \sigma$, where τ and σ are terms; and the well-formed formulas are built from the atomic formulas as described above (the non-logical axioms are given in Appendix A).

We can similarly describe the first-order theories of groups, rings, fields, ordered fields, and so on with a handful of operator and relation symbols together with the appropriate axioms.

A first-order theory is *effective* if it has a recursively enumerable set of axioms. That is, there is an algorithm which will tell us, when presented with a sentence in the theory, whether that sentence is an axiom. An effective theory is able to recursively generate its theorems. Both Peano Arithmetic and Zermelo–Fraenkel set theory are effective. In general, when we say something is effective or that some process can be effectively carried out, we mean that it can be described by an algorithm.

First-order language is of limited expressive power. For example, in Peano Arithmetic we would like to state as an axiom the full induction principle as presented in Peano's Postulates, but this involves a quantification over subsets, something which cannot be done in first-order terms. Instead we have to resort to a much weaker first-order axiom schema: for each predicate $\phi(x)$ we have the axiom $[\phi(0) \wedge (\forall a)[\phi(a) \rightarrow \phi(sa)]] \rightarrow (\forall a)\phi(a)$. Similarly, in a first-order theory of complete ordered fields we are unable to give the full completeness axiom, again because of the quantification over subsets, and so we would have to resort to a much weaker first-order axiom schema based on the idea that for each predicate $\phi(x)$ we have $(\exists u)(\forall x)[\phi(x) \rightarrow x \leq u] \rightarrow [(\exists v)[[\phi(x) \rightarrow x \leq v] \wedge (\forall u)[[\phi(x) \rightarrow x \leq u] \rightarrow v \leq u]]]$ (this being the translation of 'if $\{x : \phi\}$

has an upper bound then it has a least upper bound'). Tarski's theory of real closed fields (see Subsection 2.4.2) is rather more elegant.

In order to accommodate subset quantification (and quantification over predicates) we must pass to second-order logic. The generality in which we describe second-order theory below is much greater than is required for most applications, as later examples reveal.

Let **term**$_0$ be the class of all terms of the first-order theory. Following the same order of presentation as our description of a first-order theory above, a second-order theory comprises all the machinery of its first-order parent together with the following:

1. Primitive second-order n-ary operator symbols.

2. Primitive second-order n-ary relation symbols.

3. Second-order variables $O_0^n, O_1^n, O_2^n, \ldots, R_0^n, R_1^n, R_2^n, \ldots$, $n \in \{1, 2, 3, \ldots\}$, where O_i^n ranges over all n-ary operators **term**$_0^n \to$ **term**$_0$ and R_i^n ranges over all n-ary relations on **term**$_0$, i.e. subsets of **term**$_0^n$.

4. Second-order terms comprising (i) all terms in **term**$_0$; (ii) all expressions of the form $O_i^n(t_1, \ldots, t_n)$, where $t_1, \ldots, t_n \in$ **term**$_0$; and (iii) all expressions of the form $\Omega^n(t_1, \ldots, t_n)$, where Ω^n is a primitive second-order n-ary operator symbol. (The terms described in (ii) and (iii) are elements of **term**$_{(0,\ldots,0)}$ ($n{+}1$ zeros) – see the text following this list for an explanation of this notation).

5. Second-order atomic formulas comprising (i) the first-order atomic formulas; (ii) all expressions of the form $R_i^n(t_1, \ldots, t_n)$, where $t_1, \ldots, t_n \in$ **term**$_0$; and (iii) all expressions of the form $\Xi^n(t_1, \ldots, t_n)$, where Ξ^n is a primitive second-order n-ary relation symbol. Note in particular that the expression $R_0^1(t_0)$ means the term t_0 is an element of the unary relation (i.e. the set) R_0^1.

6. The second-order well-formed formulas comprise all second-order atomic formulas together with all formulas of the form $\neg\phi$, $\phi \land \psi$, $\phi \lor \psi$, $\phi \to \psi$, $\phi \leftrightarrow \psi$, $(\forall x_i)\phi$, $(\forall O_i^n)\phi$, $(\forall R_i^n)\phi$, $(\exists x_i)\phi$, $(\exists O_i^n)\phi$ and $(\exists R_i^n)\phi$, where ϕ, ψ are second-order well-formed formulas, x_i is a first-order variable and O_i^n and R_i^n are second-order variables.

7. Logical axioms (the same schema as for a first-order language but extended to accommodate second-order well-formed formulas – the meaning of free and bound variables being clear), rules of inference (likewise extended to allow second-order well-formed formulas) and non-logical second-order axioms. By quantifying over the unary relation variables we quantify over sets of terms and by quantifying over unary operator variables we quantify over unary predicates, allowing us to condense first-order axiom schemas into single second-order axioms.

We could go on. A $k + 1$th order theory bears the same relation to its parent kth order theory as second-order theory bears to first-order theory and the union of all nth order theories yields a theory of types (an extension of the structure which admits theories of transfinite order is possible).

More precisely, a *type* is defined recursively by (i) 0 is a type; and (ii) if $\tau_1, \ldots, \tau_n$ are types then the n-tuple $(\tau_1, \ldots, \tau_n)$ is a type. Each n-ary operator and relation (primitive or variable) and each term has a type (the type of an n-ary operator is an $n+1$-tuple of types; the type of an n-ary relation is an n-tuple of types; the class of all terms of type τ is denoted $\mathbf{term}_\tau$). Each $k + 1$th order theory introduces new operators and relations of type $(\tau_1, \ldots, \tau_n)$ (where the component types τ_i are those of the terms of the kth theory) and generates more terms $\mathbf{term}_{(\tau_1, \ldots, \tau_n, \tau_{n+1})}$ comprising all expressions of the form $O^n(t_1, \ldots, t_n)$, where O^n is an n-ary operator of type $(\tau_1, \ldots, \tau_n, \tau_{n+1})$ and t_i is a term of type τ_i.

In practice a second-order extension of a first-order theory will generally not introduce any new primitive constants and the non-logical second-order axioms will replace any first-order non-logical axiom schemas, if possible. For example, second-order Peano arithmetic introduces the usual new variables O_i^n, R_i^n (no new primitive operators or relations) and, together with the extended logical axioms and rules of inference (these being the same schema as the first-order axioms but ranging over all second-order well-formed formulas), just one new second-order axiom which is the true induction axiom (the first-order axiom schema then becoming redundant): $(\forall R_0^1)[R_0^1(0) \wedge (\forall x_0)[R_0^1(x_0) \rightarrow R_0^1(sx_0)]] \rightarrow (\forall x_0)R_0^1(x_0)$. Second-order ZF introduces a second-order axiom replacing its first-order axiom schema of Replacement (see Section 4.4), this time using quantification over an operator variable.

The above notion of a second-order theory is the widest possible. There are weaker forms of second-order logic where, instead of adding the full strength of second-order operator and relational variables, a modest new collection of variables is introduced which are interpreted as ranging only over finite subsets of $\mathbf{term}_0$ (weak second-order logic) or over arbitrary subsets of $\mathbf{term}_0$ (monadic second-order logic).

The classical conception of an axiomatic system, that apparently held by Euclid for example, is that it must have finitely many axioms, but it is a harsh fact of model theory that no finite list of first-order axioms can fully describe ZF; axiom schemas are an inevitable feature of first-order set theory (see Section 10.2 for the reason why).

The significant distinction between first- and second-order logic, originally detailed by Charles Peirce,[1] was not fully appreciated at first; the fact that the two are profoundly different only became apparent in the 1930s. By the 1960s it had become commonplace to use first-order logic to describe foundational

---

[1] Peirce, C. S., 'On the Algebra of Logic; A Contribution to the Philosophy of Notation', *American Journal of Mathematics* **7**: 180–202, (1885). See Moore and Robin [152].

issues.[1] Few mathematicians working outside of logic and set theory are overly concerned with such matters; most researchers use a higher level natural language content with the belief that their results could, if necessary, be dissected and laid out in some sort of bare logical form without any conceptual crises.

<hr>

REMARKS

1. W. V. O. Quine famously had philosophical problems with higher-order logic, describing it as 'set theory in sheep's clothing'.[2] One can define a power set operation in second-order logic and, in a sense not detailed here, go on to 'simulate' set theory within second-order logic. In fact, via this device, one can simulate higher-order logic within second-order logic (a result of Hintikka).[3] More explicitly, given any sentence ϕ of a higher-order extension of some second-order theory T one can effectively find a sentence ϕ_2 in T such that, given any model M, ϕ is satisfied in M if and only if ϕ_2 is satisfied in M. In this sense, second-order theory is the end of the line as far as expressive power goes. An argument in support of the use of higher-order logic in mathematics and philosophy is given by Shapiro [192].

2. The variations in types of logic are too rich to discuss here. Since we are discussing ZF, we shall restrict ourselves to rather more sober and tractable first-order logic, but it does no harm to be aware of higher-order alternatives. One should keep in mind that there is a profound difference in expressive power between first and second-order logic but, based on the previous remark, no such gulf between the second- and higher-order logics.

<hr>

There is a great difference between first-order logic and its higher-order cousins. The clearest indication of the relative weakness of first-order theories versus their second-order variants is given by the familiar examples of Peano Arithmetic and the theory of complete ordered fields. In each case the second-order theory is categorical: all models are isomorphic. So the set of natural numbers is completely characterized by the second-order axioms of Peano Arithmetic, but the first-order theory of Peano Arithmetic has infinitely many non-isomorphic models. Likewise the second-order theory of complete ordered fields gives us a characterization of the field of real numbers, but any first-order theory which tries to capture the properties of $\mathbb{R}$ will have infinitely many non-isomorphic models. Likewise, the first-order theory of ZF has many different models.

<hr>

[1] An interesting discussion of why first-order logic has been so favoured over its extensions appears in Chapter 1 of Potter [170]. In first-order logic we have such results as the Completeness Theorem and the Löwenheim–Skolem Theorem, neither of which is available in second-order logic.

[2] Quine [174].

[3] Hintikka, K. Jaakko., 'Reductions in the Theory of Types', in Two Papers on Symbolic Logic, *Acta Philosophica Fennica*, No. 8, Helsinki, 57–115, (1955).

2.3.5   Some remarks on the development of ZF

Zermelo's original set theory[1] took as its non-logical axioms Extensionality (see Subsection 2.5.2), Powers (Section 4.3), Unions (Section 4.2), Choice and Infinity (there exists a set Z with $\emptyset \in Z$ and such that $\{a\} \in Z$ whenever $a \in Z$ – see Section 6.2)) together with an axiom of elementary sets ($\emptyset$ is a set and if a and b are sets so are $\{a\}$ and $\{a,b\}$) and his axiom of separation (see Section 4.4). In ZF, the theory into which Zermelo's system evolved, Choice is excluded, the axiom of elementary sets is replaced by Pairing (Subsection 4.1.1), Replacement takes the place of the axiom of separation (and is used to establish that $\emptyset = \{x : \neg(x = x)\}$ is a set), the Axiom of Infinity is usually expressed differently and a new axiom, Regularity, is added (see Section 4.5). ZF, usually together with the Axiom of Choice, is apparently able to capture the modes of reasoning used in modern mathematics, and to date no paradoxes have been found to result from the axioms.

Hilbert's school of Formalism, which spawned much of this work, made a significant contribution to the development of logic. Hilbert had intended firstly to describe a symbolic formalization of all of the principles of mathematics and secondly to prove that this formalization was consistent. His famous *Entscheidungsproblem* was to find an algorithm which would decide the provability of any input symbolic sentence. Gödel's Incompleteness Theorems, which must be viewed as a very positive development in mathematics, not as an obstacle, completely undermined the latter part of Hilbert's plan in an extraordinary way. Gödel proved that undecidable sentences are present even in such simple systems as Peano Arithmetic: there are PA-undecidable sentences about the arithmetic of natural numbers. Had Hilbert achieved the impossible he would, depending on the practicability of the algorithm, have put an end to all foundational research.

REMARKS

1. The general idea of the *Entscheidungsproblem* goes back to Leibniz (seemingly triggered by his invention of a calculating machine). If we begin with a recursively enumerable collection of axioms then it is possible to generate in a well-defined way the provable sentences of the theory, so if we wish to determine whether a given sentence is *provable* then we simply have to wait for it to be generated by the algorithm (in practice, however, we could be waiting for a few billion years before this happens). If the sentence is not provable then we would be waiting indefinitely, never sure if the sentence is unprovable or if we just need to hold on for a few more years for it to appear on the list. This in itself isn't too much of a hurdle, one can instruct the machine to terminate if it generates the sentence *or its negation*. Hilbert's dream was to be able to feed any symbolically expressed conjecture into

---

[1] Zermelo, E., 'Untersuchungen über die Grundlagen der Mengenlehre I', *Mathematische Annalen*, **65**, 261–281, (1908).

a machine and have it reply 'provable' or 'refutable'. Unfortunately this assumes that all sentences expressible in the underlying language are either provable or disprovable. If the machine were set up to identify provable and refutable sentences in ZF and we were to give this hypothetical machine a ZF-undecidable statement such as the Continuum Hypothesis, it would fail to terminate, and so its workings could not properly be described as an algorithm. The precise notion of an algorithm was independently clarified by Alonzo Church and Alan Turing in 1936/1937, and their work, equivalent to the work of Kurt Gödel, made it clear that Hilbert's problem had no solution.

Hilbert had asked a related question (Hilbert's tenth problem), at first sight less ambitious than the full *Entscheidungsproblem*, but nevertheless far reaching: *Is there an algorithm which decides whether a given Diophantine equation (i.e. a polynomial equation $p(x_1, \ldots, x_n) = 0$ with integer coefficients) has a solution in integers?* It turns out that it suffices to consider natural number solutions, so we shall make this assumption below. This question also has a negative answer, as was proved by Yuri Matiyasevich in 1970, following work of Julia Robinson, Martin Davis and Hilary Putnam. Although this might initially look as if it is an isolated result in a small corner of mathematics, in fact it runs much deeper.

Suppose $\mathcal{R}$ is a set of n-tuples of natural numbers and there exists an algorithm which, when given an arbitrary n-tuple of natural numbers, will eventually halt if that n tuple is in $\mathcal{R}$. We call such sets $\mathcal{R}$ *recursively enumerable*. A stronger condition is that there exists an algorithm which tells us whether or not an input is in $\mathcal{R}$, that is, both $\mathcal{R}$ and the complement of $\mathcal{R}$ in the set of all n-tuples are recursively enumerable. Such sets are called *computable*. Importantly: *there exist recursively enumerable sets which are not computable*. For such a set $\mathcal{R}$ there exists an algorithm which will halt whenever the input is in $\mathcal{R}$ but any such algorithm will run indefinitely for at least one input that is not in $\mathcal{R}$.

The set of n-tuples $(x_1, \ldots, x_n)$ satisfying a Diophantine equation

$$f(x_1, \ldots, x_n, y_1, \ldots, y_m) = 0$$

is obviously recursively enumerable (given an arbitrary n-tuple $(x_1, \ldots, x_n)$, we can list the $n + m$-tuples $(x_1, \ldots, x_n, y_1, \ldots, y_m)$ and run through the list, simply checking in each case if $f(x_1, \ldots, x_n, y_1, \ldots, y_m) = 0$). Any such set of n-tuples is called a *Diophantine set*.[1] Matiyasevich's Theorem is the converse: *every recursively enumerable set is Diophantine*. Putting all of this together, the negative answer to Hilbert's tenth problem follows: if there was an algorithm which decided the solvability of any given Diophantine equation then every Diophantine set would be computable, equivalently every recursively enumerable set would be computable, contradicting Matiyasevich's

---

[1] Considering equations of the form $(x - a_1)(x - a_2) \cdots (x - a_n) = 0$ we see that every finite subset of $\mathbb{Z}$ is Diophantine, and by a trivial extension every finite subset of $\mathbb{Z}^n$ is Diophantine. The equation $x - my - n = 0$ tells us that the set of integers congruent to n modulo m is Diophantine. The reader is left to contemplate more complicated examples.

Theorem. This is good news for number theorists interested in Diophantine equations; the class of all Diophantine equations supplies an unlimited source of problems ready to inspire ever deeper mathematical ideas.

Now take, for example, Goldbach's conjecture. If $G(n)$ is the statement '$2n + 4$ is expressible as a sum of two primes', then the Goldbach conjecture is that $G(n)$ is true for all $n = 0, 1, 2, 3, \ldots$. But the set F of n for which $G(n)$ is *false* is recursively enumerable (in fact it is computable) since for any given n we need only run through the finite number of pairs $(a, b) \in \mathbb{N}^2$ with $a + b = 2n + 4$ and determine in each case whether a and b are both prime; as soon as we encounter a pair with this property we declare $G(n)$ to be true and if we run through all such pairs without encountering a pair of primes we declare $G(n)$ to be false. By Matiyasevich's Theorem F is Diophantine. In other words, there exists a Diophantine equation which has a solution if and only if the Goldbach conjecture is false.

More generally a formula $\phi(x_1, \ldots, x_m)$ in PA is called a Π_0-*formula* if all the quantifiers in ϕ are bounded, that is, the quantifiers appearing in ϕ are of the form $\forall x \leq N$ or $\exists x \leq N$,[1] where N is either a numeral or a variable distinct from the quantified variable x. This means that for fixed numerals $n_1, \ldots, n_m$ the truth value of the sentence $\phi(n_1, \ldots, n_m)$ can be determined by an algorithm. A Π_1-*formula* is a formula which is equivalent to a formula of the form $(\forall x_1)(\forall x_2) \ldots (\forall x_m)\phi$ where ϕ is a Π_0-formula.[2] A Π_1-sentence is also sometimes called a sentence of *Goldbach type*. By Matiyasevich's Theorem, if G is a sentence of Goldbach type, then there exists a Diophantine equation which is solvable if and only if G is false. Many familiar conjectures in mathematics can be shown to be equivalent to Goldbach type sentences. Indeed, the Riemann Hypothesis is such a sentence, i.e. one can find a Goldbach type sentence which is equivalent, say in ZF with the usual interpretation of PA, to the Riemann Hypothesis. This means there is a sense in which one can ask whether a sophisticated proposition in analysis such as the Riemann Hypothesis is 'provable in Peano Arithmetic', even though the language of Peano Arithmetic is incapable of describing the mathematical objects which appear in the original conjecture.

One last word on this subject. Since the prime numbers are recursively enumerable there must exist a polynomial f with the property that the set of all n such that $f(n, x_1, \ldots, x_m) = 0$ has a solution is precisely the set of

---

[1] $(\forall x \leq N)\psi$ is an abbreviation for $(\forall x)[x \leq N \to \psi]$ and $(\exists x \leq N)\psi$ is an abbreviation for $(\exists x)[x \leq N \wedge \psi]$. Here x must be free in ψ.

[2] This classification of formulas forms the initial part of the *Kleene–Mostowski arithmetical hierarchy* which recursively describes a collection of ever more complex formulas (and associated classes of natural numbers). A formula equivalent to a formula of the form $(\exists x_1)(\exists x_2) \ldots (\exists x_m)\phi$ where ϕ is a Π_0-formula is called a Σ_1-formula, and in general the Π_n- and Σ_n-formulas are defined recursively in terms of their simpler siblings, so that the Π_{n+1}-formulas are equivalent to Σ_n-formulas preceded by a block of universal quantifiers and the Σ_{n+1}-formulas are equivalent to Π_n-formulas preceded by a block of existential quantifiers. The intersection $\Pi_n \cap \Sigma_n$ is usually denoted by Δ_n (and we set $\Pi_0 = \Sigma_0 = \Delta_0$). Sometimes there is also a superscript, usually omitted in this context, which indicates the type of objects being quantified over, so here we have Π_n^0-formulas, Σ_n^0-formulas and Δ_n^0-formulas.

primes. In fact one can go further and find a polynomial with the property that the set of all positive integer values of the polynomial is the set of primes. Several explicit polynomials have been found with this property. One example found in 1976 has 26 variables and degree 25.[1] Other explicit examples have fewer variables but larger degree.

2. If the *Entscheidungsproblem* had turned out to have a solution then it might be pessimistic to say that it would put an end to foundational research. It isn't too much of a stretch to imagine, perhaps modulo developments in quantum computing, that for all but the most trivial of sentences the algorithm would take a long time (many years or possibly centuries) to terminate. Putting aside time and the question of insufficient memory, waiting to be told that the Riemann hypothesis is 'true' is not entirely satisfactory. We don't want a terse 'true/false' response from an oracle, we want to know *why* something is true or false, or at least to be given some insight. The next step would be to ask the computer to output the proof, but the length and nature of such a proof is unlikely to be of any use to human mathematicians. We should keep in mind that some first-order theories *are* decidable in this sense, however, the existence of a decision algorithm doesn't dissuade us from working in these theories in the 'old fashioned' way nor does it make them any less interesting.

It took years for Zermelo's original axioms to be modified, augmented and distilled into the axioms of ZF that we are about to describe. The development ran in parallel with Hilbert's highly ambitious formalist programme to find a master algorithm that would in principle decide the provability of any symbolically expressed mathematical sentence that it could be fed. This programme failed for fundamental logical reasons, but set theory and the axiomatic method continued to grow in strength.

2.4   The axiomatic method

> *There is nothing mysterious, as some have tried to maintain, about the applicability of mathematics. What we get by abstraction from something can be returned.*
>
> – R. L. WILDER[2]

2.4.1   Abstraction

In those branches of mathematics which, historically, were distilled from a huge number of examples, the axioms assume the role of a humble definition. We

---

[1] J. P. Jones, D. Sato, H. Wada and D. Wiens, 'Diophantine representation of the set of prime numbers', *Amer. Math. Monthly*, **83**, 449–464, (1976).
[2] Wilder [**227**].

might observe in the course of research that a multitude of seemingly different structures satisfy the same algebraic or topological properties; it then becomes a matter of interest to explore the extent to which these common attributes characterize the structures from which they have been abstracted. One studies in isolation, without the distraction of ad-hoc interpretation, the class of all objects satisfying a given set of axioms.

A classical example of this approach is the theory of groups. A group is defined abstractly as a set G, together with an associative operation

$$\circ : G \times G \to G$$

which possesses an identity element e (i.e. $e \circ x = x = x \circ e$ for all $x \in G$) and such that every $x \in G$ has an inverse (an element $y \in G$ such that $x \circ y = y \circ x = e$). This abstract definition represents a synthesis of many different strands of mathematics, the history of which requires a book-length account to do it justice. Its basic ideas, with hindsight, can be traced back to three major contributory strands which were woven together in the eighteenth and nineteenth centuries: number theory, the quest for general solutions of polynomial equations, and the investigation of symmetry in geometry. The description of a group in purely logical terms, as a first-order theory, is naturally the last stage in this process of abstraction, but it is not one we need to pursue here.

Group theory provides a purely mathematical means of describing symmetry in its greatest generality. Abstractly, if an object D is embedded in some ambient space A then a symmetry of D is a bijective map $\phi : A \to A$ for which D is mapped onto itself in the sense that $D = \{\phi(x) : x \in D\}$. The collection of all symmetries of D then forms a group with respect to composition. In practice one would normally restrict attention to certain types of map, dictated by the mathematical or physical context one is working in, so, for example, we might look at maps which preserve certain metric or topological qualities. Classically a symmetry is a distance preserving bijection (an *isometry*) of Euclidean space, i.e. a composition of rotations, reflections and translations, which maps the object of interest onto itself. In applications to the real world, only the physically realizable isometries of the object would be included, so, for example, all reflections would be excluded. It is a short leap from this to consider maps which preserve alternative 'non-Euclidean' metrics.

Let us give a concrete example. Consider the set of isometries of three-dimensional space which map a cube onto itself. To ensure our transformations are physically realizable we exclude reflections (which reverse orientation), leaving us only with rotations.[1] There are thirteen axes of rotational symmetry in a cube: three axes passing through the centres of opposite faces; six axes passing through the centre of opposite edges; and four axes passing through

---

[1] Of course, if we were being very precise about this being physically realizable we would have to stress that it is the cube doing the rotating and not the ambient spatial framework in which it is embedded, assuming such a framework is understood to be made of elements which conform to the laws of physics (one cannot rigidly rotate an object if it has infinite extent in any direction away from the axis of rotation because we would have to exceed light speed beyond some cylinder centred on the axis)!

opposite corners (see Figure 2.2). Rotating space through an integer multiple of 90°, 180° and 120° about these axes respectively will map the cube onto itself and hence is one of our admissible symmetries. (The 120° rotations are often forgotten due to the lazy habit we have of sketching cubes on paper as a square with two adjoining rhombuses, approximating neighbouring faces in perspective, concealing the alternative hexagonal-silhouetted corner-centred viewpoint.)

The four corner to corner axes will be permuted by any admissible symmetry and all 24 permutations of these axes are achievable by composing the rotations described above (indeed, only a small set of rotations is required to generate the whole set of symmetries). Thus the group of orientation preserving symmetries of the cube is the same group as the group of all permutations of four objects. We are fortunate in this example to have isolated a set (the corner to corner axes) on which the group acts as the full group of permutations. This is atypical.

Another example of a group is $\mathbb{Z}$ with respect to addition, which we can again view geometrically as left or right translations of a line by some multiple of a fixed unit. Other familiar examples include the non-zero rational numbers (or indeed the non-zero elements of any field) with respect to the operation of multiplication, or, as we have just seen, the set of all permutations of a set of objects with respect to composition.

The last example is important: *every* group can be realized as a set of permutations (not necessarily *all* permutations) of a collection of objects. This is known as Cayley's Theorem. The proof is easy: if G is a group with group operation $\star$ and $a \in G$ then let $f_a : G \to G$ be defined by $f_a(x) = a \star x$. f_a is injective because $f_a(x) = f_a(y)$ implies $a \star x = a \star y$, and since a has an inverse (a^{-1}) we have $x = a^{-1} \star a \star x = a^{-1} \star a \star y = y$. f_a is surjective because for all $y \in G$, $y = f_a(a^{-1} \star x)$. Thus f_a is a bijection, i.e. a permutation of G. Next observe that the composition $f_a \circ f_b$ is $f_{a \star b}$. So the group G can be viewed via the homomorphic embedding $a \mapsto f_a$ as a set of permutations of the set G with respect to composition.

In contrast to the example-motivated axioms of group theory, set theoretic axioms seem to have a more experimental nature; they capture some notions that we intuitively want sets to have and we hope that no un-set-like structures creep through the filter, possibly bringing paradoxes in their wake. The structure we try to describe via set theoretic axioms is at the outset cloaked in darkness while the prototypical examples of groups are easily comprehended. The amount of 'elbow room' allowed by, say, the axioms of ZF is surprising when first encountered; ZF gives rise to a rich variety of models just as the axioms of group theory determine a huge array of different groups.

REMARKS

1. The first-order theory of groups has one constant, 1, one unary operator (inverse), where the inverse of x is denoted x^{-1} and one binary operator $\circ$. The only relation is the equality relation which is subject to the usual axioms of identity (see axioms (E1)–(E4) in Appendix A). There are three non-logical

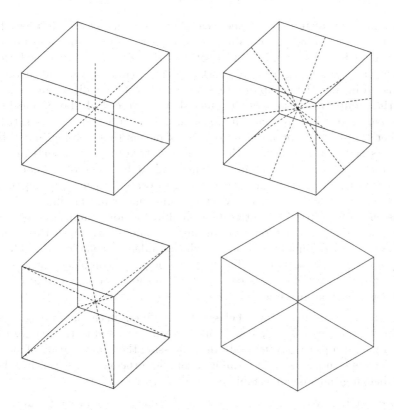

Figure 2.2   The cube and its thirteen axes of rotational symmetry. The top left diagram shows the three axes passing through the centre of opposite faces; the top right diagram shows the six axes passing through the centre of opposite edges; and the bottom left diagram shows the four diagonal axes passing through opposite corners. In the bottom right diagram the cube has been rotated to illustrate the often forgotten rotational symmetry of the cube about the diagonal axes. Any rotation of the cube about one of these thirteen axes will permute the four diagonal axes, and any permutation of these four axes can be attained by a composition of rotations. Thus the rotational symmetry group of the cube is the group of all permutations of four objects.

axioms: (i) $(\forall x)[1 \circ x = x \wedge x \circ 1 = x]$; (ii) $(\forall x)[x^{-1} \circ x = 1 \wedge x \circ x^{-1} = 1]$; and (iii) $(\forall x)(\forall y)(\forall z)[(x \circ y) \circ z = x \circ (y \circ z)]$. Fewer axioms are possible at the expense of a loss of clarity.

Consider the following sentence:

$$(\forall x)(\forall y)[x \circ y = y \circ x].$$

If we were completely new to the idea of groups we might ask ourselves if this sentence is provable or refutable. It is neither. We can find a group (that is, a model of the axioms) which satisfies the sentence (a so-called *abelian group*) and another group for which the sentence is false. The sentence is satisfied in, say, the group of integers with respect to addition and it is false in the group of rotations of the cube that we described above, and indeed one can find infinitely many non-isomorphic examples of groups satisfying either the sentence or its negation. This means that neither the sentence nor its negation is provable; the sentence is true in some models and false in others. So we see that unprovability isn't quite the esoteric notion it is sometimes painted as; our surprise at unprovability generally depends on how many models of the theory we can imagine or are familiar with. The existence of unprovable sentences in arithmetic, for example, is surprising since we generally have only one intuitive model of it. With this example in mind, and the many theories to which Gödel's Incompleteness Theorem applies, we begin to realize that it is the negation complete theories (such as Tarski's theory of real closed fields) which should be regarded as oddities, not the negation incomplete theories.

The theory of groups is nevertheless rather different in character to the theory of arithmetic described by PA. Suppose we were to add to the axioms of the theory of groups the sentence above, expressing the commutation of any pair of elements, in the hope of forming a complete theory. The resulting theory of abelian groups is not complete.

Since abelian groups are most often represented as additive groups let us now denote the group operation by $+$ instead of $\circ$ and the identity element by 0 instead of 1. A trivial example of a sentence which is undecidable in the theory of abelian groups (and indeed in the general theory of groups) is $(\forall x)\neg[x = 0]$: it is false in the trivial group $\{0\}$ and true in all other groups. If we were to adopt its negation we would obtain a complete theory, but only in the most uninteresting way: up to isomorphism this theory has only one model, the group containing just one element. In chasing completeness we would rather end up with a theory that has a wealth of non-isomorphic models.

Adopting this sentence as a new axiom we have the still incomplete theory of non-trivial abelian groups. Here we might consider the notion of divisibility. A group is *divisible* if for any $n \in \{1, 2, 3, \ldots\}$ and any element x of the group there exists a y such that x is the sum $y + y + \cdots + y$, where the y appears n times (it is convenient to abbreviate the latter sum by the symbol ny). In more informal terms, a group is divisible if every element can be split into n equal parts for any given n. As the natural numbers do not form part of the language of groups, in order to capture the notion of divisibility we have to add an axiom schema, indexed by positive natural numbers: D_n is the axiom $(\forall x)(\exists y)[x = ny]$. Perhaps the most familiar example of a divisible group is the group of rational numbers with respect to addition. The group of integers with respect to addition is *not* a divisible group. Adding this countable collection of axioms $\{D_n\}$ to our list we obtain the theory of non-trivial abelian divisible groups. We are still not there yet. This theory is again incomplete.

A group is said to be a *torsion group* if for every element x, $x \circ x \circ \cdots \circ x = 1$ for some number of occurrences of x. Thus, adopting the additive notation, an abelian group is a torsion group if for every x there exists an $n \geq 1$ such that $nx = 0$. An abelian group is torsion-free if for every $x \neq 0$ and $n \geq 0$, $nx \neq 0$. Expressing the property of torsion-freeness in first-order terms we again need an axiom schema: T_n is the axiom $(\forall x)[\neg x = 0 \rightarrow \neg nx = 0]$ (or equivalently $(\forall x)[nx = 0 \rightarrow x = 0]$). An example of a torsion-free non-trivial divisible abelian group is again $\mathbb{Q}$. An example of a non-trivial divisible abelian torsion group is $\mathbb{Q}/\mathbb{Z}$, meaning the group comprising the equivalence classes of rational numbers where two rational numbers are equivalent if they differ by an integer; it is a torsion group because, for $b \geq 1$, if one adds $\langle \frac{a}{b} \rangle$ (i.e. the equivalence class $\frac{a}{b} + \mathbb{Z} = \{n + \frac{a}{b} : n \in \mathbb{Z}\}$) to itself b times one obtains the zero element $\langle 0 \rangle$. By adopting this new countable collection of axioms $\{T_n\}$ we finally have a complete theory. The theory of non-trivial divisible torsion-free abelian groups is negation complete. It has infinitely many non-isomorphic models (in fact for each cardinal number $\alpha > \aleph_0$, there is, up to isomorphism, exactly one divisible torsion-free abelian group of cardinality α).

Thus, beginning with the basic theory of groups, we can add a recursively enumerable collection of new axioms which results in a non-trivial complete theory. In contrast, by Gödel's Incompleteness Theorem, no consistent recursive extension of PA can be complete. So in a sense a theory to which Gödel's theorem applies is 'much more incomplete' than the kind of theory, like the theory of groups, which is completable in this way. Group theory is 'completable' in many different effective ways whereas PA cannot be completed at all without stepping over the boundary into an ineffective theory.

2. The fact that any given group is isomorphic to a group of permutations does not make the theory of groups any easier to study (that is, it does not reduce it to a tractable case). Even the theory of finite groups is far from being fully understood. The most famous result in this direction is the classification of all finite *simple* groups,[1] finally announced in 2004, which represents half a

---

[1]A simple group is irreducible in the technical sense that its only normal subgroups are itself and the trivial group comprising the identity element alone. (The normal subgroups of a group G are precisely the kernels of group homomorphisms defined on G, i.e. the subgroups of the form $\{x \in G : \phi(x) = 0\}$, where $\phi : G \rightarrow H$ is a homomorphism from G to a group H.)

century of work and has a proof which currently spans tens of thousands of pages. Note that the properties of being finite and of being simple are not expressible in the first-order theory of groups.

Mathematical abstractions are usually drawn from large collections of concrete examples. For instance, the axioms of group theory are an algebraic characterization of the symmetry properties of a huge number of well understood examples, the notion of a topological space is abstracted from the properties of concrete metric spaces, the notion of a ring is abstracted from the algebraic properties of a familiar class of number systems, and so on. The axiomatic development of set theory is different in that there didn't exist a wealth of set universes from which to abstract the axioms, only an intuitive notion of what a set ought to be. The oldest example of an axiomatic system is Euclidean geometry and one of the first examples of the dramatic power of abstraction was the discovery of its non-Euclidean relatives.

2.4.2 Geometry

To Euclid and his contemporaries the axioms of geometry were self-evident truths. However, from the outset it appeared that one of these assumptions, the parallel postulate (see Figure 2.3), was less 'primitive' than the other postulates; it seemed that it might be possible to prove this statement from the other axioms.[1] For centuries mathematicians tried to find a proof, but none succeeded. By making use of further geometric properties which they regarded as obvious, their efforts merely created a catalogue of equivalent statements. For example, assuming the other postulates of Euclidean geometry hold, the parallel postulate is equivalent to each of the following (in decreasing order of intuitive immediacy):

(i) there is no upper limit to the area of a triangle;

(ii) every triangle can be circumscribed;

(iii) the sum of the interior angles of any triangle is 180°; and

(iv) Pythagoras' Theorem.

Over two thousand years would pass before the difficulties were fully understood. Nikolai Lobachevsky (1829) and Janos Bolyai (1832) independently, and using very different methods, found that no contradiction was generated by assuming the negation of the parallel postulate, i.e. by assuming that more than one parallel might pass through a point. By doing this they had shown

---

[1] For completeness we ought to mention that the other four postulates are, in Euclid's constructive language: 1. To draw a straight line from any point to any point. 2. To produce a finite straight line continuously in a straight line. 3. To describe a circle with any centre and distance. 4. All right angles are equal to one another.

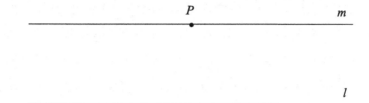

Figure 2.3 In modern accounts of Euclidean geometry Euclid's parallel postulate is usually replaced by Playfair's Axiom: if l is a line and P is a point not on l, then there exists a unique line m passing through P such that l and m are parallel. Euclid's original postulate, which translates as follows, is not quite as transparent as Playfair's: '*If a straight line falls on two straight lines in such a manner that the interior angles on the same side are together less than two right angles, then the straight lines, if produced indefinitely, meet on that side on which are the angles less than the two right angles.*' Taken alone, Playfair's axiom is stronger than the parallel postulate: it is equivalent to the combination of the parallel postulate and the assumption that parallel lines exist (John Playfair (c. 1795) credited others for this result). However, the existence of parallel lines follows from the other four Euclidean postulates (the construction of a line parallel to a given line passing through a given point is described in Proposition 31 of Book One of the *Elements*). Thus the geometry described by Euclid's first four postulates and Playfair's axiom is identical to that of Euclidean geometry.

that the parallel postulate was *unprovable*. This was a dramatic discovery: Euclidean geometry is not the 'one true geometry' – there are alternative geometric universes awaiting exploration.[1]

Non-Euclidean geometries can be modelled using Euclidean elements, but the newly embedded features will, of course, be quite distorted, certain curves or line segments playing the role of lines and possibly bounded regions playing the role of the plane, for example.

There are two varieties of non-Euclidean geometry: hyperbolic and elliptic. Hyperbolic geometry, discovered by Lobachevsky and Bolyai, allows many parallels passing through a point and has models such as the Klein model and the Poincaré disc and half-plane models (described below). These models establish the relative consistency of hyperbolic geometry with Euclidean geometry, i.e. if Euclidean geometry is consistent, then so is hyperbolic geometry. Elliptic geometry, which does not allow any parallels, arose from projective geometry and was further developed later in the far-reaching work of Riemann.

---

[1]Lobachevsky's work appeared in his *Geometriya* (not published in full until 1909). Bolyai's results were published in an appendix to a textbook written by his father Farkas Bolyai who had originally advised his increasingly obsessive son: '*For God's sake, I beseech you, give it up. Fear it no less than sensual passions because it too may take all your time and deprive you of your health, peace of mind and happiness in life*'. Gauss claimed that he had discovered the result earlier, but didn't publish it.

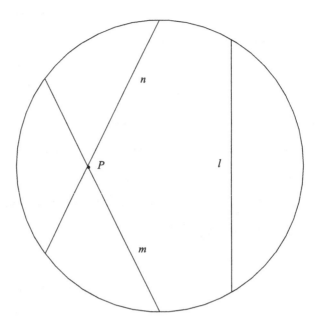

Figure 2.4 The Klein model of Bolyai–Lobachevski (hyperbolic) geometry. By defining 'point' to be a point in the interior of the disc and 'line' to be a chord of the circle forming the boundary of the disc, all of Euclid's axioms can be shown to hold except the parallel postulate. In the diagram a point P is shown not on the line l, yet two distinct lines n and m pass through P, both 'parallel' to l in the sense that they do not intersect l.

In the Klein model of hyperbolic geometry[1] the primitive notions of plane, point, and line are modelled by the interior of a disc, a point in this disc, and an open chord of the bounding circle respectively (see Figure 2.4).

In the Poincaré disc model of the hyperbolic plane (see Figure 2.5) the 'points' are interior points of a disc and the 'lines' are diameters or arcs of circles which meet the disc boundary orthogonally (this model inspired M. C. Escher's famous *Circle Limit* images).

The Poincaré half-plane model is conformally equivalent to the disc model (see Figure 2.6). In this case the 'points' are points in the upper half plane (i.e. the set of complex numbers with positive imaginary parts) and the 'lines' are semicircles centred on the real axis together with vertical straight lines.

A model of elliptic geometry is given by a sphere where *point* is interpreted as a pair of antipodal points on the sphere and *line* is a great circle on the sphere (see Figure 2.7).

---

[1] Models of hyperbolic geometry are proposed in Eugenio Beltrami's 'Teoria fondamentale degli spazii di curvatura costante', *Annali. di Mat.* ser II, **2**, 232–255, (1868). Arthur Cayley in 'A Sixth Memoire upon Quantics', *Philosophical Transactions of the Royal Society of the Royal Society of London*, **159**, 61–91, (1859) introduces a certain distance function which is given a hyperbolic geometric interpretation in Felix Klein's 'Über die sogenannte Nicht-Euklidische Geometrie', *Mathematische Annalen*, **4**, 573–625, (1871).

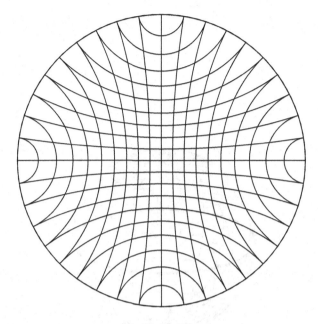

Figure 2.5 In the Poincaré disc model of the hyperbolic plane the points are the interior points of the disc and the lines are the diameters together with circular arcs which meet the boundary of the disc orthogonally. Here we show an aesthetically pleasing arrangement of circular arcs and diameters representing a collection of lines in hyperbolic space.

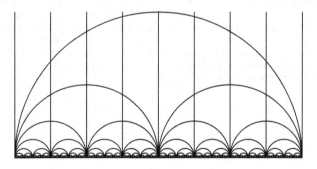

Figure 2.6 In the Poincaré half plane model of the hyperbolic plane the points are the points of the upper half plane and the lines are the vertical lines together with semicircles centred on the real axis.

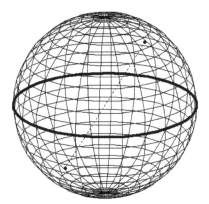

Figure 2.7   A model of elliptic geometry. Each pair of antipodal points on a sphere represent a single point and the great circles on the sphere represent lines. The diagram shows the equator and a pair of antipodal points (joined by the dashed line through the centre of the sphere) not on the equator. One can see from this that in this model no 'line' passing through the 'point' is parallel to the line represented by the equator: any great circle passing through the two points intersects the equator. This model is sometimes called the *real projective plane* as it provides a concrete way of visualizing the projective geometry which extends the real affine plane by associating with each line and all lines parallel to it a single idealized point at infinity through which the parallel lines now pass (these points at infinity collectively comprise the 'line at infinity'). The correspondence is set up by simply imagining the sphere to be sitting on the plane and projecting the various affine elements through the centre of the sphere onto its surface. (Via this correspondence the affine lines correspond to great circles on the sphere which are not equal to the equator, the affine points are antipodal pairs of points not on the equator, the points at infinity are antipodal points on the equator and the line at infinity corresponds to the equator itself.)

At first non-Euclidean geometries were seen as esoteric novelties, however, they later came to have a more tangible significance in physics. Euclidean geometry generally provides an accurate model of regions of space, but in general, models must take into account relativistic space-bending phenomena. Geometries exist which do not reflect our common perception of local regions of space, but which are nevertheless equiconsistent with the more familiar Euclidean model as well as being an accurate framework for physical space at large.

The monumental leap of imagination that gave birth to non-Euclidean geometry is easy to take for granted now that the models have been with us for so long. In modelling these geometries it took an impressive perspective shift to dare to represent the lines and points of an axiomatic geometry as anything other than the objects that inspired them. Models still have the power to surprise those who insist on tying themselves to the original intuitive inspiration of a set of axioms (one need look no further than the more exotic models of ZF or of PA). Exactly how surprisingly varied models of first-order theories can be will be revealed in Chapter 10.

The discovery of non-Euclidean geometries made mathematicians and logicians look harder at Euclidean geometry. Euclid's work was not perfect; we must not kneel before the *Elements* with blind unquestioning adoration. By modern standards of rigour there *are* flaws, but factoring in the age of the work these flaws are remarkably subtle and few in number (and most importantly they are fillable gaps, not contradictions); the work as a whole is magnificently strong even after nearly two and a half thousand years of close scrutiny.[1] Euclid was guilty of making a number of tacit assumptions – for example Proposition 1 in Book I, which describes the construction of an equilateral triangle on a given line segment, assumes that two circles passing through the centre of one another intersect in two points – something which is clearly intuitively true but which is nevertheless unprovable using Euclid's axioms (see Figure 2.8). Some other inferences also appeal to visual intuition. In addition Euclid fails to give a proper definition of several concepts, including 'betweenness'. However, it is a little unfair to judge the *Elements* by these modern formalist standards.

Euclid, as far as we can infer, intended to deduce all the geometry known to him as a logical consequence of a small collection of unproved statements. From a modern perspective he treated point and line as primitive notions. Accounts of geometry which attempt to completely banish the prejudices and faults of visual intuition (one of the first of which was published by Moritz Pasch (1882))[2] treat the subject as a type of logic.

In Hilbert's modern axiomatization of Euclidean geometry there are six primitive notions: point, line, plane, incidence, betweenness and congruence. As is well-known, a model of Euclidean (plane) geometry, as precisely reformulated by Hilbert's axioms, is given by interpreting a point as an element of $\mathbb{R}^2$ and a line as a class $\{(x, y) : ax + by + c = 0\}$ for some fixed $a, b, c \in \mathbb{R}$, a and

---

[1]For instant Euclidean gratification I recommend highly the single volume edition of Euclid [**65**], edited by D. Densmore and based on the famous translation of T. L. Heath.
[2]Moritz Pasch, *Vorlesungen über Neue Geometrie*. Berlin, Springer (1882).

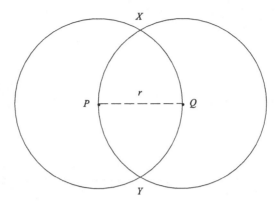

Figure 2.8 Consider two circles of equal radius r centred at P and Q, where P and Q are distance r apart. Intuitively, it is obvious that the circles intersect in two points X and Y. Indeed, this was so abundantly clear to Euclid that it did not occur to him to try to prove it or use it as an axiom. In fact it *cannot* be proved from Euclid's axioms, for there is a model which satisfies all of Euclid's original axioms yet in which two such circles do not intersect (for example, in the model in which points are elements of $\mathbb{Q}^2$ and lines are sets of the form $\{(x,y) : ax + by = 0,\ a,b,x,y \in \mathbb{Q},\ a,b \text{ not both zero}\}$, we can find two circles of radius one, centres distance one apart, which have empty intersection: $\{(x,y) \in \mathbb{Q}^2 : x^2 + y^2 = 1\} \cap \{(x,y) \in \mathbb{Q}^2 : (x-1)^2 + y^2 = 1\} = \varnothing)$. A demonstration of the mutual independence of the axioms of Hilbert's revised axiomatization of Euclidean geometry (an axiomatization which remedies this problem) may be found in Chapter II of Hilbert [**100**].

b not both zero. Other formal geometries use different primitive notions and different axioms. Mario Pieri, for example, considered a geometry where point and motion are the only primitive elements.[1]

Hilbert's axiomatization of Euclidean geometry included second-order axioms of continuity and made use of a relatively large number of primitive elements. By contrast Alfred Tarski developed an elegant first-order theory of geometry in the 1930s and 40s using just the primitive elements of point, congruence and betweenness.[2] Related to this was Tarski's work on the first-order theory of real closed fields. In place of the unwieldy schema of completeness, this theory adds to the axioms of an ordered field F the conditions:[3]

(i) every positive element of F is the square of an element of F; and

(ii) the schema: if f is a polynomial of odd degree with coefficients in F then f has a root in F.

Tarski proved that this theory is complete and, since it is also an effective theory, it is decidable, meaning that there is an algorithm which determines whether any given sentence in the underlying language is true or false. From this he

---

[1] Pieri, M. *Della geometria elementare come sistema ipotetico deduttivo*. C. Clausen (1899). For an introduction to axiomatizations of geometry see, for example, Martin [**146**].

[2] Rather than citing any individual papers I will simply point to Tarski [**211**].

[3] Naturally there are alternative axiomatizations of real closed fields. For example, an ordered field F is real closed if and only if whenever $p(x)$ is a polynomial with coefficients in F, $a < b$ in F and $p(a)p(b) < 0$, there is a $c \in F$ such that $a < c < b$ and $p(c) = 0$.

was able to deduce the closely related result that his Euclidean geometry too is complete and decidable (if R is a real closed field then R^2 is a model of the Tarski–Euclidean plane).

Gödel's Incompleteness Theorems do not apply to Tarski's Euclidean geometry or to the first-order theory of real closed fields because the requisite 'certain amount of arithmetic' is not present in either theory. We cannot define in the first-order theories of real closed fields or of Tarski's geometry the set of natural numbers: the unsurpassable hurdle is the absence of an induction principle. The definition 'the smallest subset N of $\mathbb{R}$ containing 0 and such that $n + 1 \in N$ whenever $n \in N$', for example, cannot be expressed in the underlying language of Tarski's theory.

REMARKS

1. While Lobachevsky simply negated the parallel postulate, Bolyai's approach was a little more elaborate, describing possible geometries depending on a parameter. The first concrete models of hyperbolic geometry did not appear until Beltrami's constructions nearly forty years after Bolyai and Lobachevsky's work.

2. Tarski's theory of real closed fields is a first-order theory, so besides the motivating example of $\mathbb{R}$ it also has (infinitely many) other models. We ought to give some examples. The first is easy to describe: the real algebraic numbers $\mathbb{A} \cap \mathbb{R}$. Another example is the field of computable numbers, being the set of real numbers which are approximable to any desired precision by a finite algorithm. A *definable set* in ZF is a set a for which there exists a 1-predicate $\phi(x)$ such that $a = \{x : \phi(x)\}$. The definable numbers, the set of real numbers a which are definable in this sense, is another model of Tarski's theory. Any hyperreal field $^*\mathbb{R}$ is a model, and so on. For any of these 'non-standard' models of real closed fields we form 'non-standard' models of Tarski's Euclidean geometry. The fact that the theory of real closed fields is negation complete is rather extraordinary. It means that, in order to prove or refute any given statement, one only has to show it is true or false in any one given model, and it will then be automatically be true or false respectively in all other models. Another example of a negation complete theory is, for some fixed p (prime or zero), the theory of algebraically closed fields of characteristic p. This means that to prove a first-order theorem about algebraically closed fields of characteristic 0 one need only prove it in the model $\mathbb{C}$, where we have all the wonderful tools of complex analysis at our disposal.

3. One can develop geometric theories which have finite models. Define an *affine plane geometry* to be a set (playing the role of the set of points) together with a collection of subsets of points (playing the role of the lines) satisfying the following properties: (i) for every pair of distinct points there is exactly one line containing both points; (ii) the obvious analogy of Playfair's Axiom; and

(iii) (to avoid trivialities) we insist that there exists a set of four points such that no three of them belong to the same line.

The definition of a *projective plane geometry* replaces (ii) with the condition that the intersection of two distinct lines contains exactly one point.

In both geometries we can take the set of points to be finite. Some difficult questions remain. A *finite plane* is a geometry (affine or projective) in which every line has the same number of points (n or $n + 1$, respectively). It is an open problem to prove that n must be a prime power.

One of the great feats of imagination of the last two hundred years was the construction of non-Euclidean geometries. This was achieved only by an act of abstraction that broke the shackles binding the intuitive interpretation of a notion and the object that models it. This was the primitive beginnings of model theory; a glimpse of a mathematical multiverse that sowed the seeds of Formalism.

2.4.3   Formalism

In the modern view of the axiomatic method, where the fanciful notion of 'absolute truth' has long been discarded, the primitive notions have been assigned suggestive labels but are treated as arbitrary objects to be manipulated by the axioms and rules of inference. These axioms form an initial stock of theorems and hopefully capture enough of the intuitive notions being modelled and imbue the primitive objects with sufficient properties to be of some use. It should be added that aesthetic considerations play a much more prominent role in the development of mathematics than most non-mathematicians realize; an elegant proof often informs the researcher where to look for the 'correct' path and where to find the most interesting results.

At this early stage, having not encountered any non-logical axioms, we need to exercise a bit of healthy detachment. The strings of symbols which form well-formed formulas may as well be treated as hieroglyphics or syntactical follies with no relationship to the outside world at all.

Despite the logical necessity of a linear exposition, the process of *understanding* seems a decidedly nonlinear process. One gets a fair grip of a notion before leaping ahead a few steps to sample some of the new things to come. While familiarizing ourselves with the new concepts we simultaneously revise the old partially digested material. Occasionally this means looking at applications of a proposed notion in order to gauge whether it is a successful representation of the intuitive idea it has been designed to model. We then return to the formal definition enlightened, ready for a deeper analysis. What drives all good scientists and philosophers to distraction is that the deeper you look the less you feel you know, a mechanism which fuels addiction.

REMARK

The axiomatic method is *not* a model of the thought processes of mathematicians (or anyone else for that matter), and it is not intended to be. By whatever means the brain comes up with proofs, it certainly is not by generating all possible theorems from a base of assumptions and stopping when the target is met; for all but the simplest of problems and systems this technique is hopelessly slow and little insight is given. Instead one builds on intuitive foundations, by surprisingly vague analogies, sometimes hitting on the right idea. The actual process, how the leaps are made, is still fairly mysterious. It is only after the vague initial ideas are clear and after a sketch of a proof has been written that one can reverse engineer it to see precisely what assumptions have been or need to be made. This is the axiomatic method run backwards; the (hidden) assumptions appear at the end of the journey, not at the beginning. But the fact that the naive bottom-up generation from axioms is not the way mathematics is *done* does not in any way reduce the importance of the idea of an axiomatic system. At the very least a set of axioms, if we accept them, provides a precise lens enabling us to focus on *why* a statement is true.

Strict Formalists are content to examine the logical consequences of a set of axioms without regard to their intuitive interpretation, but a fragment of the Platonistic world remains – they think of the set of all consequences of the axioms as existing in some sense, even though this infinite class cannot be exhibited. We are now prepared to present the first of the non-logical axioms of ZF.

2.5 Equality, substitution and extensionality

Mathematicians do not study objects, but relations between objects. Thus, they are free to replace some objects by others so long as the relations remain unchanged. Content to them is irrelevant: they are interested in form only.

– Jules Henri Poincaré[1]

2.5.1 Equality and substitution

We have thus far survived without a formal definition of equality. In some theories equality is taken as a primitive binary relation '$=$' subject to the following axioms:

(E1) (Reflexivity) $(\forall a)[a = a]$.

(E2) (Symmetry) $(\forall a)(\forall b)[a = b \to b = a]$.

---
[1]Poincaré [**166**].

(E3) (Transitivity) $(\forall a)(\forall b)(\forall c)[a = b \wedge b = c \to a = c]$.

(E4) (Substitutivity) If $\phi(x)$ is a predicate then

$$(\forall a)(\forall b)[a = b \to [\phi(a) \leftrightarrow \phi(b)]].$$

The fourth axiom (substitutivity) is an axiom schema, each predicate giving rise to a different axiom, and is a precise statement of the intuitive principle that if a satisfies a property ϕ and if $a = b$ then b also satisfies ϕ.

We could present ZF in this way, however we prefer to introduce equality as a defined relation, making use of membership. We say that two sets are equal if an element is a member of one of the sets if and only if it is a member of the other, in other words we use $a = b$ as an abbreviation for the well-formed formula:

$$(\forall x)[x \in a \leftrightarrow x \in b].$$

Having made this definition, the axioms (E1), (E2) and (E3) easily follow from the logical axioms, however, we cannot establish the substitution property (E4) using the logical axioms alone; we need to introduce a non-logical axiom. Before we do that let us focus on the delicacy of substituting variables in a well-formed formula.

If $a = b$, then given any statement ϕ including a we might expect to obtain an equivalent statement if we simply replace each occurrence of a in ϕ by b. However, we must not be too hasty; if the variable b is bound in ϕ this straightforward substitution will lead to difficulties. A simple example is the predicate $\phi(a)$ characterizing the statement that a is non-empty: $(\exists b)[b \in a]$. If we were to form $\phi(b)$ simply by replacing every a by b we would obtain the false sentence $(\exists b)[b \in b]$ (this contradicts the Axiom of Regularity – see Section 4.5). Instead we must replace all bound occurrences of b by a variable c, $(\exists c)[c \in a]$, then substitute b in place of a, $(\exists c)[c \in b]$, yielding the desired predicate.

In general, by '$\phi(a)$ holds for b' we mean that the formula obtained via the following two steps holds.

(i) Replace all bound occurrences of b in ϕ by a variable that does not occur in ϕ.

(ii) Replace each occurrence of a by b.

The resulting formula is denoted $\phi(b)$ and the substitution property for equality is then expressible in the expected form: if $a = b$ then $\phi(a)$ if and only if $\phi(b)$. The substitution schema (E4) was introduced as a new axiom by Fraenkel as an addition to Zermelo's original set theory. We shall instead introduce another axiom from which the substitution schema follows.

REMARK

The substitution schema has a long history, being a formal expression of the philosophical principle of the *indiscernibility of identicals*: identical objects must have the same properties. The converse is the principle of the *identity of indiscernibles*, also called *Leibniz's Law*: if two objects have the same properties then they are identical. The latter is perhaps less self-evident than the former.

Equality is introduced as an abbreviation of a simple predicate. If $\phi(x)$ is a well-formed formula we need to carefully describe how to replace x by another variable y. We first replace all bound occurrences of y in ϕ by some new variable not occurring in the formula, then we replace all occurrences of x by y, yielding $\phi(y)$. The substitution schema, a corollary of the first axiom of ZF, Extensionality, then states that if $x = y$ then $\phi(x)$ if and only if $\phi(y)$.

2.5.2 Extensionality

One can prove that the substitution schema (E4) described in the previous subsection is a consequence of the definition of equality and the following weak version of extensionality.

Weak Extensionality

For all sets a, x and y, if $x = y$ and $x \in a$ then $y \in a$.

The proof of substitution is by induction on the number of well-formed subformulas of the predicate in question and is left to the interested reader as an exercise.

The definition of equality we gave in the previous subsection might be called *inner equality*. Within ZF we could define $a = b$ instead by $(\forall x)[a \in x \leftrightarrow b \in x]$, i.e. reversing the direction of membership in the definition of inner equality, a relation we might call *outer equality*. To avoid ambiguity let us denote outer equality by the symbol '$=_{o}$' and reserve the usual symbol '$=$' for inner equality. It is almost immediate from weak extensionality that $a = b$ implies $a =_{o} b$, but what of the converse? It turns out that in ZF the converse is also true, but we need another non-logical axiom to prove it. In fact all we need is that $\{a\}$ is a set whenever a is, which follows from the Axiom of Pairing, then we have $a \in \{a\}$ which by outer equality of a and b implies $b \in \{a\}$ which in turn implies $b = a$. Having to rely on another non-logical axiom to prove this is a little awkward so, as is traditional in set theories which don't take equality to be a primitive relation, we shall instead take this equivalence as our Axiom of Extensionality.

Axiom 1: Axiom of Extensionality
For all sets a and b, $a =_o b$ if and only if $a = b$.[1]

Colloquially, this means that every set is completely determined by its elements. Weak extensionality immediately follows from this axiom and hence we have substitutivity.

Even if equality is assumed as a primitive relation then extensionality still cannot be excluded, at least not without resulting in a weaker theory.[2] If equality is introduced as a primitive relation then extensionality takes the more intuitive form $(\forall x)(\forall y)[(\forall u)[u \in x \leftrightarrow u \in y] \to x = y]$. The converse follows from the substitution schema.

The admission of atoms in the theory generally necessitates a modification of Extensionality (and Regularity, see later) so that they apply only to sets and, depending on the way atoms are characterized, other axioms also need to be altered. For example, the atomic version of Quine's theory NF (New Foundations),[3] known as NFU ('NF with urelements'), features atoms which are regarded as hybrid sets: self-membered singletons, a definition which causes some novel complications. (Quine also introduced a theory ML (Mathematical Logic)[4] which in its initial formulation turned out to be inconsistent.)

Normally atoms would be introduced as primitive 0-ary operator symbols (i.e. constants), one such atom given the special name $\emptyset$, each atom a with the property $\neg(\exists x)[x \in a]$, equality is introduced as a primitive binary relation with the usual axioms of an equivalence relation, and the predicate set(x) is defined as $x = \emptyset \lor (\exists y)[y \in x]$. Extensionality is then $(\forall x)(\forall y)[[\text{set}(x) \land \text{set}(y)] \to [(\forall u)[u \in x \leftrightarrow u \in y] \to x = y]]$.

Remarks

1. Extensionality has a rather different character from the other non-logical axioms of ZF. Its negation is to be interpreted as 'there are distinct sets with the same members' which would seem to be far more counterintuitive than the negation of the other axioms. The negation of Extensionality has been proposed and explored as part of the axiomatization of a theory of fuzzy sets.

2. In the most general sense we say that two predicates $\phi(x)$ and $\psi(x)$ are extensionally identical if their extensions are equal: $\{x : \phi(x)\} = \{x : \psi(x)\}$, i.e. $(\forall x)[\phi(x) \leftrightarrow \psi(x)]$. Two 'intensionally distinct' predicates can easily be

---

[1] More precisely, $(\forall a)(\forall b)[(\forall x)[a \in x \leftrightarrow b \in x] \leftrightarrow (\forall x)[x \in a \leftrightarrow x \in b]]$. Each non-logical axiom appearing in the text will generally be expressed in an intelligible mixture of natural language and logical symbols; a footnote, for the sake of completion but also highlighting the necessity of an abbreviated language, giving the axiom purely in primitive logical symbols.

[2] This is a result of Dana Scott, 'More on the Axiom of Extensionality', in Bar-Hillel [9].

[3] See Quine, W. V. O., 'New foundations for mathematical logic', *American Mathematical Monthly*, **44**, 70–80, (1937).

[4] See Quine [171].

extensionally identical, for example we could take $\phi(x)$ to be 'x is an odd prime expressible as a sum of two squares' and $\psi(x)$ to be 'x is an odd prime congruent to 1 modulo 4'. In ZF, every set a is identified with the extension of the predicate $x \in a$.

Extensionality tells us that sets are completely determined by their elements. In theories with atoms or in which equality is taken as a primitive relation Extensionality has to be modified slightly. Extensionality ensures that equality behaves as we expect it to, in particular it respects substitution. The remaining axioms control which classes are admitted as sets and how new sets can be formed from old. Before we meet them we shall take a closer look at what can go wrong.

3

Avoiding Russell's paradox

3.1 Russell's paradox and some of its relatives

Perhaps the greatest paradox of all is that there are paradoxes in mathematics.

— E. KASNER AND J. NEWMAN[1]

3.1.1 The consequences of the Axiom of Abstraction

In general, a logical paradox is a contradiction, usually expressed in its simplest form $\phi \leftrightarrow \neg\phi$, which reveals a theory to be inconsistent, even though the axioms of the theory *seem* to be plausible and the rules of inference *appear* to be valid. The emphasis is on the surprise of the contradiction, the paradox arising unexpectedly in what is intuitively a perfectly sound system. Zermelo, quite early on, suggested the more precise term 'antinomy' to describe the phenomenon of contradictions which arise within logical systems, demoting the description 'paradox' to those results which are merely counterintuitive but not contradictory.

Contradictions which arise in apparently sensible systems are harsh reminders of our poor intuition regarding large abstract structures. Humans, unfortunately, are very good at cherry-picking attractive sounding ideas without questioning their total lack of empirical foundation and, more to the point, without examining their logical consequences. Even extremely convincing systems can have a sting in the tail, Frege's set theory being an important example. Contradictory systems can be embraced by an individual in the full course of a lifetime without any signs of nausea simply because the system-bearer has not lived long enough, or has not analyzed the system closely enough, to reveal the contradiction. (This logical myopia is a breeding ground for all sorts of peculiar beliefs.) We can all be excused for unknowingly entertaining systems

---

[1]Kasner and Newman [**115**].

which are contradictory; perhaps the long path to the contradiction is beyond our analytical means.

Axiomatic set theory aims to present a theory of sets which is free of contradiction. The most notorious of the set theoretic paradoxes, Russell's paradox, arises if one accepts that, given any property, there exists a set comprising all sets having that property. More precisely, given a predicate $\phi(x)$ we are naively happy to believe that there exists a set that contains all sets x for which $\phi(x)$ holds and contains no set for which ϕ does not hold:

$$(\exists a)(\forall x)[x \in a \leftrightarrow \phi(x)].$$

This principle, the Axiom of Abstraction, is the troublesome 'Axiom V' of Frege's *Grundgesetze der Arithmetik* (1893).

Consider the predicate $x \notin x$. If, as the Axiom of Abstraction tells us, there exists a set a such that

$$(\forall x)[x \in a \leftrightarrow x \notin x]$$

then, setting $x = a$, we have the contradiction

$$a \in a \leftrightarrow a \notin a.$$

This is sometimes called the absolute version of Russell's paradox. Russell had arrived at the relative version (the paradox of the set of all sets) as a corollary of Cantor's Theorem, and by further inspecting the proof of the latter obtained the absolute version.[1] Since Zermelo had discovered the same contradiction independently at about the same time there is a great danger of muddling the order of discovery, or distorting history, when discussing the origins of the set theoretic paradoxes.[2] For us it is the existence of the paradox that is of importance, not who discovered it or when. It took over thirty years for set theorists to fully come to terms with its presence.

The relative version of Russell's paradox is a consequence of very low level logic applied in any set theory for which every subclass of a set is a set. Viewing this property as the root of the problem, alternative set theories in which subclasses of sets need not be sets have also been studied. Such theories are able to accommodate a universal set.[3]

REMARKS

1. Frege's original notation was wildly different from the modern notation we are using here and there are some technicalities involved in the translation.

2. If we can find a predicate $\phi(x)$ which is true of all sets then the class of all sets is $\{x : \phi(x)\}$. The predicate we usually use is $x = x$, but we still need to

---

[1] Cantor, who was no longer active in research by this time, was well aware of the problems that result from considering a universal set of all sets.

[2] For a discussion, see Garciadiego [**78**].

[3] See Forster [**70**].

prove that $x = x$ for all sets! This is easily done: by the definition of equality it reduces to the tautology $(\forall a)[a \in x \leftrightarrow a \in x]$.

At first encounter the Axiom of Abstraction seems harmless. If $\phi(x)$ is a predicate then the axiom says that the class of all sets a such that $\phi(a)$ holds is itself a set. This matches our intuitive idea of a collection (the set of all sentences including the word 'sentence', the set of all neutron stars, the set of all primes of the form $10^n + 1$). However, if one takes $\phi(x)$ to be the predicate $x \notin x$ the resulting class $\mathbf{Ru} = \{x : x \notin x\}$, the Russell class, cannot be a set: we have the contradiction $\mathbf{Ru} \in \mathbf{Ru} \leftrightarrow \mathbf{Ru} \notin \mathbf{Ru}$. Russell's paradox is an example of a logical paradox. Another class of paradoxes, the semantic paradoxes, also pose some amusing problems.

3.1.2   Logical versus semantic paradoxes: some examples

Paradoxes in general are usually partitioned into two classes: the logical paradoxes (also called mathematical paradoxes) such as Russell's paradox and the Burali-Forti paradox, and the semantic paradoxes, examples of which are given below. This division is usually credited to F. P. Ramsey.[1]

Logical paradoxes result from purely mathematical considerations while semantic paradoxes arise, broadly speaking, from a blurring of the distinction between the metalanguage employed to describe mathematics and logic and the mathematical and logical statements themselves. It should be stressed that a device which avoids Russell's paradox need not also banish the Burali-Forti paradox. For example, the Burali-Forti paradox arises in the original formulation of Quine's ML (Mathematical Logic) (1940), yet Russell's paradox is avoided in the same system.[2]

One of the oldest recorded semantic paradoxes is that of Epimenides the liar. Epimenides (a Cretan philosopher) announces 'all Cretans are liars' and we are asked to contemplate whether he is telling the truth. If he is telling the truth then by his own admission he is lying. If he is lying, then by the same reasoning he must be telling the truth. Modern versions are usually closer in form to the more direct 'I am lying' of Eubulides (fourth century BC). Consider, for example, the sentence:

THIS SENTENCE IS FALSE

If the sentence is true then it must be false, and if it is false then it must be true. It is easy to manufacture such self-referential conundrums in almost any

---

[1] Ramsey, F. P., 'The foundations of mathematics', *Proc. London Math. Soc.*, **25**, 338–84, (1926).

[2] J. B. Rosser, 'The Burali-Forti paradox', *J. Symb. Logic*, **7**, 1–17, (1942).

context; they continue to capture the imagination of philosophers (and many non-philosophers) and provide a rich source of creative ideas to this day.

An early modern semantic paradox is Richard's paradox (1905),[1] which is clearly inspired by Cantor's diagonal proof of the uncountability of $\mathbb{R}$ (different formulations of Richard's paradox have appeared over the years – the influence of Cantor's proof is transparent in the original version given here). A string is understood to be any finite sequence of symbols taken from a fixed set adequate to express any notion in English, so to ease readability we might include all digits, letters, punctuation and a blank space, for example. Having agreed on an arbitrary order of the set of basic symbols, the collection of all strings is ordered 'lexicographically'.

Formally, suppose our basic symbols $\{s_i\}$ are arbitrarily ordered by some total order $\prec$. A string s is either the empty string $\varnothing$ or a k-tuple $(s_1, \ldots, s_k)$ of symbols. By s' we mean the $k - 1$-tuple $(s_2, \ldots, s_k)$ comprising the second through kth symbols of s. The order $\prec$ extends to a total order on all strings by defining $\varnothing \prec s$ for all non-empty strings s and $s \prec t$ if:

(i) $s_1 \prec t_1$; or

(ii) $s_1 = t_1$ and $s' \prec t'$.

Denote by S the subsequence of this sequence of strings comprising all English expressions which define a unique real number and let the real number R be defined by the following sentence:

> *The real number with integer part 0 and whose nth decimal digit is equal to m plus 1 if the nth decimal digit of the number described by the nth sentence in S is neither 8 nor 9, and is equal to 1 otherwise.*

By construction R is not a member of S; R differs from the number defined by the nth sentence in S in the nth decimal place. (The slightly convoluted nature of the sentence R is to avoid technicalities involving recurring 9s.) But R is defined by a finite English expression, whence it is in S, a contradiction.

Any theory with sufficient recursive resources can refer to its own sentences in a cleverly indirect way, without blurring metalanguage and object language. By recursively enumerating the well-formed formulas of the theory we can speak unambiguously, in first-order terms, about '*the formula with (Gödel) number n*'. This is the trick behind Gödel's Incompleteness Theorems (see Section 11.3 and Appendix C); via a diagonalization argument we prove the existence of a sentence which (indirectly) asserts its own unprovability, the sentence '*the Gödel number of this sentence is the Gödel number of an unprovable sentence*' substituting the '*I am lying*' of the liar paradox. Gödel explicitly mentions the analogy in his 1931 paper[2] – '*...the analogy of this result with the antinomy of Richard is immediately evident; there is also a close relation to the liar paradox*'.

---

[1] Jules Richard, 'Les principes des mathématiques et le problème des ensembles', *Revue Générale des Sciences Pures et Appliquées*, (1905).

[2] Gödel, K., 'Uber formal unetscheidbare Sätze der Principia Mathematica und verwandter, Systeme I.', *Monatshefte für Mathematik und Physics*, **38**, 173–198, (1931).

Berry's paradox (named by Bertrand Russell after G. G. Berry, librarian of the Bodleian Library, Oxford) appears in a footnote to Whitehead and Russell's *Principia Mathematica* and can be viewed as a simplification of Richard's paradox. It results from an attempt to classify the positive integers according to the smallest number of syllables of English needed to describe them. According to this classification we happily identify 1, 2, 3, 4, 5, 6, 8, 9 and 10 as being describable in one syllable (simply by announcing their names); 7 is describable in two syllables but not in one syllable; 11 is describable in three syllables in several different ways ('eleven'; 'ten plus one'; 'the fifth prime'); and so on – we can see that by some ingenuity one can describe large numbers in far fewer syllables than are required to give their full name. Let us suppose that this classification has been carried out, then among all those positive integers not describable in fewer than 22 syllables there will be a smallest number, B say. Assuming the correctness of the classification, it is impossible to describe B in fewer than 22 syllables, however the sentence:

the least integer not describable using fewer than twenty-two syllables

describes B in just twenty-one syllables, a contradiction.

Berry's paradox, like the paradox of the liar, has also been formalized to give rise to incompleteness results (in the form given here it is a little too vague to produce anything of logical impact, relying on an imprecise notion of being 'describable').[1]

In his allusion to Berry's paradox,[2] Gregory Chaitin defines a computer program (in an arbitrarily chosen language) to be *elegant* if it is the shortest that produces a given output. Chaitin's Theorem is that it is not possible in general to determine whether or not a given program is elegant.

Another popular semantic paradox which borrows ideas from both Berry's paradox and the liar paradox is the Grelling–Nelson paradox of heterologicality (1908).[3] A predicate P is called *heterological* if the sentence 'the predicate "P" is P' is false. Thus the predicate 'a sentence fragment with fewer than eight words' is heterological since the sentence 'the predicate "a sentence fragment with fewer than eight words" is a sentence fragment with fewer than eight words' is false. The paradox is revealed when we ask if the predicate 'a heterological predicate' is heterological. Clearly, if it is we infer that it is not and if it is not we infer that it is.

The appearance of a paradox in a new theory should not condemn the whole theory to oblivion. Instead one must first investigate whether the cause of the problem can be excised without severing any important arteries, leaving something useful behind. Cantor, influenced by Aristotelian ideas on the potential infinite, reacted to the paradoxes known to him, the Burali-Forti paradox for

---

[1]See, for example, George Boolos' 'A new proof of Gödel's Incompleteness Theorem', *Notices of the American Mathematical Society*, **36**, 388–90, (1989).

[2]See Chaitin, G., 'The Berry Paradox', *Complexity*, **1**, 1, 26–30, (1995).

[3]Named after Kurt Grelling and Leonard Nelson, 'Bemerkungen zu den Paradoxien von Russell und Burali-Forti', *Abhandlungen der Fries'shen Schule* (Neue Serie) 2, 300-334, (1907/1908).

example, by drawing a firm distinction between the type of infinite amenable to the methods of his new transfinite theory and his notion of the 'absolute infinite' which he treated as truly inaccessible.[1]

The self-reference of Russell's paradox is a feature, possibly very well-hidden or indirect, that is close to all paradoxes, however the exclusion of *all* statements involving self-reference as a strategy for avoiding internal contradiction is a drastic proposal which is likely to strangle the life out of a theory before it has even begun. Consequently a more sober solution is called for.

REMARKS

1. One can replace the self-reference of the liar paradox with circularity between different statements. Consider the following two sentences.

   A: Sentence B is true.

   B: Sentence A is false.

   Assigning a truth value to either sentence results in a paradox. This is sometimes known as *Jourdain's paradox* after Philip Jourdain.

2. In an attempt to create a non-self-referencing non-circular paradox Stephen Yablo[2] considered an infinite sequence of statements (A_n), $n = 1, 2, \ldots$.

   A_n: for all $k > n$, A_k is false.

   There is no consistent way of assigning a truth value to all of the statements in this sequence. If A_n is true then $A_{n+1}, A_{n+2}, \ldots$ are all false. But if A_{n+1} is false this means that there is an A_k with $k > n + 1$ which is true, a contradiction. If A_1 is false, then there is an A_n with $n \geq 2$ which is true, which we have already seen leads to contradiction.

---

[1] In the historical philosophical literature one can find extensive discussions on the Aristotelian notions of 'actual' versus 'potential', much of it trying to determine what Aristotle really meant by the two terms and how we are to understand the distinction now. It is not a dichotomy I am particularly comfortable with, largely because the notion of an 'actual' anything is so hopelessly vague, and 'potential' doesn't fare any better. Aristotle himself states that he uses the terms differently in different contexts. Most of his discussion appears in *Metaphysics* (particularly Book IX), but there is also mention of actuality and potentiality in *On the Soul* and elsewhere. The discussion in Book IX:6–9 of *Metaphysics* comes closest to giving a precise definition, '*Actuality means the existence of the thing, not in the way which we express by 'potentially'; we say that potentially, for instance, a statue of Hermes is in the block of wood and the half-line is in the whole, because it might be separated out, and even the man who is not studying we call a man of science, if he is capable of studying. Otherwise, actually.*' Both notions seem to involve consideration of whether something presently exists or may possibly exist some time in the future, via some unmentioned process. The introduction of temporal notions into mathematics sits very uneasily with most mathematicians; if a mathematical statement is true today then it was true yesterday and will continue to be true tomorrow.

[2] Yablo, S., 'Paradox without self-reference', *Analysis*, **53** (4), 251–252, (1993).

3. Gregory Chaitin tells an amusing story about his proof of the incompleteness theorem in a lecture dated 27 October 1993 (University of New Mexico): '*In early 1974 I was visiting the Watson Research Center and I got the idea of calling Gödel...Gödel answered the phone. I said "Professor Gödel, I'm fascinated by your incompleteness theorem. I have a new proof based on the Berry paradox that I'd like to tell you about". Gödel said "it doesn't matter which paradox you use".*'

4. The statement of the Grelling–Nelson paradox can seem artificial and awkward. An alternative, a clever modification of the liar paradox known as *Quine's paradox*, is less clumsy and also has the benefit of not directly referring to itself. Consider the sentence:

   'Yields falsehood when preceded by its quotation' yields falsehood when preceded by its quotation.

The sentence is indirectly declaring that it is false. But if it is false then what it states is true.

Russell's paradox and the Burali-Forti paradox arise from logical assumptions and are known as logical or mathematical paradoxes. Another class of paradoxes, the semantic paradoxes, arise from variations on self-reference or from an interplay between the object language and metalanguage. Semantic paradoxes can be resolved by a close examination of the types of language being used. Logical paradoxes are more difficult to resolve and require a careful axiomatic excision.

3.1.3 Resolutions

The semantic paradoxes arise because natural language is able to refer effortlessly to other expressions in the same language. In the more precise framework of set theory it is necessary to make a clear distinction between the language used to talk about sets (the object language) and the language used to talk about the object language (the metalanguage). Formulating the statements of the theory using an exact syntax, such contradictions are immediately eliminated (i.e. one can clearly recognize in a given semantic paradox where metalanguage notions have invaded the object language) and the logical paradoxes, although not resolved, are easily expressed and analyzed (i.e. one can see which assumptions lead to the contradiction).

After the shock of the paradoxes there was a natural focus on devising ways of axiomatically isolating the concept of *set* in such a way that all *known* routes to internal contradiction are unattainable. Once this had been achieved, after the dust had settled, theorists began to look at the theory anew and asked to what extent the new restrictions had been stronger than necessary. Such research seeks to describe the elusive borderline that separates a paradox-free theory from its unwelcome neighbours.

Most of 'mainstream' mathematical research is not directly affected by the devices designed to avoid paradoxes. Such taming principles tend to have influence only when the boundaries of extreme generalization are deliberately prodded to test the strength of a theory. This is not to say that such logical matters are irrelevant to mathematical research; whichever field one works in it is not difficult to produce a statement whose resolution relies critically on some deep set theoretic postulate (the Continuum Hypothesis or the Axiom of Choice, for example).

Paradoxes of the Russell type can be avoided by replacing the full schema of comprehension (i.e. that for all predicates $\phi(x)$, $\{x : \phi(x)\}$ is a set) by a weaker version, the schema of separation, which states that for any set a the class $a \cap \{x : \phi(x)\}$ is a set. In ZF this schema of separation is a corollary of the Axiom Schema of Replacement (see Section 4.4). This weakening of full comprehension to separation successfully banishes the known paradoxes, but in turn handicaps the theory to such an extent that further principles need to be postulated.

The axioms of ZF are generally accepted as a correct formalization of those principles that mathematicians apply when working intuitively with sets, but in light of the independence of the Continuum Hypothesis and the Axiom of Choice from ZF some might still regard ZF as lacking in some important ingredients. By Gödel's First Incompleteness Theorem there are (infinitely many) undecidable statements expressible in the language of ZF. Whether some of these 'ought' to be included as truths in an extension of ZF is a matter for philosophical debate. The mathematical principles encompassed by ZF plus the Axiom of Choice are nevertheless those that are assumed by most working mathematicians, possibly unknowingly.

Modern mathematics has, ever since the existence of non-Euclidean geometries came to light, drifted away from more absolute forms of Platonism. Mathematics draws inspiration from its own structures as well as from the new findings of physics, so that the elucidations are many times removed from the concrete intuitive and counterintuitive objects of the Universe about us. The fact that a lot of mathematics is an abstraction of an abstraction of an abstraction makes an accurate popular exposition close to impossible. This progressive departure from physical ties is justified by an unstoppable and continuing barrage of powerful abstract results which bring a deep new understanding of the old material, at the same time opening the gates to vast new landscapes of ideas.

REMARKS

1. There is a misunderstanding in some quarters that if set theory (ZF, say) was found to be inconsistent then all of mathematics would come tumbling down, as if it formed a foundation of mathematics in the structural engineering sense. It would not – mathematics would carry on exactly as it has done

for the last few centuries, uninjured and unperturbed (somewhat analogous to the way the building of a metropolis would be unhindered by setbacks in theoretical physics; if a model of the subatomic world turns out to be incorrect, buildings don't suddenly fall down). Set theorists would go about understanding what lies behind the newfound contradiction (and there are many other set theories to compare the 'broken' ZF to, so much of the conceptual work may have been done already), perhaps eventually finding a new modification that seems to work.

In truth the vast majority of mathematicians would know very little about the finding, beyond overheard gossip, and would continue with their research untroubled.

2. There is a reductive sense that all mathematics and indeed all ideas, being 'just' brain activity, cannot escape the finite confines of a local region of physical space – each idea, particular to the individual, is just a large but finite number of particles jostling about in some organized neurological maze. As a naive argument for ultrafinitism this backfires a little, since all mathematical ideas are revealed to be equally fictitious. Reducing everything to this disconnected level the number two or a tesseract is no more or less real than the set $P(P(P(P(P(P(P(\mathbb{N}))))))$ or the thought of a pineapple. Everything in mathematics is imaginary, even the objects that we pretend to be based on 'real life' notions (triangles, small finite sets, lines) are once abstracted, and notions like infinite sets are, so to speak, twice abstracted – abstractions of abstractions. Where do we set the boundary? Thrice abstracted? This seems arbitrary. Back to sorites again.

Semantic paradoxes generally play on a blurring of object and metalanguage and are avoided by a precise definition of the well-formed formulas of the theory. By imitating this blurring in the formal framework one can devise self-referential analogues of statements such as 'this sentence is false' which lead in a cleverly indirect way to deep incompleteness results. To tame the logical paradoxes one must choose appropriate axioms which prevent certain classes, such as the Russell class, from being sets.

3.2   Introducing classes

One of the endlessly alluring aspects of mathematics is that its thorniest paradoxes have a way of blooming into beautiful theories.
                                                         – PHILIP J. DAVIS[1]

The idea of a collection of all objects having some fixed property seems to be a 'hardwired' feature of our intuition and is used freely throughout mathematics.

---
[1] Davis, P. J., 'Number', *Scientific American*, **211**, 51–59, (Sept 1964).

We need to limit the notion (without abandoning it completely) in such a way that Russell's paradox is unattainable. In ZF the approach is as follows.

With each predicate $\phi(x)$ we associate the class symbol $\{x : \phi(x)\}$. We interpret $\{x : \phi(x)\}$ principally as the collection of all sets x satisfying the property ϕ. The membership relation $\in$ is extended to allow class symbols in place of variables so that $Y \in \{x : \phi\}$ if and only if Y is a *set* with property ϕ. Classes which are not sets, called *proper classes*, are not members of any class. The Russell paradox is then resolved by proving that the Russell class $\mathbf{Ru} = \{x : x \notin x\}$ is a proper class and hence is not an element of any class, including itself.

We can adopt this strategy provided we admit the new class symbols into our language in a harmonious manner. Either we treat this collection of entities as a fresh assortment of primitive objects and provide revised axioms and rules for the extended well-formed formulas – a potentially messy procedure – or, what is much more economical, we introduce classes as defined terms. To do this we simply need to exhibit a means of reducing any formula that contains class symbols to an equivalent formula that does not. Before doing that we need to describe the valid ways in which class symbols may appear in the formulas of the theory. We shall call the formulas formed by these rules *generalized well-formed formulas*. We stress that generalized well-formed formulas are simply convenient abbreviations of well-formed formulas.

(i) If a and b are variables then $a \in b$ is a generalized well-formed formula.

(ii) If $\phi(x)$ and $\psi(y)$ are generalized well-formed formulas and a and b are variables then $a \in \{x : \psi(x)\}$, $\{x : \phi(x)\} \in b$ and $\{x : \phi(x)\} \in \{x : \psi(x)\}$ are generalized well-formed formulas.

(iii) If ϕ and ψ are generalized well-formed formulas then so too are $\neg\phi$, $\phi \wedge \psi$, $\phi \vee \psi$, $\phi \rightarrow \psi$ and $\phi \leftrightarrow \psi$.

(iv) If ϕ is a generalized well-formed formula and x is a variable then $(\exists x)\phi$ and $(\forall x)\phi$ are generalized well-formed formulas.

(v) A string of symbols is a generalized well-formed formula if and only if it can be formed using the four rules above.

Each generalized well-formed formula ψ is equivalent to a unique well-formed formula ψ^* according to the following intuitive rules based on the interpretation of $\{x : \phi(x)\}$ as the collection of all sets x satisfying ϕ:

ψ	ψ^*
$a \in b$	$a \in b$
$a \in \{x : \phi(x)\}$	$\phi^*(a)$
$\{x : \phi(x)\} \in a$	$(\exists y)[y \in a \wedge (\forall z)[z \in y \leftrightarrow \phi^*(z)]]$
$\{x : \phi(x)\} \in \{x : \eta(x)\}$	$(\exists y)[\eta^*(y) \wedge (\forall z)[z \in y \leftrightarrow \phi^*(z)]]$
$\neg\phi$	$\neg\phi^*$
$\phi \wedge \eta$	$\phi^* \wedge \eta^*$
$(\forall x)\phi$	$(\forall x)\phi^*$

It follows from the expression of $\vee$, $\rightarrow$, $\leftrightarrow$ and $\exists$ in terms of $\neg$, $\wedge$ and $\forall$ that for generalized well-formed formulas ϕ and ψ the well-formed formula $(\phi \vee \psi)^*$ is equivalent to $\phi^* \vee \psi^*$; $(\phi \rightarrow \psi)^*$ is equivalent to $\phi^* \rightarrow \psi^*$; $(\phi \leftrightarrow \psi)^*$ is equivalent to $\phi^* \leftrightarrow \psi^*$; and $((\exists x)\phi)^*$ is equivalent to $(\exists x)\phi^*$.

If $\phi(x)$ is a generalized well-formed formula then $(\forall x)[\phi(x) \leftrightarrow \phi^*(x)]$, so $(\forall x)[x \in \{y : \phi\} \leftrightarrow x \in \{y : \phi\}]$ or in other words $\{y : \phi\} = \{y : \phi^*\}$,[1] an equality which assures us that we have not accidentally created any genuinely new entities in the process of introducing class symbols. Via the tautology $x \in a \leftrightarrow x \in a$ we obtain the equality $a = \{x : x \in a\}$; *every set is a class.*

Although the above technique shows that it is possible to develop set theory without explicitly mentioning classes, using the defining condition ϕ alone instead of the class $\{x : \phi(x)\}$, such a version of the formalism is almost unworkable in practice and, among other things, hides many otherwise intuitively clear issues behind an unnecessary fog of symbols. We shall denote classes by upper case Roman letters, A, B, C, understanding that any formulas involving them are abbreviations of statements regarding their defining predicates.

The underlying conceptual difference between sets and classes is that of a collection of objects versus an extension of a property (an ideal class comprising all objects having the said property). The ideal object $\{0, 1, 2, \ldots\}$ of all natural numbers, for example, is first presented to us in the guise of a very precisely defined class ω and an axiom (the Axiom of Infinity, see Section 6.2) is needed to 'promote' ω to set status.

For Frege the notion of an extension of a property (i.e. a class) was critical to any foundation of arithmetic because, for example, the empty set, used to define 0 and ultimately all sets, is meaningful as the extension of a never attained condition ($x \neq x$, for instance) but the idea of a 'collection of no objects' was regarded with suspicion.

REMARKS

1. Notice that in the definitions of $\{x : \phi(x)\} \in a$ and $\{x : \phi(x)\} \in \{x : \eta(x)\}$ the class $\{x : \phi(x)\}$ appearing on the left-hand side (the member) is necessarily a set.

2. Given a predicate $\phi(x)$ we must ask whether $\{x : \phi(x)\}$ is a set. It may be possible to prove this from prior assumptions. Alternatively it might be possible to prove that the assumption that it is a set leads to contradiction (as we find, for example, by taking $\phi(x)$ to be the predicate $x \notin x$). If the assumption that $\{x : \phi(x)\}$ is a set can neither be proved nor disproved then, if the property ϕ seems to be of some interest, we might consider adopting as an axiom: $\{x : \phi(x)\}$ is a set. With the exception of Extensionality and Regularity, all non-logical axioms of ZF are of this form.

---

[1] Equality of class symbols is defined in the obvious way: $\{x : \phi\} = \{x : \psi\}$ means $(\forall x)[\phi(x) \leftrightarrow \psi(x)]$.

3. It should be pointed out that we haven't yet provided a concrete instance of a set. We shall continue the suspense; at the moment we are just describing some properties we want sets to have and some ways of avoiding paradoxes.

Intuitively a class is simply a collection of objects satisfying some property. If $\phi(x)$ is a predicate then $\{x : \phi(x)\}$ is a class. Classes are incorporated in ZF as defined terms, so that well-formed formulas including class symbols are abbreviations for well-formed formulas without class symbols. The rules for eliminating class symbols in this way are based on the intuitive interpretation of classes. Most of the non-logical axioms of ZF declare that certain relatively simple and mathematically natural predicates have sets as their extensions.

3.3   Set hierarchies

> *We lay down a fundamental principle of generalization by abstraction: 'The existence of analogies between central features of various theories implies the existence of a general theory which underlies the particular theories and unifies them with respect to those central features....'*
>
> – E. H. MOORE[1]

3.3.1   Building a universe of sets

The non-logical axioms to come provide a means of constructing new sets from old and also identify some classes as sets. For example, one axiom identifies the class of natural numbers (in the form of ω) as a set and another says that if x is a set then so is $P(x)$. In combination these axioms lead us to the higher infinities. We tend to view the universe of sets as a recursively generated class of objects. In a non-atomic set theory the starting point for this recursive hierarchy is the empty set.

It is customary in most axiomatic developments of pure set theory to define the empty set as the class $\{x : x \neq x\}$. We could use any never-attained predicate in place of $x \neq x$ to the same effect, but the assertion $(\forall x)[x = x]$ has the advantage of being a purely logical truth (it follows from the tautology $y \in x \leftrightarrow y \in x$) whereas, say, the fact that there are no x for which $x \in x$ is a consequence of a *non-logical* axiom (Regularity). A non-logical axiom (Replacement) is needed to prove that the class $\{x : x \neq x\}$ is a set. Set theories which admit an empty set and atoms; atoms but no empty set; or an

---

[1] Moore [**151**]. Preface.

empty set and no atoms have all been subjected to intensive study – only the baseless case of excluding atoms *and* the empty set has been avoided.

REMARK

I am guilty of using the word 'constructing' casually here. As mentioned earlier, the operation of taking the power set of an infinite set is not constructive at all in any algorithmic sense of the word. We will return to this when we discuss the relevant axiom.

Perhaps the most important two sets in set theory are the empty set and the set of natural numbers. As soon as these sets are made available to us, we can generate a vast set hierarchy. In the case of the natural numbers, a special axiom is required to promote the class ω to set status. The empty class could similarly be granted set status by a new axiom, or alternatively we can define it as a particular class in terms of a never-attained property and use existing axioms to prove that the said class is a set.

3.3.2 Type theory and other alternatives

Type theories try to construct a foundation of mathematics recursively, building their components from a simple base. One can imagine a crude prototype of this approach – a tower of sets built with the avoidance of internal contradiction in mind. At the lowest level we find all atoms, at the next level all sets of atoms, next all sets of sets of atoms, then sets of sets of sets of atoms, and so on. One could label these collections of elements in the obvious way, the lowest level comprising type 0 objects, the next level comprising type 1 objects, and so on. If this tower were to begin with a solitary atom, then each level would have only one object.

This simple rigid structure is, for most purposes, not enormously useful by itself; of course we need to add further features in order to actually do mathematics. A natural first cosmetic adjustment makes the types cumulative, so that the objects of one level (the type $n + 1$ sets) are sets of objects of type less than or equal to n rather than sets of objects of type n only. But neither the simple hierarchy nor this slight generalization allow for any infinite sets unless infinitely many atoms are assumed at the outset or alternatively, by appealing to another external device, allowing an ordinal hierarchy, so that the type β sets have as members sets of type α for $\alpha < \beta$. Both options seem to grossly violate the idea of a wholly bottom up approach to foundations, appealing to fairly sophisticated notions that ought to be emergent features of the theory that is being constructed. To evade this criticism and to transform the basic notion into a fully fledged type theory, further elaborations must incorporate a complex hierarchy of well-formed formulas, subject to a host of restrictive or corrective axioms. History has taught us a lesson that in constructing such theories there is a danger that one must invent increasingly unnatural and ad hoc axioms to patch up technical problems. This is not to say that, as a foundation of mathematics, type theory is doomed to failure, it is not, but there are some difficult hurdles that must be overcome.

The classical paradoxes are resolved in such type hierarchies as follows: a class $\{x : \phi\}$ is a set if and only if there is an ordinal γ_ϕ for which every set that satisfies ϕ is of type no greater than γ_ϕ. No such ceiling exists for the Russell property $x \notin x$, for example, because within this hierarchy $x \notin x$ is true of all objects. Likewise the class of all sets $\{x : x = x\}$ is not a set.

The basic idea behind the theory of types is a natural one and, on the face of it, it has the sort of structure that one might hope for in a foundation of mathematics. However, as Russell developed the theory into the ramified theory of types[1] he was forced to introduce further assumptions and to extend the basic foundation until the resulting theory became overly complicated, and certainly had no legitimate claim to be any more natural than the much simpler and more intuitive set theory of Zermelo. Some of the complications involved in constructing a theory of types are hinted at in Subsection 2.3.4 (we will not spell out the details here).

Owing to its awkwardness and relative complexity in comparison with set theories such as ZF, Russell's theory of types has not been directly successful in mathematical circles, but it was nonetheless a highly influential stepping stone to new approaches and spawned many fruitful ideas, variations and simplifications. Modern type theory, initiated by Alonzo Church in the 1940s, plays an important role in logic and in computer science, but a description of its principles, its relationship to set theory and its deep connection with category theory is beyond the scope of this text. The most famous reference to the Russell–Whitehead theory is in Gödel's epoch-making paper of 1931, 'On formally undecidable propositions of Principia Mathematica and related systems I'.[2]

The ramified theory of types is a form of *no-class theory*, one of three approaches suggested by Russell in 1906[3] in response to the problem of paradox avoidance. The other two ideas were *limitation of size* and the so-called *Zig-Zag theory*. Russell's main concern was, oversimplifying a little, the problem of identifying those concepts (*predicative functions*) whose extensions are sets. Zig-Zag theory has at its core the assumption that if ϕ is predicative then so is $\neg\phi$ (an approach which is incompatible with any axiom of subsets, i.e. subclasses of sets needn't be sets). 'Limitation of size' refers, broadly speaking, to any principle which classifies predicative properties according to the number of objects with that property, for example one such principle might be 'if A is less numerous or equally as numerous as B and B is a set, then A is a set' (the meaning of numerous will be appropriately defined by the other principles of the theory). We know through his letters to Hilbert and Dedekind that Cantor sketched some

---

[1] Russell develops his theory in 'Mathematical logic as based on the theory of types', *Amer. J. Math.*, **30**, 222–62, (1908) and it reaches maturation in Whitehead and Russell [**225**].

[2] Now published in book form, see Gödel [**83**].

[3] Russell, B., 'Les paradoxes de la logique', *Rev. Métaphys. Morale*, **14**, 627–50, (1906).

ideas based on limitation of size principles during the 1890s. Russell subsequently worked on another such theory but abandoned it (and Zig-Zag theory) in favour of the ramified theory of types. The late 1920s saw renewed interest in the idea of limitation of size, apparently encouraged by von Neumann.

REMARKS

1. On the appearance of the mysterious 'I' at the end of the title of Gödel's paper: Gödel had planned a part II which was to provide fine details of the proof of the second incompleteness theorem he had outlined in part I, but it never materialized since everyone seemed to be satisfied with the outline. (Of course since then the details have been fully spelled out.) Gödel worked in a general system he called P, the results of which apply to *Principia Mathematica* and many other systems.

2. Zig-Zag theory aims to find some criteria of admissible predicate complexity, allowing simple predicates and disallowing complex predicates, but Russell was unable to find an adequate distinction. To see why Zig-Zag theory is incompatible with any consistent set theory in which subclasses of sets are always sets consider the predicate $x \neq x$. Its extension is the empty class. We want the empty class to be a set, but the extension of the negation of the predicate $x \neq x$ is the class V of all sets, which by the zig-zag principle must also be a set. But every class is a subclass of V, so according to this theory there are subclasses of a set which are not sets.

3. Von Neumann introduced an axiom of limitation of size to his class theory which states that a class is a proper class if and only if it can be mapped *onto* V. This is a powerful axiom which implies, among other things, the Axiom of Replacement, the Axiom of Unions and the Axiom of Choice (in fact it implies the Axiom of Global Choice since it entails the existence of a mapping of the proper class of all ordinal numbers onto the class of all sets – it well-orders the entire set universe!).

The theory of types is an alternative approach to the foundations of mathematics initiated by Bertrand Russell and developed in Russell and Whitehead's Principia Mathematica. It, and other approaches such as Zig-Zag theory and limitation of size, has not been as successful (as a foundation of mathematics) as the set theoretic approach we are familiar with today. Both the theory of types and set theory can be thought of as having a hierarchical structure, but the set theoretic universe is not defined by this hierarchy.

3.3.3   The von Neumann cumulative hierarchy

The most intuitively pleasing starting place for a set hierarchy is the empty set, and, assuming the levels of the hierarchy are naturally indexed by the ordinal numbers in the obvious way, for each ordinal α the corresponding level V_α is populated by sets whose elements all lie in $\cup_{\beta<\alpha} V_\beta$. Having defined the level V_α, we need to specify clearly which sets are to be included in $V_{\alpha+1}$. If $V_{\alpha+1}$ comprises *all* subsets of V_α, then the resulting structure is von Neumann's cumulative hierarchy of sets:

$$
\begin{aligned}
V_0 &= \varnothing \\
V_{\alpha+1} &= P(V_\alpha) \\
V_\beta &= \bigcup_{\alpha<\beta} V_\alpha, \text{ if } \beta \text{ is a limit ordinal.}
\end{aligned}
$$

Although this construction would seem to be nothing more than an atomless version of the simple type hierarchy we described above, there is a crucial conceptual difference: von Neumann's hierarchy is a stratification 'after the event'. Sets are first moulded by the axioms of ZF, and the ordinal numbers are properly defined, then we see how the sets are arranged in relation to one another. In particular it can be proved that *all* sets are accounted for in the hierarchy, whereas the type hierarchies are used to *define* the class of all sets (the statement that the class of sets is precisely all those sets falling in the hierarchy providing a form of the Axiom of Regularity) and the ordinal numbers are either plucked from nowhere in the style of Cantor's principles of generation or must carefully be crafted from the unfolding hierarchy itself, sometimes with great technical difficulty.

Every set in the von Neumann hierarchy is an element of some V_α. The smallest ordinal γ for which V_γ contains a given set is called the *rank* of that set. This seemingly innocent position marker turns out to be very useful.

REMARK

If, instead of taking the full power set at each successor stage of the construction, one instead takes the set of subsets which are 'definable' from sets in the previous level, i.e. those sets a such that there exists a predicate $\phi(x_1,\ldots,x_{n+1})$ and elements $a_1, a_2, \ldots, a_n$ of the previous level for which $a = \{x : \phi(x, a_1, \ldots, a_n)\}$, then the resulting hierarchy forms Gödel's constructible universe. We shall describe this in more detail in Section 11.1.

The von Neumann hierarchy generates sets by starting with the empty set, then successively taking the power set, forming the union of all lower levels at each limit ordinal. It follows from the Axiom of Regularity that all sets appear somewhere in this hierarchy.

4

Further axioms

4.1 Pairs

> *We used to think that if we knew one, we knew two, because one and one are two. We are finding that we must learn a great deal more about 'and'.*
>
> – SIR ARTHUR EDDINGTON[1]

4.1.1 Unordered pairs

The collection comprising an unordered pair of sets a and b, denoted $\{a, b\}$, is formally defined as the class $\{x : x = a \vee x = b\}$. The properties $\phi \vee \phi \leftrightarrow \phi$ and $\phi \vee \psi \leftrightarrow \psi \vee \phi$ of disjunction entail the facts that neither repeating elements nor changing the order of presentation of elements in the n-tuple $\{a_1, a_2, ..., a_n\} = \{x : x = a_1 \vee x = a_2 \vee ... \vee x = a_n\}$ will give rise to a different class. In particular the axiom below tells us that the singleton $\{a\} = \{a, a\}$ is a set whenever a is a set.

The predicate 'A is a set', used in the following axiom, simply means

$$(\exists x)[x = A]$$

or, replacing equality by its definition,

$$(\exists x)(\forall y)[y \in x \leftrightarrow y \in A].$$

In particular '$\{x : \phi(x)\}$ is a set' means $(\exists x)(\forall y)[y \in x \leftrightarrow \phi(y)]$. The latter is one instance of the Axiom Schema of Abstraction. Our axioms are designed to restrict the collection of predicates $\phi(x)$ for which this sentence is true. For example, to avoid Russell's paradox, the instance of Abstraction corresponding to the predicate $x \notin x$ will be false.

---

[1]Quoted in Mackay [141].

Axiom 2: Axiom of Pairing
For all sets a and b, $\{a, b\}$ is a set.[1]

It turns out that Pairing is redundant, i.e. one can deduce Pairing from the other axioms of ZF (see Subsection 4.4.3). Retrospective redundancy of this kind is a common feature of the development of any axiomatic theory; we shall continue to call Pairing an axiom regardless.

REMARK
Now that $\{a\}$ is a set whenever a is a set we can generate a sequence of sets a, $\{a\}$, $\{\{a\}\}$, $\{\{\{a\}\}\}$, etcetera. However, we don't yet have sufficient machinery to prove that these sets are distinct, and we certainly can't gather them all together to form a new set.

Unordered pairs are defined naturally as certain classes, but we need an axiom to promote these classes to set status. However, the later requirements of the theory will call for a new axiom which is more powerful that the Axiom of Pairing, making it redundant. To define relations we also need to define ordered pairs.

4.1.2   Ordered pairs

The question arises how to define an *ordered* pair. The intuitive idea is clear, we readily see that (a, b) and (b, a) are different, but how do we encode this order dependence in our primitive language? All we ask is that (a, b) is defined in such a way that $(a, b) = (c, d)$ if and only if $a = c$ and $b = d$. The most commonly given definition is, in terms of unordered pairs, $(a, b) = \{\{a\}, \{a, b\}\}$. This formal trick is due to Kazimierz Kuratowski.[2] The first purely set theoretic encoding of ordered pairs was given by Norbert Wiener[3] who defined (a, b) as $\{\{\{a\}, \varnothing\}, \{\{b\}\}\}$.

These set theoretic descriptions of ordered pairs provide a means of embedding the theory of relations in the theory of sets without having to introduce new axioms. Using Kuratowski's definition the sentences 'a is the first element of the ordered pair P' and 'b is the second element of the ordered pair P' are expressible in simple terms as 'a is an element of all elements of P' and 'b is an element of exactly one element of P'' respectively. Using the same definition, we do not need to postulate that (a, b) is a set if a and b are sets since it follows immediately from the Axiom of Pairing.

---

[1] $(\forall a)(\forall b)(\exists c)(\forall d)[d \in c \leftrightarrow (\forall e)[e \in d \leftrightarrow e \in a] \vee (\forall e)[e \in d \leftrightarrow e \in b]]$.

[2] Kuratowski, K., 'Sur la notion d'ordre dans le théorie des ensembles', *Fund. Math.*, **2**, 161–71, (1921).

[3] Wiener, N., 'A simplification of the logic of relations', *Proc. Camb. Phil. Soc.*, **17**, 387–90, (1914).

For $n \geq 3$ the ordered n-tuple, $(a_1, a_2, \ldots, a_n)$ is defined (recursively) as the class $((a_1, a_2, \ldots, a_{n-1}), a_n)$.

REMARKS

1. Wiener's definition was motivated by the types of *Principia Mathematica* and the unusually ornate form of the definition is forced by the insistence that each member of a set must be of the same type.

2. Felix Hausdorff also proposed a (somewhat unsatisfactory) definition of ordered pairs: if a and b are sets then choosing two distinct objects α and β, neither equal to a or b, we define (a, b) to be the set $\{\{a, \alpha\}, \{b, \beta\}\}$.

There are several different ways of giving a purely set theoretic definition of ordered pairs, all of them carrying with them some extra properties that ordered pairs do not intuitively have. So long as the definitions have the properties that ordered pairs intuitively <u>do</u> have, this normally doesn't bother us.

4.1.3   Benacerraf's problem

Explicit set theoretic formulations of the idea of an ordered pair bring with them 'accidental consequences', properties which we would not normally associate with the intuitive idea that is being modelled. For example, using Kuratowski's definition we have the relation $\{a, b\} \in (a, b)$. We need to learn to tolerate this; any arbitrary definition of some intuitive notion in terms of sets will inevitably carry with it such peculiar side-effects.

This issue of models doing more than what they are required to do, and more generally the existence of different models of the same intuitive object, sometimes referred to as Benacerraf's identification problem, is the cause of much discussion amongst certain philosophers of mathematics. In his original article '*What numbers could not be*'[1] Paul Benacerraf discusses, among other things, the set theoretic definitions of 3 (as, say $\{\{\{\emptyset\}\}\}$ or as $\{\emptyset, \{\emptyset\}, \{\emptyset, \{\emptyset\}\}\}$), highlighting that this identification is performed for the purposes of a theory and is not intended to declare what the number 3 'really is'.

Although it is most visible in the foundational developments beginning with Dedekind in the late nineteenth century, in its most general form the problem is as old as mathematics itself, arguably as old as abstraction and the development of even the most rudimentary language.

Benacerraf's *epistemological* problem, a challenge to mathematical platonism, seems to have generated a lot of comment. There are signs, however, that the arguments have become a little tired (Michael Potter describes it as a

---

[1] *Phil. Rev.*, **74**, 47–73, (1965).

'painful cliché' to begin an article on mathematical epistemology with a statement of Benacerraf's problem).[1] The basic idea is this. Benacerraf hypothesizes that the best theory of knowledge is the causal theory, where a chain of events, facts or beliefs, each following from its predecessor, leads from the perception of something to knowledge of it. According to Platonism, however, mathematical objects have no material or temporal content, while mathematicians do, so there is a barrier to any true knowledge of mathematical objects since we can have no causal interaction with them, no chain can stretch from us in the material world to the objects in their Platonic world.[2]

The problem of introducing such accidental relations as $\{a, b\} \in (a, b)$ is avoidable by introducing each new entity as a primitive term together with a supply of axioms governing its behaviour. This extra baggage often burdens the theory with such inelegance that the oddities associated with the supposedly problematic original definition seem innocent in comparison. For example, Kuratowski's definition gives a clear position of a given ordered pair in the set theoretic hierarchy, whereas an axiomatic treatment of ordered pairs as new primitive objects calls for an additional and arguably far more artificial means of determining their hierarchical placement. Mathematicians are exposed to this phenomenon continually and perceive no difficulty.

REMARK

To replace modelled objects by primitives is a backward step – Wiener's original motivation in defining an ordered pair as a set, for example, was precisely to reduce the number of primitive elements in *Principia Mathematica*.

Models of intuitive mathematical objects will inevitably have additional properties not usually associated with the object being modelled. This raises some philosophical issues, however, mathematically it is not a problem at all. We continue to introduce axioms which allow us to form new sets from old in familiar ways.

---

[1] See Michael Potter, 'What is the problem of mathematical knowledge?' in Leng, Paseau and Potter [**138**].

[2] Variations based on differing hypotheses lead to similar conclusions which can be roughly summarized in the question: 'how can material beings have any knowledge of purely Platonic objects?' Perhaps more pressing is the mystery of how we gain knowledge of the material world by invoking Platonic notions that are not part of it. Put another way, how is it that imaginative flights of fancy such as infinite sets (or indeed extra-astronomically large finite sets), real numbers, the widthless objects of geometry, and so on, can tell us so much, with such economy of language, about the finite real world?

4.2   Unions

*So much of modern mathematical work is obviously on the border-
line of logic, so much of modern logic is symbolic and formal, that
the very close relationship of logic and mathematics has become obvious to every instructed student. The proof of their identity is, of
course, a matter of detail: starting with premises which would be
universally admitted to belong to logic, and arriving by deduction at
results which as obviously belong to mathematics, we find that there
is no point at which a sharp line can be drawn, with logic to the left
and mathematics to the right.*

– BERTRAND RUSSELL[1]

Ordered n-tuples of sets are sets, as one can prove by induction. It is easy to fall
into the trap of assuming that a similar argument holds for unordered n-tuples,
but careful scrutiny of any such naive inductive argument soon reveals the need
for properties of unions of sets. The inductive step *if* $\{a_1, \ldots, a_n\}$ *and* a_{n+1} *are
sets then* $\{a_1, \ldots, a_n, a_{n+1}\}$ *is a set* is the sticking point. No problem occurs in
the case of ordered n-tuples because of the recursive way n-tuples are formally
defined; $(a_1, \ldots, a_n)$ is, despite notational appearances, an ordered pair of sets
but $\{a_1, \ldots, a_n\}$ is not an unordered pair of sets ($n \geq 3$).

The following axiom provides the necessary machinery. For any class A we
define $\cup A$ to be the class $\{x : (\exists y)[y \in A \wedge x \in y]\}$, for example, we have
$\cup\{\{1, 2, 3\}, \{a, b\}, \{x, y, z\}\} = \{1, 2, 3, a, b, x, y, z\}$.

Axiom 3: Axiom of Unions
For all sets a, $\cup a$ is a set.[2]

Dual to the notion of unions is that of intersection:

$$a_1 \cap \cdots \cap a_n = \{x : x \in a_1 \wedge \cdots \wedge x \in a_n\}.$$

The general intersection $\cap A$ is defined to be the class $\{x : (\forall y)[y \in A \to x \in y]\}$.
In some set theories $\cap A$ is defined only if A is non-empty. If we examine the
case $A = \emptyset$ we find, owing to vacuous satisfaction of the defining condition, that
$\cap\emptyset = V$, the class of all sets: $\cap\emptyset = \{x : (\forall y)[y \in \emptyset \to x \in y]\}$, but as $y \in \emptyset$ is
false for all y the implication $y \in \emptyset \to x \in y$ is true, so $(\forall y)[y \in \emptyset \to x \in y]$ is
true for *all* sets x, hence $\cap\emptyset = V$.

If a theory is designed to equate $\cap\emptyset$ with $\emptyset$, as some do, it is because of
a total inadmission of proper classes, or as a result of other logical restrictions
intended to force $\cap$, like $\cup$, to be a unary operator $V \to V$.[3]

---

[1]Russell [**184**]. Chapter 18.
[2]$(\forall a)(\exists b)(\forall c)[c \in b \leftrightarrow (\exists d)[d \in a \wedge c \in d]]$.
[3]See, for example, the theory (with urelements – or more precisely, which does not exclude
the possibility of admitting urelements) given in Suppes [**206**].

---

1. At the moment, since we are still unable to collect together infinitely many sets, we can only say that if $a_1, \ldots, a_n$ are sets, then so is $\{a_1, \ldots, a_n\}$. The class of all sets V, at this point, could consist only of sets of finite cardinality.

2. If $B \backslash A$ denotes the complement of A in B then it is easy to prove from the definitions that $B \backslash \cup A = \cap B \backslash A$ and $B \backslash \cap A = \cup B \backslash A$.

A new axiom is needed to grant set status to the union of a set. In particular this means that, via the Axiom of Pairing, unordered n-tuples of sets are sets, so the Axiom of Unions might be regarded as an extension of the Axiom of Pairing, although it does not replace it (it cannot tell us that $\{a\}$ is a set if a is a set, for example, nor can we replace Pairing with an axiom that only says the latter). A much more powerful generator of new sets is the power set operation.

4.3   Power sets

> *The method of 'postulating' what we want has many advantages; they are the same as the advantages of theft over honest toil.*
>
> – BERTRAND RUSSELL[1]

The statement 'A is a subclass of B', denoted $A \subseteq B$, means every element of A is an element of B, that is, it is an abbreviation for the well-formed formula $(\forall x)[x \in A \to x \in B]$. The expression $A \subset B$ means A is a subclass of B and $A \neq B$.

If a is a set then the power set of a, $P(a)$, is the class $\{x : x \subseteq a\}$. One can demonstrate that the axioms and rules of inference we have given so far are insufficient to prove that $P(a)$ is a set for all sets a, so we adopt this assertion as our fourth non-logical axiom.

Axiom 4: Axiom of Powers
For all sets a, $P(a)$ is a set.[2]

If a is a finite set with n elements then the power set of a has 2^n elements. More generally, if a is a set then 2^a, the set of all functions $f : a \to \{0, 1\}$, has the same cardinality as $P(a)$ and we sometimes see the notation 2^a in place of $P(a)$, treating the bijection mapping $f \in 2^a$ to the subset $\{x : f(x) = 1\}$ of a as an identity.

---

[1] Russell [184]. Chapter 7.
[2] $(\forall a)(\exists b)(\forall c)[c \in b \leftrightarrow (\forall d)[d \in c \to d \in a]]$.

It is the Axiom of Powers which introduces a curious element of flexibility into the metatheory of ZF. Viewed externally, the power set of ω in one model of ZF may differ dramatically from the power set of ω in another model, as we shall discuss later, and it is this lack of absoluteness – the availability of 'spare' subsets of ω outside small models – which is exploited in the independence results of Cohen.

There is a natural inclination among some mathematicians to modify ZF (or its extensions) so that the Axiom of Powers does not admit pathological sets of real numbers. One axiom which has been suggested to play such a 'taming' role (and which is used as an alternative to the Axiom of Choice) is the game theoretic flavoured Axiom of Determinacy which we shall meet in Subsection 9.3.4.

REMARK

There are only countably many 1-predicates $\phi(x)$ and consequently there are only countably many associated class symbols $\{x : \phi(x)\}$. Fixing some set X, there are only countably many definable subsets of X, these being precisely $X \cap \{x : \phi(x)\}$ as $\phi(x)$ ranges over all 1-predicates (that this intersection is always a set follows from Replacement – see the next section). This means, for any infinite X, by Cantor's Theorem most sets in $P(X)$ cannot be given a description in the form $\{x : \phi(x)\}$, for a 1-predicate ϕ.

Consider, for example, the ZF-definable subsets of $\mathbb{N}$. Enumerating the 1-predicates (ϕ_n), some of the classes $\{x : \phi_n(x)\}$ will be forced to be empty by the non-logical axioms of ZF, many will be identical, and many non-identical classes will have equal intersections with $\mathbb{N}$, so we can expect a lot of repetitions, but the point is we can list the definable subsets $X_n = \mathbb{N} \cap \{x : \phi_n(x)\}$ of $\mathbb{N}$. At this point the natural inclination is to 'diagonalize'. Define a subset D of $\mathbb{N}$ by $n \in D$ if and only if $n \notin X_n$, so that D differs from all definable subsets and hence is not definable. The problem is, we cannot define D or indeed *definability itself* in the first-order theory of ZF; we cannot say 'there exists a 1-predicate ϕ such that $a = \{x : \phi(x)\}$'. So in an external sense we can see that not all subsets of $\mathbb{N}$ are definable, and we can even 'construct' an example of an undefinable set by diagonalization, but this is out of reach of the theory itself.

What we are anticipating here is the potential richness of different models of ZF. We shall see later that, paradoxical as it may sound, there are *countable* models of ZF.

It should be added that one can interpret the above result as saying that the Axiom of Powers is 'too strong' and that we should instead confine ourselves to more conservative classes of subsets, such as the class of definable subsets, or some other recursively defined collection. This is a fruitful avenue of research. One can still do a lot of classical mathematics assuming such restricted power set axioms.

The Axiom of Powers tells us that the class of all subsets of a set is again a set. This powerful axiom together with the Axiom of Unions helps us to generate a vast set hierarchy; as soon as we admit the class of natural numbers as a set we can generate an infinite class of different sizes of infinities. We now need an axiom which defines sets in terms of properties, but which avoids the problems caused by the Axiom of Abstraction.

4.4   Replacement

Perhaps the most surprising thing about mathematics is that it is so surprising. The rules which we make up at the beginning seem ordinary and inevitable, but it is impossible to foresee their consequences. These have only been found out by long study, extending over many centuries.

– E. C. TITCHMARSH[1]

4.4.1   An axiom schema

Searching for a weaker alternative to the paradox-friendly Axiom of Abstraction, Zermelo proposed in 1908 the Axiom of Separation. This asserts that the class of all objects in a *set* satisfying a given property is also a set. In other words, if a is a set and A is a class, then $a \cap A$ is a set. Zermelo's original statement of this axiom incorporated a rather vague notion of 'definiteness' of a property:

For any set x and any predicate ϕ which is meaningful ('definite') for all elements of x, there exists a set y that contains just those elements a of x which satisfy the condition $\phi(a)$.[2]

The meaning of 'definiteness' remained in an uncomfortably poorly defined state for over a decade.

Fraenkel and Skolem independently sought to clarify Zermelo's idea of Separation and their two solutions, both devised around 1922, turned out to be essentially equivalent. Definite statements for Skolem are those which satisfy a precise definition of formula (what we have called a well-formed formula).[3] In Fraenkel's formulation,[4] in modern terms, a predicate $\phi(u, v)$ *induces a function*

---

[1]Titchmarsh [**213**]. Last chapter.

[2]Zermelo, E., 'Untersuchungen über die Grundlagen der Mengenlehre: I', *Math. Annalen*, **65**, 261–281, (1908).

[3]Skolem, T., 'Einige Bemerkungen zur axiomatischen Begründung der Mengenlehre', *Wiss. Vortöge gehalten auf dem 5. Kongress der Skandinav. Mathematiken in Helsingfurs*, (1922).

[4]Fraenkel, A., 'Zu den Grundlagen der Cantor–Zermeloschen Mengenlehre', *Math. Annalen*, **86**, 230–237, (1922). See also Fraenkel's 'Über den Begriff "definit" und die Unabhängigkeit des Auswahlaxioms', *Sitzungsb. d. Preuss. Akad. d. Wiss., Physik.-math. Klasse*, 253–257, (1922).

if, for all u, whenever $\phi(u, v)$ and $\phi(u, w)$ both hold, then $v = w$.[1] Fraenkel's Axiom asserts that if $\phi(x, y)$ induces a function and a is a set then the class $\{y : (\exists x)[\phi(x, y) \wedge x \in a]\}$, the 'range' of the induced function, is a set. Zermelo later aimed to clarify the notion of definiteness, but it was not as successful as Skolem's construction.[2] Hermann Weyl also deserves mention as independently simplifying definiteness.[3]

Fraenkel's idea gives us the fifth axiom of ZF, an axiom schema comprising an infinite bundle of axioms, one axiom for each function inducing predicate $\phi(x, y)$.

Axiom 5: Axiom Schema of Replacement
If a predicate $\phi(x, y)$ induces a function then for all sets a,
$\{y : (\exists x)[x \in a \wedge \phi(x, y)]\}$ is a set.[4]

The Axiom of Replacement is strictly stronger than Zermelo's axiom schema of separation. Richard Montague proved that no finite extension of Zermelo's original set theory is as strong as the extension of the same theory by the addition of the full Replacement schema.[5]

The reader may be uncomfortable with this infinite cluster of axioms and pine for a finite alternative. Unfortunately no such animal exists. In other words, given any finite collection of axioms $\mathfrak{A}_1, \ldots, \mathfrak{A}_n$ expressible in the language of ZF (we might take these axioms to be each of the other axioms of ZF together with a finite number of instances of Replacement, for example), ZF will always be able to prove something that $\mathfrak{A}_1 + \cdots + \mathfrak{A}_n$ cannot. We shall give an argument for this in Subsection 10.3.1.

Gödel–Bernays set theory achieves a finite axiomatization by introducing class variables, the Schema of Replacement reduced to a single axiom of the form 'if a class C induces a function f_C and a is a set then $\{f_C(x) : x \in a\}$ is a set'. More generally, a second-order extension of ZF would express Replacement in a single sentence by making use of an operator variable.

That Replacement implies Separation is straightforward: suppose $\psi(x)$ is a predicate, let A be the class $\{x : \psi(x)\}$ and let y be a variable that does not occur in $\psi(x)$, then applying Replacement to the predicate $\phi(x, y)$ defined by

---

[1]Equivalently, this means that the class $\{(u, v) : \phi(u, v)\}$ is a single-valued class (see Subsection 5.1.2).

[2]Zermelo, E., 'Über den Begriff der Definitheit in der Axiomatik', *Fundamenta Mathematicae*, **14**, 339–344, (1929). This paper is criticized by Skolem in 'Einige Bemerkungen zu der Abhandlung von E. Zermelo: "Über den Begriff der Definitheit in der Axiomatik"', *Fundamenta Mathematicae*, **15**, 337–341, (1930).

[3]Weyl, H., 'Über die Definitionen der mathematischen Grundbegriffe', *Mathematisch-naturwissenschaftliche Blätter*, **7**, 93–95, (1910).

[4]$(\forall a)(\forall b)(\forall c)[\phi(a, b) \wedge \phi(a, c) \rightarrow (\forall d)[d \in b \leftrightarrow d \in c]] \rightarrow (\forall a)(\exists b)(\forall c)[c \in b \leftrightarrow (\exists d)[d \in a \wedge \phi(d, c)]]$.

[5]Montague, R., 'Semantic closure and non-finite axiomatizability I', *Infinitistic Methods: Proceedings of the Symposium on Foundations of Mathematics* (Warsaw, 1959), New York: Pergamon, 45–69, (1961).

$\psi(x) \wedge x = y$ we obtain the result that for all sets a,

$$\begin{aligned} \{y : (\exists x)[x \in a \wedge \phi(x, y)]\} &= \{y : (\exists x)[x \in a \wedge \psi(x) \wedge x = y]\} \\ &= \{y : y \in a \wedge \psi(y)\} \\ &= a \cap A \end{aligned}$$

is a set. (ϕ induces a function, for if $\phi(x, a)$ and $\phi(x, b)$ both hold then $\psi(x) \wedge x = a$ and $\psi(x) \wedge x = b$ both hold, and in particular $a = b$.)

Although Fraenkel and Skolem are credited with the invention of the Axiom of Replacement, it was von Neumann who was to recognize the importance of Replacement in the theory and application of ordinal numbers and in the construction of the cumulative hierarchy of sets.[1]

REMARKS

1. What, then, can Replacement do that cannot be done in its absence? Let us consider two examples. In each example we use the ordinal class ω. The formal definition and arithmetic of ordinals will be described later, but for now it will suffice to think of ω as the ordered set of natural numbers and keep in mind the definition of ordinals from the introduction. Note also that ω is well-ordered (this being part of its definition) which means that mathematical induction is valid over ω.

Let $\phi(x, y)$ be the predicate $x \in \omega \wedge y = \omega + x$. Then ϕ clearly induces a function. By Replacement (applied to the set ω) we have that

$$\{\omega, \omega + 1, \omega + 2, \ldots\}$$

is a set and hence its union with ω,

$$\omega + \omega = \{0, 1, 2, 3, \ldots \omega, \omega + 1, \omega + 2, \ldots\}$$

is a set. Without Replacement we cannot prove that $\omega + \omega$ is a set, so the axiom is crucial for the theory of ordinals.

The second example was given by Skolem as a reason to adopt Replacement. Let $X_0 = \{\omega\}$. This is a set. Define X_1 to be $\{\omega, P(\omega)\}$. By Pairing (or Unions) and Powers this is also a set. Next $X_2 = \{\omega, P(\omega), P(P(\omega))\}$ is a set, again by Unions and Powers. For $n \in \omega$ define

$$X_{n+1} = \{\omega\} \cup \{y : y = P(x) \wedge x \in X_n\}.$$

If X_n is a set then X_{n+1} is a set. By induction all X_n are sets, but without Replacement we cannot say that

$$\begin{aligned} \{\omega, P(\omega), P(P(\omega)), P(P(P(\omega))), \ldots\} &= \{y : (\exists n)[n \in \omega \wedge y \in X_n\} \\ &= \cup\{y : n \in \omega \wedge y = X_n\} \end{aligned}$$

---

[1] Von Neumann develops his set theory (which was to evolve into Gödel–Bernays set theory) in 'Eine Axiomatisierung der Mengenlehre', *J. Reine Angew. Math.*, **154**, 219–40, (1925).

is a set. We cannot use the Axiom of Unions here because, without Replacement, we cannot assert that $\{y : n \in \omega \wedge y = X_n\}$ is a set.

2. Replacement tells us that if A is equipollent to a set then A is a set. In some expositions the Axiom of Replacement takes the form: if $f : V \to V$ and x is a set, then $f''x = \{y : (\exists a \in x)[f(a) = y]\}$ is a set.

The Axiom Schema of Replacement tells us that for any definable function f : $A \to B$ the image $\{f(x) : x \in a\}$ of a set $a \subseteq A$ under f is always a set. Using this powerful axiom we can construct increasingly complex sets from existing sets and we can give certain abstract classes set status. Importantly, one such class is the empty set.

4.4.2   The empty set: our first concrete set

We found earlier that the underlying logic itself implies a non-empty universe of discourse. Some authors, perhaps nervous of this quirk of logic, or if they use an alternative logic, like to add an axiom that says that the universe of discourse is non-empty, usually by stating that the empty class $\{x : x \neq x\}$ is a set. The Axiom of Infinity, to be introduced in Subsection 6.2, states that the class ω is a set, and the empty set plays a role in ω's definition.

If A and B are classes then $A - B$ is the class $\{x : x \in A \wedge x \notin B\}$. Via Replacement (in fact, Separation suffices), for any set a and any class A, the class $a - A = a \cap \{x : x \notin A\}$ is a set. For all sets a we have

$$
\begin{aligned}
a - a &= a \cap \{x : x \notin a\} \\
&= \{x : x \in a \wedge x \notin a\} \\
&= \{x : x \neq x\} \\
&= \varnothing
\end{aligned}
$$

(the penultimate equality simply follows from the tautology $x \in a \wedge x \notin a \leftrightarrow x \neq x$, or equivalently $x = x \leftrightarrow x \in a \vee x \notin a$), so by the above we conclude that $\varnothing$ is a set. We use the empty set, which is our first explicit example of a set, to build all other sets.[1]

The notion of the empty set at first provoked the same sort of controversy that the number zero enjoyed centuries earlier. It is difficult to appreciate the source of the discomfort nowadays, perhaps it was simply due to the fusion-collection distinction, or to more philosophical problems. Even in the twentieth century Zermelo called the empty set 'improper'.[2]

---

[1] It is customary to make a light quip at this point about making something from nothing, although liberal use of the axioms and rules of inference together with the definition of $\varnothing$ cannot seriously be described as 'nothing'.

[2] Zermelo, E., 'Ünter suchungen über die Grundlagen der Mengenlehre I', *Math. Ann.* **65**: 261–81, (1908).

---

1. The notation $A - B$ is unlikely to intrude on the same notation in arithmetic. In case there is a clash, the set theoretic difference is denoted by $A \backslash B$ instead.

2. The task of trying to comprehend the difficulties encountered by the mathematicians of another age is as insurmountable as any other historical endeavour, but is further compounded by modern presentations of the subject of their toil, since resolved, in a crisp anxiety-free order dictated by logic. This order is, more often than not, far from the chronological order of discovery and in many cases begins with axiomatized hindsight-laden preliminaries, going on to deftly weave past decades worth of obstacles in a few pages. This is a gift for the student of mathematics, but of little use to the historian.

With the Axiom of Replacement we are able to prove that the empty set, initially defined as an abstract class, is indeed a set. This is the initial seed of the set theoretic universe. Using the axioms of ZF, from the empty set (and from ω, which is also promoted to set status by another axiom) one generates all other sets. Given the apparent power of Replacement it is not too surprising that it makes one of our earlier set-making axioms redundant: the Axiom of Pairing.

4.4.3   The redundancy of Pairing

We are now in a position to see why the Axiom of Pairing is redundant. Suppose a and b are sets. Replacement, as described above, tells us that $\varnothing$ is a set, so by the Axiom of Powers $\{\varnothing\}$ is a set, and hence so is $P(\{\varnothing\}) = \{\varnothing, \{\varnothing\}\}$. The predicate $\phi(x, y)$ defined by

$$[x = \varnothing \wedge y = a] \vee [x = \{\varnothing\} \wedge y = b]$$

induces a function and so by Replacement $\{y : x \in \{\varnothing, \{\varnothing\}\} \wedge \phi(x, y)\} = \{a, b\}$ is a set.

Here we used the Axiom of Powers to prove the existence of a two element set, and from that deduced the Axiom of Pairing. By a similar argument we can show that as soon as a model of Replacement contains some finite set with n elements, then it will contain all sets of the form $\{a_1, \ldots, a_n\}$ where $a_1, \ldots, a_n$ are in the model. In particular, iterating this, if a is in the model, the model must also contain each of the sets $\{a\}$, $\{\{a\}\}$, $\{\{\{a\}\}\}$, and so on.

REMARK

The Axiom of Replacement is an extremely strong one, and its full strength is unlikely to be needed in typical mathematical settings. It is a 'set theorist's axiom', critically used within the subject of set theory but only weakly used outside of set theory.

The Axiom of Pairing has now been demoted as a corollary of the more powerful Axiom of Replacement in combination with the Axiom of Powers. By introducing such a powerful set-making axiom there is a danger that certain classes which we intuitively prefer not to be sets might be granted set status. The other axioms of the theory should further control the types of sets that are allowed in the theory.

4.5 Regularity

...by natural selection our mind has adapted itself to the conditions of the external world. It has adopted the geometry most advantageous to the species or, in other words, the most convenient. Geometry is not true, it is advantageous.

– JULES HENRI POINCARÉ[1]

4.5.1 Banishing membership loops and ensembles extraordinaire

An effective approach to avoiding paradoxes is to posit that no set be permitted to be an element of itself and more generally to avoid 'membership loops', i.e. finite collections of sets $\{a_1, \ldots, a_n\}$ for which $a_1 \in a_2 \in \cdots \in a_n \in a_1$. To prevent such curiosities Zermelo introduced the following axiom, also known as the Axiom of Foundation.

Axiom 6: Axiom of Regularity
If $a \neq \emptyset$ then there exists an $x \in a$ such that $x \cap a = \emptyset$.[2]

Although the Axiom of Regularity as we have come to recognize it is due to

---

[1]Poincaré [**167**].

[2]$\neg(\forall b)[b \in a \leftrightarrow \neg(\forall c)[c \in b \leftrightarrow c \in b]] \rightarrow (\exists x)[x \in a \wedge (\forall b)[b \in x \wedge b \in a \leftrightarrow \neg(\forall c)[c \in b \leftrightarrow c \in b]]]$.

Zermelo,[1] von Neumann[2] and Mirimanoff[3] had anticipated the need for such an axiom.

The version of Regularity quoted above is sometimes called the weak form of Regularity, the strong form asserting the same except with the set a replaced by a *class*, however, the distinction is unnecessary, for it turns out that the weak and strong forms of Regularity are equivalent in ZF (see Section 6.5). Membership loops $a_1 \in \cdots \in a_n \in a_1$ are forbidden by Regularity because for any such loop the set $a = \{a_1, \ldots, a_n\}$ has the Regularity-incompatible property $x \cap a \neq \emptyset$ for all $x \in a$.

The class $V = \{x : x = x\}$ is, via the tautology $(\forall x)[x \in a \leftrightarrow x \in a]$ (which simply asserts that every set is equal to itself), the class of all sets. V is easily seen to be a proper class by an application of Regularity: if V were a set then we would have $V \in V$. Regularity also gives rise to the following, which is an induction principle of sorts.

If for all a, $a \subseteq A$ implies $a \in A$, then $A = V$.

The proof is easy: if A is not equal to V then the non-empty class $V - A$ possesses, by Regularity, an element x such that $x \cap (V - A) = \emptyset$, that is, for all y, if $y \in x$ then $y \notin V - A$, which in turn implies $y \in A$. So $x \subseteq A$ and, by the hypothesis, $x \in A$, a contradiction. Therefore $A = V$. This theorem assures us that if, for all sets a, a satisfies a property ϕ whenever all elements of a satisfy ϕ, then *every* set has property ϕ (let A be the class $\{x : \phi\}$, the hypothesis is then a straightforward rewording of the statement that $a \subseteq A$ implies $a \in A$, and the conclusion $A = V$ simply asserts that all sets have property ϕ).

In 1917 Mirimanoff[4] first pondered the question of whether set theory could accommodate *ensembles extraordinaires*, by which he meant infinite descending membership chains $\cdots \in a_2 \in a_1 \in a$. Under the assumption of Regularity, no such chain exists: invoking the induction result just mentioned, we simply need to show that if each element of a set a has no infinite descending $\in$-chain then a itself has no infinite descending $\in$-chain. The contrapositive statement – if a has an infinite descending $\in$-chain then there exists an element of a which has an infinite descending $\in$-chain – is transparent. The reader might be nervous of the appearance of the natural number indexed sequence $\cdots \in a_3 \in a_2 \in a_1 \in a_0$ above. To remedy this we will return to this result in Subsection 6.2.2 after defining ω.

The empty set and the class of all sets V are the two most extreme classes in the sense that any class A satisfies $\emptyset \subseteq A \subseteq V$. The Russell class $\mathbf{Ru} = \{x : x \notin x\}$ is, by Regularity, *equal to V*. The proof is immediate.

---

[1] Zermelo, E., 'Über Grenzzahlen und Mengenbereiche', *Fundamenta Mathematicae*, **16**, 29–47, (1930).

[2] von Neumann, J., 'Eine Axiomatisierung der Mengenlehre', *J. Reine Angew. Math.*, **154**, 219–240, (1925) and von Neumann, J., 'Über eine Widerspruchsfreiheitsfrage in der axiomatischen Mengenlehre', *J. Reine Angew. Math.*, **160**, 227–241, (1929).

[3] Mirimanoff, D., 'Les antinomies de Russell et de Burali-Forti et le problème fondamental de la théorie des ensembles', *Enseignement Math.*, **19**, 37–52, (1917).

[4] Mirimanoff, D., 'Les antinomies de Russell et de Burali-Forti et le problème fondamental de la théorie des ensembles', *Enseignement Math.*, **19**, 37–52, (1917).

1. Regularity prevents the snake from eating its own tail and by doing so simultaneously blocks the absolute and relative forms of Russell's paradox (and indeed makes the focal class of each form of the paradox identical). However, it is only with this purpose in mind that it is introduced; self-membered sets, membership loops and *ensembles extraordinaire* simply don't appear in typical mathematical settings, and Regularity is rarely invoked even in metamathematical results. In this sense it is the weakest (the least necessary) of the axioms of ZF.

2. Regularity is strictly stronger than the clean statement denying self-membership: $(\forall x)[x \notin x]$. The latter does not block, for example, the sandwiched membership chain $x \in y \in x$ (it is difficult, however, to imagine an intuitive example of distinct aggregates x and y satisfying this).

3. Assuming Regularity it is possible to redefine the ordered pair (a, b) as $\{a, \{a, b\}\}$, i.e. to remove the braces around the first set in Kuratowski's definition. (I leave the reader to check the details.)

The Axiom of Regularity excludes from the theory sets which are members of themselves and more generally membership loops such as $a_1 \in a_2 \in \cdots \in a_n \in a_1$. It also excludes infinite descending membership chains $\cdots \in a_3 \in a_2 \in a_1$. The axiom attacks Russell's paradox at its source: by adopting Regularity the Russell class becomes the proper class of all sets. Some alternatives to Regularity have been also been explored.

4.5.2   Alternatives to Regularity

The Axiom of Regularity has consequences that appear to be very natural and it is automatically satisfied in most mathematical settings, nevertheless set theories which contradict it have been investigated in detail. Regularity arguably has less of the 'indisputable' quality of the other axioms and was, for a variety of reasons, one of the last axioms to have been adopted by set theorists. For example, in a 1963 note outlining his proof of the independence of the Continuum Hypothesis from ZF, Cohen still needs to explicitly state that the Axioms of Zermelo–Fraenkel set theory are to exclude the Axiom of Choice *but include the Axiom of Regularity*.[1]

Regularity is a safe assumption in the sense that it cannot introduce any contradictions into the theory which are not already present in its absence (von

---

[1]Cohen, P., 'The independence of the Continuum Hypothesis', *Proc. N.A.S.*, **50**, 1143–1148, (1963).

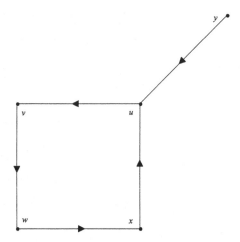

Figure 4.1   A directed graph representing the non-well-founded set $\{u, v, w, x, y\}$ with $u \in v \in w \in x \in u$ and $y \in u$.

Neumann proved that if ZF without Regularity is consistent then so is ZF), but removing Regularity does not seem to handicap set theory as severely as the removal of other axioms of ZF. Most of mathematics is unaffected by the omission of Regularity purely because the sets one encounters in non-exotic settings are easily seen to be free of membership loops without having to appeal to any special axioms in order to force them to be so.

In some applications of set theory to computer science a modification of ZF which allows membership loops is useful. The resulting theory is called *non-well-founded* set theory. Its sets may be defined within ZF minus Regularity, but in addition to extensionality a new axiom is needed to distinguish between non-well-founded sets. The existence of atoms is also often assumed. A non-well-founded set is commonly pictured as a directed graph, a collection of vertices representing the members of the set joined by directed edges indicating membership. For example the non-well-founded set $\{u, v, w, x, y\}$ with $u \in v \in w \in x \in u$ and $y \in u$ is represented by the directed graph shown in Figure 4.1. The so-called Anti-foundation Axiom then asserts that *every* directed graph depicts exactly one non-well-founded set. We shall not pursue non-well-founded set theory any further here.[1]

REMARK

Different non-well-founded set theories give alternative notions of equality between non-well-founded sets. The axiom alluded to above is Aczel's Antifoundation Axiom. An example of a non-well-founded set in this theory is a *Quine atom*, a set a with $a = \{a\}$. Such a set is represented by a single vertex and a loop joining the vertex to itself.

---

[1] For details see, for example, Aczel [**2**].

By modifying the other axioms of ZF, Regularity can be excluded and one can study an alternative universe of non-well-founded sets. These non-well-founded sets can be pictured as directed graphs, the vertices representing elements and the directed edges indicating membership. Choosing to adopt Regularity, we have only one more axiom of ZF to describe. Before we meet that we need some technical machinery to help us precisely define the ordinal numbers.

5

Relations and order

5.1 Cartesian products and relations

At the age of eleven, I began Euclid, with my brother as my tutor.
This was one of the great events of my life, as dazzling as first love.
I had not imagined there was anything so delicious in the world.
From that moment until I was thirty-eight, mathematics was my
chief interest and my chief source of happiness.

– BERTRAND RUSSELL[1]

5.1.1 Functions and relations

The clarification of the concept of function, even when confined to the subject of analysis, had become a matter of desperate importance in the eighteenth and nineteenth centuries. Fourier's representation of piecewise continuous functions $[a, b] \to \mathbb{R}$ by trigonometric series and, later, Weierstrass and Riemann's examples of continuous nowhere differentiable functions and functions which are nowhere continuous yet integrable forced a generation of mathematicians (who were accustomed to treating functions of real variables as visualizable geometric entities) to reluctantly revise their understanding. A precise and general definition of function which is not tied to analysis is expressible within the set-theoretic framework.

Functions and relations saturate every part of mathematics. Even in our humble introduction we had to introduce these ideas at an early stage in order to construct models of the basic number systems. Now that the logical prerequisites have been brought into focus it is worth reviewing the formal definitions.

The *cartesian product* of two classes $A \times B$ is the class of all ordered pairs (a, b) with $a \in A$ and $b \in B$, i.e. $\{x : (\exists a)(\exists b)[a \in A \land b \in B \land x = (a, b)]\}$. If $A_1, \ldots, A_n$ are classes then the n-fold cartesian product $A_1 \times \cdots \times A_n$ is the set of n-tuples $(a_1, \ldots, a_n)$ with $a_i \in A_i$. As $(a_1, \ldots, a_n)$ is formally the ordered

---

[1]Russell [189].

pair $((a_1, \ldots, a_{n-1}), a_n)$, we lose no generality in confining our discussion to binary relations and functions of one variable. If a and b are sets then their cartesian product $a \times b$ is also a set. Indeed, if $x \in a$ and $y \in b$ then both $\{x\}$ and $\{x, y\}$ are elements of $P(a \cup b)$ and so $\{\{x\}, \{x, y\}\}$, i.e. (x, y), is an element of $P(P(a \cup b))$. Therefore $a \times b \subseteq P(P(a \cup b))$. From the Axiom of Unions $a \cup b$ is a set and two applications of the Axiom of Powers tell us that $P(P(a \cup b))$ is a set. From the Axiom of Replacement every subclass of a set is a set, so we conclude that $a \times b$ is a set.

A function $F : A \to B$ is a subclass F of $A \times B$ with the property

$$(\forall a)[a \in A \to (\exists! b)[(a, b) \in F]],$$

and if $(a, b) \in F$ we would normally write $F(a) = b$. The expression $(\exists! x)\phi$ is a shorthand for 'there exists a unique x such that $\phi(x)$', which means firstly that there exists an x such that $\phi(x)$ and secondly that for all x, y, if $\phi(x)$ and $\phi(y)$ then $x = y$. More formally, then, $(\exists! x)\phi$ is an abbreviation for the formula

$$(\exists x)\phi(x) \land (\forall x)(\forall y)[[\phi(x) \land \phi(y)] \to x = y].$$

We use the abbreviation A^2 for the class $A \times A$ (and A^n for the n-fold cartesian product $A^{n-1} \times A$ of n copies of A). A^{-1} denotes the class $\{(x, y) : (y, x) \in A\}$, that is, A^{-1} comprises all the ordered pairs in A but with their order reversed. If A contains elements that are not ordered pairs then $(A^{-1})^{-1} \neq A$. A $relation$ on a class A is simply a subclass of A^2. If we say 'R is a relation' without mention of another class we mean $R \subseteq V^2$.

As all classes are subclasses of the class V, we have $V^2 \subset V$. Consideration of cardinality alone shows that the strict inclusion $A^2 \subset A$ can hold only if A is infinite. The simplest example of a set a with $a \times a \subset a$ is the union $a = \cup_{i \in \omega} a_i$ where a_i is defined recursively by

$$
\begin{aligned}
a_0 &= \varnothing \\
a_{i+1} &= \{(x, y) : x \in a_n,\ y \in a_m,\ 0 \leq n \leq i,\ 0 \leq m \leq i\},
\end{aligned}
$$

so

$$
\begin{aligned}
a_0 &= \varnothing, \\
a_1 &= \{\varnothing\}, \\
a_2 &= \{\varnothing, (\varnothing, \varnothing)\}, \\
a_3 &= \{\varnothing, (\varnothing, \varnothing), (\varnothing, (\varnothing, \varnothing)), ((\varnothing, \varnothing), \varnothing), ((\varnothing, \varnothing), (\varnothing, \varnothing))\}, \\
&\ \vdots
\end{aligned}
$$

In this example the difference between $a \times a$ and a is as small as possible, only the absence of the empty set in $a \times a$ obstructing equality.

For non-empty A the inclusion $A \subseteq A^2$ $never$ holds. Suppose $A \subseteq A^2$ and choose, by Regularity, an element x of A which is minimal in the sense that for any membership chain $x_1 \in x_2 \in \cdots \in x$ none of the x_i are elements of

A, then $x = (a, b) = \{\{a\}, \{a, b\}\}$ for some $a, b \in A$. In particular there is an $a \in A$ with $a \in \{a\} \in x$, contradicting minimality. Although this argument seems to critically rely on the Kuratowski definition of ordered pairs, the proof applies to any set theoretic definition of the ordered pair (a, b) in which the set a is embedded, i.e. where there is a finite (possibly empty) sequence of elements $a_1, a_2, \ldots a_n$ with $a \in a_1 \in a_2 \in \cdots \in a_n \in (a, b)$.

The injections $f_x : A \to A^2$, $x \in A$, defined by $f_x(a) = (a, x)$ or $f_x(a) = (x, a)$ naturally embed a copy of A in A^2 and it is this type of embedding that is intended, as described in the introduction, when we casually write '$\mathbb{R} \subset \mathbb{C}$', $\mathbb{C}$ being regarded as $\mathbb{R}^2$ with certain algebraic operations.

REMARK

$(A^{-1})^{-1}$ is precisely the set of ordered pairs in A, so it is the 'largest relation' that is a subclass of A. If R is a relation then it is symmetric if and only if $R^{-1} = R$. $R \cap R^{-1}$ is the largest symmetric subrelation of a relation R and $R \cup R^{-1}$ is the smallest symmetric superrelation of R.

Now that the language and sufficient axioms of ZF have been set up, we can give a rigorous definition of cartesian products, relations and functions. We shall look at various modifications, generalizations and particular cases of these notions.

5.1.2   Generalizations

Functional notions enjoy a smooth generalization to classes in general simply by 'ignoring' those elements of the class which are not ordered pairs. For example, we say that a class A is *single-valued* if for all u, v, w the equality $v = w$ holds whenever $(u, v) \in A$ and $(u, w) \in A$ and A is *injective* if A and A^{-1} are both single-valued. The familiar notions of domain, range, restriction, image and composite are similarly extended as follows.

(i) The domain of a class A is the class $\{x : (\exists y)[(x, y) \in A]\}$.

(ii) The range of a class A is the class $\{y : (\exists x)[(x, y) \in A]\}$.

(iii) The restriction of A to B is $A \cap (B \times V)$.

(iv) The image of B under A, denoted $A(B)$, is the range of the restriction of A to B. One sometimes finds the notation $A''B$ in place of $A(B)$.

(v) The composite of A with B, denoted $A \circ B$, is the class

$$\{(x, y) : (\exists z)[(x, z) \in B \wedge (z, y) \in A]\}.$$

If A is single-valued then a simple application of Replacement proves that the image of any set under A is a set. (The result fails if A is not single-valued. Consider the class $A = \{a\} \times V$. The image of the set $\{a\}$ under A is the proper class V.) Indeed, it is a straightforward exercise to prove that the class of sets is stable with respect to all of the constructions we have just introduced: if a is a set then a^{-1}, the domain of a, and the range of a are all sets. Corollaries: $A \times B$ is a set if and only if $B \times A$ is a set; and if A is a proper class and $B \neq \varnothing$ then $A \times B$ and $B \times A$ are proper classes.

For a class A, $A'b$ is the class:

$$\{x : (\exists y)[x \in y \wedge (b, y) \in A] \wedge (\exists! y)[(b, y) \in A]\}.$$

In other words, if A is a function and if b is in the domain of A then $A'b$ is the value of A at b in the usual sense; if b is not in the domain of A then $A'b = \varnothing$; if b is in the domain of A but there are two different ordered pairs in A with first entry b then again $A'b = \varnothing$; and if b is in the domain of A and (b, y) is the only ordered pair in A with first entry b then $A'b = y$. We will make use of this generalization of function later in the context of transfinite recursion.

It is easy to take the formal definitions for granted. They emerged after a difficult period of revision – the syllogistic reasoning of old being inadequate for the demands of the new theory. Frege, and others, extended formal logic from its Aristotelian roots in such a way that it was capable of handling relations and functions and hence captured the (semi)-intuitive reasoning being employed in the new mathematics.

REMARK

R is a transitive relation if and only if $R \circ R \subseteq R$. Relations satisfying $R = R \circ R$ have the strong 'interpolation property' that whenever aRc there exists a b such that aRb and bRc (transitive reflexive relations trivially have this property, but a relation satisfying $R = R \circ R$ needn't be reflexive, consider $<$ on $\mathbb{Q}$ for example).

This suggests an algebraic approach to relations. We define I to be the constant relation $\{(x, x) : x \in V\}$, so R is reflexive if and only if $I \subseteq R$. One can then characterize all relational properties (function, injective function, surjective function, bijection, equivalence relation etc.) as short algebraic expressions in terms of the relation and possibly I.

The study of relation algebra, as it came to be known, began in the nineteenth century (Augustus De Morgan and Charles Peirce were the pioneers and their work was extended by Ernst Schröder). This continued with the more abstract work of Alfred Tarski and others.

An arbitrary class can be regarded as a generalized function in a natural way by examining the occurrence and uniqueness of certain ordered pairs in the class. Next we look at relations and ordering in general and isolate certain types of relation that are modelled on some familiar prototypes.

5.2 Foundational, well-ordered and well-founded relations

> *To new concepts correspond, necessarily, new signs. These we choose in such a way that they remind us of the phenomena which were the occasion for the formation of the new concepts.*
>
> – DAVID HILBERT[1]

5.2.1 Foundational relations

Let R be a relation on a class A. The class of R-*predecessors* of a given $a \in A$ is the class $R^{-1}(a) = \{x : xRa\}$ (when R is a strict order this is also called the R-*initial segment* determined by a). If S is a non-empty subset of A and $m \in S$ has the property $R^{-1}(m) \cap S = \emptyset$, then m is an R-*minimal element* of S. If *every* non-empty subset of A has an R-minimal element then R is a *foundational relation*.

Note that the definition neatly excludes from the class of foundational relations all relations R for which there exists an a with aRa, for in such a case the singleton $\{a\}$ is a non-empty set with no R-minimal element. Indeed, more generally, the defining condition of a foundational relation tells us immediately that, in common with the membership relation, there are no 'infinite descending foundational chains' $\ldots a_3 R a_2 R a_1$ and no 'foundational loops' $a_1 R a_2 \ldots R a_n R a_1$ because in these cases the sets $\{a_1, a_2, \ldots\}$ and $\{a_1, \ldots, a_n\}$, respectively, would have no R-minimal elements. If R is a foundational relation on A then one can prove that every non-empty sub*class* of A has an R-minimal element.[2]

Foundational relations are sufficiently well structured to admit an induction principle. If R is a foundational relation on A, B is a subclass of A, and, for all $x \in A$, $A \cap R^{-1}\{x\} \subseteq B$ implies $x \in B$, then $A = B$. In terms of a property ϕ (setting $B = \{x : \phi\} \cap A$) this means that if, for all $x \in A$, x has property ϕ whenever every element of the R-initial segment $A \cap R^{-1}(x)$ has property ϕ then every element of A has property ϕ.

An instructive example of a foundational relation is the following.[3] Define F on the class of all finite sets by aFb if and only if a has fewer elements than b. F is foundational because in any non-empty collection C of finite sets there exists a set a which has an equal number or fewer elements than all sets in C; any such a is F-minimal.

There are two pertinent observations to make about this relation: firstly, the F-minimal set a need not be unique – there may be more than one set in C with the same cardinality as a; secondly, for any finite set a with at least two

---

[1]Hilbert's address *Mathematische Probleme* was printed in *Göttinger Nachrichten*, (1900) and in *Archiv der Mathematik und Physik* (1901). An English translation appears in the *Bulletin of the American Mathematical Society* (1902).

[2]We shall return to this result, and the induction principle of the following paragraph, in our discussion of rank in Subsection 6.5.2.

[3]We use the notion of finiteness again here, some time before we have properly defined it, however, the comments in the introduction will suffice for the time being.

elements the set of F-predecessors of a is a *proper class* – for example, if a is a pair, then $F^{-1}(a) = \{\{x\} : x \in V\}$. This motivates the two natural candidates for stricter forms of foundational relation described in Subsections 5.2.2 and 5.2.3 below.

This relation does not extend naturally to a foundational relation on *all* sets: if we define I on the class of all sets by aIb *if and only if there exists an injection from a onto a proper subset of b* then the restriction of I to the class of finite sets coincides with the foundational relation F, however I itself is not foundational: the set of natural numbers is related to itself. Indeed aIa for all Dedekind infinite sets a by definition. The same is true of the subset relation $\subset$: on the class of all sets V, $\subset$ is not foundational – it is easy to manufacture an infinite descending $\subset$-chain $\dots a_3 \subset a_2 \subset a_1$ by taking, say, a_n to be the complement of $\{0, 1, 2, 3, \dots, n\}$ in the set of natural numbers, but restricted to the class of finite sets $\subset$ is a foundational relation (it is also well-founded – see Subsection 5.2.3 below – since $\subset^{-1}(a) = P(a) - a$).

REMARKS

1. In our example of cardinal comparison of finite sets the corresponding induction principle is: if $\phi(F)$ is true whenever $\phi(E)$ is true for all sets E of smaller cardinality than F, then $\phi(G)$ is true for all finite sets G. The induction principle corresponding to the subset relation on the class of all finite sets is: if $\phi(F)$ is true whenever $\phi(E)$ is true for all proper subsets of F, then $\phi(G)$ is true for all finite sets G.

2. We can define a foundational relation $\prec$ on all of $P(\mathbb{N})$ as follows. Define $\min(\emptyset) = 0$ and define $A \prec B$ if and only if $\min(A) < \min(B)$. The corresponding induction principle is: if $\phi(F)$ is true whenever $\phi(E)$ is true for all subsets E of $\mathbb{N}$ with $\min(E) < \min(F)$, then $\phi(G)$ is true for all subsets G of $\mathbb{N}$. I make no judgement on the usefulness of these induction principles.

A foundational relation on a class is a relation with the property that every subset of the class has a minimal element with respect to the relation. This property imposes enough order on a class for it to satisfy some elementary induction principles and to banish foundational loops and infinite descending foundational chains. The minimal element need not be unique, and the class of relational predecessors of an element need not be a set.

5.2.2 Well-ordering

If a foundational relation R on A has the property that every non-empty subset of A has a *unique* R-minimal element, then R is said to be a *well-ordering* (A is *well-ordered* by R). The most immediate example of a well-ordering is $<$ on the set of all natural numbers.

The uniqueness of the minimal element forces a well-ordering R to be a linear ordering, i.e. R is transitive and for all $x, y \in A$ with $x \neq y$ we have exactly one of xRy or yRx. The proof is fairly straightforward: suppose W is a well-ordering on a class A and let $x, y \in A$ be distinct elements. If x and y were not W-comparable then the set $\{x, y\}$ would have two W-minimal elements, x and y, contradicting the uniqueness property. So we have xWy or yWx. If both xWy and yWx then $\{x, y\}$ would have no W-minimal element, contradicting the foundational property. Now consider three elements $x, y, z \in A$. Suppose xWy, yWz and $\neg xWz$. From what we have just proved this means xWy, yWz and zWx, and furthermore x, y and z are distinct. It follows that $\{x, y, z\}$ has no W-minimal element, again contradicting the foundational property, and we conclude that W is transitive.

REMARKS

1. Every countable set has a well-ordering, but we should emphasize that there are many different such orderings. Also recall that every countable ordinal can be represented as a subset of $\mathbb{Q}$ with respect to the usual ordering on $\mathbb{Q}$.

   The usual ordering on $\mathbb{N}$ is a well-ordering, but there are (uncountably many) other well-orderings. For example, we might order all even numbers before all odd numbers: $0 \prec 2 \prec 4 \prec 6 \prec \cdots 1 \prec 3 \prec 5 \cdots$ (yielding an ordering with order type $\omega + \omega$), or fixing some n, order all multiples of n before all numbers which are congruent to 1 mod n, then all numbers which are congruent to 2 mod n and so on, yielding an ordering of order type ωn. To describe a reordering of $\mathbb{N}$ of order type ω^2 we might place zero and all powers of two at the beginning, followed by all numbers of the form $2^k + 1$ that haven't already been used, then numbers of the form $2^k + 2$ that haven't already been used, and so on: $0 \prec 1 \prec 2 \prec 4 \prec 8 \prec 16 \prec \cdots \prec 3 \prec 5 \prec 9 \prec 17 \prec \cdots 6 \prec 10 \prec 18 \prec 34 \prec \cdots$.

2. The foundational induction principle applied to a well-ordering of an arbitrary set leads to the notion of transfinite induction, which we discuss in the next chapter.

Using the relation $<$ on the natural numbers as a prototype, a well-ordering is a foundational relation for which every subset has a unique minimal element. Well-orderings are automatically linear.

5.2.3   Well-founded relations

If R is a foundational relation on A with the property that, for all $a \in A$, $R^{-1}(a)$ is a *set* then R is said to be a *well-founded relation*. The prototypical example of a well-founded relation is the membership relation $\in$ on the class of all sets V: that $\in$ is foundational follows from the Axiom of Regularity – $a \cap x = a \cap \in^{-1}(x)$ and so there exists an $x \in a$ with $a \cap \in^{-1}(x) = \emptyset$; and the well-founded property is immediate from $\in^{-1}(a) = \{x : x \in a\} = a$. $\in$ is not a well-ordering – one can easily find $\in$-incomparable sets.

Well-orderings need not be well-founded. The relation R on the proper class $\mathcal{O}n \times \{0, 1\}$ defined by

$(\alpha, 0)R(\beta, 0)$ if $\alpha \leq \beta$;

$(\alpha, 0)R(\beta, 1)$ for all α, β; and

$(\alpha, 1)R(\beta, 1)$ if $\alpha \leq \beta$,

is a well-ordering but, for any α, $R^{-1}((\alpha, 1))$ is a proper class. We picture this ordering as two copies of $\mathcal{O}n$, all members of one copy preceding the members of the second copy.

REMARK

Sets on which $\in$ *is* a well-ordering and a transitive relation will be our models of the ordinal numbers, which we come to in the next chapter.

Using the membership relation on the class of all sets as a prototype, a well-founded relation is a foundational relation with the property that the class of relational predecessors of every element is a set. Well-founded relations need not be well-orderings and well-orderings need not be well-founded. A natural question arises: which classes admit foundational relations, well-founded relations or well-orderings?

5.3   Isomorphism invariance

> *Everything is vague to a degree you do not realize till you have tried to make it precise, and everything precise is so remote from everything that we normally think, that you cannot for a moment suppose that is what we really mean when we say what we think.*
>
> – BERTRAND RUSSELL[1]

---
[1]Russell [**183**].

If R_1 and R_2 are relations on classes A_1 and A_2, respectively, then an *iso-morphism* $\phi : A_1 \to A_2$ is a bijection with the property aR_1b if and only if $\phi(a)R_2\phi(b)$ for all $a, b \in A_1$. The class of all relational systems is partitioned into isomorphism equivalence classes – (A_1, R_1) and (A_2, R_2) being equivalent if and only if there exists an isomorphism $A_1 \to A_2$.

Isomorphisms map minimal elements to minimal elements and initial segments to initial segments, that is, if $\phi : A_1 \to A_2$ is an order isomorphism and if $a \in A_1$ is R_1-minimal then $\phi(a)$ is R_2-minimal in A_2 and $\{\phi(x) : xR_1a\}$ is an R_2-initial segment of A_2 for all $a \in A_1$. From this it follows that if R_1 is respectively a foundational relation, well-founded relation or well-ordering then R_2 is the same. Consequently, if in a given isomorphism equivalence class there is a relational system that is foundational then every relational system in that equivalence class is foundational, if there is a relational system in the equivalence class that is well-founded then all systems in the class are well-founded and (importantly for the theory of ordinal numbers) if one system is well-ordered, all are well-ordered.

If a bijection $f : A \to B$ exists between two classes A and B and S is a relation on B then a relation R on A of the same type as S is induced simply by defining aRb if and only if $f(a)Sf(b)$. While this has its uses (in particular, it proves that all countable sets have a well-ordering) it does not provide any information as to what types of ordering may be defined on a given class in the first place.

The fact that the membership relation is well-founded tells us immediately that well-founded relations (and hence foundational relations) exist on any class, but the question of whether a class has a well-ordering has no such simple resolution. In fact the hypothesis that every *set* can be well-ordered, the Well-Ordering Theorem, is equivalent to the Axiom of Choice and hence, from the work of Cohen, is undecidable in ZF. That is, neither the assertion nor the negation of the assertion that every set can be well-ordered can be proved from the axioms of ZF.

Remark

Isomorphisms obviously preserve reflexivity, symmetry and transitivity, and so if an isomorphism equivalence class contains an equivalence relation, then all relations in that class are equivalence relations. Two equivalence relations on sets of the same cardinality will be equivalent provided their respective equivalence classes are of compatible cardinalities. More precisely, for each cardinal κ each equivalence relation must have the same number of equivalence classes of cardinality κ. So on a finite set of n elements the number of equivalence relations (up to isomorphism) is equal to the number of different partitions of n (the nth *Bell number*).

Since the membership relation is well-founded, well-founded relations can be defined on any class, however, the existence of a well-ordering of every set cannot be proved without appealing to the Axiom of Choice. Indeed, the assumption that every set has a well-ordering is equivalent to the Axiom of Choice.

6

Ordinal numbers and the Axiom of Infinity

6.1 Ordinal numbers

> *Guided only by their feeling for symmetry, simplicity, and generality, and an indefinable sense of the fitness of things, creative mathematicians now, as in the past, are inspired by the art of mathematics rather than by any prospect of ultimate usefulness.*
>
> – ERIC TEMPLE BELL[1]

6.1.1 The emergence of ordinality

Cantor described an ordinal number, where M is an arbitrary well-ordered set, as '*the general concept which results from M if we abstract from the nature of its elements while retaining their order of precedence...*'.[2] Making this precise, the classical view of ordinal numbers begins by defining the notion of an order type – the set of all ordered sets which are order isomorphic to some fixed ordered set – and then isolates the ordinal numbers as the collection of all order types of *well-ordered* sets.

This approach is intuitively pleasing to a certain degree, yet one might object to something as 'simple' as 1 being modelled by the class of all single element sets, which is, after all, a proper class. Some would provocatively adopt an extremist position, arguing that the apparent simplicity we perceive in the Platonic idea of ordinal numbers is an illusion; an unfortunate result of brainwashing in infancy. From a constructivist point of view the most immediate objection to the classical definition of ordinal numbers is that the entire set theoretical universe must be defined before we can extract any ordinals from it.

---

[1] Bell [13].

[2] See Philip Jourdain's translation of Cantor's papers of 1895 and 1897 in Cantor [29]. The translation given here differs a little from Jourdain's.

Von Neumann's alternative (formulated in the early 1920s)[1] is to define ordinal numbers to be sets a which are well-ordered by the membership relation and which are *transitive* in the sense that $x \in a$ implies $x \subset a$. For example, $\{\emptyset, \{\emptyset\}, \{\emptyset, \{\emptyset\}\}\}$ is the ordinal number 3. Every well-ordered set a is order isomorphic to exactly one of von Neumann's ordinal sets.

REMARK

There is a layman's misunderstanding that all mathematics is born pristine and precise – everything exquisitely defined, clear as crystal. This is generally not so. Most original mathematical ideas, like all good ideas, are formed out of a struggle and emerge as vague hints towards the final polished idea. Cantor's ideas are no exception. He refers to '*definite and separate objects of our intuition or our thought*' when defining a set, and there are similar 'psychological' allusions elsewhere in his work. This is not too precise, but it is a good thing to draw attention to the intuitive or even imaginary nature of sets (and *all* other mathematical objects) once in a while. Working with them, they become familiar enough to seem real, rather than idealized (consistent) generalizations of physically observable things. The consistency is crucial. Mathematics is a *consistent* ideal extension of the real world; it has its feet firmly on the ground.

Cantor's notion of ordinal numbers has a vague psychological quality. The definition was later improved, treating an ordinal number as an equivalence class of well-ordered sets under the relation of order isomorphism. This also has its problems; one must define the entire universe first in order to extract the equivalence classes. Von Neumann's definition takes the best aspects of both ideas. Each ordinal number is a concrete set, precisely a well-ordered transitive set, and the successor of an ordinal number is defined by a simple set theoretic operation.

6.1.2   The modern realization of ordinal numbers

A is an *ordinal class* if it is transitive (i.e. if $x \subset A$ whenever $x \in A$) and well-ordered by $\in$. We call ordinal sets *ordinal numbers* or simply *ordinals*. The membership relation well-orders the class $\mathcal{O}n$ of all ordinal numbers, and $\mathcal{O}n$ is clearly transitive, so $\mathcal{O}n$ is an ordinal class. If $\mathcal{O}n$ were a set then we would have $\mathcal{O}n \in \mathcal{O}n$, contradicting Regularity, therefore $\mathcal{O}n$ is a proper class.

If A and B are distinct ordinal classes, then either $A \in B$ or $B \in A$. The proof relies on the following four results.

(i) $X \notin X$ for all classes X.

---

[1] Neumann, J. von, 'Zur Einführung der transfiniten Zahlen', *Acta Szeged*, **I**, 199–208, (1923).

(ii) For all classes A and B, $A \cap B \subseteq A$ and $A \cap B \subseteq B$.

(iii) If A and B are ordinal classes, then $B \subset A$ if and only if $B \in A$.

(iv) If A and B are ordinal classes, then $A \cap B$ is an ordinal class.

(i) follows immediately from Regularity; (ii) is trivial; (iii) is, in one direction, an immediate consequence of the definition of ordinals, and in the converse direction follows from the properties of $\in$ (in particular, from the fact that $\in$ is a well-founded well-ordering); and (iv) is easily proved directly.

Let us assume A and B are distinct ordinal classes and $A \cap B \subset A$ and $A \cap B \subset B$. By (iii) and (iv) we have $A \cap B \in A$ and $A \cap B \in B$ and so $A \cap B \in A \cap B$, contradicting (i). Therefore, by (ii), $A \cap B = A$ or $A \cap B = B$, so either $B \subset A$ or $A \subset B$, which by (iii) means $A \in B$ or $B \in A$.

It follows that among all the ordinal classes only one is not an ordinal number: if A is an ordinal class then either $A \in \mathcal{O}n$ or $A = \mathcal{O}n$. The Burali-Forti paradox evaporates – $\mathcal{O}n$ is *not* an ordinal number.

The class of ordinal numbers is naturally partitioned into two subclasses. The class of *successor ordinals* is the class containing the empty set (newly labelled 0) together with all ordinals of the form $\alpha \cup \{\alpha\}$ (a set which we shall call $\alpha + 1$). All other ordinals are *limit ordinals*. The ordinal class ω is defined to be the class of all ordinal numbers α such that α and all ordinal numbers less than α (i.e. all members of α) are successor ordinals. That ω is an ordinal class follows firstly from the fact that $\omega \subseteq \mathcal{O}n$ (we haven't yet ruled out $\omega = \mathcal{O}n$) and secondly that if $x \in \omega$ then for all $y \in x$, y and all members of y are successor ordinals, i.e. $x \subseteq \omega$.

The elements of ω form a model of the natural numbers, i.e. ω satisfies Peano's Postulates:

(i) $0 \in \omega$;

(ii) for all i, $i + 1 \in \omega$;

(iii) for all i, $i + 1 \neq 0$;

(iv) for all i, j, $i + 1 = j + 1$ if and only if $i = j$; and

(v) if $0 \in A$ and if, for all i, $i \in A$ implies $i + 1 \in A$, then $\omega \subseteq A$.

The first three properties follow trivially from the definition of ω, as does the implication $i = j \to i+1 = j+1$ of (iv). The converse implication of (iv) follows from the fact that $\in$ is foundational, for if $i + 1 = j + 1$, i.e. $\{i, \{i\}\} = \{j, \{j\}\}$, then $i = j$ or $i \in j$, but $j = i$ or $j \in i$, each ruling out $i \in j$: $i \in j = i$ or $i \in j \in i$. The induction principle (v) is proved as follows: suppose A is a class satisfying $0 \in A$ and $i \in A \to i + 1 \in A$, and suppose there is an element of ω not in A, then since ω is well-ordered there must be a smallest such element m. We know that $0 \in A$, so $m \neq 0$, but all non-zero members of ω are of the form $\alpha + 1$, so $m = m' + 1$ for some ordinal m'. By the minimality of m, $m' \in A$, but this implies $m = m' + 1 \in A$, a contradiction.

Thus the principle of finite induction, a trivial corollary of (v) above, is revealed to be a theorem of ZF: if $A \subseteq \omega$ and $0 \in A$ and if for all i, $i \in A$ implies $i + 1 \in A$, then $A = \omega$.

We say a set is *finite* if it is equipollent to an element of ω and is *infinite* otherwise.

REMARKS

1. The von Neumann successor operator $a \mapsto a \cup \{a\}$ is arguably the most natural one. If we were to look at or imagine a set of objects then, besides each element of the set, one instantly perceives another object, the set itself, and since this cannot be equal to any of its members it represents something new in the universe, a new object that can be added to the original collection. This presents a challenge against naive ultrafinitism – a theory which is able to collect together any finite collection of objects and call it another object is able to define a new larger set. The inability to collect a finite number of objects together imposes a huge constraint on the theory.

2. The condition of transitivity implies that the membership relation is a transitive relation on the class of ordinals: if $x \in y$ and $y \in z$ then by transitivity of z we have $x \in y \subseteq z$, so $x \in z$. Conversely, if $\in$ is a transitive relation on A and $x \in A$ then for all $y \in x$ we have $y \in A$. In other words $x \subseteq A$. Therefore the statements 'A is a transitive class' and '$\in$ is a transitive relation on A' are equivalent.

3. Since $\in$ is the order relation on $\mathcal{O}n$, we use $<$ interchangeably with $\in$, and the set theoretic union is the supremum operator.

The class $\mathcal{O}n$ of all ordinal numbers is a proper class. An ordinal class is either equal to $\mathcal{O}n$ or is an ordinal number. The ordinal class ω is the class of all ordinal numbers α such that α and all ordinals less than α are successor ordinals. A set is finite if it is equipollent to a member of ω. ω is the smallest class containing the empty set and the successor of each of its elements. Alternative models of the natural numbers are also possible.

6.1.3   Modelling the natural numbers

Modelling the natural numbers as the class ω, i.e. the smallest class containing $\emptyset$ which is stable under the application of the von Neumann successor operator $a \mapsto a \cup \{a\}$, is standard in ZF, however there are various other ways of producing a copy of the natural numbers; one simply specifies a successor operator on the class of all sets, that is, an injection $s : V \to V - \{\emptyset\}$ (see the remarks at the end of this subsection), and the 'alternative' ω is the smallest class containing $\emptyset$

with the property that $s(x) \in \omega$ whenever $x \in \omega$. Note, however, that by using an alternative successor we must also make some further fairly drastic changes to the theory, for example one must also revise the general definition of ordinal numbers so that it is compatible with the successor operator, finiteness can no longer generally be defined in terms of equipollence with elements of ω, and there are likely to be problems in describing the arithmetic and structure of the ordinals (problems which are absent in the von Neumann representation).

For instance, the successor operator $s(a) = \{a\}$ yields a copy of the natural numbers in the form $\varnothing$, $\{\varnothing\}$, $\{\{\varnothing\}\}$, $\{\{\{\varnothing\}\}\}$,... (the Zermelo model) in contrast to the standard representation $\varnothing$, $\{\varnothing\}$, $\{\varnothing, \{\varnothing\}\}$, $\{\varnothing, \{\varnothing\}, \{\varnothing, \{\varnothing\}\}\}$,... that we get using the von Neumann successor. Ultimately it does not matter which successor operator is used, however, there is a strong case for using the von Neumann successor, which has numerous technical advantages over the other choices. In particular the general definition of von Neumann ordinals requires little effort – an ordinal is a transitive set well-ordered by the membership relation, the successor operator is then used to partition $\mathcal{O}n$ into successor and limit ordinals and to define ω, and the set of all ordinals is linearly ordered by the simple membership relation.

By contrast, a Zermelo ordinal would have to be defined in a way which recalls Cantor's original principles of generation: we begin with the empty set, generate the successors $\{\varnothing\}$, $\{\{\varnothing\}\}$, $\{\{\{\varnothing\}\}\}$..., then form the union of all sets so constructed, generating the new ordinal ω, then form all successors of ω: $\{\omega\}$, $\{\{\omega\}\}$,... etc. By relying so critically on this recursion, the general definitions (ordinals, successor ordinals, limit ordinals, order) in the theory of Zermelo ordinals will be considerably less elegant than their von Neumann counterparts. It is easy to see why the von Neumann construction of ordinals has been favoured. Indeed, I would argue that it is almost inevitable that on close inspection of what would be required to construct 'Zermelo ordinals' one would be led, by the need for a transitive ordering, to various modifications, eventually hitting upon the von Neumann construction.

There is also an obvious intuitive advantage to the von Neumann model of the natural numbers: each element has exactly the same cardinal value as the intuitive ordinal it is intended to represent, whereas all but one of the elements in Zermelo's model is a singleton (meaning that the definition of finiteness we have just given will have to be abandoned).

Having finally found a set theoretic model of the class of natural numbers, ω, knowing how to interpret ordered pairs as sets $(a, b) = \{\{a\}, \{a, b\}\}$ and having the other axioms of ZF at hand, we have enough machinery to define set theoretic models of all of the classical number systems. It is worth pausing to appreciate how elaborate these sets are. How, for example, are we to represent the complex number i as a set? Recall i is to be interpreted firstly as an ordered pair of real numbers $(0, 1)$, that is, it is the set $\{\{0\}, \{0, 1\}\}$, where 0 and 1 are real numbers. But a real number is, for us, a Dedekind cut, so each of the real numbers appearing in $\{\{0\}, \{0, 1\}\}$ is an abbreviation for a certain ordered pair of *sets* of rational numbers (or just a set of rational numbers, depending on how we define Dedekind cuts). An *individual* rational number is an infinite set of ordered pairs (n, m), where n, m are integers, $m \neq 0$. An individual integer is an infinite set of ordered pairs of natural numbers. So, unravelling this, the humble complex number i is to be represented by an extremely complicated infinite set. Fortunately one never needs to unpeel the identities of set theoretic embedded classical objects in this way!

REMARKS

1. Let us denote by 0 the empty set. If $s : V \to V - \{0\}$ is an injection then it is a valid successor operator; we just need to show that the elements $0, s0, ss0, sss0, \ldots$ are pairwise distinct. By choice of codomain none of $s0, ss0, sss0, \ldots$ are equal to 0. If $s^n 0 = s^m 0$, with, say, $n \geq m$, then since s is injective we have $s^{n-1}0 = s^{m-1}0$, and hence $s^{n-2}0 = s^{m-2}0$, and so on, so that $s^{n-m}0 = 0$, implying $n = m$ as claimed.

2. With just one more axiom to go, we are very nearly able to model all of the classical structures of mathematics. It should be abundantly clear at this point that these models and the axioms that led to them do not serve a pedagogical function for the newcomer to mathematics. No one thinks of individual numbers intuitively as infinite sets – these are models of intuitive ideas, in this case expressed in terms of the basic underlying notion of sets. The point is to show that a sparse foundation (a machine language, if you wish) is powerful enough to generate in encoded form all of classical mathematics and much beyond. What we have shown is that it is possible to encode everything in the same language; to represent everything in terms of sets.

Different successor operators will give rise to different models of the natural numbers, however, von Neumann's successor enjoys numerous technical advantages over its rivals. In order to generate the full hierarchy of ordinals and to admit infinite sets we need one more axiom: the class ω needs to be promoted to set status. Once this is done we have the machinery to build all of the basic number systems described in the introduction.

6.2   The Axiom of Infinity

> *There is a concept which corrupts and upsets all others. I refer not to Evil, whose limited realm is that of ethics; I refer to the infinite.*
> — JORGE LUIS BORGES[1]

---

[1]'Avatars of the Tortoise' (1932), translated by James E. Irby, from *Labyrinths* [**26**], copyright ©1962, 1964 by New Directions Publishing Corp. Reprinted by permission of New Directions Publishing Corp.

6.2.1 The status of ω

As ω is an ordinal class it must either be equal to $\mathcal{O}n$ itself (yielding the classical finitist's universe of choice: all sets are finite; there are no limit ordinals) or it must be a member of $\mathcal{O}n$. The status of ω as a set or otherwise cannot be resolved by the axioms stated thus far (see Subsection 10.4.2) so we postulate as the last of the non-logical axioms of ZF that ω is a set.[1] As we shall see later, this is just one of a host of axioms of infinity which postulate the existence of ever larger cardinal numbers.

Axiom 7: Axiom of Infinity
ω is a set.

In this form the axiom is deceptively brief. If we adopt the common notation Mx, meaning 'x is a set' (M for *Menge*), then the axiom can be expressed even more economically in just two symbols: $M\omega$. However, if we mechanically unravel its constituent parts, writing out each definition in full, stopping when all but the basic logical symbols have been replaced and ignoring all possible simplifications, the result vomited forth is the magnificently unreadable:
$(\exists a)(\forall b)[b \in a \leftrightarrow (\forall c)[c \in b \to (\forall d)[d \in c \to d \in b]] \land (\forall c)(\forall d)[[c \in b \land d \in b] \to [c \in d \lor (\forall e)[e \in c \leftrightarrow e \in d] \lor d \in c]] \land (\forall c)[(\forall d)[d \in c \leftrightarrow d \in b] \lor (\forall d)[d \in c \leftrightarrow (\forall e)[e \in d \leftrightarrow e \in b]] \to (\forall d)[d \in c \to (\forall e)[e \in d \to e \in c]] \land (\forall d)(\forall e)[[d \in c \land e \in c] \to [d \in e \lor (\forall f)[f \in d \leftrightarrow f \in e] \lor e \in d]] \land [(\forall d)[d \in c \leftrightarrow (\forall e)\neg[e \in d \leftrightarrow e \in d]] \lor (\exists d)[(\forall e)[e \in d \to (\forall f)[f \in e \to f \in d]] \land (\forall e)(\forall f)[[e \in d \land f \in d] \to [e \in f \lor (\forall g)[g \in e \leftrightarrow g \in f] \lor f \in e]] \land (\forall e)[e \in c \leftrightarrow (\forall f)[f \in e \leftrightarrow f \in d] \lor (\forall f)[f \in e \leftrightarrow (\forall g)[g \in f \leftrightarrow g \in d]]]]]]]]]].$ The bulk of this monster comprises repeated statements that certain sets are ordinal numbers together with expressions of equality. Defining ω as the intersection of all inductive classes which contain $\varnothing$ and using von Neumann's axiom of infinity turns out to be a much simpler alternative (see Subsection 6.2.2 below).

The Formalist sensibly dodges all controversy over interpretation here; the axiom simply states that a certain well-defined class is a set – a finite string of symbols has been added to the class of true statements in the theory which, in combination with the other axioms, has an interesting web of consequences

---

[1]It is at this point that the Finitist will depart, preferring to assume the negation of the axiom: ω is a proper class. One can and of course one should study both variants, but the urge to amputate the theory, insisting that only the smaller theory is worth studying, does strike me as peculiarly conservative given the apparent consistency of the wider theory (the theory one obtains by replacing the Axiom of Infinity by its negation is rather close to Peano Arithmetic – see the remark to Subsection 10.4.2 for a more precise statement). The motivation seems to come from a dogmatic assumption that mathematics should be confined to modelling discrete physical structures or classical computational procedures as we currently understand them, introducing an unwelcome temporal element into mathematics thereby depicting 'completed' infinite sets as the end results of impossible supertasks. Viewing mathematics as a logically consistent and timeless game of the imagination, this censorship comes across as positively Orwellian, an act of thought policing! Insisting on confining ourselves to 'physically realizable mathematics' has the curious and unfortunate effect of suffocating physics itself, which often successfully draws on the esoteric and 'useless' mathematics of the past.

apparently (but, if we confine our reasoning to the system itself, not provably) free of contradiction.

At last we have described all of the axioms of ZF. Each of them is expressible in a finite number of symbols. One of them (Replacement) is indexed by a recursively enumerable set of well-formed formulas, so we can imagine winding a machine or getting a computer to progressively grind through the steps of an algorithm, generating the provable sentences of ZF.

This approach is generally too slow to do anything useful, but in principle this machine can be set up and, given any provable sentence, it will eventually churn the sentence out, or the Universe may die first.

Automated theorem proving these days employs far more sophisticated techniques than this exhaustive trawling, but there is still much research to do.

Since only one of the ordinal classes is a proper class we either have $\omega = \mathcal{O}n$, which would result in a classical finitist's universe, or ω must be a set. Neither alternative can be proved from the other axioms of the theory so to choose one way or the other would require a new axiom. Opting for the richer alternative we choose the axiom: ω is a set. Although the natural statement of this axiom seems to be elementary, if we unravel its meaning in pure logical symbols the statement is much longer than all the other axioms. Some alternative ways of stating the Axiom of Infinity have been explored.

6.2.2   Applications and alternatives

In Section 4.5 we promised to look again at the impossibility of *ensembles extraordinaire* following our definition of ω. We are now in a position to do this. If f is a function with domain ω then there exists an $n \in \omega$ such that $f(n+1) \notin f(n)$. This is a simple application of Regularity. As ω is assumed to be a set then (by Replacement) the image $a = \{f(n) : n \in \omega\}$ is also a set. Regularity tells us that there exists an $f(n) \in a$ such that $a \cap f(n) = \emptyset$ and, since $f(n+1) \in a$, $f(n+1) \notin f(n)$. The non-existence of infinite descending membership chains follows: if such a chain (a_n) exists, with $a_{n+1} \in a_n$ for all n, then $f(n) = a_n$ defines a function with domain ω such that $f(n+1) \in f(n)$ for all n, a contradiction. Clearly Regularity is very much at the heart of the matter here.

Zermelo and von Neumann introduced axioms of infinity in terms of their successor operators. Thus Zermelo postulated the existence of a set Z with the properties $\emptyset \in Z$ and $x \in Z \to \{x\} \in Z$ and von Neumann postulated the existence of a set W with the properties $\emptyset \in W$ and $x \in W \to x \cup \{x\} \in W$ (a set with the latter property is said to be *inductive*). Gödel also introduced an axiom of infinity in the form 'there is a non-empty set which has no inclusion maximal element'.[1] This means there exists a set a with the property that if $x \in a$ then there exists a $y \in a$ with $x \subset y$ (strict inclusion).

Most accounts of ZF use von Neumann's axiom, defining ω as the smallest inductive class which contains $\emptyset$, i.e. the intersection of all such inductive classes. As we do not have to repeatedly state that a set is an ordinal number, and since ω plays no role, the axioms of Zermelo, von Neumann and Gödel are substantially easier to express in pure logical terms than the ω-based axiom we gave in Subsection 6.2.1.

Zermelo's axiom of infinity is simply $(\exists a)[(\exists b)[b \in a \wedge (\forall c)[c \in b \leftrightarrow \neg(\forall d)[d \in c \leftrightarrow d \in c]]] \wedge (\forall b)[b \in a \rightarrow (\exists c)[c \in a \wedge (\forall d)[d \in c \leftrightarrow (\forall e)[e \in d \leftrightarrow e \in b]]]]]$.

Von Neumann's axiom of infinity is the slightly longer $(\exists a)[(\exists b)[b \in a \wedge (\forall c)[c \in b \leftrightarrow \neg(\forall d)[d \in c \leftrightarrow d \in c]]] \wedge (\forall b)[b \in a \rightarrow (\exists c)[c \in a \wedge (\forall d)[d \in c \leftrightarrow (\forall e)[e \in d \leftrightarrow e \in b] \vee (\forall e)[e \in d \leftrightarrow (\forall f)[f \in e \leftrightarrow f \in b]]]]]]$.

Gödel's axiom of infinity is the shortest of the four we have mentioned (using $(\exists b)[b \in a]$ for 'a is non-empty' rather than the literal $a \neq \emptyset$, i.e. $\neg(\forall b)[b \in a \leftrightarrow \neg(\forall c)[c \in b \leftrightarrow c \in b]])$ we have $(\exists a)[(\exists b)[b \in a] \wedge (\forall b)[b \in a \rightarrow (\exists c)[c \in a \wedge \neg(\forall d)[d \in b \leftrightarrow d \in c] \wedge (\forall d)[d \in b \rightarrow d \in c]]]]$.

These examples underline the importance of using higher level abbreviations rather than restricting ourselves to pure logical symbols – forcing oneself to work continually in 'machine code' is beyond perverse!

REMARK

Gödel's axiom is the only one of the axioms of infinity that we have mentioned that is not linked in any way to a natural successor operation, and indeed the set whose existence it declares need not contain the empty set. Here is an informal argument that Gödel's axiom implies that ω is a set (the converse is clear since ω satisfies the property of Gödel's axiom). Let a be the set whose existence is asserted by the axiom, then choose $x_0 \in a$. By assumption there exists an $x_1 \in a$ such that $x_0 \subset x_1$. By the same assumption there exists an x_2 with $x_0 \subset x_1 \subset x_2$, and so on. That is, there exists an injective function $f : \omega \rightarrow a$, so that ω is equipollent to a subclass of a set, implying via Replacement that ω is a set.

Each successor operator s gives rise to a different Axiom of Infinity in the form: there exists a non-empty set a such that $s(x) \in a$ whenever $x \in a$. One can then define the set of natural numbers to be the smallest such set containing the empty set. Gödel's Axiom of Infinity differs from these, declaring the existence of a non-empty set which has no inclusion maximal element. The principles of induction and recursion, which have proved to be so useful in countable settings, can be extended to powerful transfinite principles which allow us to prove results and build constructions on arbitrary well-ordered sets.

---

[1]See, for example, in Gödel's Collected Works [**82**], 'The consistency of the axiom of choice and of the generalized continuum hypothesis with the axioms of set theory' (1940).

6.3   Transfinite induction

> *Mathematics, rightly viewed, possesses not only truth, but supreme*
> *beauty – a beauty cold and austere, like that of sculpture, without*
> *appeal to any part of our weaker nature, without the gorgeous trap-*
> *pings of painting or music, yet sublimely pure, and capable of a stern*
> *perfection such as only the greatest art can show.*
>
> — BERTRAND RUSSELL[1]

6.3.1   Generalizing induction

We would like to extend finite induction, that is induction on ω, to a principle
which applies to all ordinals: transfinite induction. There are many approaches
to this,[2] for example, one can prove that if A is a class of ordinal numbers
having the property that $\alpha \in A$ whenever $\alpha \subseteq A$, then $A = \mathcal{O}n$. The proof is
straightforward (compare the induction result in Subsection 4.5.1). If A is not
equal to $\mathcal{O}n$ then the class $\mathcal{O}n - A$ has a $\in$-minimal element α, so $(\mathcal{O}n - A) \cap \alpha =$
$\varnothing$. Since $\alpha \subset \mathcal{O}n$ it follows that $\alpha \subseteq A$, which by the hypothesis means
$\alpha \in A$, a contradiction. Re-expressing this in terms of a property ϕ, setting
$A = \{x : x \in \mathcal{O}n \wedge \phi(x)\}$, we have the following: to prove $\phi(\alpha)$ for all ordinals
α it suffices to prove that if $\phi(\beta)$ for all $\beta < \alpha$ then $\phi(\alpha)$. The specialization of
the latter to ω yields complete induction.

A version of transfinite induction which is even closer in spirit to the familiar
finite induction is easily found; this is a mixture of complete induction and
normal induction performed on limit and successor ordinals respectively.

Principle of transfinite induction

   If

   (i) ϕ is a predicate;

   (ii) γ is a limit ordinal;

   (iii) $\phi(0)$ holds;

   (iv) for every $\alpha < \gamma$, $\phi(\alpha)$ implies $\phi(\alpha + 1)$; and

   (v) for every limit ordinal $\beta < \gamma$, $\phi(\beta)$ whenever $\phi(\delta)$ for all $\delta < \beta$,

   then $\phi(\alpha)$ for all $\alpha < \gamma$.

---

[1]Russell [**182**]. Chapter IV (The Study of Mathematics).
[2]Suppes [**206**] gives four different formulations of transfinite induction (and six formulations of transfinite recursion).

REMARK

We have already seen that with every foundational relation comes a principle of induction, so focussing on the particular case of the well-ordering $\in$ on $\mathcal{O}n$ we instantly get transfinite induction. The reason for splitting the predecessor condition between the two cases of successors and limit ordinals is a purely practical one. In most applications it is simply easier to apply in this form, as we shall see.

There are many ways of extending induction to arbitrary well-ordered sets. If a property holds for the successor of an ordinal whenever it holds for an ordinal, if it holds for a limit ordinal whenever it holds for all ordinals less than that limit ordinal, and if it holds for zero, then it is true for all ordinals. As soon as a well-ordering has been defined on a set, one can use this principle of transfinite induction to prove that the set has certain properties.

6.3.2 Applications

Hilbert had hoped to use (finite) induction to prove the consistency of various formal systems, a dream that was derailed by Gödel's Second Incompleteness Theorem. Nevertheless Hilbert's idea managed to survive incompleteness, albeit in a slightly indirect and mutated form – an 'externally realized' transfinite induction as far as the countable ordinal ε_0 (to be defined in Subsection 7.2.2) was employed by Gerhard Gentzen in 1936 in his argument for the consistency of Peano Arithmetic:[1] ordinals smaller than ε_0 are naturally assigned to proofs in Peano Arithmetic and the existence of a proof of internal contradiction is shown to entail the existence of an infinite descending sequence of ordinals, itself a contradiction.

Interestingly, Gentzen's proof operates in a theory which is not stronger and obviously not weaker than Peano Arithmetic, the two systems are simply incomparable, yielding a solid counterexample to the common misconception that any theory to which Gödel's Second Incompleteness Theorem applies can only be proved consistent in a strictly stronger theory. Similar techniques have been used for other systems.

REMARKS

1. The theory Gentzen worked in was Skolem's finitistic arithmetic theory *primitive recursive arithmetic* augmented with ε_0-induction. Primitive recursive arithmetic is a quantifier-free formalization of arithmetic which is usually adopted as a suitable background theory for Finitist mathematics.

---

[1]Gentzen, G., 'Die Widerspruchsfreiheit der reinen Zahlentheorie', *Mathematische Annalen*, **112**, 132–213, (1936).

2. Gentzen's remarkable result was the beginning of a new branch of proof theory known as *ordinal analysis*. Ordinal analysis has now successfully been used to prove the consistency of a wide range of theories (relative, of course, to the consistency of the theory in which the consistency proof is being formulated), but theories which allow very large cardinals via a power set operation and replacement seem to be out of reach of the technique at present.

One of the most well known 'metatheoretic' uses of transfinite induction was Gentzen's proof of the consistency of Peano Arithmetic in which ordinals are assigned to proofs and the existence of a contradiction in the theory entails the existence of an infinite descending sequence of ordinals. Transfinite induction is used to prove results about well-ordered sets. Transfinite recursion is used to construct infinite sets with some desired properties.

6.4   Transfinite recursion

> *By and large it is uniformly true in mathematics that there is a time lapse between a mathematical discovery and the moment when it is useful; and that this lapse of time can be anything from 30 to 100 years, in some cases even more; and that the whole system seems to function without any direction, without any reference to usefulness, and without any desire to do things which are useful.*
>
> – JOHN VON NEUMANN[1]

6.4.1   Generalizing recursion

Given an element x of a set a and a function $h : a \to a$, there exists a unique sequence (x_n) such that $x_0 = x$ and $x_{n+1} = h(x_n)$ for all $n \in \omega$. This much taken for granted statement, the principle of (finite) recursion, was to receive its first formal treatment in the work of Dedekind in 1888.[2] Several variations of the principle are possible, for example one sometimes needs recursion with a parameter: if a is a set, (h_n) is a sequence of functions $a \to a$, and $x \in a$, then there exists a unique sequence (x_n) in a such that $x_0 = x$ and $x_{n+1} = h_n(x_n)$ for all $n \in \omega$.

Dedekind's contribution must not be underestimated. Recursive sequences had been used for centuries before the arrival of set theory, so it was inevitable that an over-familiarity was to taint the subject, in its formative years, with some hidden assumptions. It took Dedekind's insight to point out that there *is*

---

[1] *The Role of Mathematics in the Sciences and in Society* (1954), in volume VI of von Neumann [221].
[2] Dedekind [43].

something to be proved here, while others had used the principle blind to the fact that there was an issue at all.

With the principle of finite recursion one can confidently give recursive definitions of operations on ω, including the familiar addition, multiplication and exponentiation. However, the method of recursive definition is not confined to the natural numbers, it extends to a powerful general principle known as transfinite recursion, which applies to arbitrary ordinal numbers. This generalized recursion, which plays a much underappreciated but central role in set theory, was first stated and proved by von Neumann[1] who stressed the essential role played by the Axiom of Replacement in its proof. The proof does not, as is often claimed, require the Axiom of Choice; the confusion seems to arise because of the presence of a well-ordered set. Choice is not needed to prove the principle, but in *applying* transfinite recursion we work with a well-ordering of the set on which the recursion is being applied and we may need to invoke the well-ordering theorem in order to assert the existence of such an ordering.

Like transfinite induction, transfinite recursion has many variations; we choose a form which is most easily recognized as a direct generalization of finite recursion and which features a mixture of conditions on limit and successor ordinals similar to those employed in the version of transfinite induction presented in the previous section.

Let λ be a limit ordinal and suppose $h : \lambda \to \lambda$ and $g : P(\lambda) \to \lambda$. For any successor ordinal $\alpha + 1 < \lambda$ we define $f(\alpha + 1) = h(f(\alpha))$ and if $\beta < \lambda$ is a limit ordinal we define $f(\beta) = g(\{f(\gamma) : \gamma < \beta\})$. Having defined the recursion conditions h and g and fixed $f(0)$, f is then determined on all of λ.

This is more than required, of course. We could dispense with h and define $f(\beta) = g(\{f(\gamma) : \gamma < \beta\})$ for all $\beta < \lambda$, not just limit ordinals. The reason we do not give this more economical version of recursion here is that it is rather awkward to use in practice and can seem a little unnatural especially in the context of defining ordinal addition, multiplication and exponentiation as introduced in the next chapter.

In summary:

Principle of transfinite recursion

If λ is a limit ordinal, $h : \lambda \to \lambda$, $g : P(\lambda) \to \lambda$ and $\delta < \lambda$ then there is a unique f such that:

(i) f is a function $\lambda \to \lambda$;

(ii) $f(0) = \delta$;

(iii) $f(\alpha + 1) = h(f(\alpha))$ for all $\alpha + 1 < \lambda$; and

(iv) $f(\beta) = g(\{f(\gamma) : \gamma < \beta\})$ for all limit ordinals $\beta < \lambda$.

---

[1]von Neumann, J., 'Über die Definition durch transfinite Induktion und verwandte Fragen der allgemeinen Mengenlehre', *Math. Annalen*, **99**, 373–391, (1928).

This formulation of transfinite recursion admits a natural generalization: h and g may be replaced by arbitrary classes H and G, where $H(x)$ and $G(x)$ are interpreted as $H'x$ and $G'x$ as defined in Subsection 5.1.2. This is not a generalization for the sake of generalization; the recursion conditions can be given in terms of some property and there may be no *a priori* guarantee that the associated classes G and H determine functions.

REMARK

Note that the principle of transfinite recursion is not simply an existence result; it does, at least in principle, give us a way to construct the function f from g and h. We will give a few examples shortly.

The principle of recursion extends to a general principle which applies to arbitrary ordinals. We have already seen examples of this in the analysis that motivated the invention of the ordinal numbers in the first place – recall Cantor's iterated derivation operator, and Baire's functions. This recursion can also be used to prove some counterintuitive geometric results.

6.4.2   Geometric applications (a miscellany)

Transfinite induction can be used together with transfinite recursion and the Well-Ordering Theorem to prove a host of surprising geometric results. A few examples:[1]

1. There exists a subset A of $\mathbb{R}^2$ with the property that for every horizontal cross-section H of $\mathbb{R}^2$, $H \cap A$ is dense in H (i.e. for every $h \in H$ and all $\varepsilon > 0$ there is an $x \in H \cap A$ such that the distance between h and x is less than ε) and yet for every vertical cross-section V of $\mathbb{R}^2$, $V \cap A$ has exactly one element. A sketch of the construction: well-order $\mathbb{R}^2$ as $\{(x_\lambda, y_\lambda) : \lambda < \gamma\}$. Set $P_0 = \mathbb{R}^2$. Take the first point (x_0, y_0) in P_0 according to this well-ordering. Construct the horizontal copy of $\mathbb{Q}$ passing through (x_0, y_0), $A_0 = \{(x_0 + q, y_0) : q \in \mathbb{Q}\}$, and construct the set of all vertical lines passing through the points comprising A_0. We define P_1 to be P_0 minus all points lying on this set of vertical lines and on the horizontal line passing through (x_0, y_0). The well-ordering of $\mathbb{R}^2$ will then present us with a unique 'smallest' member of P_1, (x, y). As before, construct the horizontal copy of $\mathbb{Q}$ passing through (x, y), $A_1 = \{(x+q, y) : q \in \mathbb{Q}\}$, and construct the set of all vertical lines passing through the points comprising A_1. We define P_2 to be P_1 minus all points lying on this set of vertical lines and on the horizontal line passing through (x, y). Repeating this we see how to construct P_α for successor ordinals α. We then define, for limit

---

[1] Proofs of these statements can be found in Ciesielski [**33**], Chapter 6.

ordinals β, $P_\beta = \cap_{\delta<\beta}P_\delta$ and so A_δ is defined for all ordinals $\delta < \gamma$. The set in question is then the (disjoint) union $\cup_{\delta<\gamma}A_\delta$.

2. $\mathbb{R}^3$ is a disjoint union of circles.[1] By contrast $\mathbb{R}^2$ is not a disjoint union of circles.

3. There is a subset A of $\mathbb{R}^2$ that intersects every one of its translates $\{x+y : x \in A \wedge y \in \mathbb{R}^2 \wedge y \neq (0,0)\}$ in a single element.

4. There is a countable partition (X_i) of $\mathbb{R}^2$ such that the distance between any two different points of any given X_i is irrational.[2] The same result, much more difficult to prove, is true in $\mathbb{R}^n$ for $n > 2$.

5. There is a subset of $\mathbb{R}^2$ that intersects *every* line in exactly two points.[3] A sketch of the construction of such a set is as follows. For $x \neq y$ in $\mathbb{R}^2$ let $[xy]$ be the set of points $\{\lambda x + (1 - \lambda)y : \lambda \in \mathbb{R}\}$ comprising the unique line passing through x and y. Fix a well-ordering of $\mathbb{R}^2$ so that $\mathbb{R}^2$ is indexed by an ordinal segment: $\mathbb{R}^2 = \{a_\gamma : \gamma < \alpha\}$. Let $A_1 = \{a_0, a_1\}$ and $L_1 = [a_0 a_1]$. We define sets A_δ, L_δ and a class of ordinal numbers $\{\gamma_\delta\}$ by: γ_δ is the least ordinal such that $a_{\gamma_\delta} \notin L_\delta$; $A_{\delta+1} = A_\delta \cup \{a_{\gamma_\delta}\}$ and $A_\beta = \cup_{\delta<\beta}A_\delta$ for limit ordinals β; and $L_\delta = \cup[a_i a_j]$ where a_i, a_j range over distinct pairs in A_δ. The set $\cup_{\delta<\alpha}A_\delta$ has the stated property.

REMARK

'Well-order $\mathbb{R}^2$' we casually said above, which of course we can do if we assume the Axiom of Choice. However, there are infinitely many wildly different orders available. This is already abundantly clear when choosing a well-ordering of a countable set A. We could choose an ordering of A of order type ω (and there are uncountably many of those – one for each permutation of A), but we could equally well choose any other countable ordinal. The instinct is generally to choose the ordering to have order type corresponding to the smallest ordinal which is equipollent to the set in question (i.e. the cardinal number of the set), but depending on the result you are trying to prove and the structure (topological, algebraic, etc.) of the set it may be appropriate to use an alternative larger ordinal.

Assuming the continuum has a well-ordering we can make use of that well-ordering and use transfinite recursion to construct a host of surprising geometric results. Recursion is also used to define the set theoretic hierarchy.

---

[1] Andrzej Szulkin, '$\mathbb{R}^3$ is the disjoint union of circles', *The American Mathematical Monthly*, **90**, No. 9, 640–641, (Nov., 1983).

[2] Credited to Paul Erdős and András Hajnal (1969).

[3] S. Mazurkiewicz, 'O pewnej mnogsci plaskiej, ktora ma z kazda prosta dwa i tylkp dwa punkty wspolne', *C. R. Varsovie*, **7**, 382–384, (1914).

6.5 Rank

> *Man is a rational animal – so at least I have been told. Through-*
> *out a long life, I have looked diligently for evidence in favour of this*
> *statement, but so far I have not had the good fortune to come across*
> *it, though I have searched in many countries spread over three con-*
> *tinents.*
>
> – BERTRAND RUSSELL[1]

6.5.1 The cumulative hierarchy and Regularity

Recall from Subsection 3.3.3 that the von Neumann recursive set hierarchy is generated as the union $V = \cup V_\alpha$, where the set V_α, for ordinal α, is defined (by transfinite recursion) as follows:

$$
\begin{aligned}
V_0 &= \varnothing; \\
V_{\alpha+1} &= P(V_\alpha); \text{ and} \\
V_\beta &= \bigcup_{\alpha<\beta} V_\alpha, \text{ if } \beta \text{ is a limit ordinal.}
\end{aligned}
$$

The universe of sets is often pictured as a widening tower of nested classes forming a V-shaped wedge (see Figure 6.1).

If $\alpha < \beta$ then $V_\alpha \in V_\beta$ and $V_\alpha \subset V_\beta$. A set a is said to be *well-founded* if there exists an ordinal α such that $a \in V_\alpha$. It is easily proved that if x is well-founded for all $x \in a$ then a itself is well-founded. In combination with the Axiom of Regularity this tells us that *every set is well-founded*: the contrapositive of '*if, for all $x \in a$, x is well-founded then a is well-founded*' is '*if a is not well-founded then there is an $x \in a$ such that x is not well-founded*'. If a set a_0 is not well-founded, then we infer from the latter statement the existence of an *ensemble extraordinaire* of non-well founded sets $\cdots \in a_3 \in a_2 \in a_1 \in a_0$, contradicting Regularity.

So in the presence of Regularity the union V of the nested sets V_α is the entire universe of sets. Even in the absence of Regularity the set hierarchy can still be constructed by the recursion. One approach to the theory is to begin with this construction and use the statement that every set is well-founded as an axiom of regularity.

If α is a limit ordinal then any set in V_α can also be found in some V_β with $\beta < \alpha$. It follows that for each set a the smallest ordinal β for which $a \in V_\beta$ is a successor ordinal. This particular ordinal β we call the rank of a, rank(a). Note in particular that if $a \in b$ then rank(a) < rank(b) and the rank of an ordinal number α is α itself. The rank of a set can be regarded as a measure of its structural complexity.

---

[1] Russell [**187**]. Chapter 7 (An Outline of Intellectual Rubbish).

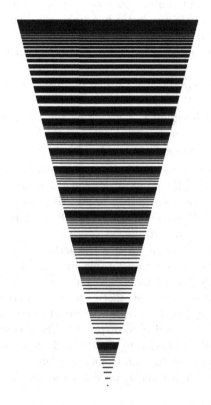

Figure 6.1 The hierarchy of sets begins simply enough with $V_0 = \varnothing$, $V_1 = \{\varnothing\}$, $V_2 = \{\varnothing, \{\varnothing\}\}$, $V_3 = \{\varnothing, \{\varnothing\}, \{\varnothing, \{\varnothing\}\}\}$, but expands hyperexponentially thereafter (V_4 has 16 elements, V_5 has 65 536 elements, and to write the number of elements of V_6 we need 19 729 digits). Once the pure finite sets are exhausted (all such sets comprising the level V_ω) we begin to climb through the infinite ordinal stages. The union V of all of the levels is, thanks to the Axiom of Regularity, the universe of all sets. The diagram is a symbolic representation of the initial part of this hierarchy up to V_{ω^2}. Each horizontal line represents one level in the hierarchy.

REMARK

Proofs of theorems which employ high ranking superstructures might be said to be more abstract than those that use only lower ranked sets. However, the main measure of the perceived difficulty of a proof is drawn from another less well-defined hierarchy: if the proof of theorem A relies, among other things, on theorem B then we would rank it no lower, in terms of complexity, than the proof of theorem B. This is an elastic hierarchy that can change dramatically, not only in height but in order, with new developments in mathematics; proofs that use the full force of some well-developed theory sometimes have elementary alternatives.

Perhaps the most well-known example of this kind of restructuring is the prime number theorem: if $\pi(n)$ is the number of primes not exceeding n then $\lim_{n\to\infty} \frac{\pi(n)\ln n}{n} = 1$. Roughly speaking this tells us that there are approximately $\frac{n}{\ln n}$ primes equal to or less than n. A much closer approximation to $\pi(n)$ is obtained by the logarithmic integral Li defined by $\text{Li}(x) = \int_2^x \frac{dt}{\ln(t)}$. This was proved using complex analysis in 1896, independently, by Jacques Hadamard[1] and Charles-Jean de la Vallée Poussin[2]. So, at the time we would have claimed that this is a high ranking result that relies critically on a number of results from complex analysis. However, in 1949 Paul Erdős[3] and Atle Selberg[4] found elementary proofs of the same result.

One could argue that, by a forensic inspection of the underlying logical details, all proofs ought to be translatable into an elementary combinatorial argument, after all any proof can be broken down into a long string of basic symbols. However, this is a very unrealistic strategy – in practice great ingenuity and insight is needed to isolate the right elements and eliminate the complex abstract machinery that is used in most proofs. There is also no guarantee that 'elementary' proofs will be easier to understand than their abstract parents, nor is it necessarily the case that an elementary proof will provide any special insight into the result in question; it may have philosophical but not practical or pedagogical value.

By the Axiom of Regularity the cumulative set hierarchy, defined using transfinite recursion, is the class of all sets. This hierarchy begins with the simplest set, the empty set, and creates progressively more complex sets as one moves upward. Using the notion of rank, which labels where a set first appears in this hierarchy, we can extend several results from sets to general classes.

[1] Hadamard, J., 'Sur la distribution des zéros de la fonction $\zeta(s)$ et ses conséquences arithmétiques', *Bull. Soc. Math France*, **24**, 199–220, (1896).

[2] Published in *Mém. Couronnés Acad. Roy. Belgique*, **59**, 1–74, (1899).

[3] Erdős, P., 'Demonstration élémentaire du théorème sur la distribution des nombres premiers', *Scriptum 1, Centre Mathématique*, Amsterdam, (1949).

[4] Selberg, A., 'An elementary proof of the prime number theorem', *Ann. Math.*, **50**, 305–313, (1949).

6.5.2 Applications

The equivalence of the weak and strong forms of the Axiom of Regularity, mentioned in Subsection 4.5.1, comes as an easy application of rank, for if we let $R(A) = \{\operatorname{rank}(x) : x \in A\}$ (which is non-empty if A is non-empty) then, using only the weak form of Regularity, the class of ordinal numbers $R(A)$ has a minimal element α.[1] Let $x \in A$ with $\operatorname{rank}(x) = \alpha$. Suppose there is a set y such that $y \in A \cap x$, then $\operatorname{rank}(y) < \operatorname{rank}(x) = \alpha$, and $y \in A$, contradicting the minimality of α, hence $A \cap x = \emptyset$, proving the result.

Now assume R is a foundational relation on a class A and B is a non-empty subclass of A. We prove that B has an R-minimal element.[2] Let B_0 be the class of all sets of minimal rank m in B (B_0 is a set since it is a subclass of the set V_m). If B_n is a set then so is

$$B_{n+1} = \bigcup_{y \in B_n} \{x : x \text{ has minimal rank in } R^{-1}\{y\} \cap B\}.$$

Thus $b = \cup_{n \in \omega} B_n$ is a non-empty subset of B. Assume B has no R-minimal element and let $x \in b$. Since $B \cap R^{-1}\{x\}$ is non-empty it contains an element y of minimal rank. As $x \in B_n$ for some n, $y \in B_{n+1}$ and so $y \in b$, i.e. yRx. So $b \cap R^{-1}\{x\}$ is non-empty for all $x \in b$, contradicting the fact that R is foundational.

This settles the first of two claims made in Subsection 5.2.1. The second claim is the induction principle for foundational relations: if R is a foundational relation on A, B is a subclass of A and for all $x \in A$, $A \cap R^{-1}\{x\} \subseteq B$ implies $x \in B$, then $B = A$. This is a direct corollary of the first result. Suppose $B \subset A$, then the non-empty class $A - B$ has an R-minimal element x, but by minimality $A \cap R^{-1}\{x\} \subseteq B$, whence $x \in B$, a contradiction.

REMARK

The general induction principle for foundational relations is sometimes called *Noetherian induction* after the notion of a Noetherian space, a topological space with the property that every strictly decreasing chain of closed subsets is finite (i.e. the class of closed subsets satisfies the 'descending chain condition'). This is an important idea in algebraic geometry.

The Axiom of Regularity and various induction results hold for general classes, not just for sets. In the next chapter we come to one of the motivating applications of transfinite recursion (besides the description of the set hierarchy): the formal definition of the arithmetic operations on the ordinal numbers.

---

[1] Since $R(A)$ is non-empty it has an element β. If β is minimal in $R(A)$ then we are done. Assume otherwise, then $\beta \cap R(A)$ is a non-empty *set* and so by weak regularity has a minimal element γ. It is easy to see that γ is then a minimal element (*the* minimal element) of $R(A)$.

[2] This proof is a paraphrasing of a proof appearing in Chapter 9 of Takeuti and Zaring [**207**].

7

Infinite arithmetic

7.1 Basic operations

*Most human beings have an almost infinite capacity for taking things
for granted.*

– ALDOUS HUXLEY[1]

7.1.1 Ordinal addition

If α is an ordinal number we define $\alpha + 1$ to be the ordinal number $\alpha \cup \{\alpha\}$.
The ordinal sum $\alpha + \beta$ can then defined by transfinite recursion as follows:

$$\begin{aligned}
\alpha + 0 &= \alpha; \\
\alpha + (\beta + 1) &= (\alpha + \beta) + 1; \text{ and} \\
\alpha + \beta &= \bigcup_{\gamma < \beta} (\alpha + \gamma), \text{ if } \beta \text{ is a limit ordinal.}
\end{aligned}$$

To spell out exactly how the principle of recursion applies (in the form
stated in the previous chapter) we describe the recursive framework, for some
fixed ordinal δ, of the operation of right addition by δ.

Let α be a limit ordinal. If we define $g : P(\alpha) \to \alpha$ by $g(a) = \cup a$ and
$h : \alpha \to \alpha$ by $h(\gamma) = \gamma \cup \{\gamma\}$, then transfinite recursion assures us of the
existence of a function $f_\delta^\alpha : \alpha \to \alpha$ such that $f_\delta^\alpha(0) = \delta$, $f_\delta^\alpha(\gamma + 1) = h(f_\delta^\alpha(\gamma))$
and, for limit ordinals β, $f_\delta^\alpha(\beta) = g(\{f_\delta^\alpha(\gamma) : \gamma < \beta\})$. This function f_δ^α is
precisely the desired addition operation '$+\delta$' on α, i.e. for any ordinals $\gamma, \delta < \alpha$
we have $\gamma + \delta = f_\delta^\alpha(\gamma)$. If $\alpha < \beta$ are limit ordinals then f_δ^α is the restriction of
f_δ^β to α so we can think of $f_\delta = \cup f_\delta^\alpha$ (taking the union over all limit ordinals
α) as the '$+\delta$' operator $\mathcal{O}n \to \mathcal{O}n$. The formal details of transfinite recursion
as it applies to multiplication and exponentiation as defined below are similar
so we omit them.

---

[1]'Variations on a Philosopher' in Huxley [**105**].

303

Ordinal addition restricts to the familiar natural addition on ω. Beyond ω addition is not commutative and we do not have a right cancellation law, for example $1+\omega = 0+\omega$, but $1 \neq 0$. Furthermore, right addition may not preserve strict inequality, as we see from the same example: $0 < 1$ but $0 + \omega = 1 + \omega$.

REMARKS

1. This definition is equivalent to the order type sum described in Subsection 1.11.2 in the sense that the ordinal $\alpha + \beta$ is among the ordered sets in the order equivalence class defined by the ordered sum of the order types α and β. Similarly for multiplication below. The benefit of the present view of ordinal numbers is that we have a canonical representative set for each equivalence class.

2. An easy exercise in transfinite induction (left to the reader) is to prove that $\beta < \gamma$ if and only if $\alpha + \beta < \alpha + \gamma$ (perform the induction on γ). From this we quickly deduce the validity of left cancellation (i.e. $\alpha + \beta = \alpha + \gamma$ if and only if $\beta = \gamma$). This in turn leads to associativity (another transfinite induction, this time on γ in the expression $(\alpha + \beta) + \gamma = \alpha + (\beta + \gamma)$).

3. If $\alpha + 1 = 1 + \alpha$ then α must be a successor ordinal. If $\alpha \geq \omega$ then $\alpha = \omega + \beta$ for some β and we have $\omega + \beta + 1 = 1 + \omega + \beta$ but since $1 + \omega = \omega$, by left cancellation we have the contradiction $1 = 0$. Therefore the set of ordinals α which additively commutes with 1 is precisely the set of finite ordinals (the set ω).

4. Consider the class of ordinals which additively commute with ω, i.e. let us solve the ordinal equation $\alpha + \omega = \omega + \alpha$. As ω is clearly a solution, so is any multiple ωn for $n < \omega$ (including 0). We show that all the solutions are of this form. First suppose there is a solution larger than ωn for all $n < \omega$. Such a solution would have to be of the form $\alpha = \omega^2 + \beta$ for some ordinal β. But then $\omega^2 + \beta + \omega = \omega + \omega^2 + \beta$, and using the fact that $\omega^2 = \omega + \omega^2$ and left cancellation we obtain the contradiction $\omega = 0$. So all solutions are less than ω^2, and hence they are of the form $\omega n + m$ for some $n, m < \omega$. Then $\omega n + m + \omega = \omega + \omega n + m$. Since the ordinal on the left is a limit ordinal, so is the ordinal on the right, meaning $m = 0$, and the resulting equation holds for all $n < \omega$. Therefore the set of ordinals which additively commutes with ω is $\{\omega n : n < \omega\}$.

Ordinal addition is formally defined by transfinite recursion, finally giving a concrete clarification of Cantor's original idea. Further operations are similarly defined.

7.1.2   Ordinal multiplication

The ordinal product $\alpha\beta$ can be defined recursively as follows:

$$\begin{aligned}
\alpha 0 &= 0; \\
\alpha(\beta + 1) &= \alpha\beta + \alpha; \text{ and} \\
\alpha\beta &= \bigcup_{\gamma < \beta} \alpha\gamma, \text{ if } \beta \text{ is a limit ordinal.}
\end{aligned}$$

When restricted to ω ordinal multiplication is the familiar product of natural numbers. Beyond ω multiplication is not commutative: $2\omega = \omega$ yet $\omega 2 = \omega + \omega \neq \omega$; we do not have a right cancellation law: $1\omega = 2\omega$ but $1 \neq 2$; nor do we have a right distributive law: $(1 + 1)\omega \neq \omega + \omega$.

REMARKS

1. Left distributivity follows almost immediately from the definition. We prove that $\alpha(\beta + \gamma) = \alpha\beta + \alpha\gamma$ by transfinite induction on γ. The case $\gamma = 0$ is trivial. Suppose $\alpha(\beta + \gamma) = \alpha\beta + \alpha\gamma$, then $\alpha(\beta + \gamma + 1) = \alpha(\beta + \gamma) + \alpha$ by definition and then by assumption this is in turn equal to $\alpha\beta + \alpha\gamma + \alpha$ which by the definition of multiplication is equal to $\alpha\beta + \alpha(\gamma + 1)$. This proves that if left distributivity holds for γ then it holds for $\gamma + 1$. Assume γ is a limit ordinal and that for all $\delta < \gamma$ we have $\alpha(\beta + \delta) = \alpha\beta + \alpha\delta$, then the equality $\alpha(\beta + \gamma) = \alpha\beta + \alpha\gamma$ follows directly from the definition of right multiplication by a limit ordinal.

2. Left cancellation for multiplication (i.e. if $\alpha > 0$ and $\alpha\beta = \alpha\gamma$ then $\beta = \gamma$) is proved by transfinite induction, rather like the corresponding result for addition, first by establishing monotonicity, that is, if $\alpha > 0$ then $\beta \leq \alpha\beta$ and if $\alpha > 0$ and $\beta > 1$ then $\alpha < \alpha\beta$. Once left cancellation is established we obtain associativity (again by transfinite induction).

Ordinal multiplication is formally defined by transfinite recursion. Like addition, it coincides with the usual operation on the natural numbers, but at ω and beyond we generally lose commutativity.

7.1.3   Ordinal subtraction and division

Although neither ordinal addition nor multiplication is commutative, it is possible to define a 'one-sided' subtraction and division for ordinal numbers.

If $\alpha < \beta$ then there is a unique ordinal number γ such that $\alpha + \gamma = \beta$, γ being the 'one sided difference' between β and α. There need not be a γ such that $\gamma + \alpha = \beta$, for example $1 < \omega$ but there is no ordinal γ with $\gamma + 1 = \omega$.

For any two ordinal numbers $\alpha > 0$ and β there is a uniquely determined pair of ordinal numbers γ and δ such that $\beta = \alpha\gamma + \delta$, where $\delta < \alpha$. If $\delta = 0$, then γ is the 'one sided quotient' of β by α. There may not exist a γ and $\delta < \alpha$ with $\beta = \gamma\alpha + \delta$, for example there is no γ or $\delta < 2$ for which $\omega = \gamma 2 + \delta$.

Two special cases of this one-sided ordinal division are particularly illuminating. Setting $\alpha = \omega$ we get a representation $\beta = \omega\gamma + \delta$, $\delta < \omega$ (repeating this division leads us to the Cantor normal form, which we come to in the next section), and we see that the limit ordinals are precisely those ordinals of the form $\omega\gamma$ for some non-zero ordinal γ. Alternatively, setting $\alpha = 2$ we see that, in common with the natural numbers, every ordinal number is either of the form 2γ or $2\gamma + 1$, i.e. either 'even' or 'odd'.

REMARKS

1. The existence of the one-sided difference of α and β, where $\alpha \leq \beta$, is assured. Let $\delta = \cup\{\gamma : \alpha + \gamma \leq \beta\}$. It is easy to see that δ is the required difference. (The uniqueness is obvious by left cancellation.)

2. Defining $\gamma = \cup\{\mu : \alpha\mu \leq \beta\}$ we have $\alpha\gamma \leq \beta$. If δ is the unique ordinal with $\beta = \alpha\gamma + \delta$ then δ is clearly less than α (otherwise $\gamma + 1$ would be among those ordinals μ with $\alpha\mu \leq \beta$).

3. In imitation of the usual definition of prime numbers, we might define an ordinal number p to be 'prime' if $p \geq 2$ and if $p = ab$ implies either a or b is equal to p. The ordinal primes less than ω are clearly the usual primes. ω itself is a prime, as is $\omega + 1$, but since $\omega + n = n(\omega + 1)$, $\omega + n$ is not prime for any $2 \leq n < \omega$. The next prime after $\omega + 1$ is $\omega^2 + 1$. Uniqueness of prime factorization cannot be properly formulated owing to noncommutativity, but even in the noncommutative sense we lose uniqueness since $(\omega + 1)\omega = \omega^2$, nevertheless every ordinal is expressible as a finite product of primes. Most inappropriately, a prime can be 'composite', for example $\omega = 2\omega$. Modifications of the definition are possible, but none are particularly satisfactory. One can recover some of the familiar structural results (at least, for successor ordinals) by using the Hessenberg natural sum and product instead (see the remarks to Subsection 7.2.4).

4. For any fixed ordinal α we can define addition and multiplication on $\mathcal{O}n_\alpha = \{\gamma : \gamma < \alpha\}$ to be the remainder of the operation upon one sided division by α. For $n < \omega$ we clearly have $\mathcal{O}n_n = \mathbb{Z}_n$. The algebra $\mathcal{O}n_\omega$ is the usual algebra of natural numbers (and in general, if α is a limit ordinal, $\mathcal{O}n_\alpha$ is simply the algebra of ordinals less than α). $\mathcal{O}n_{\omega+1}$ is the algebra obtained by appending to the natural numbers the symbol ω subject to the rules $0\omega = \omega 0 = 0$ and $n\omega = \omega = \omega n$ (for $1 \leq n \leq \omega$).

*Ordinal numbers have a 'one-sided' notion of subtraction and division. By re-
peating the division algorithm we are able to represent ordinal numbers to the
base of some fixed ordinal number, just as in the special case of natural numbers.*

7.2 Exponentiation and normal form

> *As far as the laws of mathematics refer to reality, they are not cer-
> tain; and as far as they are certain, they do not refer to reality.*
>
> – ALBERT EINSTEIN[1]

7.2.1 Ordinal exponentiation

The ordinal α^β is defined recursively as follows:

$$\alpha^0 = 1;$$
$$\alpha^{\beta+1} = \alpha^\beta \alpha;$$
$$\alpha^\beta = \begin{cases} \bigcup_{\gamma<\beta} \alpha^\gamma & \text{if} \quad \beta \text{ is a limit ordinal and } \alpha \neq 0; \text{ and} \\ 0 & \text{if} \quad \beta \text{ is a limit ordinal and } \alpha = 0. \end{cases}$$

The observation that $2 < 3$ but $2^\omega = 3^\omega = \omega$ shows that if $\alpha < \beta$ and γ is
a limit ordinal then it need not be the case that $\alpha^\gamma < \beta^\gamma$. Also, since ordinal
multiplication is not commutative, we lose one of the familiar laws of powers: it
can happen that $(\alpha\beta)^\gamma \neq \alpha^\gamma\beta^\gamma$ as is shown, for example, by $(\omega 2)^2 \neq \omega^2 2^2$. We
do have, however, that $\gamma^\alpha\gamma^\beta = \gamma^{\alpha+\beta}$ and $(\gamma^\alpha)^\beta = \gamma^{\alpha\beta}$ for all ordinals α, β, γ.

REMARKS

1. The validity of the identity $\gamma^\alpha\gamma^\beta = \gamma^{\alpha+\beta}$ is immediate by transfinite induction on β.

2. The validity of the identity $(\gamma^\alpha)^\beta = \gamma^{\alpha\beta}$ follows by transfinite induction on
   β, using the identity of the previous remark for the successor part of the
   induction.

---

*Ordinal exponentiation is formally defined by transfinite recursion. Not all of
the laws of powers we are familiar with through natural exponentiation hold for
ordinal exponentiation. Passing to limits of exponential towers we are able to
define classes of large ordinals.*

---

[1]Einstein [**62**].

7.2.2   Ordinal limits and Cantor's epsilon numbers

A workable general notion of a limit of an indexed collection of ordinal numbers
is available,[1] and from this we define continuity for ordinal functions. Briefly, if
$\{\alpha_\beta\}_{\beta<\gamma}$ is a collection of ordinals with γ a limit ordinal, then it has limit α if
and only if there exists an ordinal $\mu < \gamma$ such that for every β with $\mu \le \beta < \gamma$,
$\alpha_\beta \le \alpha$ and for every $\delta < \alpha$ there exists an α_β, $\mu \le \beta$, with $\delta < \alpha_\beta$. When this
is the case we write $\alpha = \lim_{\beta<\gamma} \alpha_\beta$.

Let f be an ordinal function on an ordinal α. Then f is continuous if and
only if for every limit ordinal $\lambda < \alpha$, $f(\lambda) = \lim_{\xi<\lambda} f(\xi)$. It is easy to see
that, for fixed α, the functions $f(\beta) = \alpha + \beta$, $f(\beta) = \alpha\beta$ and $f(\beta) = \alpha^\beta$ are
continuous in β. On the other hand, none of these three functions are continuous
as functions of α (fixing β), since

$$\lim_{\beta<\omega} (\beta + 1) = \omega \ne \omega + 1;$$

$$\lim_{\beta<\omega} (\beta 2) = \omega \ne \omega 2; \text{ and}$$

$$\lim_{\beta<\omega} \beta^2 = \omega \ne \omega^2.$$

Every ordinal number can be viewed naturally as a topological space, where
the open sets are possibly infinite unions of sets of the form $\{\beta : \beta < \gamma\}$,
$\{\beta : \gamma < \beta\}$ or $\{\beta : \gamma < \beta < \delta\}$, yielding the so-called *order topology*. The
associated notion of convergence is precisely that described above.

As a brief aside, we ought to mention that a much more general theory of
limits is available in arbitrary topological spaces. This uses what used to be
called *Moore–Smith sequences*,[2] now generally called *nets*.[3] By a *directed set*
I we mean a set together with a reflexive and transitive relation $\preceq$ with the
property that for all $i, j \in I$ there is a $k \in I$ with $i \preceq k$ and $j \preceq k$. A *net* in
a topological space X is a function $f : I \to X$. f converges to $x \in X$ if for
every open neighbourhood U of x there is an $\alpha \in I$ with $f(\beta) \in U$ whenever
$\alpha \preceq \beta$. The most familiar example of net convergence in classical mathematics
is given by integration: the set P of all finite partitions of an interval $[a, b]$ of real
numbers into subintervals is naturally directed by the subpartition relation. For
a Riemann integrable $f : [a, b] \to \mathbb{R}$ the net $s : P \to \mathbb{R}$ mapping the partition
p to the Riemann (upper or lower) sum of f associated with p has net limit
$\int_a^b f \, dx$.

For every *set* a of ordinals $\cup a$ is the smallest ordinal equal to or greater than
all members of a. That $\cup a$ is transitive and well-ordered by $\in$ is immediate: if
$x \in \cup a$ then $x \in \alpha$ for some ordinal α in a, hence $x \subseteq \alpha$. If $y \in x$ then $y \in \alpha$
and so $y \in \cup a$, i.e. $x \subseteq \cup a$; if $x, y \in \cup a$ then x and y are ordinals so either
$x \in y$, $x = y$ or $y \in x$. $\cup a$ is a set by the Axiom of Unions.

---

[1] I have been careful not to use the word 'sequence' here, this implying a set indexed by ω.

[2] Introduced in Moore, E. H. and Smith, H. L., 'A general theory of limits', *Amer. J. Math.*,
44, 102–121, (1922).

[3] This terminology was introduced by John L. Kelley in 'Convergence in topology', *Duke
Math J.*, **17**, 277–283, (1950).

The increasing sequence (a_n)

$$1, \omega, \omega^\omega, \omega^{\omega^\omega}, \omega^{\omega^{\omega^\omega}}, \ldots$$

recursively defined by $a_0 = 1$ and $a_{n+1} = \omega^{a_n}$ has an ordinal limit denoted by ε_0, that is, ε_0 is the smallest ordinal $\cup\{1, \omega, \omega^\omega, \omega^{\omega^\omega}, \omega^{\omega^{\omega^\omega}}, \ldots\}$ which dominates all terms in the sequence (a_n).

It is easily seen that $\omega^{\varepsilon_0} = \varepsilon_0$. Cantor called every ε satisfying the ordinal equation $\omega^\varepsilon = \varepsilon$ an *epsilon number*, ε_0 being the smallest such solution, ε_1 the next smallest, and so on. The class of epsilon numbers is a proper class indexed by $\mathcal{O}n$: $\varepsilon_0, \varepsilon_1, \ldots, \varepsilon_\omega, \varepsilon_{\omega+1}, \ldots, \varepsilon_{\omega 2}, \ldots$ etc. Generally $\varepsilon_{\alpha+1}$ is the limit of the sequence (a_n) defined by $a_0 = 0$, $a_{n+1} = \varepsilon_\alpha^{a_n}$, i.e. the sequence $0, 1, \varepsilon_\alpha, \varepsilon_\alpha^{\varepsilon_\alpha}, \varepsilon_\alpha^{\varepsilon_\alpha^{\varepsilon_\alpha}}, \ldots$ and, for β a limit ordinal, ε_β is the ordinal $\cup\{\varepsilon_\alpha : \alpha < \beta\}$.

REMARKS

1. If f and g are continuous ordinal functions on some ordinal α then $(f + g)(x) = f(x) + g(x)$ and $(fg)(x) = f(x)g(x)$ need not be continuous. For example, let f be the identity function $f(x) = x$ and g the constant function $g(x) = 2$, then neither $(f + g)(x) = x + 2$ nor $(fg)(x) = x2$ is continuous at ω.

2. If we give the first uncountable ordinal ω_1 the usual order topology, the result is an interesting topological space. It has the property that every infinite sequence has a convergent subsequence (it is *sequentially compact*) and yet it is not a compact topological space – it is easy to find an open cover which has no finite subcover.

3. ε_0 is countable (it is a countable union of countable sets). Indeed, for the same reason, any epsilon number with a countable base ordinal will again be countable. Since $\omega^{\omega_1} = \cup_{\alpha<\omega_1}\omega^\alpha = \omega_1$ and ε_α is countable if and only if α is countable, ω_1 is equal to the epsilon number ε_{ω_1}. More generally every uncountable cardinal ω_κ is equal to the epsilon number $\varepsilon_{\omega_\kappa}$.

4. An ordinal α is *recursive* if there is a recursively defined well-ordering of $\mathbb{N}$ with order type α. ω is recursive, the successor of a recursive ordinal is recursive and all predecessors of a recursive ordinal are recursive. The smallest non-recursive ordinal (equal to the set of all recursive ordinals) is a countable ordinal called the *Church–Kleene ordinal*.

Let us denote by PA_0 Peano Arithmetic (see Appendix A). By Gödel's Second Incompleteness Theorem, although we can formulate a sentence $\text{Con}(\text{PA}_0)$ in the language of PA_0 which is true if and only if PA_0 is consistent, it cannot be proved or refuted in PA_0. Let us assume that PA_0 is consistent, and consider the augmented theory $\text{PA}_1 = \text{PA}_0 + \text{Con}(\text{PA}_0)$. Again by the Incompleteness Theorem $\text{Con}(\text{PA}_1)$ is not provable in PA_1, so we consider $\text{PA}_2 = \text{PA}_1 +$

Con(PA$_1$), and more generally, for natural n, PA$_{n+1}$ = PA$_n$ + ConPA$_n$. But we can define PA_ω to be the theory $\cup_{n<\omega}$PA$_n$, and the Incompleteness Theorem still applies to it, leading us to PA$_{\omega+1}$ = PA$_\omega$ + Con(PA$_\omega$), and so on, so that for any ordinal α less than the Church–Kleene ordinal κ the theory PA$_\alpha$ is still incomplete and in particular is unable to prove its own consistency, but since PA$_\kappa$ is no longer an effective theory, the Incompleteness Theorem no longer applies.

Note that, since Con(PA) is neither provable nor refutable in PA, as well as augmenting PA by Con(PA) we can also consistently (assuming PA is consistent) augment PA by ¬Con(PA). The resulting theory, PA+¬Con(PA) is a peculiar one: it is a consistent theory which 'thinks' it is inconsistent.

An epsilon number is a solution of the ordinal equation $\omega^\varepsilon = \varepsilon$. The class of epsilon numbers is a proper class indexed in order of size by the ordinal numbers. The first epsilon number, ε_0, is the limit of the sequence $\omega, \omega^\omega, \omega^{\omega^\omega}, \omega^{\omega^{\omega^\omega}}, \ldots$ Each epsilon number beyond ε_0 is similarly constructed as a limit of a sequence of exponential towers or as a union of epsilon numbers.

7.2.3 Picturing ordinals

Visualizing even the humble countable ordinals in terms of well-ordered sets can be a complicated business. Normally we tend to favour the 'ordinals as topological spaces' approach, successor ordinals being isolated points and limit ordinals being limit points. For example, we might picture ω, $\omega + \omega$, $\omega + \omega + \omega$, ω^2, $\omega^2 + \omega^2$ and $\omega^2 + \omega^2 + \omega^2$ as illustrated in Figure 7.1 (the 'points' being replaced by vertical lines for clarity).

This becomes less feasible for larger ordinals, where we might resort instead to choosing a representative natural mathematical set, for example the ordinal ω^ω might be imagined as representing the lexicographically ordered set of all polynomials with natural coefficients, i.e. the order defined by

$$a_n x^n + \cdots + a_1 x + a_0 \prec b_m x^m + \cdots + b_1 x + b_0$$

if and only if

(i) $m > n$; or

(ii) $m = n$ and $a_n < b_n$; or

(iii) $m = n$, $a_n = b_n$ and $a_{n-1}x^{n-1} + \cdots + a_1 x + a_0 \prec b_{n-1}x^{n-1} + \cdots + b_1 x + b_0$.

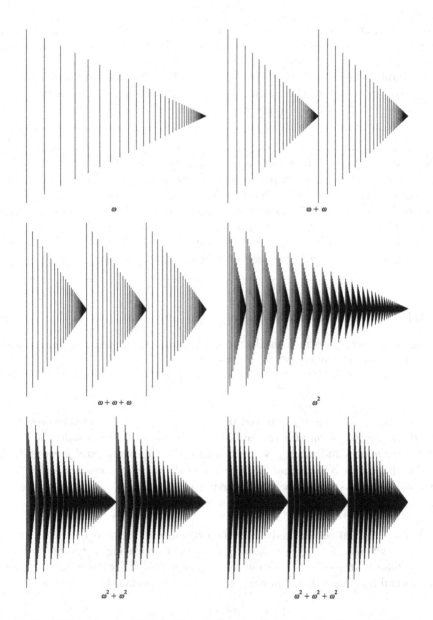

Figure 7.1 The smaller infinite ordinals can easily be pictured as certain infinite arrangements of lines, each isolated line representing a successor ordinal and the limits of these lines representing limit ordinals. Depicted above are the ordinals ω, $\omega + \omega$, $\omega + \omega + \omega$, ω^2, $\omega^2 + \omega^2$ and $\omega^2 + \omega^2 + \omega^2$. Beyond a certain point, however, this technique becomes less feasible.

We know that every countable ordinal, and this of course includes ε_0 (and indeed ε_α for any countable α), can be embedded in $\mathbb{Q}$ with respect to the usual ordering, so for these large ordinals it is not the existence of a planar representation that is in doubt but the difficulty of successfully visualizing it. Nevertheless, assuming the continuum is a set, there will of course be a point beyond which all ordinals cannot be represented in this planar form.

The smaller infinite ordinals can be visualized as certain limiting configurations of lines or points. However, once we reach the heights of ω^ω and beyond, such geometric representations become less feasible. Instead we might think of representative mathematical sets with familiar orderings, but even these examples will soon run out. Nevertheless, from an algebraic point of view, the class has some familiar characteristics: like the natural numbers, each ordinal number can be written uniquely to the base of a fixed ordinal number $\beta > 1$.

7.2.4   Normal form

If β is an arbitrary ordinal number greater than 1, then *every* ordinal number $\delta > 0$ has a unique representation in the form

$$\delta = \beta^{\alpha_0}\gamma_0 + \beta^{\alpha_1}\gamma_1 + \cdots + \beta^{\alpha_n}\gamma_n,$$

with $\alpha_0 > \alpha_1 > \cdots > \alpha_n \geq 0$ and $0 < \gamma_0, \gamma_1, \ldots, \gamma_n < \beta$. This expression is called the *normal form* of δ to the base β. When $\beta = 2$ we obtain the *dyadic normal form* of δ and when $\beta = \omega$ we obtain the *Cantor normal form* of δ.

The least $\gamma > \beta$ such that $\beta^\gamma = \gamma$ is called the *salient ordinal* for β. That such ordinals exist is easily established; the equation $\beta^\gamma = \gamma$ is satisfied by the least upper bound γ of the sequence of ordinals (β_n) defined by $\beta_0 = 1$; $\beta_{n+1} = \beta^{\beta_n}$.

If δ_0 is the salient ordinal for β then its normal form to base β is simply $\delta_0 = \beta^{\delta_0}$, whereas for any $\delta < \delta_0$ all powers of β appearing in the normal form of δ to base β are strictly less than δ_0. For such 'small' δ we can further reduce the normal form as follows: having expressed δ in normal form to base β,

$$\delta = \beta^{\alpha_0}\gamma_0 + \beta^{\alpha_1}\gamma_1 + \cdots + \beta^{\alpha_n}\gamma_n,$$

we then express each of the powers α_0 to α_n of β in normal form to base β; then write all the powers of β appearing in these subexpressions in normal form to base β; and so on. The fact that δ is smaller than the salient ordinal of β guarantees that this process will terminate in a *finite* number of steps. The resulting expression is called the *complete normal form* of δ to base β;

alternatively we say that δ has been written in *superbase* β. To take a simple finite example, the complete normal form of 100 to base 2 is:

$$2^{2^{2^1}+2^1} + 2^{2^{2^1}+2^0} + 2^{2^1}.$$

If the base β is a finite ordinal then the salient ordinal is ω, so that a reduction of an ordinal α to complete normal form to base β is possible only if α is a finite ordinal. If the base is ω, the salient ordinal is ε_0, so every ordinal less than ε_0 can be rendered in complete normal form to base ω. One sometimes finds transfinite induction employed up to ε_0 (most famously in Gentzen's proof of the consistency of PA, but also in the proof of Goodstein's Theorem, which we come to in the next section), the representation of each smaller ordinal in complete normal form being reduced to a finite configuration of natural terminal powers of ω.

REMARKS

1. The representation of an ordinal δ to base β follows from repeated one-sided division by β. Firstly there exist unique γ_0 and $\alpha_0 < \beta$ with $\delta = \beta\gamma_0 + \alpha_0$. But $\gamma_0 = \beta\gamma_1 + \alpha_1$ for unique γ_1 and $\alpha_1 < \beta$, and so on, with $\gamma_{i+1} < \gamma_i$, so after a finite number of divisions the representation is complete.

2. The salient ordinal of ε_α is $\varepsilon_{\alpha+1}$.

3. The Cantor normal form can be used to define alternative operations of ordinal addition and multiplication. Although we shall have no use for this, it is worth describing. Given two arbitrary ordinal numbers δ_1 and δ_2 with Cantor normal form representations

$$\delta_1 = \sum \omega^\alpha c_\alpha; \text{ and}$$
$$\delta_2 = \sum \omega^\alpha d_\alpha,$$

we form their *natural* (or *Hessenberg*) sum $\delta_1 \# \delta_2 = \sum \omega^\alpha (c_\alpha + d_\alpha)$. This does not generally coincide with $\delta_1 + \delta_2$ (it is equal or greater than the usual sum, for example $\omega + 1 = \omega\#1 = 1\#\omega > 1 + \omega$). One can also form natural products in the obvious way, again treating δ_1 and δ_2 as polynomials in a neutral symbol ω with natural coefficients.[1]

These alternative operations appear at first glance to be rather artificial, however, there is an alternative approach to them which avoids the Cantor normal form, revealing the operations to be a little more deserving of their name 'natural'. In a sense the natural operations yield the most generous possible extensions of the arithmetical operations on ω which continue to

---

[1]The natural sum is due to Gerhard Hessenberg (*Grundbegriffe der Mengenlehre*, Vandenhoeck & Ruprecht 1906). Hausdorff [96] also credits the natural product to Hessenberg, although it does not appear in the former reference.

be commutative and associative (and which ensure that multiplication distributes over addition). To find $\delta_1 \# \delta_2$ we form the disjoint union of a pair of ordered sets X and Y with order types δ_1 and δ_2, $X \cup Y$ is partially ordered by declaring $a \prec b$ if:

(a) $a, b \in X$ and $a <_X b$; or

(b) $a, b \in Y$ and $a <_Y b$.

The set of ordinal numbers representing all of those well-orderings of $X \cup Y$ which restrict to $\prec$ on X and on Y has a least upper bound (in fact a maximal element) which is the natural sum $\delta_1 \# \delta_2$. The natural product is similarly defined as the ordinal number representing the maximal well-ordered extension of the partial order on $X \times Y$ defined by $(x_1, y_1) \prec (x_2, y_2) \leftrightarrow x_1 <_X x_2 \wedge y_1 <_Y y_2$.

The normal form of an ordinal δ to some base $\beta > 1$ is a sum of powers of β, each power multiplied (on the right) by an ordinal coefficient less than β. By expressing the exponents appearing in this sum again as a linear sum of powers of β, and repeating this procedure, if δ is small enough (smaller than the salient ordinal for β) then this procedure will terminate in a finite number of steps, resulting in the complete normal form of δ to base β. The complete normal form is an ingredient of a surprising theorem discovered in the mid-1940s.

7.2.5   Goodstein's Theorem

The complete normal form has been introduced here because it plays a leading role in a significant result proved by Reuben L. Goodstein in 1944.[1] This theorem is surprising in two respects. Firstly, it is hard to believe based on the 'empirical' evidence, underlining our poor intuition concerning such matters; and secondly, although the theorem is entirely about sequences of natural numbers, the proof uses infinite ordinals. Most importantly, nearly forty years after Goodstein's proof appeared, Laurence Kirby and Jeff Paris[2] showed that this is a theorem about natural numbers which cannot be proved in the first-order theory of Peano Arithmetic, so in a sense, the use of infinite ordinals cannot be avoided. (However, it should come as no surprise that Goodstein's Theorem *is* provable in the 'set theory in disguise' that is second-order Peano Arithmetic.)

A *Goodstein sequence* is generated by the following recursive procedure:

Start with any natural number and write it in superbase 2.

---

[1] Goodstein, R. L., 'On the restricted ordinal theorem', *Journal of Symbolic Logic*, **9**, 33–41, (1944).

[2] Kirby, L. and Paris, J., 'Accessible independence results for Peano arithmetic', *Bulletin of the London Mathematical Society*, **14**, 285–93, (1982).

Replace all 2s with 3s, subtract 1. Rewrite the new number in superbase 3.
Replace all 3s with 4s, subtract 1. Rewrite the new number in superbase 4.
Replace all 4s with 5s, subtract 1. Rewrite the new number in superbase 5.
Replace all 5s with 6s, subtract 1. Rewrite the new number in superbase 6.
And so on.

The Goodstein sequences beginning with 1, 2 and 3 are easily seen to decrease to zero, these being:

$1, 0$
$2, 2, 1, 0$
$3, 3, 3, 2, 1, 0$

respectively.

Once we get beyond these initial examples, however, we find what appear to be rapidly increasing sequences. For example:

Start with 7. Write in superbase 2:
$7 = 2^{2^1} + 2^1 + 2^0$.
Change 2 to 3 then subtract 1:
$3^{3^1} + 3^1 + 3^0 - 1 = 30$.
Write in superbase 3:
$30 = 3^{3^1} + 3^1$.
Change 3 to 4 and subtract 1:
$4^{4^1} + 4^1 - 1 = 259$.
Write in superbase 4:
$259 = 4^{4^1} + 4^0 \cdot 3$.
Change 4 to 5 and subtract 1:
$5^{5^1} + 5^0 \cdot 3 - 1 = 3127$.
Write in superbase 5:
$3127 = 5^{5^1} + 5^0 \cdot 2$.
Change 5 to 6 and subtract 1:
$6^{6^1} + 6^0 \cdot 2 - 1 = 46\,657$.
Write in superbase 6:
$456\,657 = 6^{6^1} + 6^0$.
Change 6 to 7 and subtract 1:
$7^{7^1} + 7^0 - 1 = 823\,543$.
Write in superbase 7.
$823\,543 = 7^{7^1}$.
Change 7 to 8 and subtract 1:
$8^{8^1} - 1 = 16\,777\,215$.
Etcetera.

The first eleven terms of the Goodstein sequence beginning with 4 are

$$4, 26, 41, 60, 83, 109, 139, 173, 211, 253, 299, \ldots ,$$

which is not a rapid increase by any stretch of the imagination, however it does begin to expand quite dramatically as we move further along the sequence: a maximum is reached at $3 \times 2^{402\,653\,210} - 1$. Appearances are deceptive – Goodstein's Theorem tells us that no matter which number we start with, the generated Goodstein sequence will eventually descend to zero.

The proof of Goodstein's Theorem is surprisingly simple. If the number we start with is m_0 and the Goodstein sequence generated is (m_k) then we are at each stage in the recursion writing m_k in superbase $k + 2$. Suppose running parallel to this we form another sequence (n_k) where n_k is obtained from the complete normal form of m_k to base $k + 2$ by replacing each instance of $k + 2$ with ω. n_k is then an ordinal in Cantor normal form. For example, if we perform this substitution on the above sequence generated from 7 we obtain the parallel sequence:

$$n_0 = \omega^{\omega^1} + \omega^1 + \omega^0;$$
$$n_1 = \omega^{\omega^1} + \omega^1;$$
$$n_2 = \omega^{\omega^1} + \omega^0 \cdot 3;$$
$$n_3 = \omega^{\omega^1} + \omega^0 \cdot 2;$$
$$n_4 = \omega^{\omega^1} + \omega^0;$$
$$n_5 = \omega^{\omega^1}$$
. . ...

This is evidently a strictly decreasing sequence of ordinals. It is not difficult to prove that the parallel sequence (n_k) is *always* a strictly decreasing sequence of ordinals and hence must eventually hit zero.[1] Now one simply observes that $m_k \leq n_k$ for all k so that m_k too must eventually reach zero.

REMARKS

1. One can define a relation E on ω recursively so that the system (ω, E) is isomorphic to $(\varepsilon_0, \in)$. This means that induction up to ε_0 can be interpreted in the language of Peano Arithmetic (or indeed in primitive recursive arithmetic) via this identification.

2. We know from Gödel's First Incompleteness Theorem that there exist unprovable statements in Peano Arithmetic, but finding natural candidates is a challenge. Gentzen's proof of the relative consistency of Peano Arithmetic with primitive recursive arithmetic plus ε_0-induction can be interpreted as the first 'natural' independent statement (in the form 'ε_0-induction is independent of Peano Arithmetic'). The next example was found by Paris and Harrington[2] in the form of the strengthened finite Ramsey Theorem. Let n, k and m be positive integers. Then there exists an N such that if we

---

[1] A full proof is given, for example, in Potter [**170**] (Section 13.1).

[2] Paris, J. and Harrington, L., 'A Mathematical Incompleteness in Peano Arithmetic.' In *Handbook for Mathematical Logic* (ed. J. Barwise). Amsterdam, Netherlands: North-Holland, 1977.

colour each of the n-element subsets of $[1, N] = \{1, 2, \ldots, N\}$ with one of the k colours 1 to k then there exists a subset X of $[1, N]$ with at least m elements so that all n-element subsets of X have the same colour and the number of elements of X is at least equal to the smallest element of X. The Paris–Harrington Theorem states that this cannot be proved in Peano Arithmetic (for the same reason as Gentzen's result – it implies the consistency of Peano Arithmetic). The independence of Goodstein's Theorem was the third natural example of a statement independent of Peano Arithmetic.

Goodstein sequences are sequences of natural numbers generated by a recursion based on complete normal forms. At first sight it appears that all but the initial few Goodstein sequences are rapidly increasing. However, it turns out that all Goodstein sequences eventually reach 0. This cannot be proved in Peano Arithmetic, but it can be proved in ZF, using infinite ordinals. Now that ordinal arithmetic operations and limits have been formally defined, it is natural to examine infinite sums and products of ordinals. The novelty here is that the indices over which the sum or product is taken may range over arbitrary ordinals.

7.2.6 Appendix: infinite ordinal sums and products

Infinite sums of ordinals are defined by the following recursion scheme:

$$\sum_{\beta < 0} \alpha_\beta = 0;$$

$$\sum_{\beta < \gamma + 1} \alpha_\beta = \left(\sum_{\beta < \gamma} \alpha_\beta \right) + \alpha_\gamma; \text{ and}$$

$$\sum_{\beta < \gamma} \alpha_\beta = \bigcup_{\delta \in \gamma} \left(\sum_{\beta < \delta} \alpha_\beta \right), \text{ if } \gamma \text{ is a limit ordinal,}$$

and infinite products of ordinals are defined by:

$$\prod_{\beta < 0} \alpha_\beta \;=\; 1;$$

$$\prod_{\beta < \gamma+1} \alpha_\beta \;=\; \left(\prod_{\beta < \gamma} \alpha_\beta \right) \cdot \alpha_\gamma; \text{ and}$$

$$\prod_{\beta < \gamma} \alpha_\beta \;=\; \begin{cases} \bigcup_{\delta \in \gamma} \left(\prod_{\beta < \delta} \alpha_\beta \right) & \text{if} \quad \gamma \text{ is a limit ordinal and } \alpha_\kappa \neq 0 \\ & \qquad \text{for all } \kappa < \gamma; \text{ and} \\ 0 & \text{if} \quad \alpha_\kappa = 0 \text{ for some } \kappa < \gamma. \end{cases}$$

REMARKS

1. If $\alpha_\beta = \alpha$ for all β then $\Sigma_{\beta<\gamma+1}\alpha_\beta = \alpha\gamma$, as falls out of the definition. So in particular we have $n+n+n+\cdots = \omega$ for non-zero $n \in \omega$ and $\omega+\omega+\omega+\cdots = \omega^2$.

2. If $\alpha_\beta = \alpha$ for all β then $\Sigma_{\beta<\gamma+1}\alpha_\beta = \alpha^\gamma$. So in particular $n \cdot n \cdot n \cdots = \omega$, for $1 < n \in \omega$ and $\omega \cdot \omega \cdot \omega \cdots = \omega^\omega$.

Infinite ordinal sums and products are formally defined by transfinite recursion. Application of these definitions to certain simple collections of ordinals yields results which match our intuition. Now that the ordinal numbers have been explored we take a closer look at cardinal numbers.

8

Cardinal numbers

8.1 Cardinal theorems and Cantor's paradox

We all have a tendency to think that the world must confirm to our prejudices. The opposite view involves some effort of thought, and most people would die sooner than think – in fact, they do so.

– BERTRAND RUSSELL[1]

8.1.1 The Cantor–Bernstein Theorem

At the centre of the theory of cardinal numbers is the notion of equipollence of sets and critical in the application of equipollence is the Cantor–Bernstein Theorem.

The Cantor–Bernstein Theorem

If a set x is equipollent to a subset of a set y and if y is equipollent to a subset of x then x is equipollent to y.

The Cantor–Bernstein Theorem is intuitively highly plausible. One might even mistakenly think that it is obvious, based on an extrapolation from the trivial finite case. However, a glance at its history reveals that the route to full understanding was not smooth; the transparent proofs known to us today make it very easy for us to forget these early difficulties.

Cantor first conjectured in 1882, and a year later proved,[2] the theorem in the special case of subsets of $\mathbb{R}^n$, however, he assumed the Continuum Hypothesis in the proof. Over the next twelve years Cantor was to share his conjecture with several sources. The proofs Cantor outlined in this period assumed the

---

[1] Russell [185].

[2] Cantor, G., *Grundlagen einer allgemeinen Mannigfaltingkeitslehre. Ein mathematisch-philosophischer Versuch in der Lehre des Unendlichen.* Leipzig: Teubner, 1883. A translation (Foundations of a general theory of manifolds: a mathematico-philosophical investigation into the theory of the infinite) can be found in Ewald [66].

Well-Ordering Theorem. Developments over the remaining part of the nine-teenth century gradually stripped away these further assumptions. It turns out, however, that Dedekind had proved the conjecture fairly early on, in 1887, just four years after Cantor had posed the problem to him in a letter. This lay unannounced in Dedekind's notebooks until it was discovered in 1932.

Cantor published the conjecture as an open problem in 1895.[1] Burali-Forti proved the theorem for countable sets in 1896, without making any further as-sumptions. A year later Bernstein proved the general result assuming a weak form of the Axiom of Choice. Finally, in 1899, Dedekind proved the full con-jecture again, and this time sent it to Cantor. The published proofs of the theorem which appeared in the early years of the twentieth century are either based on Dedekind's proof or manage to remove the use of Choice in Bernstein's argument.

Many different proofs of the Cantor–Bernstein Theorem are now known. Honouring the promise made in Subsection 1.6.9, we present Birkhoff and Mac Lane's short elementary Choice-free proof.

Suppose $f : a \to b$ and $g : b \to a$ are injections and let us make the harmless assumption that a and b are disjoint. The functions f, g and the set $a \cup b$ form a dynamical system: the orbit of an element x of a is the sequence $x, f(x), g(f(x)), f(g(f(x))), \dots$ and the orbit of an element x of b is the sequence $x, g(x), f(g(x)), g(f(g(x))), \dots$. We call an element $x \in a \cup b$ an *ancestor* of $y \in a \cup b$ if y is in the orbit of x. Let a_O be the set of all elements in a with an odd number of ancestors, a_E the set of all elements in a with an even number of ancestors and a_I the set of all elements in a with an infinite number of ancestors. The set a is the disjoint union of a_O, a_E and a_I. Similarly we partition b into b_O, b_E and b_I. f maps a_E onto b_O and a_I onto b_I and g^{-1} maps a_O onto b_E, therefore the function $h : a \to b$ defined by

$$h(x) = \begin{cases} f(x) & \text{if} \quad x \in a_E \cup a_I, \\ g^{-1}(x) & \text{if} \quad x \in a_O, \end{cases}$$

is a bijection. This concludes the proof.

If the injections f and g are simple enough, allowing us to easily describe the orbit of any given element and hence the sets $a_E \cup a_I$ and a_O, then it should be possible to exhibit the bijection h.

For example, if we fix k distinct primes $p_1, \dots, p_k$ and we define natu-ral injections $f : \mathbb{N} \to \mathbb{N}^k$ and $g : \mathbb{N}^k \to \mathbb{N}$ by $f(n) = (n, 0, \dots, 0)$ and $g((m_1, \dots, m_k)) = p_1^{m_1} p_2^{m_2} \cdots p_k^{m_k}$, then $\mathbb{N}_O = \cup_n \Lambda_n$ where Λ_0 is the set of all natural numbers of the form $p_1^{m_1} \cdots p_k^{m_k}$ with $m_2, \dots, m_k$ not all zero and $\Lambda_{n+1} = \{p_1^t : t \in \Lambda_n\}$. The value of $h(n)$ (for the induced bijection $h : \mathbb{N} \to \mathbb{N}^k$) is then easily found by inspecting the prime factorization of n. Taking the two primes $p_1 = 2$ and $p_2 = 3$, for instance, the induced enumeration of $\mathbb{N}^2$ begins: $(0,0), (1,0), (2,0), (0,1), (4,0), (5,0), (1,1), (7,0), (3,0), (0,2), (10,0) \dots$

---

[1]Cantor, G., 'Beiträge zur Begründung der transfiniten Mengenlehre I', *Math. Ann.*, **47**, 481–512, (1895).

In the same way one can generate novel permutations of an infinite set from a pair of self-injections. For example, the injections $\mathbb{N} \to \mathbb{N}$ given by $f(n) = 2n$ and $g(n) = 3n$ yield the bijection $h : \mathbb{N} \to \mathbb{N}$

$$h(n) = \begin{cases} \frac{n}{3} & \text{if } n = 2^a 3^b m \text{ with } a < b \text{ and neither 2 nor 3 divide } m \\ 2n & \text{otherwise.} \end{cases}$$

REMARKS

1. Recall the proof of the equipollence of $[0,1)$ and $[0,1)^2$ where we appealed to the Cantor–Bernstein Theorem after exhibiting the two injections $f : [0,1) \to [0,1)^2$ and $g : [0,1)^2 \to [0,1)$ defined by $f(x) = (x,0)$ and

$$g(0.a_1 a_2 a_3 \ldots, 0.b_1 b_2 b_3 \ldots) = 0.a_1 b_1 a_2 b_2 a_3 b_3 \ldots$$

(where neither $0.a_1 a_2 a_3 \ldots$ nor $0.b_1 b_2 b_3 \ldots$ end in a string of 9s). Examining the above proof of the Cantor–Bernstein Theorem we can define a bijection $h : [0,1) \to [0,1)^2$ recursively as follows. To determine $h(x)$ write x as a decimal $0.a_1 b_1 a_2 b_2 a_3 b_3 \ldots$ (without a tail of 9s). If all a_is are 9s or if all b_is are 9s then $h(x) = (x,0)$. If not all a_is are 9s and not all b_is are 9s and the b_is are not all zero, then $h(x) = (0.a_1 a_2 a_3 \ldots, 0.b_1 b_2 b_3 \ldots)$. In the remaining case, where all b_is are zero, we then define $h(x)$ to be equal to $h(0.a_1 a_2 a_3 \ldots)$.

2. For a fixed set X consider the set $\mathcal{H}(X)$ of all injective functions $X \to X$. The above construction associates with any two such injections f and g two bijections $X \to X$ depending on the order in which one takes f and g in the proof. Agreeing to denote h as defined earlier by $f \square g$ (so the bijection with f and g interchanged is $g \square f$) we define an operation $\square$ on $\mathcal{H}(X)$. The reader is invited to study the algebraic properties of the system $(\mathcal{H}(X), \square)$.

3. There are several other proofs of the Cantor–Bernstein Theorem which employ similar back-and-forth techniques to the one presented here. If the original injections are given explicitly and if they are sufficiently simple, then each such proof should be able to generate an explicit bijection between the two sets in question.

To establish the equipollence of two sets we simply need to prove that there exists an injection from each into the other. In some cases these injections will be simple enough to explicitly describe the bijection. The assignment of cardinality can only apply to sets; if we try to extend it to proper classes we soon meet a paradox.

8.1.2 Cantor's paradox

Cantor's Theorem states that for any set a, there is no injection from $P(a)$ into a. As we saw in Subsection 1.10.2, the proof of Cantor's Theorem does not require the Axiom of Choice.

This easy to prove result was initially the cause of much consternation owing to its extraordinary corollary: *if there is an infinite set then there are (infinitely many) different sizes of infinity*, a fact that some found difficult to swallow – indeed, many found the treatment of any infinite aggregate as a manipulable whole to be problematic.

Just as Cantor's principles of generation, when applied to the 'set' of all ordinals, gave rise to the Burali-Forti paradox of the largest ordinal, Cantor's Theorem when naively applied to the 'set' of all sets leads to Cantor's paradox of the largest cardinal number: if the class V of all sets were a set then $P(V)$, being a collection of sets, would be a subset of V, contradicting Cantor's Theorem. Therefore V is not a set.

REMARK

We should perhaps restate that the objects in the language of ZF are sets – it is not a class theory; statements about classes are abbreviations for statements about sets. The statement 'V is not a set' when fully unravelled is simply the statement $\neg(\exists a)[y \in a \leftrightarrow (\forall z)[z \in y \leftrightarrow z \in y]]$, and more generally '$\{x : \phi(x)\}$ is a proper class' simply means $\neg(\exists a)[y \in a \leftrightarrow \phi(y)]$.

If the class V of all sets was a set then by the Axiom of Powers $P(V)$ would be a set and hence would be a subset of V, contradicting Cantor's Theorem. This paradox, Cantor's paradox, illustrates why it is necessary to distinguish between sets and proper classes. Another boundary which is surprisingly difficult to draw is that between finite and infinite sets.

8.2 Finite and infinite sets

> *Only two things are infinite, the Universe and human stupidity, and I'm not sure about the former.*
>
> – ANONYMOUS[1]

To give a satisfactory definition of finiteness is, as has been stressed already, a curiously difficult problem. By defining finite sets to be those sets which are equipollent to elements of the smallest inductive set ω we certainly capture the essence of finiteness, but it is arguably too technical a definition for what should

---

[1]This quote is usually attributed to Albert Einstein, however, it is difficult to source. Frederick S. Perls claims in two of his books that Einstein said something of this flavour to him. Of course the sentiment, which has often been repeated, predates Einstein.

be a fairly natural and simple property. It is more instructive to attempt to find a definition which does not make reference to such sophisticated machinery. We seek a definition which mirrors more faithfully our natural enumerative sense.[1]

Dedekind[2] (and independently, Peirce)[3] proposed the most well-known 'non-numerical' definition of finiteness in the late 1880s. Recall from the introduction that a *Dedekind finite* set is one which is not equipollent to any of its proper subsets. Simple examples suggest that this definition is intuitively sound, but the Axiom of Choice is needed to prove that every Dedekind finite set is finite in the sense described above.

By fixing a well-ordering of a given infinite set a we extract a countable subset $X = \{x_n : n \in \omega\}$ of a by defining x_0 to be the smallest element of a and, for $n \geq 1$, x_n to be the smallest element in $a - \{x_0, x_1, \ldots, x_{n-1}\}$. The function $f : a \to a$ which fixes all elements of $a - X$ and which maps x_n to x_{n+1} is a non-surjective injection, implying a is Dedekind infinite (this is a generalization of the construction of a bijection $[0, 1] \to [0, 1)$ presented in Subsection 1.6.8).

The converse, that Dedekind infinite sets are infinite, is most easily proved in contrapositive form. Suppose a is a finite set (i.e. equipollent to an element of ω), b is a proper subset of a and $f : a \to b$ is a bijection. For a subset c of a denote by $f[c]$ the image set $\{f(t) : t \in c\}$. Let $X = \{x : x \subseteq a \text{ and } f[x] \subset x\}$. X is non-empty ($f[b]$ cannot be equal to b because f is an injection, so $b \in X$) and, since a is finite, X too is finite so it has an inclusion minimal element m. By minimality, and since $f[m] \subset m$, $f[m] \notin X$. But $f[f[m]] \subset f[m]$, so $f[m] \in X$, a contradiction.

Because of this dependence on the Axiom of Choice, some texts speak both of infinite cardinals (cardinal numbers of sets which are not equipollent to any member of ω) and *transfinite cardinals* (cardinal numbers of Dedekind infinite sets).

Dedekind's definition of infinite sets seems to be popular because it spotlights in the property of proper self-injection one of the oldest documented 'counter-intuitive' characteristics of infinite sets. Tarski's definition of finiteness, on the other hand, highlights the intuitive: a set a is Tarski finite if every non-empty subset of $P(a)$ has a $\subseteq$-minimal element. This is equivalent, without the need for Choice, to ordinary finiteness. It is also equivalent to the definition proposed by P. Stäckel:[4] a is finite if and only if there is a well-ordering W of a such that the inverse order W^{-1} is also a well-ordering of a. Tarski finiteness is arguably more intuitive than Dedekind finiteness because it instantly captures 'exhaustibility' – given an infinite set a consider the collection of subsets of a formed by successive depletion, say $a_0 = a$; a_{n+1} obtained by removing one element from a_n, then $\{a_n : n \in \omega\}$ has no $\subseteq$-minimal element.

In *Principia Mathematica* Whitehead and Russell provide an induction in-

---

[1]This section draws on, among other sources, Chapter 4 of Suppes [**206**] which in turn follows Tarski's survey 'Sur les ensembles finis', *Fundamenta Mathematicae*, **6**, 45–95, (1924).
[2]Dedekind [**43**].
[3]See Volume III of Hartshorne and Weiss [**94**].
[4]Stäckel, P., 'Zu H. Webers Elementarer Mengenlehre', *Jahresber. d.d. M.-V.*, **16**, 425, (1907).

spired definition of finiteness: a set a is finite if and only if it is an element of every set F with the properties:

(i) $\emptyset \in F$; and

(ii) if $b \in F$, $b \subseteq a$ and $x \in a$, then $b \cup \{x\} \in F$.

A comparable definition was given by Sierpinski:[1] a set a is finite if and only if it is an element of every set F such that:

(i) $\emptyset \in F$;

(ii) if $x \in a$ then $\{x\} \in F$; and

(iii) if $b \in F$ and $c \in F$ then $b \cup c \in F$.

Kuratowski pursued this theme,[2] defining a to be finite if and only if $P(a)$ is the only set F with the properties:

(i) $F \subseteq P(a)$;

(ii) $\emptyset \in F$;

(iii) if $x \in a$ then $\{x\} \in F$; and

(iv) if $b \in F$ and $c \in F$, then $b \cup c \in F$.

In his 1924 survey Tarski is careful to indicate when the equivalence of two of his surveyed definitions of finiteness depends on the Axiom of Choice – his list of definitions appears in an order whereby each statement freely implies the next but the converse requires Choice. Tarski's list gives an excellent illustration of the characteristic way the Axiom of Choice fuses together what would in its absence be a catalogue of inequivalent properties.

REMARKS

1. To recap. In ZF a set is finite if it is equipollent to an element of ω. That is, it is finite if and only if it is equipollent to an ordinal α such that α and all ordinals less than α are successor ordinals.

2. For the classical Finitist the boundary between finite and infinite sets and between sets and proper classes is *the same boundary*.

---

[1] Sierpinski, W., 'L'axiome de M. Zermelo et son rôle dans la théorie des ensembles et l'analyse', *Bull. Acad. Sc. Cracovie*, 97–152, (1918).

[2] Kuratowski, C., 'Sur la notion de l'ensemble fini', *Fundamenta Mathematicae*, **1**, 129–131, (1920).

There are many alternative definitions of finiteness, and their equivalence often critically relies on the Axiom of Choice. In ZF we carefully define the ordinal ω as the smallest inductive class containing the empty set, declare ω to be a set by the Axiom of Infinity, and say a set is finite if it is equipollent to one of the members of ω.

8.3 Perspectives on cardinal numbers

> *'Tis strange – but true; for truth is always strange; Stranger than fiction.*
>
> – LORD BYRON[1]

8.3.1 Classical and modern definitions of cardinal numbers

Cantor's approach to cardinal numbers is reflected in his notation. He used a single bar over a set, $\bar{a}$, to indicate that the 'specific nature' of the elements of a is irrelevant, but the ordering is to be retained, while a second bar, $\bar{\bar{a}}$, indicates that both the nature and the ordering of the elements are to be ignored. Single bars were used for ordinals, double bars for cardinals.

Frege, in 1884,[2] and Russell, independently in 1903,[3] made Cantor's notion of cardinality sharp and accessible to analysis by defining the cardinal number of a set a to be the class of all sets that are equipollent to a. This is now regarded as the 'classical' approach to cardinality.

Within the classical framework the definitions of the three cardinal arithmetic operations are intuitively clear: selecting disjoint members x and y of the equivalence classes m and n respectively, the cardinal number $m + n$ is the set of all sets which are equipollent to the union $x \cup y$; the product mn is the set of sets equipollent to the cartesian product $x \times y$; and the exponential m^n is the set of all sets equipollent to the set x^y of all functions $y \to x$.

In modern set theory it is more common to define the cardinal number of a to be the smallest ordinal number that is equipollent to a. In particular, in ZF the proper class of cardinal numbers is the subclass of $\mathcal{O}n$ comprising all α such that α is not equipollent to any ordinal less than α. For example, ω is a cardinal number but the ordinal sum $\omega + 1$ is not.

The class of all *infinite* cardinal numbers is order isomorphic to $\mathcal{O}n$. We denote the order isomorphism from $\mathcal{O}n$ onto the class of infinite cardinal numbers by $\aleph$ and write $\aleph_\alpha$, rather than $\aleph(\alpha)$, for the value of $\aleph$ at α. Clearly $\aleph_0 = \omega$; $\aleph_1$ is the smallest uncountable ordinal; $\aleph_2$ is the smallest ordinal greater than $\aleph_1$ but not equipollent to $\aleph_1$, and so on. Generally, $\aleph_\alpha$ is the smallest ordinal

---

[1] *Don Juan* (1823), Canto XIV.
[2] Frege [**74**].
[3] Russell [**181**].

greater than (and not equipollent to) all $\aleph_\beta$ with $\beta < \alpha$. The arithmetic of alephs is simple: any given sum or product is equal to the largest summand or multiplicand; only cardinal exponentiation leads into new territory.

Cantor considered the attractive question of determining the position of $2^{\aleph_0}$, the cardinality of $\mathbb{R}$, in the hierarchy of alephs. Since the cardinality of $\mathbb{R}$ is greater than $\aleph_0$ the simplest possibility is that conjectured by the Continuum Hypothesis, namely $2^{\aleph_0} = \aleph_1$. That such a technical statement can be expressed so economically is a good advertisement for the notation.

The classical and modern definitions of cardinal numbers are not equivalent in the absence of the Axiom of Choice. If a set a has no well-ordering then the equipollence equivalence class containing a (a classical cardinal number) will not include any of our modern cardinal numbers; a is not equipollent to any ordinal number. From Cohen's independence results it is known that the question of whether or not such a set exists is undecidable in ZF.

REMARK

If $P(a)$ has a well-ordering then so does a (embed a in $P(a)$ via $x \mapsto \{x\}$). Thus if a has no well-ordering then neither do any of the sets $P(a)$, $P(P(a))$, $P(P(P(a)))$ and so on. So the existence of one non-well-orderable set entails the existence of infinitely many such sets of different (classical) cardinalities.

Modern cardinal numbers are certain types of ordinal number – they are precisely those ordinal numbers which are not equipollent to any of their members. Assuming the Axiom of Choice this definition will assign a unique cardinal number to each set. Now that cardinal numbers are identified as a subclass of the ordinal numbers we need to stress that cardinal addition, multiplication and exponentiation will not generally coincide with their ordinal counterparts.

8.3.2   Cardinal versus ordinal operations

If α and β are (modern) cardinal numbers then the cardinal sum $\alpha + \beta$ is the smallest ordinal number equipollent to the ordinal sum $\alpha + \beta$, so that, for example, if κ is an infinite cardinal then the cardinal sum $\kappa + 1$ is equal to κ. Similarly the cardinal product $\alpha\beta$ is the smallest ordinal equipollent to the ordinal product $\alpha\beta$ (equivalently, $\alpha\beta$ is the smallest ordinal equipollent to the cartesian product $\alpha \times \beta$).

The cardinal number α^β is the smallest ordinal equipollent to the set of functions $\beta \to \alpha$ and generally does not coincide with the smallest ordinal equipollent to the ordinal α^β, for example $2^\omega = \omega$ (ordinal exponentiation), but $2^{\aleph_0} \neq \aleph_0$ (cardinal exponentiation).

The infinite cardinal sum $\sum \alpha_i$ is the smallest ordinal equipollent to the infinite ordinal sum $\sum \alpha_i$, but the infinite cardinal product $\Pi_{i\in\Lambda}\alpha_i$ is the smallest ordinal equipollent to the infinite cartesian product $\times_{i\in\Lambda}\alpha_i$, not the smallest ordinal equipollent to the infinite ordinal product $\Pi\alpha_i$, so, for example, the cardinal product $1 \cdot 2 \cdot 3 \cdot 4 \cdots$ is equal to $2^{\aleph_0}$ (see the remark at the end of this subsection), while the ordinal product $1 \cdot 2 \cdot 3 \cdot 4 \cdots$ is equal to ω.

There is little chance of confusion when using the same symbols for cardinal and ordinal operations. We could, for example, use $+_c$ and $+_o$ for cardinal and ordinal addition, respectively, so that $\kappa +_c 1 = \kappa \neq \kappa +_o 1$ for infinite κ, but in practice no such notational distinction is made. The context is usually sufficient to avoid confusion – in particular the appearance of an aleph, as in the Continuum Hypothesis: $2^{\aleph_0} = \aleph_1$, indicates that cardinal exponentiation is intended.

REMARK

The infinite cartesian product $\times_{i \in \Lambda} a_i$ of a collection of sets $\{a_i\}_{i \in \Lambda}$ is the set of functions on Λ whose value at i is in the set a_i. If a_i is a set with cardinal number α_i then the infinite cardinal product of all α_i as i ranges over Λ is equal to the cardinal number of $\times_{i \in \Lambda} a_i$.

If $\alpha_i = 2$ for all i then there is a natural bijection between $P(\Lambda)$ and the infinite cartesian product $\times_{i \in \Lambda} \alpha_i$ given by $i \in X$ if and only if $\alpha_i = 1$ (recall that $2 = \{0, 1\}$). So the cardinal product $\Pi_{i \in \Lambda} 2$ is equal to the cardinal number $2^{|\Lambda|}$, where $|\Lambda|$ is the cardinal number of Λ, while the ordinal product $\Pi_{i \in \Lambda} 2$, assuming now that Λ is an ordinal, is the ordinal 2Λ.

If $2 \leq \alpha_i < \aleph_0$ for all $i \in \Lambda$ then the cardinal product $\Pi_{i \in \Lambda} \alpha_i$ will have lower bound $2^{|\Lambda|}$, as indicated in the previous paragraph, and it will have upper bound $\aleph_0^{|\Lambda|}$. However, if Λ is infinite we have $\aleph_0^{|\Lambda|} \leq (2^{\aleph_0})^{|\Lambda|} = 2^{\aleph_0 |\Lambda|} = 2^{|\Lambda|}$, so we have $\Pi_{i \in \Lambda} \alpha_i = 2^{|\Lambda|}$. (Note that although $\aleph_0 < 2^{\aleph_0}$ is a strict inequality, we have $\aleph_0^{|\Lambda|} = (2^{\aleph_0})^{|\Lambda|}$ if $|\Lambda| \geq \aleph_0$.) A special case of this is the result stated earlier that the cardinal product $1 \cdot 2 \cdot 3 \cdots$ is equal to $2^{\aleph_0}$.

Although a moderate amount of care is needed to ensure that ordinal and cardinal operations are not confused, in most cases the intended operation is clear in context. We now turn to the interesting phenomenon of inaccessibility.

8.4 Cofinality and inaccessible cardinals

> *'Ought the word 'infinite' to be avoided in mathematics?' Yes; where it appears to confer a meaning upon the calculus; instead of getting one from it.*
>
> – LUDWIG WITTGENSTEIN[1]

---

[1] Wittgenstein [229].

8.4.1 Cofinality

Suppose α and β are ordinal numbers with $\beta \leq \alpha$ and there exists a strictly increasing function $f : \beta \to \alpha$ with the property that every element of α is equal to or less than the image $f(\gamma)$ of some element γ of β. If these conditions hold we say that α is *cofinal* with β. Some elementary results concerning cofinality are immediate: α is cofinal with 0 if and only if $\alpha = 0$; and every (non-zero) successor ordinal is cofinal with 1, indeed if $\beta + 1 \leq \alpha + 1$ then $\alpha + 1$ is cofinal with $\beta + 1$ – the required increasing function is obtained by defining $f(\beta) = \alpha$ and $f(\gamma) = \gamma$ for $\gamma < \beta$. In other words every successor ordinal δ is cofinal with all non-zero successor ordinals less than δ.

The first limit ordinal ω is not cofinal with any of its members. It is natural to ask whether this type of extreme dominance can be observed elsewhere in $\mathcal{O}n$; are there other ordinals which are not cofinal with any smaller ordinal? Since every non-zero successor ordinal is cofinal with 1, any such ordinal must be a limit ordinal, but this condition alone is not sufficient, after all $\omega + \omega$ is cofinal with ω. We begin to realize that if such an ordinal exists it must be very large indeed.

For ease of discussion it is convenient to introduce the notion of the *character of cofinality* $\mathrm{cf}(\alpha)$ of α, by which we mean the smallest ordinal $\beta \leq \alpha$ such that α is cofinal with β. The character of cofinality of any ordinal is a cardinal number. Non-trivial lower bounds can be found for the character of cofinality of certain cardinal expressions, for example, if κ and λ are cardinal numbers such that $\kappa \geq \aleph_0$ and $\lambda \geq 2$ then $\kappa < \mathrm{cf}(\lambda^\kappa)$.

REMARKS

1. It can be shown that if $\beta < \alpha$ and there exists a function $f : \beta \to \alpha$, not necessarily increasing, such that every element of α is either equal to or smaller than $f(\delta)$ for some $\delta \in \beta$, then there is a $\gamma \leq \beta$ such that α is cofinal with γ. In particular, if β is equipollent with α then there is a $\gamma \leq \beta$ such that α is cofinal with γ. It then follows from the transitivity of cofinality that no ordinal less that $\mathrm{cf}(\alpha)$ is equipollent to $\mathrm{cf}(\alpha)$. In other words, $\mathrm{cf}(\alpha)$ is a cardinal number.

2. Since the character of cofinality is a cardinal number, the character of cofinality of any countable limit ordinal must be ω (in particular cases it is easy to exhibit an increasing sequence of ordinals approaching the given ordinal with the required property, for example the map $n \mapsto \omega + n$ shows that $\mathrm{cf}(\omega + \omega) = \omega$; the map $n \mapsto \omega n$ shows that $\mathrm{cf}(\omega^2) = \omega$; the map $n \mapsto \omega^n$ shows that $\mathrm{cf}(\omega^\omega) = \omega$ and the sequence $\omega, \omega^\omega, \omega^{\omega^\omega}, \ldots$ shows that $\mathrm{cf}(\varepsilon_0) = \omega$).

An ordinal α is cofinal with a smaller ordinal β if one can reach, and exceed, every element of α by a β-indexed increasing collection of elements of α. If α is a successor ordinal this is trivially possible, but if α is a limit ordinal it may not be – ω is the smallest example of an ordinal which is not cofinal with any smaller ordinal. Beyond ω such unreachable limit ordinals are forced to be extraordinarily large.

8.4.2   Inaccessibility

An ordinal α is *regular* if $\mathrm{cf}(\alpha) = \alpha$ (for example, ω), and is *singular* if $\mathrm{cf}(\alpha) < \alpha$ (for example, $\aleph_\omega$, which has character of cofinality ω). Assuming the Axiom of Choice (to avoid complications we will be making this assumption throughout this section – without it we cannot generally compare two cardinal numbers!) one can prove that $\aleph_\alpha$ is always regular if α is a successor ordinal.[1] A cardinal α is regular if for every non-empty family $\{\alpha_i\}_{i \in I}$ of cardinals with $\alpha_i < \alpha$, for all i, and $\mathrm{Card}(I) < \alpha$, we have the strict cardinal inequality $\sum_{i \in I} \alpha_i < \alpha$.

A natural question arises: does there exist a regular cardinal $\aleph_\alpha$ with α a limit ordinal? Such cardinals are called *weakly inaccessible*. The cardinal $\aleph_\alpha$ is *inaccessible* if, in addition to being weakly inaccessible, $\aleph_\beta < \aleph_\alpha$ implies $2^{\aleph_\beta} < \aleph_\alpha$.

If $\aleph_\alpha$ is weakly inaccessible then $\aleph_\alpha = \alpha$: by definition, α is a limit ordinal and $\mathrm{cf}(\aleph_\alpha) = \aleph_\alpha$. Since $\aleph_\alpha$ is cofinal with α for every limit ordinal α (via the increasing function $\beta \mapsto \aleph_\beta$) we have $\aleph_\alpha = \mathrm{cf}(\aleph_\alpha) \leq \alpha \leq \aleph_\alpha$, whence $\aleph_\alpha = \alpha$, as claimed. The converse is false; it is easy to find 'accessible' α with $\aleph_\alpha = \alpha$. The smallest ordinal α such that $\aleph_\alpha = \alpha$ (i.e. the smallest fixed point of the aleph function) is the enormous $\alpha = \lim_{n < \omega} \alpha_n$, where $\alpha_0 = \aleph_0$ and $\alpha_{n+1} = \aleph_{\alpha_n}$. The ordinal α is clearly cofinal with ω, via the function mapping n to α_n, and hence is singular.

Every inaccessible cardinal is weakly inaccessible, but it is consistent with ZF that there exist weakly inaccessible cardinals which are not inaccessible – the two classes coincide if we assume the Generalized Continuum Hypothesis. The notions of inaccessibility outlined here were introduced by Sierpinski and Tarski in 1930.[2]

A *Grothendieck universe*, originally introduced in the context of algebraic geometry, is a set U such that: (i) the membership relation is transitive on U; (ii) $\omega \in U$; (iii) if $x \in U$ then $P(x) \in U$; and (iv) for all $a \in U$ and all functions $f : a \to U$ the union $\cup_{x \in a} f(x)$ is an element of U. In ZFC one can prove that every Grothendieck universe is of the form V_κ for some inaccessible cardinal

---

[1] This is a result of Felix Hausdorff ('Grundzüge einer Theorie der geordneten Mengen', *Mathematische Annalen*, **65**, 435–505, (1908)).

[2] Sierpinski, W. and Tarski, A., 'Sur une propriété caractéristique des nombres inaccessibles', *Fund. Math.*, **15**, 292–300, (1930).

κ and conversely every such V_κ is a Grothendieck universe. Consequently the existence of Grothendieck universes is equivalent to the existence of inaccessible cardinals. Note that some texts omit (ii) or replace it by $\varnothing \in U$, permitting the empty set or V_ω to be Grothendieck universes.

A cardinal number $\aleph_\alpha$, with α a limit ordinal, is inaccessible if it is not cofinal with any smaller ordinal and if it cannot be reached by cardinal exponentiation on smaller cardinals. Assuming the Generalized Continuum Hypothesis cardinal exponentiation is completely determined, but as soon as the set theoretic conditions are weakened a whole range of different values can consistently assumed.

8.4.3   Cardinal exponentiation and Silver's Theorem

The value of κ^λ can, if λ is small enough relative to $\mathrm{cf}(\kappa)$, be calculated from the values of τ^λ for $\tau < \kappa$. If κ and λ are cardinal numbers such that $1 \leq \lambda < \mathrm{cf}(\kappa)$ and $\kappa \geq \aleph_0$, then

$$\kappa^\lambda = \left(\sum_{\tau < \kappa} \tau^\lambda \right) \kappa,$$

where the sum is understood to range over all cardinal numbers less than κ. This theorem is the end result of progressive investigations of Bernstein, Hausdorff and Tarski. Bernstein's original theorem is the special case: $\aleph_n^{\aleph_0} = 2^{\aleph_0} \cdot \aleph_n$ for all $n \in \omega$.

Assuming the Generalized Continuum Hypothesis, we have the following more general result. For an arbitrary infinite cardinal κ:

$$\kappa^\lambda = \begin{cases} 1 & \text{for} \quad \lambda = 0, \\ \kappa & \text{for} \quad 1 \leq \lambda < \mathrm{cf}(\kappa), \\ \kappa^+ & \text{for} \quad \mathrm{cf}(\kappa) \leq \lambda \leq \kappa, \\ \lambda^+ & \text{for} \quad \kappa < \lambda, \end{cases}$$

where by α^+ we mean the smallest cardinal strictly larger than α.[1] The special case $\kappa^{\mathrm{cf}(\kappa)} = \kappa^+$ forms the basis of the Singular Cardinals Hypothesis, which proposes that this is true for all singular κ, even in the absence of the Generalized Continuum Hypothesis.

For an embarrassment of years it had been assumed that the behaviour of κ^λ is governed only by the above results. However, starting in the 1970s, some surprising new inequalities began to emerge. At the forefront was Silver's Theorem.[2]

Let κ be a singular cardinal with $\mathrm{cf}(\kappa) > \omega$, and assume that $2^\lambda = \lambda^+$ if $\omega \leq \lambda < \kappa$, then $2^\kappa = \kappa^+$.

---

[1] Proofs of these two theorems can be found, for example, in Hajnal and Hamburger [88].
[2] Silver, J. H., 'On the singular cardinal problem', *Proc. Int. Congr. Math.*, Vancouver, 1, 265–286, (1974).

Informally, Silver's Theorem says that if the Generalized Continuum Hypothesis holds for all cardinals less than κ then it holds for κ. This kind of result is not available for singular cardinals of countable cofinality (like $\aleph_\omega$) or for regular cardinals. Naturally there is a flood of research on this topic. PCF theory (PCF stands for 'possible cofinalities'), an invention of the astonishingly productive Saharon Shelah, provides upper bounds for 2^κ, κ singular. For example, in 1989 Shelah proved that if $2^{\aleph_n} < \aleph_\omega$ for all $n < \omega$ then $2^{\aleph_\omega} < \aleph_{\omega 4}$.[1]

For any uncountable regular cardinal κ it is consistent with ZF that the Generalized Continuum Hypothesis holds *below* κ but fails at κ itself (i.e. $2^\kappa > \kappa^+$). For instance, it is consistent with ZF that $2^{\aleph_0} = \aleph_1$, $2^{\aleph_1} = \aleph_2$, $2^{\aleph_2} = \aleph_3$ and $2^{\aleph_3} = \aleph_{873\,906}$. We shall return to this in the next chapter.

REMARK

The Singular Cardinals Hypothesis, being a consequence of the Generalized Continuum Hypothesis, is consistent with ZFC. The negation of the Singular Cardinals Hypothesis is also consistent with ZFC, however the proof assumes the existence of a certain inaccessible cardinal.

For arbitrary cardinals κ and λ the possible values of κ^λ are determined by set theoretic assumptions. These restrictions can be weak. For example, it is consistent with ZF that $2^{\aleph_0} = \aleph_\alpha$, where $\aleph_\alpha$ is any infinite cardinal greater than $\aleph_0$ which does not have countable character of cofinality. Postulating the existence of larger and larger cardinals strengthens a theory.

8.4.4   Applications to models of ZF

A more precise discussion of models will be given later. For now, a model of an axiomatic theory is simply a class in which all of the axioms are satisfied, where all quantified variables are restricted to the said class. If μ is an inaccessible cardinal then the class V_μ is a model of ZF, for if we apply the axioms of ZF to sets in V_μ, any new sets formed will also lie in V_μ. Thus if it were possible to prove in ZF the existence of an inaccessible cardinal, it would be possible to prove in ZF that ZF has a model. But this implies that ZF (which we assume to be consistent) is able to prove its own consistency, contrary to Gödel's Second Incompleteness Theorem. An incompleteness-free argument for the ZF-unprovability of the existence of inaccessible cardinals, making use of the phenomenon of absoluteness, is given at the end of Section 10.4.

It is an entertaining experiment to posit the existence of hyper-inaccessible cardinals, the latter being inaccessible cardinals μ which dominate a set of inaccessible cardinals with cardinality μ. It can then be shown that if μ_0 is the

---

[1] A proof can be found in T. J. Jech, 'Singular cardinal problem: Shelah's Theorem on $2^{\aleph_\omega}$', *Bull. London Math. Soc.*, **24**, 127–131, (1992).

smallest hyper-inaccessible cardinal then V_{μ_0} is a model of ZF containing inaccessible cardinals but no hyper-inaccessible cardinals. We might then say that a cardinal μ is hyper-hyper-inaccessible if there are μ hyper-inaccessible cardinals smaller than μ, in which case, if μ_1 is the least hyper-hyper-inaccessible cardinal, V_{μ_1} is a model of ZF containing hyper-inaccessible cardinals but no hyper-hyper-inaccessible cardinals, and so on.

New axioms asserting the existence of ever larger cardinals result in increasingly stronger systems. Such additional axioms of infinity imitate at a higher level the jump from the finite to the infinite provided by the ordinary axiom of infinity of ZF and bring with them a similar abundance of new sets. Recent research in set theory has been fuelled to a significant degree by the study of these large cardinal axioms. A huge literature now exists detailing the consequences of the existence of all manner of cardinals: inaccessible, Mahlo, reflecting, weakly compact, subtle, ineffable, remarkable, Ramsey, measurable (see Subsection 8.4.5 below), strong, Woodin, Shelah, superstrong, supercompact, extendible, Vopěnka, n-huge, rank-into-rank... This list is not exhaustive. The ordering is by increasing strength, so, for example, the existence of Shelah cardinals entails the existence of Woodin cardinals. The fact that all the large cardinal axioms so far examined are linearly ordered by strength in this way is a curiosity; there is no particular reason that they should be.

REMARK

The terminology 'hyper-inaccessible' is used differently by different authors. One natural approach is the following. Let $I_1(\alpha)$ be the αth inaccessible cardinal, where α ranges over the ordinal numbers > 0. The ordinals such that $I_1(\alpha) = \alpha$ (the fixed points of I_1) are the 1-inaccessible cardinals. The class of 1-inaccessible cardinals form an ordered class so we can define $I_2(\alpha)$ to be the αth 1-inaccessible cardinal. The fixed points of I_2 are the 2-inaccessible cardinals. Generally we define $I_{\beta+1}(\alpha)$ to be the αth β-inaccessible cardinal and for limit ordinals δ, $I_\delta(\alpha)$ is the αth cardinal which is a fixed point of I_β for all $\beta < \delta$. One can then define α to be hyper-inaccessible if it is α-inaccessible.

The existence of inaccessible cardinals is not provable in ZF. There now exist a host of axioms of infinity, each one postulating the existence of ever larger cardinals and each strengthening the theory it is added to. One such class of cardinals is the class of measurable cardinals.

8.4.5   Measurable cardinals

Of the species of large cardinals just listed I shall single out one for special attention. In order to introduce this we need a little measure theory.

A measure attempts to abstract those properties possessed by measures of length, area and volume in Euclidean space. The simplest concrete case is

Lebesgue measure on $\mathbb{R}$ (which generalizes easily to natural measures on $\mathbb{R}^n$). A *cover* of a subset A of $\mathbb{R}$ is a collection $\mathcal{C}$ of intervals with $A \subset \cup_{I \in \mathcal{C}} I$. The *outer measure* $m^*(A)$ of A is defined to be the greatest lower bound of the set of numbers $\{\sum_{I \in \mathcal{C}} |I| : \mathcal{C}$ is a finite or countable cover of $A\}$. Here $|I|$ is the length of the interval I, i.e. $|b - a|$, where a and b are the end-points. If $\sum_{I \in \mathcal{C}} |I|$ diverges for all covers $\mathcal{C}$ of A then we write $m^*(A) = \infty$, otherwise $m^*(A)$ is a non-negative real number.

A set A is *(Lebesgue) measurable* if, for every set $E \subseteq \mathbb{R}$,

$$m^*(E) = m^*(E \cap A) + m^*(E - A).$$

This definition, due to Constantin Carathéodory, is difficult to motivate in this short space. An equivalent alternative is: A is measurable if, for each $\varepsilon > 0$, A can be squeezed between a closed set F, i.e. the complement of an open set in $\mathbb{R}$, and an open set G so that $F \subset A \subset G$ and $m^*(G - F) < \varepsilon$.

If A is measurable then the complement of A in $\mathbb{R}$ is also measurable, the intersection of two measurable sets is measurable and the union of a sequence of measurable sets is measurable. More generally, abstracting these properties, we say that a non-empty class of subsets of a set X is a *σ-algebra* if it contains the union of any sequence of its members, the set theoretic difference of any two of its members and X itself. Thus the class S of measurable sets is a σ-algebra of subsets of $\mathbb{R}$. The restriction of m^* to S is *Lebesgue measure*. In general, abstracting the properties of Lebesgue measure, an extended real-valued non-negative function m on a σ-algebra S is called a *measure* if $m(\emptyset) = 0$ and $m(A) = \sum m(A_i)$ whenever (A_i) is a pairwise disjoint sequence of members of S (i.e. $A_i \cap A_j = \emptyset$ whenever $i \neq j$).

The connections between measure theory and inaccessible cardinals are plentiful and deep. It turns out that if a κ-additive[1] measure can be defined on the power set of a set X of cardinality κ in such a way that the measure of every single element set is zero and the measure of X is 1, then κ is forced to be astonishingly large. Such a cardinal is said to be *measurable* and the assertion that measurable cardinals exist has many interesting consequences, the study of which was initiated by Stanisław Ulam in the early 1930s.[2]

From the work of Alfred Tarski and William Hanf[3] measurable cardinals are hyper-inaccessible: each measurable cardinal μ exceeds a set of inaccessible cardinals of cardinality μ. In particular, if there is a measure $m : P(\mathbb{R}) \to [0, 1]$

---

[1]This means that the measure of the union of a collection $\{S_\alpha : \alpha < \kappa\}$ of disjoint sets is equal to the sum of the measures of the sets. For example Lebesgue measure is $\aleph_0$-additive but not c-additive (if it were c-additive then the measure of every Lebesgue measurable set would be zero).

[2]See Ulam, S., 'Zur Masstheorie in der allgemeinen Mengenlehre', *Fundamenta Mathematicae*, **16**, 140–50, (1930).

[3]See Hanf, W., 'Incompactness in languages with infinitely long expressions', *Fundamenta Mathematicae*, **53**, 309–324, (1964) and Tarski, A., 'Some problems and results relevant to the foundations of set theory', *Logic, Methodology and the Philosophy of Science* (Proceedings of the 1960 International Congress, Stanford), Stanford University Press, Stanford, 125–135 (E. Nagel, *et al.* eds.).

which vanishes on singletons but does not vanish on all sets, then the Continuum Hypothesis is (spectacularly) false. One also gets an impression of the extraordinary magnitude of measurable cardinals from the fact that their existence is strong enough to contradict the Axiom of Constructibility (see Section 11.1), an axiom which is unperturbed by more sober axioms of infinity. Each new axiom of infinity is motivated by an invisible informal principle that each class of cardinals should be dwarfed by higher cardinals just as the finite are dominated by the infinite.

REMARKS

1. The extended real number system is used in measure theory as a matter of convenience to avoid a multitude of clumsy special cases; it simply augments $\mathbb{R}$ by two symbols $-\infty$ and ∞ with the order properties one might expect and *partially* extends the algebraic operations to $\mathbb{R} \cup \{-\infty, \infty\}$. This structure should not be confused with any profound extension of $\mathbb{R}$, it is merely an inelegant necessity.

2. Lebesgue measure on $\mathbb{R}^n$ uses as its 'intervals' n-cells, i.e. cartesian products $C = I_1 \times I_2 \times \cdots \times I_n$ of n intervals and the basic volume used to define the outer measure is obtained by taking the product $|C| = |I_1| \cdot |I_2| \cdots |I_n|$. From the measure one obtains a natural integral (the Lebesgue integral). An alternative approach begins by extending the Riemann integral to the Lebesgue integral and defining the measure via the integral. There is much literature on this subject.

If one can define a certain probability measure on the set of all subsets of a set X then the cardinality of X is a measurable cardinal. Measurable cardinals are hyper-inaccessible.

9

The Axiom of Choice and the Continuum Hypothesis

9.1 The Axiom of Choice

Zermelo regards the axiom as an unquestionable truth. It must be confessed that, until he made it explicit, mathematicians had used it without a qualm; but it would seem that they had done so unconsciously. And the credit due to Zermelo for having made it explicit is entirely independent of the question whether it is true or false.

– BERTRAND RUSSELL[1]

9.1.1 Strong versus weak Choice

The Axiom of Choice is one of the most interesting and most discussed axioms to have emerged in mathematics since the status of Euclid's parallel postulate was resolved in the mid-nineteenth century. The question of whether it should be adopted as one of the standard axioms of set theory was the cause of more philosophical debate among mathematicians in the twentieth century than any other question in the foundations of mathematics.

Axiom of Choice
For every set a there exists a function f such that,
for all $x \in a$, if $x \neq \emptyset$ then $f(x) \in x$.

The function f is called a *choice function*; it selects exactly one element from each set in a. Choice functions had been used without full recognition by Cantor and others long before Zermelo distilled the idea.

A much stronger form of Choice asserts that there is a universal choice function, that is, a function F such that, for all sets x, if $x \neq \emptyset$ then $F(x) \in$

---

[1]Russell [**184**].

x. The strong form of the Axiom of Choice includes a quantification on a class symbol (the proper class *F*) and so cannot be expressed in the first-order language of ZF. One can nevertheless construct a model of ZF in which such a universal choice function is defined. The Axiom of Global Choice, as this strong variant of Choice is sometimes called, is a consequence of Gödel's Axiom of Constructibility (see Chapter 11). Global Choice, which is expressible in such class theories as Gödel–Bernays set theory, is equivalent to the existence of a well-ordering of the universe *V* of all sets. From here on by the 'Axiom of Choice', unless stated otherwise, we will mean the weaker version displayed above.

The main application of Choice in the early development of set theory was to problems involving well-ordering, indeed the axiom was first introduced by Zermelo (who coined the name 'Axiom of Choice') in the first decade of the twentieth century in order to prove the Well-Ordering Theorem. In his original paper Zermelo used the following version of Choice.[1] *If a is a set of non-empty, pairwise disjoint sets, then there is a set x whose intersection with any member of a has exactly one element.* It was partly in reaction to the criticism directed at his original proof that Zermelo was prompted to develop his axiomatization of set theory.

Choice is sometimes called the 'multiplicative axiom'. This name, due to Russell, comes from the fact that Choice implies that a product of non-zero cardinal numbers is non-zero; put another way, the cartesian product of an arbitrary collection of non-empty sets is non-empty – a particularly convincing case for the intuitive soundness of the axiom. In a letter to Philip Jourdain (March 1906) Russell wrote:

> *As for the multiplicative axiom, I came upon it so to speak by chance. Whitehead and I make alternate recensions on the various parts of our book, each correcting the last recension made by the other. In going over one of his recensions, which contained a proof of the multiplicative axiom, I found that the previous proposition used in the proof had surreptitiously assumed the axiom. This happened in the summer of 1904. At first I thought probably a proof could easily be found; but gradually I saw that, if there is a proof, it must be very recondite.*

Despite such plausible consequences as the cartesian product result, Intuitionists reject the Axiom of Choice because of its purely existential nature; it provides no means of constructing a choice function, merely asserting that such a function exists.

---

[1] Zermelo's original statement of the Axiom of Choice is in 'Beweis dass jede Menge wohlgeordnet werden kann', *Mathematische Annalen*, **59**, 514–16, (1904) and in its revised form in 'Untersuchungen über die Grundlagen der Mengenlehre I', *Mathematische Annalen*, **65**, 261–81, (1908).

REMARK

The equivalence of the Axiom of Choice and Russell's multiplicative axiom is fairly transparent. First we assume Choice. Let $\{x_i : i \in I\}$ be a family of non-empty sets. By Choice there exists a choice function, f say, for $\{x_i : i \in I\}$. Let $F(i) = f(x_i) \in x_i$, then F is an element of the cartesian product $\times_{i \in I} x_i$ of all x_is, hence that product is non-empty.

Conversely, let a be a non-empty set of non-empty sets. By the multiplicative axiom the cartesian product $\times_{x \in a} x$ is non-empty, which means there exists a function on a with the property that $f(x) \in x$, i.e. a choice function.

Objections to the Axiom of Choice, either the strong or the weak version, are typically either philosophical, based on the intuitive temporal implausibility of making an infinite number of choices, or on the non-constructive nature of the axiom, or are based on a peculiar identification of continuum-based models of physics with the physical objects being modelled; properties of the model which are implied by the Axiom of Choice are deemed to be counterintuitive because the physical objects they model don't have these properties. Motivated by these objections, or just for curiosity, several alternatives to Choice have been explored.

9.1.2   Alternatives to Choice

The most obvious way of restricting the Axiom of Choice is to add assumptions about the cardinality of the sets involved. In the case where the collection of sets on which the choice function is to be defined is finite, the axiom becomes redundant – it is an easy theorem of ZF. The difficulty occurs when the collection of sets is infinite; the weakest variant of Choice which cannot be proved in ZF is that in which one must make a choice from countably many pairs.

The Axiom of Countable Choice (often employed in analysis – even when it is not strictly needed) states that for every sequence (a_n) of non-empty sets there exists a sequence (x_n) such that $x_n \in a_n$ for all $n \in \omega$. A good impression of the comparative strength of full choice over Countable Choice can be formed by studying the consequences of the latter. Some of these results are exactly what we would expect, for example, it follows from the Axiom of Countable Choice that every cardinal number is comparable with $\aleph_0$. Other consequences, however, are surprising. Robert Solovay[1] proved the existence of a model of ZF in which Countable Choice holds and yet in which every set of real numbers is Lebesgue measurable, in notable contrast to the more familiar setting where the full Axiom of Choice is employed to construct non-Lebesgue measurable subsets of $\mathbb{R}$ (such as the Vitali set, for example – see Section 9.3). Countable Choice, like full choice, implies that the union of countably many countable sets

---

[1] R. M. Solovay, 'A model of set theory in which every set of reals is Lebesgue measurable', *Ann. Math.*, **92**, Ser. 2, 1–56, (1970).

is countable (compare this with the result of Feferman and Levy who proved in 1963 that in the absence of Choice there is a model of set theory in which the continuum is a countable union of countable sets).[1] Countable Choice is not strong enough to prove the existence of a well-ordering of $\mathbb{R}$, nevertheless it is sufficient to prove some basic theorems in measure theory.[2]

Another refinement of Choice is the Axiom of Countable Dependent Choice (first explicitly formulated by Paul Bernays in 1942)[3] which posits the existence of countable relational chains of elements: if R is a relation on a set a such that for all $x \in a$ there exists a $y \in a$ with xRy, then for any $x_0 \in a$ there exists a sequence (x_n) in a such that $x_n R x_{n+1}$ for all $n \in \omega$. The Axiom of Countable Dependent Choice implies the Axiom of Countable Choice but is not equivalent to it. Nevertheless it is equivalent to a multitude of interesting statements in other parts of mathematics, for example the Baire Category Theorem: in every complete metric space the intersection of a countable family of open dense subsets of the space is itself dense in the space. Consequently it has a fairly stable status as a principle of interest. Dependent choices are used in model theory, and had been used, especially by analysts, long before Bernays' concrete statement of the principle.

REMARKS

1. Suppose a is an infinite set and let F_n be the collection of all subsets of a of cardinality n for $n = 1, 2, 3, \ldots$. Since a is infinite, F_n is non-empty for all n. By the Axiom of Countable Choice there exists a sequence (a_n) of sets in a with $a_n \in F_n$. The countable union of finite sets $\cup_{n=1}^{\infty} a_n$ is a countable subset of a. Therefore the Axiom of Countable Choice implies that every infinite set has a countable subset. It follows that, under the assumption of Countable Choice, every cardinal number is comparable with $\aleph_0$ and also that every infinite set is Dedekind infinite.

2. A surprising amount of classical mathematics is accessible using Countable Choice in place of the Axiom of Choice. Familiar analytical results such as the equivalence of continuity and sequential continuity of a function $f : \mathbb{R} \to \mathbb{R}$ at a point (i.e. if $a_n \to a$ then $f(a_n) \to f(a)$) are equivalent to the existence of a choice function on countable subsets of $\mathbb{R}$.

   Alternative cardinal restrictions are easy to formulate. Generalizing Countable Choice one could study the consequences of the axiom $\mathrm{AC}(\kappa)$: every set of cardinality κ or less has a choice function.

   Instead of restricting the cardinality of the set on which the choice function is being defined one can restrict the cardinality of its members. Let C_α be the

---

[1]Feferman, S. and Levy, A., 'Independence results in set theory by Cohen's method', *Not. Amer. Math. Soc.*, **10**, 593, (1963).

[2]The first line of Oxtoby [**161**]: '*The notions of measure and category are based on that of countability*'.

[3]Bernays, P., 'A system of axiomatic set theory III', *J. Symbolic Logic*, **7**, 65–89, (1942).

statement: *if a is a set all members of which have cardinality α, then there exists a choice function for a.* C_1 follows trivially from ZF (the map $\{x\} \mapsto x$ defines a choice function on any collection of singletons). The statements C_α are not independent of one another. One can prove, for example, that C_2 implies C_4. The study of the statements C_n (n finite) and their relation to one another was initiated by A. Mostowski in 1945.[1]

3. One proof that Countable Dependent Choice implies Countable Choice is as follows. Assume (X_n) is a sequence of non-empty sets. We further assume without loss of generality[2] that the X_n are disjoint. We need to show that there exists a choice function which selects one element from each X_n, i.e. a sequence (x_n) with $x_n \in X_n$. Since we are assuming Countable Dependent Choice we need to find an appropriate relation R on the disjoint union $X = \cup X_n$. Let xRy if and only if there exists an n such that $x \in X_n$ and $y \in X_{n+1}$. Let $x_0 \in X_0$. Then by Countable Dependent Choice there exists a sequence (x_n) of elements in X with $x_n R x_{n+1}$. By induction $x_n \in X_n$, so we have constructed the desired sequence.

The most obvious variations of the Axiom of Choice are those that restrict the cardinality of the sets in question. Other variations impose relational restrictions between the sets. When the early set theorists tried to prove the Axiom of Choice they invariably ended up showing it is equivalent to some other statement that they were unable to prove. This collection of equivalent statements has grown to an enormous size. One of its striking features is that some of the statements seem intuitively obvious while others are either wildly counterintuitive or evade any kind of evaluation.

9.1.3 An equivalent of Choice

The Axiom of Choice is consistent with ZF[3] and is quite robust in the sense that it remains consistent with extensions of ZF by the vast majority of the higher axioms of infinity. Reinhardt cardinals (we do not give the definition here), which can be defined in Gödel–Bernays set theory but not in an unaugmented ZF, provide an exception to this rule: Kenneth Kunen[4] proved that the existence of Reinhardt cardinals is inconsistent with Gödel–Bernays set theory plus Choice. A bewildering number of statements, drawn from all branches of

---

[1] A. Mostowski, 'Axiom of choice for finite sets', *Fund. Math.*, **33**, 137–168, (1945).

[2] If the X_n are not disjoint we can replace each element x of X_n with, say, (x, n), or use any number of other elementary tricks to ensure that the modified sets are disjoint.

[3] Gödel, K., 'The consistency of the axiom of choice and the generalized continuum hypothesis', *Proc. Nat. Acad. Sci. USA*, **24**, 556–7, (1938).

[4] Kunen, K., 'Elementary embeddings and infinite combinatorics', *J. Symbolic Logic*, **36**, 407–413, (1971).

mathematics, can be shown to be equivalent to the Axiom of Choice; indeed volumes have been filled on this subject alone.[1]

To take a simple example within set theory itself (there are so many to choose from our choice is almost arbitrary), Tarski proved that Choice is equivalent to the assertion that $a + b = ab$ for all infinite cardinal numbers a and b;[2] in fact it is equivalent to the simpler assertion that if u is an infinite cardinal number then $u^2 = u$. The proof of the latter is not too difficult (it is convenient to use Choice in the form of the Well-Ordering Theorem – see Section 9.2): We prove that (i) implies (ii) where:

(i) for every infinite set x there exists an injection $x^2 \to x$ (in particular, since an injection $x \to x^2$ is easily found, x^2 is equipollent to x);

(ii) every set has a well-ordering.

Let a be a set. As every finite set (and indeed every countable set) has a well-ordering, we will assume that a is infinite. There is an ordinal γ for which no injection $\gamma \to a$ exists. (This is sometimes called Hartogs' Theorem after Friedrich Hartogs[3] – see the remarks at the end of this subsection.) Let us assume, for the avoidance of arbitrary choice, that α is the smallest such ordinal. α is called the *Hartogs number* of a. Denote by $\hat{a}$ the set $\{(\alpha, x) : x \in a\}$. $\hat{a}$ has been designed to be equipollent to a and disjoint from α. By (i) there is an injection $f : (\hat{a}\cup\alpha) \times (\hat{a}\cup\alpha) \to \hat{a}\cup\alpha$. For all $z \in \hat{a}$ and $\beta \in \alpha$, $f(z, \beta)$ is either an element of $\hat{a}$ or an ordinal less than α. If there exists a $z \in \hat{a}$ such that, for every $\beta < \alpha$, $f(z, \beta) \in \hat{a}$ then, letting $h(\beta) = f(z, \beta)$, h is an injection $\alpha \to \hat{a}$ which implies the existence of an injection $\alpha \to a$, a contradiction. Consequently for every $z \in \hat{a}$ there is a $\beta < \alpha$ such that $f(z, \beta) \in \alpha$; let β_z be the least such ordinal. Define $g(z) = f(z, \beta_z)$. Since f is an injection it follows that g is an injection $\hat{a} \to \alpha$, inducing an injection $a \to \alpha$ giving the desired well-ordering of a.

The converse ((ii) implies (i)) can be proved via transfinite induction.

Although the Axiom of Choice implies $u = u + u$ for all infinite cardinals u, the converse, which was an open problem raised by Azriel Levy in 1963,[4] turns out to be false as Gershon Sageev proved a decade later.[5]

---

[1]See Rubin and Rubin [**179**].

[2]Tarski, A., 'Sur quelques théorèmes qui équivalent à l'axiome du choix', *Fund. Math.*, **5**, 147–54, (1924).

[3]Hartogs, F., 'Über das Problem der Wohlordnung', *Mathematische Annalen*, **76**, 438–443, (1915).

[4]Levy, A., 'The Fraenkel–Mostowski method for independence proofs in set theory', John W. Addison Jr, Leon Henkin, Alfred Tarski (eds.), *The Theory of Models* (Proceedings of the 1963 International Symposium at Berkeley), North-Holland, Amsterdam, 221–228, (1965).

[5]Gershon Sageev, 'An independence result concerning the axiom of choice', *Annals of Mathematical Logic*, **8**, 1–184, (1975).

REMARKS

1. Hartogs' Theorem. Let X be a set. Then $P(X^2)$ is a set. The class of all well-orderings of subsets of X is a subclass of $P(X^2)$ and hence is a set. Each such well-ordered subset Y of X corresponds to an injection $\alpha \to X$, where α is the ordinal number of Y (and so the class β of all such ordinals α is a set, by Replacement). It is easy to see that β is itself an ordinal, and we claim that it is the smallest ordinal which cannot be injected into X. If β could be injected into X then we would have $\beta \in \beta$, contradicting Regularity, and if $\alpha \in \beta$ then by the definition of β there exists an injection of α into X.

2. Since the proof of Hartogs' Theorem does not appeal to the Axiom of Choice, the Hartogs number of a set X exists whether or not X has a well-ordering. If we assume Choice then the Hartogs number of X is the smallest cardinal number strictly greater than the cardinal number of X.

   Let us assume that the Axiom of Choice is false so that there exists a set X which does not have a well-ordering (X is necessarily infinite). We denote by $|X|$ the classical cardinal number of X. Let α be the Hartogs number of X. No generality is lost by assuming that X and α are disjoint. This means that $|X \cup \alpha| > |\alpha|$ and $|X \cup \alpha| > |X|$. Each inequality must be strict: if $|X \cup \alpha| = |\alpha|$ there is a bijection $X \cup \alpha \to \alpha$ and in particular an injection $X \to \alpha$, implying X can be well-ordered; if $|X \cup \alpha| = |X|$ then there is an injection $\alpha \to X$, contradicting the fact that α is the Hartogs number of X. This means $X \cup \alpha$ is the disjoint union of two sets of strictly smaller cardinality. Since if we assume the Axiom of Choice the sum of two cardinal numbers at least one of which is infinite is equal to the largest summand, we have shown that the Axiom of Choice is equivalent to the statement: if X is infinite and $|X| = |A| + |B|$ then $|X| = |A|$ or $|X| = |B|$.

The most well-known equivalent of the Axiom of Choice is the Well-Ordering Theorem, which states that every set has a well-ordering. Choice is also equivalent to the statement that every infinite set is equipollent to its cartesian square (we have already seen some concrete examples of this equipollence, without having to appeal to Choice, in the case of $\mathbb{Z}$ and $\mathbb{R}$). One of the reasons that the Axiom of Choice is so widely adopted is that it is so useful, and the contortions one must make to prove a statement without it, if this is possible, are often painful.

9.1.4 Applications of Choice

Many well-known theorems rely critically on the Axiom of Choice in their proofs. Some Platonists have cited the fruitfulness of Choice (or any other abundantly useful axiom for that matter) as evidence for its 'truth' and have taken as further ammunition that many branches of mathematics are considerably simplified if Choice is assumed; it is, they claim, a natural assumption. General topology in particular becomes something of a nightmare in its absence, and we have already seen how cardinal arithmetic is almost irretrievably handicapped without Choice.

The Axiom of Choice may also be employed to provide simpler shorter proofs and stronger statements of theorems for which it is otherwise not required. For example, although we do not need the Axiom of Choice to prove the Cantor–Bernstein Theorem, with Choice the proof is instantaneous: any two cardinal numbers are comparable with respect to the relation 'there exists an injection from α to β' and the class of cardinal numbers is linearly ordered by this relation. However, even Platonists convinced of the validity of Choice nevertheless have a tendency to avoid using it if possible, generally regarding proofs which do not use it as being more informative than those that do.

Thanks to the work of Gödel and Cohen, the role played by the Axiom of Choice in mathematics is no longer shrouded in mystery, at least not to the same extent as it was in the early development of set theory.

REMARK

Before Gödel and Cohen the suspicions about the Axiom of Choice were partly mathematically motivated. There was a worry that the axiom might be inconsistent with the other axioms of set theory. Once its independence was established, the arguments for or against Choice necessarily became purely philosophical. The acceptance or rejection of Choice is a philosophical decision, as is the acceptance or rejection of any independent statement. Mathematically, the study of models of set theory in which Choice holds or fails is no more peculiar than the study of commutative versus non-commutative groups or of Hausdorff versus non-Hausdorff topological spaces.

The Axiom of Choice turns the Cantor–Bernstein Theorem into a triviality as it induces a linear ordering on the cardinal numbers. It also vastly simplifies many other parts of mathematics – often being indispensable. However, some of the geometric results one can prove with the Axiom of Choice are, to some, so wildly counterintuitive as to be regarded as evidence that Choice should not be adopted after all. One such result is the Banach–Tarski paradox.

9.1.5   The Banach–Tarski paradox

Most notoriously, the Axiom of Choice implies the Banach–Tarski paradox[1] which tells us that there is a decomposition of the unit ball

$$\{(x, y, z) \in \mathbb{R}^3 : x^2 + y^2 + z^2 \leq 1\}$$

into a finite number of subsets (five being the minimum)[2] which can by a composition of rotations and translations in $\mathbb{R}^3$ be reassembled to form two balls of unit radius.[3] The basic idea behind the proof is the following. Consider the collection of all finite strings formed by the symbols a, b, a^{-1}, b^{-1}, repetitions permitted, and denote by 1 the empty string. Using the intuitive rules:

(i) $1s = s1 = s$ for all strings s; and

(ii) $aa^{-1} = a^{-1}a = 1 = bb^{-1} = b^{-1}b$,

every string is reducible to either the empty string 1 or to a string in which no occurrences of a and a^{-1} are adjacent and no occurrences of b and b^{-1} are adjacent (let us call such strings *reduced strings*). Two reduced strings are multiplied together by simple concatenation and the string so produced is reduced, as described by the above rules, to a reduced string. So for example $(abab^{-1}a^{-1}b)(b^{-1}a^3ba) = abab^{-1}a^2ba$. (As usual, a^3 and a^2 are abbreviations for aaa and aa, respectively.) The set of all reduced strings with this multiplication is the non-commutative group F_2 (the *free group* on two elements).

If we denote by S_x the set of all elements of F_2 beginning with the symbol x ($x \in \{a, a^{-1}, b, b^{-1}\}$) then F_2 is the disjoint union:

$$F_2 = \{1\} \cup S_a \cup S_{a^{-1}} \cup S_b \cup S_{b^{-1}}.$$

Denoting by yS_x the set $\{ys : s \in S_x\}$ we have two alternative disjoint decompositions of F_2:

$$F_2 = aS_{a^{-1}} \cup S_a$$
$$F_2 = bS_{b^{-1}} \cup S_b.$$

A concrete representation of F_2 is obtained by letting a and b be orthogonal rotations in $\mathbb{R}^3$, each rotation by an irrational multiple of π radians. From here on, by F_2 we mean this concrete group of rotations, interpreting ab, say, as 'apply the rotation b followed by the rotation a'. The F_2-orbit of a point x on the sphere is the set of all points obtained by applying the rotations of F_2 to x, i.e. $\{\rho(x) : \rho \in F_2\}$.

The group of rotations F_2 partitions the unit sphere of $\mathbb{R}^3$ into F_2-orbits. Using the Axiom of Choice we select one point from each F_2-orbit. This selection

---

[1] Stefan Banach and Alfred Tarski, 'Sur la décomposition des ensembles de points en parties respectivement congruentes', *Fundamenta Mathematicae*, **6**, 244–277, (1924).
[2] That five is minimal is due to R. M. Robinson, 'On the Decomposition of Spheres', *Fund. Math.*, **34**, 246–260, (1947).
[3] Full details of the Banach–Tarski decomposition may be found in Wagon [**222**].

forms a subset K of the sphere. The five part disjoint decomposition of F_2 above when applied to K yields a partition of the sphere into the subsets K, $S_a K$, $S_{a^{-1}} K$, $S_b K$, $S_{b^{-1}} K$, where $S_x K$ is the set of all points obtained by applying the rotations in S_x to the set K. The alternative decompositions of F_2 show that by rotating the subset $S_{a^{-1}} K$ by the rotation a and rotating the subset $S_{b^{-1}} K$ by the rotation b we can put the pieces together to form two copies of the sphere $a S_{a^{-1}} K \cup S_a K$ and $b S_{b^{-1}} K \cup S_b K$, however, there is another piece, K, left over. We can cleverly remedy this as follows. Let

$$L = a^{-1} K \cup a^{-2} K \cup a^{-3} K \cup \ldots$$

and partition the sphere into four parts:

$$[S_a K \cup K \cup L] \cup [(S_{a^{-1}} K) - L] \cup S_b K \cup S_{b^{-1}} K.$$

Applying the rotation a to the second component $(S_{a^{-1}} K) - L$ we obtain the set $a S_{a^{-1}} K - (K \cup L)$, whose union with the first component is the sphere. We form another sphere from the second two components as before. Hence, skipping some fine details re the centre point and other easily resolved technicalities concerning fixed points of rotations, by radial extension we obtain a paradoxical decomposition of the ball. There is a similarity here, in the manner in which Choice is used, with the construction of the Vitali sets, described in Section 9.3, which was freely acknowledged as an inspiration in Banach and Tarski's work. It is trivially the case that the Banach–Tarski decomposition is impossible if we further require the subsets to be Lebesgue measurable since Lebesgue measure on $\mathbb{R}^3$ is translation and rotation invariant.

The Banach–Tarski paradox is sometimes called the Hausdorff–Banach–Tarski paradox in honour of a related paradox discovered by Felix Hausdorff a decade earlier:[1] there is a disjoint decomposition of the unit sphere in $\mathbb{R}^3$ into four subsets A_1, A_2, A_3, C such that C is countable and A_1, A_2, A_3 and $A_2 \cup A_3$ are congruent. In the same paper Hausdorff proves that there is a countable partition of $[0, 1]$ which can by translation be reassembled to form $[0, 2]$. Corollary: there is no non-trivial, translation invariant measure defined on all bounded subsets of $\mathbb{R}$.

REMARKS

1. The free group on n elements is defined in the obvious way as the set of 'words' in n symbols $a_1, \ldots, a_n$ and their inverses $a_1^{-1}, \ldots, a_n^{-1}$ with concatenation product and the usual cancellation rules. The free group F_1 is isomorphic to the additive group of all integers. The set of generators may be infinite. Every group is the homomorphic image of some free group.

---

[1] Hausdorff, F., 'Bemerkung über den Inhalt von Punktmengen', *Mathematische Annalen*, **75**, 428–433, (1914).

2. The Banach–Tarski paradox gives us a measure theoretic limitation in three-dimensional space. A function $\mu : P(\mathbb{R}^3) \to [0, \infty]$ is a finitely additive invariant measure if it satisfies the following: (i) the measure of a subset of $\mathbb{R}^3$ does not change if we move the subset by an isometry (distance preserving map); (ii) the measure of the disjoint union of two sets is the sum of the measures of each set; and a normalizing condition (iii) the measure of the unit ball is, say, $4\pi/3$ (or 1, or whichever constant one prefers). The Banach–Tarski paradox tells us that no such measure exists. Finitely additive invariant measures *do* exist in lower dimensions, so no direct analogue of the Banach–Tarski paradox exists in dimensions 1 or 2. Recall that Lebesgue measure is *countably additive* (the measure of a countable pairwise disjoint collection of sets is the sum of the measures), as are measures in general, a substantially stronger condition than finite additivity.

One can take a ball in $\mathbb{R}^3$, partition it into a finite number of subsets, then rotate and translate each of these subsets to form two solid balls each with the same radius as the original. This surprising application of the Axiom of Choice, the Banach–Tarski paradox, is paradoxical only in a fairly weak sense. We are surprised by it only because we tend to identify $\mathbb{R}^3$ with physical space and such a decomposition in real physical space is impossible. All the Banach–Tarski paradox really tells us is how exotic subsets of $\mathbb{R}^3$ can be if we are granted the Axiom of Choice to create them. The subsets involved in the decomposition are, not surprisingly, conceptually very distant from the kind of subsets we would normally imagine a physical sphere being divided into. Using similar techniques we can construct other paradoxical decompositions.

9.1.6 Further paradoxical decompositions

An interior point of a set $X \subseteq \mathbb{R}^n$ is a point $x \in X$ for which there exists an open n-ball B_r of some positive radius r centred at x with $B_r \subseteq X$. Two subsets X and Y of $\mathbb{R}^n$ are congruent by finite decomposition if there exists a partition of X into finitely many sets $X = A_1 \cup \cdots \cup A_n$ and orientation preserving surjective isometries (distance preserving functions) $\sigma_1, \ldots, \sigma_n$ of $\mathbb{R}^n$ to itself such that Y is the disjoint union $\sigma_1 A_1 \cup \cdots \cup \sigma_n A_n$. By $\sigma_i A_i$ we mean the set $\{\sigma_i(a) : a \in A_i\}$. The condition of orientation preservation is demanded for 'realism', the idea being that the permissible isometries are those that could be performed on a physical object, ruling out reflections.

The generalized Banach–Tarski Theorem tells us that in a Euclidean space of dimension at least three, two arbitrary bounded sets with interior points are congruent by finite decomposition. In particular any two balls of arbitrary size are congruent by finite decomposition. This is sometimes expressed, purely for dramatic purposes, in terms of physical objects ('the sun and the pea'). There are different ways of stating the paradox, all of them painting a vivid picture of $\mathbb{R}^3$ as something far removed from anything we are likely to encounter in the physical Universe.

Why dimension three? The answer is group theoretic; the group of isometries of $\mathbb{R}^n$ for $n \geq 3$ is sufficiently rich (it contains a copy of the free group F_2, for instance) to generate paradoxical decompositions while the group of isometries for $n \leq 2$ is too simple for Banach–Tarski style duplication. Nevertheless, weaker paradoxical decompositions are also available in two dimensions; for example, any disc is congruent by finite decomposition to a solid square of equal area. This surprising result, posed as a challenge by Alfred Tarski in 1925, was eventually proved by Miklós Laczkovich.[1] The decomposition can be achieved using translations alone. In contrast to the minimal five subsets of the Banach–Tarski paradox, the Laczkovich decomposition partitions the disc into approximately 10^{50} subsets.

The following simple example should help to demystify the phenomenon by illustrating how a set can be congruent by finite decomposition to some of its proper subsets. Let $C = \{e^{i\theta} : 0 \leq \theta < 2\pi\}$ be the unit circle in $\mathbb{C}$. Fix some irrational number γ, let $\Gamma = \{e^{n\pi\gamma i} : n \in \mathbb{N}\}$ and denote by Γ' the complement of Γ in C, so that C is partitioned as $C = \Gamma \cup \Gamma'$. The anticlockwise rotation R of the complex plane about the origin by $\gamma\pi$ radians (equivalently multiplication by $e^{\gamma\pi i}$) maps Γ to Γ minus the point 1 and Γ' onto Γ', in particular C is congruent by finite decomposition to $R(\Gamma) \cup \Gamma'$, i.e. C minus the point 1, and in general to $R^n(\Gamma) \cup \Gamma'$, i.e. C with n points removed.

Those who have spent some time studying mathematics but who have not spent too long modelling objects of physical space by pieces of $\mathbb{R}^3$ do not regard the conclusion of the Banach–Tarski paradox (however one may choose to state it) as paradoxical. It also needs to be stressed that paradoxical decompositions do not always use the Axiom of Choice. For example, Stefan Mazurkiewicz and Wacław Sierpinski proved in 1914[2] that there is a non-empty subset X of $\mathbb{R}^2$ (necessarily of measure zero) which has two disjoint measurable subsets each of which can be split into finitely many parts and rearranged without distortion to form a partition of X. Although this has less counterintuitive impact than the Banach–Tarski decomposition it nevertheless has the potential to plant seeds of doubt in the minds of those who reject Banach–Tarski on purely intuitive grounds (and hence who reject Choice).

REMARKS

1. The pieces (if we can call them that) in Laczkovich's decomposition are non-measurable – not the type of mosaical fragments that one can easily imagine.

2. Tangential to the present topic, but something equally surprising, and yet another example of how strange $\mathbb{R}^3$ really is, is Smale's paradox, discovered

[1] Laczkovich, M., 'Equidecomposability and discrepancy: a solution to Tarski's circle squaring problem', *J. Reine Angew. Math.*, **404**, 77–117, (1990).

[2] Mazurkeiwicz, S. and Sierpinski, W., 'Sur un ensemble superposables avec chacune de ses deux parties', *C. R. Acad. Sci. Paris*, **158**, 618–619, (1914).

in 1958.[1] It is possible to turn a sphere inside out in $\mathbb{R}^3$ *smoothly*; that is, self-intersections are allowed but there is no creasing or pinching of the sphere as it turns inside out. This process is known as sphere eversion.

For group theoretic reasons the most impressive paradoxical decompositions occur in dimension at least three, but there are also interesting decompositions in lower dimensions. For example, one can partition a disc into finitely many subsets and rigidly rearrange these subsets to form a square of the same area as the original disc. Even without the Axiom of Choice it is possible to construct some counterintuitive subsets of the plane, so Choice cannot be held responsible for all that is counterintuitive in geometry. Some more radical alternatives to the Axiom of Choice obstruct such constructions more effectively.

9.1.7 More alternatives to Choice

The Axiom of Choice had been tacitly assumed for years before it came into focus following Zermelo's closer examination of the Well-Ordering Theorem. Opponents of Choice on the whole tended to be influenced by psychological dissonance rather than logic. Many of the reasons given for the rejection of the Axiom of Choice were based on an unwillingness to accept its supposedly counterintuitive consequences; the opponents did not believe it and tried to conclude that it must therefore be false. The adoption of such intuition based prejudice is disastrous in any science. Although the Axiom of Choice has some apparently counterintuitive consequences, so does its negation (only more so). The reader is invited to compare the supposed counterintuitive Banach–Tarski decomposition with, say, the existence of an infinite yet Dedekind finite set, or the failure of trichotomy for sets. Bare intuition alone is a hopeless guide when it comes to matters as remote from our immediate physical experience as infinite sets, instead our intuition must be tuned by submersion in a set theoretic universe.

It is of course worthwhile to examine alternatives to Choice, even if one is Platonically inclined to believe that Choice is true. Some of these alternative axioms are directly motivated by the need to block the Banach–Tarski paradox and similar counterintuitive decompositions, some are fuelled by constructivist ideas, others are similarly motivated but nevertheless are compatible with Choice. Focusing on sober measure theoretic consequences, one approach is to replace the Axiom of Choice with an assertion that a certain class of subsets of $\mathbb{R}$ must be measurable or be subject to some other restriction. Jan Mycielski and Hugo Steinhaus proposed such an alternative in the early 1960s in the form of the Axiom of Determinacy[2] (we shall meet this 'game theoretic' axiom in

---

[1]Smale, Stephen, 'A classification of immersions of the two-sphere', *Trans. Amer. Math. Soc.*, **90**, 281–290, (1958).

[2]Mycielski, J. and Steinhaus, H., 'A mathematical axiom contradicting the axiom of choice', *Bull. Acad. Polon. Sci.*, **10**, 1–3, (1962).

more detail in Subsection 9.3.4 so we postpone its statement until then). If the Axiom of Determinacy holds then, among other unfamiliar consequences, every set of real numbers is measurable (and hence Determinacy is incompatible with the Axiom of Choice).

The full Axiom of Determinacy has not been a particularly popular substitute for Choice among mathematicians. More recently some set theorist have instead been attracted, both on the basis of its consequences and on the fact that it follows from the existence of certain large cardinals, to a restrictive form of determinacy concerning *projective* sets. The class of projective sets is generated from the class of closed subsets of $\mathbb{R}$ by countable union, complementation and continuous transformation – under the assumption of projective determinacy all projective sets are Lebesgue measurable. This weaker determinacy is relatively consistent with the Axiom of Choice, the Banach–Tarski decomposition is still valid and furthermore the Continuum Hypothesis remains undecidable.

REMARK

Many of the more otherworldly consequences of the Axiom of Choice were discovered in the decades between its first clear statement and Gödel's proof of its relative consistency with the other axioms. In these years the understandable nervousness about Choice led to much explicit cataloguing of those results which required Choice in their proofs and those which did not, so the project of understanding the geography of the two different worlds, mathematics with Choice and mathematics without Choice, was already under way long before Cohen's independence result appeared.

Axioms which replace the Axiom of Choice, such as the Axiom of Determinacy, tend not to be inspired by intuitively immediate properties of naive sets. The focus has moved from the intuitive soundness of the axioms themselves to the plausibility of their consequences. Unfortunately even equivalent statements can have different degrees of plausibility. Compare, for example, the Axiom of Choice with the Well-Ordering Theorem, trichotomy and Zorn's Lemma.

9.2 The Well-Ordering Theorem, trichotomy and Zorn's Lemma

In all affairs it's a healthy thing now and then to hang a question mark on the things you have long taken for granted.
– BERTRAND RUSSELL[1]

---

[1]'The Recrudescence of Puritanism', in Russell [186].

9.2.1 The Well-Ordering Theorem

Every countable set has a well-ordering. Having proved that $\mathbb{R}$ is uncountable, Cantor soon asked whether it too may be well-ordered. Initially Cantor felt that such an ordering must be possible, and went on to suggest that *every* set has a well-ordering (the Well-Ordering Theorem). At first this was treated as an axiom,[1] but by 1895 Cantor was beginning to believe that it should be provable from the other principles of his theory.[2]

The Axiom of Choice guarantees the existence of a well-ordering of $\mathbb{R}$ but it does not explicitly define one. Indeed, as has been mentioned several times already, the Axiom of Choice implies the Well-Ordering Theorem (as was proved by Zermelo). The intuitive argument is immediate. Suppose a is a set and f is a choice function for $P(a)$. Let $a_0 = f(a)$, $a_1 = f(a - \{a_0\})$, $a_2 = f(a - \{a_0, a_1\})$, $a_3 = f(a - \{a_0, a_1, a_2\})$,..., $a_\omega = f(a - \{a_i : i \in \omega\})$, etc. and for each ordinal α less than the Hartogs number of a, $a_\alpha = f(a - \{a_\gamma : \gamma < \alpha\})$.

Conversely, the Well-Ordering Theorem implies the Axiom of Choice. This is fairly elementary – if $\{a_i\}_{i \in \Lambda}$ is a set of sets then by the Well-Ordering Theorem $a = \cup_{i \in \Lambda} a_i$ has a well-ordering W, so each subset a_i of a has a W-minimal element x_i yielding the choice function $f : \{a_i\}_{i \in \Lambda} \to a$, $f(a_i) = x_i$.

Critics who believed the Well-Ordering Theorem to be false desperately tried to find an error in Zermelo's proof. Some attacked its use of transfinite induction, a use of ordinals which was still regarded with suspicion. This criticism was expertly countered by Zermelo in 1908 when he published his second proof,[3] this time eliminating the issues that had been subject to doubt in the first proof. The Axiom of Choice remained as the only conceivable weak point in the proof, so it was this freshly exposed target at which the critics aimed their shots.

REMARK

Zermelo's 1908 proof of the Well-Ordering Theorem is less intuitive than the earlier proof. He begins *'Although I still fully uphold my "Proof that every set can be well-ordered", published in 1904, in the face of the various objections that will be thoroughly discussed ... the new proof that I give below of the same theorem may yet be of interest, since, on the one hand, it presupposes no specific theorems of set theory and, on the other, it brings out, more clearly than the first proof did, the purely formal character of the well-ordering, which has nothing at all to do with spatiotemporal arrangement.'*

The argument is largely framed in terms of 'Θ-chains' by which he means the following. Given a set X and a choice function f on X, a Θ-chain is a subset

---

[1]Cantor, G., *Grundlagen einer allgemeinen Mannigfaltigkeitslehre. Ein mathematische-philosophischer Versuch in der Lehre des Unendlichen.* Leipzig: Teubner 1883. In Ewald [66].

[2]Cantor, G., 'Beiträge zur Begründung der transfiniten Mengenlehre I', *Math Ann.*, **47**, 481–512, (1895).

[3]Zermelo, E., 'Untersuchungen über die Grundlagen der Mengenlehre', *Math. Ann.*, **65**, 261–281, (1908). Translated in van Heijenoort [219].

Θ of $P(X)$ such that: (i) $M \in \Theta$; (ii) if $x \in \Theta$ then all elements of x are in Θ as is the set $x \backslash \{f(x)\}$; and (iii) if $A \in \Theta$ then $\cap A \in \Theta$.

The Well-Ordering Theorem states that every set has a well-ordering. It was in proving this principle that Zermelo brought focus on the equivalent statement: the Axiom of Choice. His set theory was initially created as a response to criticisms of his proof.

9.2.2 Trichotomy

Cantor's Law of Trichotomy tells us that for all sets a and b there exists an injection $a \to b$ or an injection $b \to a$. This intuitively highly plausible statement is equivalent to the less plausible Well-Ordering Theorem which in turn is equivalent to Zorn's Lemma (see below), which at first glance seems to be an opaque statement impervious to any decision as to its plausibility.

The importance of trichotomy for the classical theory of cardinal numbers should be obvious: without it two cardinal numbers need not be comparable, an unfortunate characteristic of any construction that aims to be a measure of quantity, finite or otherwise.

REMARK

Let X be a set which has no well-ordering. By Hartogs' Theorem (which, recall, is a theorem of ZF – it does not require Choice) let α be the Hartogs number of X, the smallest ordinal which cannot be injected into X. It is easy to prove (compare the remarks to Subsection 9.1.3) that α is precisely the set of ordinals which are the order types of well-ordered subsets of X. α cannot be injected into X by definition and X cannot be injected into α since it is not well-orderable. Thus trichotomy fails for the two sets X and α. This proves that trichotomy implies the Well-Ordering Theorem. The converse is obvious.

Trichotomy tells us that the class of cardinal numbers is linearly ordered. More precisely, given two sets there exists an injection from one of the sets into the other.

9.2.3 Zorn's Lemma

Although Zorn's Lemma appears to be more complicated than its equivalent siblings the Axiom of Choice, trichotomy and the Well-Ordering Theorem, it is used far more frequently, often implicitly, in the modern mathematical literature. For its statement we need the two simple notions of *chain* and *maximal element*.

A set b is a chain in a if $b \subseteq a$ and if for all $x, y \in b$, $x \subseteq y$ or $y \subseteq x$; and x is a maximal element of a if $x \in a$ and for all $y \in a$, $x \subseteq y$ implies $x = y$.

Zorn's Lemma

If $a \neq \emptyset$ and if $\cup b \in a$ whenever b is a chain in a, then a has a maximal element.

Zorn's Lemma is named in honour of Max Zorn who described the principle in 1935,[1] but it, and some closely related maximal principles, has a much older history (Zorn was unaware, at the time, of the earlier results). The idea was anticipated by a number of prominent mathematicians. Hausdorff's maximal principle is an early incarnation of Zorn's Lemma derived from the Axiom of Choice: every non-empty partially ordered set contains a maximal totally ordered subset.[2] Kuratowski's principle[3] is also recognizable as a predecessor of Zorn's Lemma (the chains are well-ordered in Kuratowski's principle). For this reason Zorn's Lemma is occasionally called the Kuratowski–Zorn Lemma. Note the title of Kuratowski's paper – he is avoiding the use of transfinite ordinals, a theme which Zorn continued.[4]

Many variant formulations of Zorn's Lemma are known. It has proved to be more popular than its equivalent principles because it is easily and immediately applicable in a myriad of different mathematical settings.

REMARK

Without digressing to a cornucopia of different results in almost all branches of mathematics it is difficult to convey precisely how useful and far reaching Zorn's Lemma is. One such application (which, as mentioned in Subsection 1.7.5, turns out to be equivalent to Choice) is that every vector space has a basis. The proof is as follows. Suppose V is a vector space. Let $\mathcal{L}$ be the set of all linearly independent subsets of V. $\mathcal{L}$ is non-empty since it contains all singletons $\{v\}$ ($v \in V$) and it is partially ordered by inclusion $\subseteq$. If C is a chain in $\mathcal{L}$ then $\cup\{X : X \in C\}$ is an upper bound of C in $\mathcal{L}$. We infer from Zorn's Lemma that $\mathcal{L}$ has a maximal element B. B must be a basis for $\mathcal{L}$, for if the complement of the linear span of B in V contained a vector v, then the disjoint union $B \cup \{v\}$ would be an element of $\mathcal{L}$, contradicting maximality.

Another application, also mentioned in the introduction in the discussion of non-standard extensions of $\mathbb{R}$ in Subsection 1.7.11, is that every ring with a multiplicative identity has a maximal ideal (the condition of having a multiplicative identity is essential).

---

[1]Max Zorn, 'A remark on method in transfinite algebra', *Bull. Amer. Math. Soc.*, **41**, no 10, 667–670, (1935).

[2]Hausdorff, F., *Grundzüge der Mengenlehre*, (1914).

[3]C. Kuratowski, 'Une méthode d'élimination des nombres transfinis des raisonnements mathématiques', *Fundamenta Mathematicae*, **3**, 76–108, (1922).

[4]Other principles are described by Salomon Bochner ('Fortsetzung Riemannscher Flachen', *Mathematische Annalen*, **98**, 406–421, (1928)) and R. L. Moore (see Moore [**154**]).

A third well-known application is that every field has a maximal algebraic extension, its algebraic closure.

Zorn's Lemma is not as easy to state as the Well-Ordering Theorem or trichotomy, but the framework it describes applies to a wide range of mathematical settings. With it one can instantly prove the existence of certain algebraic or topological objects. Zorn's Lemma is just one of many maximal principles.

9.2.4   The Teichmüller–Tukey Lemma

We mention one more maximal principle equivalent to Choice,[1] due independently to Oswald Teichmüller (1939)[2] and John Tukey (1940).[3] This principle is expressed in terms of the notion of *finite character*. A set a is of finite character if:

(i) a is non-empty;

(ii) if $x \in a$ then every finite subset of x is also a member of a; and

(iii) if every finite subset of a set b is a member of a then $b \in a$.

The Teichmüller–Tukey Lemma then asserts that any set of finite character has an inclusion-maximal element.

Remark

Let us see how the Teichmüller–Tukey Lemma follows from Zorn's Lemma. Suppose a is a set of finite character. We need to prove that a has an inclusion-maximal element. Let $b = \{a_i\}_{i \in I}$ be a chain in a where I is some indexing set. Let x be a finite subset of $\cup b$. Since x is finite and b is a chain there exists an index $j \in I$ such that $x \subseteq a_j$, but $a_j \in a$ so by property (ii) $x \in a$. However, x was an arbitrary finite subset of $\cup b$, so by property (iii) we have $\cup b \in a$. Then by Zorn's Lemma a has an inclusion-maximal element.

Like Zorn's Lemma, the Teichmüller–Tukey Lemma infers the existence of a maximal element from some simple inclusion conditions. Again, although it seems materially different to Choice, it is equivalent to it. The second most contentious assumption in the history of set theory is the Continuum Hypothesis.

---

[1]This is the principle favoured by Bourbaki [28].

[2]Teichmüller, O., 'Braucht der Algebraiker das Auswahlaxiom?', *Deutsche Math.*, **4**, 567–77, (1939).

[3]Tukey, J. W., 'Convergence and Uniformity in Topology', *Annals of Mathematics Studies*, **2**, Princeton University Press, (1940).

9.3   The Continuum Hypothesis

Every transfinite consistent multiplicity, that is, every transfinite set,
must have a definite aleph as its cardinal number.

– GEORG CANTOR[1]

9.3.1   Lost infinities?

Cantor's Theorem tells us that, for any set a, the cardinality of $P(a)$ is strictly greater than the cardinality of a. Are we missing anything when we leap from a set to its power set? Are there cardinalities strictly between that of a and that of $P(a)$? Put another way, does there exist a set b such that a is equipollent to a subset of b, b is equipollent to a subset of $P(a)$, yet b is neither equipollent to a nor to $P(a)$? In the case where a is finite the answer is affirmative provided a has at least two elements. But what if a is infinite?

The assertion that for every infinite set a there are no intermediate cardinalities between that of a and that of $P(a)$ is the Generalized Continuum Hypothesis, the Continuum Hypothesis (first conjectured by Cantor in 1878)[2] being the special case where a is countable. In stating either hypothesis there is no loss of generality in assuming that b is a subset of $P(a)$. That is, if we say b is equipollent to a subset of $P(a)$ we may as well replace b with that subset, since the specific nature of the elements of a set is irrelevant to all questions pertaining to its cardinality.

As we saw in the introduction, it is easily established that there is a bijective correspondence between $\mathbb{R}$ and $P(\omega)$. Thus the Continuum Hypothesis is the conjecture that the cardinality $2^{\aleph_0}$ of $\mathbb{R}$ is $\aleph_1$.

In his famous address at the International Congress of Mathematicians in Paris, 6–10 August 1900, David Hilbert took '*a look to the future, at the probable directions of mathematics of the new century*'. Ten problems were discussed in the lecture and a further thirteen were included in the published proceedings of the congress. Taking pride of place at the top of the list was the question of the resolution of the Continuum Hypothesis.[3] The attempt to prove or disprove this assertion became one of the driving forces in the early development of axiomatic set theory and mathematical logic. Inevitably several failed attempts at a solution followed (although these investigations weren't entirely unfruitful, see Section 9.3.3). For example, Hilbert himself discussed in 1925 an 'outline of a proof', but the 'proof' was sparse on details, and according to Levy,[4] somewhat amusingly, Zermelo claimed that 'no one understood what he meant'. When Gödel proved that the Continuum Hypothesis was consistent with set theory there was perhaps some hope that it was provable. Cohen put a stop to this

---

[1]Letter to Richard Dedekind (1899).

[2]Cantor, G., 'Ein Beitrag zur Mannigfaltigkeitslehre', *J. Reine Angew. Math.*, **84**, 242–58, (1878).

[3]We met Hilbert's tenth problem earlier in the remarks to Subsection 2.3.5. An account of Hilbert's problems is given in Yandell [**232**].

[4]Levy, P., 'Remarques sur un théorème de Paul Cohen', *Rev. Métaphys. Morale*, **69**, 88–94, (1964).

when he proved that the *negation* of the Continuum Hypothesis is also consistent with ZF. The question is truly undecidable, at least within the context of ZF.

After Cohen, researchers have either sought plausible new axioms of set theory which decide the Continuum Hypothesis or instead have adopted an attractive multiverse view of set theory, recognizing that for any given undecidable statement one can produce a model in which that statement is true and another model in which that statement is false and denying that there should be any Platonic preference for one model over the other. The former investigations tend to lean towards a belief that the Continuum Hypothesis is *false*. For example, Gödel, later in life, submitted a paper to Tarski claiming a proof of $2^{\aleph_0} = \aleph_2$ based on new axioms, intended for publication in the *Proceedings of the National Academy of Sciences*, but withdrew after an error was discovered by Martin and Solovay. There are also modern set theoretic principles which imply $2^{\aleph_0} = \aleph_2$.[1] Cohen was also of the opinion that the Continuum Hypothesis is false, justifying his bold claim in the conclusion to his *Set Theory and the Continuum Hypothesis* by contrasting the construction of $\aleph_1$, generated simply as the set of countable ordinals, with the construction of the continuum, which is generated by the more powerful Axiom of Powers, the heights of which can surely not be reached by tamer principles of construction.

The Generalized Continuum Hypothesis is clearly significantly stronger than the Continuum Hypothesis. Indeed, in 1926 Adolf Lindenbaum and Alfred Tarski[2] conjectured that it is strong enough to imply the Axiom of Choice. There is a sound intuition behind this claim; the Generalized Continuum Hypothesis entails an abundance of bijective correspondences between sets – potentially enough to well-order every set. This conjecture was finally proved by Sierpinski in 1947.[3]

REMARK

Sierpinski's proof that the Generalized Continuum Hypothesis implies the Axiom of Choice makes use of his upper bound on the Hartogs number of a set. Recall that Hartogs' Theorem tells us that if x is an infinite set then there exists a cardinal number α such that no injection exists from α to x. The smallest such number, we shall label it β, is the Hartogs number of x. Sierpinski proved that β is equal to or less that the cardinal number of $P(P(P(x)))$.

---

[1] See for example Foreman, M., Magidor, M. and Shelah, S., 'Martin's maximum, saturated ideals and non-regular ultrafilters I', *Ann. Math.*, **127**, 1–47, (1988) and Woodin, W. H., 'The continuum hypothesis II', *Not. Amer. Math. Soc.*, **48**, 681–90, (2001).

[2] Lindenbaum, A. and Tarski, A., 'Communication sur les reserches de la théorie des ensembles', *Comptes Rendus des Séances de la Societé des Sciences et des Lettres de Varsovie, Classe III, Sciences Mathématiques et Physiques*, **19**, 299–330, (1926).

[3] Sierpinski, W., 'L'hypothèse généralisée du continu et l'axiome du choix', *Fundamenta Mathematicae*, **34**, 1–5, (1947).

Cantor's Theorem tells us that the cardinality of $P(\mathbb{N})$ is strictly larger than the cardinality of $\mathbb{N}$, but it does not tell us anything about possible intermediate cardinalities. The Continuum Hypothesis states that there are none. More generally, the Generalized Continuum Hypothesis states that there are no intermediate cardinalities between a and $P(a)$ for arbitrary infinite sets a. Like the Axiom of Choice, the Continuum Hypothesis and its generalization are independent of the axioms of ZF, however, the Generalized Continuum Hypothesis is strong enough to imply the Axiom of Choice. Some generous weakenings of the Generalized Continuum Hypothesis are also consistent with ZF.

9.3.2 Weakening the Generalized Continuum Hypothesis

It is natural to investigate the extent to which the Generalized Continuum Hypothesis can be weakened to allow intermediate cardinalities. We ask for which $F : \mathcal{O}n \to \mathcal{O}n$ the statement $(\forall \alpha)[\mathrm{Card}(2^{\aleph_\alpha}) = \aleph_{F(\alpha)}]$ is consistent with ZFC. Since $2^{\aleph_\alpha} \leq 2^{\aleph_\beta}$ whenever $\alpha \leq \beta$, and since the character of cofinality of $2^{\aleph_\alpha}$ exceeds $\aleph_\alpha$ for all α, the function F must, at the very least, satisfy the following:

(i) $(\forall \alpha)(\forall \beta)[\alpha \leq \beta \to F(\alpha) \leq F(\beta)]$;

(ii) $(\forall \alpha)[\mathrm{cf}(\aleph_{F(\alpha)}) > \aleph_\alpha]$.

Robert Solovay conjectured that no further properties of F are needed,[1] a conjecture which was reinforced when William Easton proved in 1964 that if F satisfies both (i) and (ii) above then there is a model of ZFC in which $2^{\aleph_\alpha} = \aleph_{F(\alpha)}$ holds for all *regular* cardinals $\aleph_\alpha$.[2] Examination of the singular case spawned Shelah's PCF theory.[3]

There is a feeling among many mathematicians that the Generalized Continuum Hypothesis is a drastic oversimplification that is far too convenient to be plausible. Intuitive arguments against it begin with the fact that $\aleph_{\alpha+1}$ is the cardinality of the set of all (order isomorphism types of) well-orderings of $\aleph_\alpha$. It is easily shown that $2^{\aleph_\alpha}$ is equipollent to the set of (order isomorphism types of) *all* orderings of $\aleph_\alpha$. Yet well-ordering is an extremely severe constraint, so much so that one would expect the two classes, all well-orderings versus all orderings, to be separated by a cardinal difference rather than one of equipollent proper inclusion. In other words one would expect $2^{\aleph_\alpha}$ to be strictly greater than $\aleph_{\alpha+1}$.

---

[1] See R. Solovay, '$2^{\aleph_0}$ can be anything it ought to be', *The Theory of Models*, Proceedings of the 1963 International Symposium at Berkeley, North-Holland Amsterdam (Addison, Henkin and Tarski, eds.), Addison (1965).

[2] The result appears in Easton's PhD thesis, later published as 'Powers of regular cardinals', *Annals of Mathematical Logic*, **1**, 139–178, (1970).

[3] See Shelah [**193**].

Other arguments point with disbelief, even in the case of the Continuum Hypothesis, to geometric interpretations. For example the Continuum Hypothesis is equivalent to the following startling statement due to Sierpinski:[1]

> $\mathbb{R}^2$ is the union of two sets of which one is at most countable along any line parallel to the x-axis and the other is at most countable along any line parallel to the y-axis.

Paul Erdős proved the following generalization:[2]

> Decompose the set of all lines in $\mathbb{R}^2$ into two arbitrary disjoint sets L_1 and L_2. Then, assuming the Continuum Hypothesis, there exists a decomposition of the plane into two sets S_1 and S_2 such that each line of L_i intersects S_i $(i = 1, 2)$ in at most a countable set.

The Platonist will insist that the Continuum Hypothesis is either true or false and is forced to interpret its undecidability as proof that ZF (and related systems) provides an inadequate description of their Platonic 'reality'; there ought to exist additional or alternative axioms of set theory which decide the Continuum Hypothesis in an intuitively persuasive fashion. The axioms of ZF have little influence on the possible location of $2^{\aleph_0}$ in the cardinal hierarchy, the only restriction (as revealed above) being that $2^{\aleph_0}$ is not an aleph of countable cofinality. In particular we have to exclude the attractive value of $\aleph_\omega$ for $2^{\aleph_0}$. Any value for $2^{\aleph_0}$ (of course exceeding $\aleph_0$) which is not of countable cofinality is consistent with ZFC, but for the Platonist, arbitrary choices such as $2^{\aleph_0} = \aleph_{12\,369}$ are abhorrent.

Although Gödel rejected the comparison as superficial, there is an undeniable similarity between the story of the Continuum Hypothesis and that of Euclid's parallel postulate. Of course important differences may be exhibited between the underlying theories, perhaps the most dramatic being that the first-order theory of Euclidean geometry, as realized by Tarski, is complete, whereas Gödel's Incompleteness Theorems apply to ZF and its augmentation by the Continuum Hypothesis (or its negation).

Gödel's main point about the differences between complementary extensions of set theory and Euclidean geometry versus non-Euclidean geometry, besides arguments motivated by his strong Platonism, was that there is a kind of strength asymmetry in some complementary augmentations of set theory. For example, let ϕ be the statement that there exists an inaccessible cardinal. Then the augmented theory ZF+ϕ can prove that ZF is consistent (i.e. that ZF has a model) but one cannot prove in ZF+$\neg\phi$ that ZF is consistent, so the addition of ϕ seems to be a much stronger assumption than its negation.

Unlike the large cardinal axioms, the Continuum Hypothesis does not give rise to any new number theoretic theorems. Gödel believed that alternative assumptions concerning the cardinality of $\mathbb{R}$ might entail such new results about

---

[1] W. Sierpinski, 'Sur un théorème équivalent a l'hypothèse du continu $(2^{\aleph_0} = \aleph_1)$', *Bull. Int. Acad. Polon. Sci. Lett. Cl. Sci. Math. Nat. Ser. A. Sci. Math.*, 1–3, (1919).

[2] Erdős, P., 'Some Remarks on Set Theory IV', *Michigan Math. J.*, **2**, 169–173, (1954).

ℕ and, disturbed by the consequences of the Continuum Hypothesis, he encouraged an experimental approach to its resolution, treating properties implied by it or its negation as 'evidence' for or against it. Alas there is no unanimity about what is and what is not plausible and we are once again led to the unavoidable problem of basing our choice of axioms on the unfortunate prejudices of intuition.

REMARKS

1. In a mathematical line of thought that can be traced back to Sierpinski's example, C. Freiling[1] introduced a set theoretic axiom (the *Axiom of Symmetry*) which turns out to be equivalent to the negation of the Continuum Hypothesis. Let $\mathcal{C}$ be the set of countable subsets of the interval $[0, 1]$. The axiom states that for every function $f : [0, 1] \to \mathcal{C}$ there exist an x and y in $[0, 1]$ such that $x \notin f(y)$ and $y \notin f(x)$.

2. The statement that ZF is consistent can be encoded as a number theoretic sentence via a Gödel numbering (see later). By Gödel's Second Incompleteness Theorem this sentence is unprovable in ZF, but it is provable if we assume the existence of inaccessible cardinals. It is in this sense that the existence of large cardinals proves number theoretic statements that are otherwise unprovable.

The Generalized Continuum Hypothesis is surprisingly delicate; one can modify it to assign a value to each $2^{\aleph_\alpha}$ in a barely restricted fashion and still not conflict with the axioms of ZF or with Choice. Some Platonists are unsettled by this, and would prefer to seek an intuitively plausible axiom which forces a more rigid cardinal structure. Many early attempts to prove the Continuum Hypothesis, although doomed to failure, led to some interesting spin-off results in descriptive set theory.

9.3.3 Some properties of the continuum

The efforts to prove the Continuum Hypothesis by means describable in ZF, which we now know to be futile, nevertheless still provide valuable inspiration for those seeking alternative set theoretic axioms. Informative results come from descriptive set theory – the study of the point sets of analysis in terms of the complexity of their defining constructions.

A *perfect* subset P of $\mathbb{R}$ is one which is equal to its derived set $P^{(1)}$ (see Subsection 1.11.7). Every perfect set has cardinality $2^{\aleph_0}$, so if it could be

---

[1] Freiling, C., 'Axioms of symmetry: throwing darts at the real number line', *The Journal of Symbolic Logic*, **51** (1), 190–200, (1986).

proved that every uncountable set of real numbers has a perfect subset, then every such set must have cardinality $2^{\aleph_0}$ and there can be no set of real numbers with cardinality between $\aleph_0$ and $2^{\aleph_0}$, resolving in the affirmative the Continuum Hypothesis. Cantor had hoped that this approach would be successful, however the dream of proving the Continuum Hypothesis by such methods was crushed long before Gödel and Cohen's results appeared; Felix Bernstein showed in 1908,[1] using the Axiom of Choice, that there is a decomposition of $\mathbb{R}$ into two disjoint uncountable sets neither of which contains a perfect subset. The target result can be partially rescued by imposing further conditions on the set in question. The Cantor–Bendixson Theorem[2] tells us that every uncountable closed subset of $\mathbb{R}$ is the disjoint union of a perfect set and an at most countable set. One can say much more. The class of *Borel sets* in $\mathbb{R}$ is the smallest σ-algebra containing all open subsets of $\mathbb{R}$. In other words, it is the smallest superset of the class of open subsets of $\mathbb{R}$ which contains all countable unions of its members and all complements of its members. It can be proved that every uncountable Borel set has a perfect subset and so has cardinality $2^{\aleph_0}$.

The existence of non-Borel sets (implicit in the Bernstein decomposition) is a simple consequence of the fact that the family of all Borel subsets of $\mathbb{R}$ has the same cardinality as $\mathbb{R}$. More explicit examples of non-Borel sets may be described (again using the Axiom of Choice). One of the oldest constructions is the following: consider the uncountable partition $\{X_i\}$ of $\mathbb{R}$ given by the set of all disjoint translations of $\mathbb{Q}$, so that each X_i is of the form $\{x + q : q \in \mathbb{Q}\}$ for some $x \in \mathbb{R}$. The Axiom of Choice asserts the existence of a set V which has exactly one element in common with each X_i. Any such V is called a *Vitali set* (after Giuseppe Vitali)[3] and is easily seen to be non-Borel.

There are more exotic perfect sets, one of the most well-known examples being Cantor's ternary set (see Figure 9.1). To construct it we start with the interval $I_0 = [0,1]$, remove the open middle third, leaving $I_1 = [0,\frac{1}{3}] \cup [\frac{2}{3},1]$, then remove the open middle third of each of these two intervals, leaving $I_2 = [0,\frac{1}{9}] \cup [\frac{2}{9},\frac{1}{3}] \cup [\frac{2}{3},\frac{7}{9}] \cup [\frac{8}{9},1]$, then remove the open middle third of each of these four intervals, leaving I_3, and so on. Cantor's ternary set is the set of points common to all I_is, i.e. $\cap_{n=1}^{\infty} I_n$, or put another way it is the set of all real numbers in $[0,1]$ which can be written in base 3 without using the ternary digit 1. It is easy to see from the latter representation that, despite its initially sparse appearance, the Cantor set is uncountable: each number in the interval $(0,1]$ has a unique non-terminating binary representation $0.b_1 b_2 b_3 \ldots$.[4] Let $t_i = 2b_i$, then the ternary representation $0.t_1 t_2 t_3 \ldots$ defines a unique element of the Cantor set. This mapping, extended to map 0 to 0, is an injection of $[0,1]$ into the Cantor set.

---

[1]Bernstein, F., 'Zur Theorie der trigonometrischen Reihen', *Berichte Vernhandl. Königl. Sächs. Gesell. Wiss. Leipzig, Math.-phys. Kl.*, **60**, 325–38, (1908).

[2]Bendixson, I., 'Quelques théorèmes de la théorie des ensembles de points', *Acta Math.*, **2**, 415–29, (1883). Bendixson's paper is based on a letter sent to Cantor.

[3]Vitali, Giuseppe, 'Sul problema della misura dei gruppi di punti di una retta', *Bologna, Tip. Gamberini e Parmeggiani*, (1905).

[4]For example, the 1-free ternary representations of 1 and $\frac{1}{3}$ are $0.\dot{2}$ and $0.0\dot{2}$, respectively.

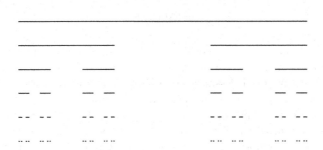

Figure 9.1 The first few stages of the construction of the Cantor set. Although the fine dust that results from the limit of this construction might seem to be a sparse set, it is equipollent to the original line. The Cantor set is a nowhere-dense, uncountable perfect set.

One of the topological properties of the real line is that every pairwise disjoint set of open intervals is at most countable. Mikhail Souslin had hoped in 1920 that this might be used as part of an alternative characterization of the continuum.[1] We briefly describe the details.

A *line* is an ordered set $(L, \prec)$ without a greatest or least element with the property that if $x \prec z$ in L then there exists a $y \in L$ with $x \prec y \prec z$. $(L, \prec)$ is *complete* if every non-empty subset of L with an upper bound has a least upper bound in L. (Equivalently every non-empty subset of L with a lower bound has a greatest lower bound in L.) A subset D of L is *dense* in L if for all $x \prec y$ in L there exists a $d \in D$ with $x \prec d \prec y$. A set of the form $\{x : a \prec x \prec b\}$ for $a, b \in L$ is an *open interval*.

A *Souslin line* is a complete line which possesses no countable dense subset and in which every pairwise disjoint set of open intervals is at most countable. The Souslin Hypothesis states that there are *no* Souslin lines. Assuming the Souslin Hypothesis, if every pairwise disjoint set of open intervals of a complete line is at most countable then that line is order isomorphic to $\mathbb{R}$. This is a strengthening of Cantor's result that every dense complete ordered set with a countable dense subset and without endpoints is order isomorphic to $\mathbb{R}$. It turns out, however, that this hypothesis is independent of ZFC.[2]

---

[1] Souslin, M., 'Problème 3', *Fundamenta Mathematicae*, **1**, 223, (1920). 'Souslin' is also rendered from the Cyrillic as 'Suslin'.

[2] The relevant papers are: Jech, T., 'Non-provability of Souslin's hypothesis', *Comment. Math. Univ. Carolinae*, **8**, 291-305, (1967); Jensen, R., 'Souslin's Hypothesis is incompatible with $V = L$', *Not. Am. Math. Soc.*, **15**, 935, (1968); and Solovay, R. M. and Tennenbaum, S., 'Iterated Cohen extensions and Souslin's problem', *Ann. of Math.* (2), **94**, 204-245, (1971).

1. Clearly each set I_n in the construction of the Cantor set is uncountable,
   being a union of intervals (and one can see from the figure that each I_n is
   equipollent to its predecessor without any sophisticated prior results, by a
   simple geometric projection of each component interval of I_n onto half of the
   segment above it). What might be uncertain is if this constancy of cardinality
   continues to hold at the limit, after all it is easy to construct a sequence of
   uncountable sets whose intersection is countable, finite, or empty.

2. The Cantor set $\mathcal{C}$ is a universal object for compact metric spaces in the
   following sense. *Every* compact metric space is a continuous image of $\mathcal{C}$ (the
   continuous map is not uniquely defined). For example, the map $f : \mathcal{C} \to [0,1]$
   sending $\Sigma a_n/3^n$ to $\Sigma a_n/2^{n+1}$ is a continuous surjection (here, $a_n \in \{0,2\}$
   for all n). $\mathcal{C}$ is homeomorphic to the countable product of Cantor sets $\mathcal{C}^\omega$,
   so by mapping $(x_n) \in \mathcal{C}^\omega$ to $(f(x_n))$ we see that the Hilbert cube $[0,1]^\omega$
   is the continuous image of the Cantor set. Every compact metric space is
   homeomorphic to a closed subset of the Hilbert cube so the result follows from
   the fact that every closed subset of $\mathcal{C}$ is a continuous image of $\mathcal{C}$. One can
   prove some interesting analytical theorems using this representation theorem
   alone.

3. The Cantor set is totally disconnected, i.e. its connected components are
   one-point sets; in fact, it is the unique totally disconnected compact metric
   space in the sense that any such space is homeomorphic to the Cantor set.

---

*The early set theorists sought topological or measure theoretic arguments for the
Continuum Hypothesis. Although their efforts failed, the mathematics generated
was of independent interest. Another novel way of looking at complex sets is in
terms of game strategies.*

9.3.4   Determinacy

Investigation of the properties of subsets of $\mathbb{R}$ inspired a useful and natural
approach to the theory of games between two players and this in turn has led
to interesting alternative axioms of set theory. Let us consider the following
example of a theoretical infinite game. Fix in advance of the game a subset X
of $[0,1]$ and reveal it to both participants. Two players (a pair of 'supertaskers',
each possessing a supernatural speed of thought allowing them to make an
infinite number of moves in a finite amount of time) take turns in selecting 0
or 1, player 1 choosing s_1, player 2 choosing s_2, player 1 choosing s_3, and so
on until together they have generated an infinite sequence (s_n) which is to be

interpreted as the binary expansion of a real number $0.s_1 s_2 s_3 \ldots$ in $[0, 1]$. If $0.s_1 s_2 s_3 \ldots \in X$ player 1 wins, otherwise player 2 wins.

A *strategy* for player 1 is a function s which for any string of $2n$ binary digits as input generates a single binary digit as output. This tells player 1 which move to make at each stage of the play on the basis of all of the previous moves. We say that s is a *winning strategy* if player 1 wins no matter which moves player 2 makes. A winning strategy for player 2 is similarly defined (on odd strings of binary digits).

The set X is said to be *determined* if there exists a winning strategy for one of the players. A question that naturally presents itself is to describe those sets which are determined. For some simple sets it is quite easy to prove determinacy, for example, every countable subset X of $[0, 1]$ is determined because player 2 is guaranteed to win if he adopts the following diagonalizing strategy: first he enumerates X as a sequence (r_n) of binary numbers:

$$r_1 = 0.r_{11} r_{12} r_{13} r_{14} \ldots$$
$$r_2 = 0.r_{21} r_{22} r_{23} r_{24} \ldots$$
$$r_3 = 0.r_{31} r_{32} r_{33} r_{34} \ldots$$

$\ldots$

If player 1 chooses $x_1, x_3, x_5, x_7, \ldots$ on his turns, player 2 (who need not pay any attention to the moves of his opponent) then chooses $\bar{r}_{12}, \bar{r}_{24}, \bar{r}_{36}, \bar{r}_{48}, \ldots$ where $\bar{r}_i = 0$ if $r_i = 1$ and $\bar{r}_i = 1$ if $r_i = 0$. At the end of the game we see that the number formed by the players' choices of binary digits

$$0.x_1 \bar{r}_{12} x_3 \bar{r}_{24} x_5 \bar{r}_{36} \ldots$$

differs from all r_i hence is not in X, and player 2 wins.

By assuming progressively more powerful large cardinal axioms one can prove the determinacy of increasingly complex subsets of $[0, 1]$. Of course the idea of this being a game is simply a useful way of thinking about the axiom. The existence of a winning strategy for one of the players, when unravelled, is a cold statement concerning the existence of a certain function having a certain property just as, say, the Axiom of Choice is a statement concerning the existence of a choice function. Do not let the seductive game analogy distract from the highly abstract nature of the axiom.

The Axiom of Determinacy says that *every* subset of $[0, 1]$ is determined. M. Davis proved in 1964 that this axiom implies that every set of real numbers is either countable or has a perfect subset and so it is inconsistent with the Axiom of Choice.[1]

As Determinacy contradicts the Axiom of Choice, it is not an instantly attractive axiom for mathematicians. However, putting this aside, it does impose some pleasingly sober properties on the real numbers (such as the aforementioned perfect set result and the fact that every subset of $\mathbb{R}$ becomes measurable, rendering impossible the Banach–Tarski decomposition). Modifications of

---

[1]Davis, M., 'Infinite games of perfect information', *Advances in Game Theory* (L. S. Shapley and A. W. Tucker, eds.), Princeton University Press, 85–101, (1964).

Determinacy which are consistent with Choice but which weaken the sobering effects just mentioned have been proposed in the form of the Axiom of Definable Determinacy which restricts attention to the class of all *constructible* real numbers (constructibility in general is defined later) and in the form of Projective Determinacy, restricting instead to projective sets. Such restricted Determinacy principles have received favourable opinion among set theorists. Advocates (Hugh Woodin, for example) stress that Projective Determinacy is attractive only in light of its consequences and of principles which imply it. Presented as a bare axiom it is difficult to make a clear assessment of its plausibility. One such piece of 'evidence' was provided by Donald Martin and John Steel[1] who proved that Projective Determinacy is a consequence of the existence of infinitely many Woodin cardinals.

REMARK

In the literature the Axiom of Determinacy is often stated instead in terms of a game where the players alternately choose integers rather than binary digits (or decimal digits). In this case each game is represented by an element of the *Baire line*,[2] the set of all sequences of natural numbers ordered by a certain natural partial order.[3]

The game theoretic Axiom of Determinacy acts as a replacement for the Axiom of Choice and blocks the selection of the type of complex sets that make the Banach–Tarski decomposition possible. Various modifications have been suggested and explored. Replacements for the Continuum Hypothesis have also been explored, one weaker alternative being Martin's Axiom.

9.3.5 Martin's Axiom

We close this chapter with a brief description of another alternative axiom of set theory, Martin's Axiom.[4] This new axiom was first formulated by Martin on inspection of Solovay and Tennenbaum's construction of a model in which Souslin's Hypothesis holds, greatly simplifying a technically difficult iteration of forcing (we briefly discuss a special case of the original method of forcing in Chapter 11); it is an example of a 'forcing axiom'. Martin's Axiom is consistent with ZFC and the negation of the Continuum Hypothesis and it is only in this setting that it is of any interest, as will be revealed below. Indeed, we can regard Martin's Axiom as a weakening of the Continuum Hypothesis.

---

[1] Martin, D. and Steel, J., 'Projective determinacy', *Proc. Natl. Acad. Sci. U.S.A.*, **85**(18), 6582–6586, (September 1988).

[2] After Baire, R., 'Sur la représentation des fonctions discontinues II', *Acta Math.*, **32**, 97–176, (1909).

[3] See Potter [**170**], Chapter 7, for an exposition.

[4] Named after Donald Martin. See Martin, D. and Solovay, R., 'Internal Cohen extensions', *Ann. Math. Logic*, **2**, 143–178, (1970).

The statement of Martin's Axiom requires a few preliminary definitions. As usual, by a *partial order* on a set A we mean a reflexive, transitive and antisymmetric relation $\leq$. A *filter* in $(A, \leq)$ is a non-empty subset F of A with the property that for every $a, b \in F$ there is a $c \in F$ with $c \leq a$ and $c \leq b$ and such that for every $a \in F$ and $b \in A$, if $a \leq b$ then $b \in F$. A subset D of $(A, \leq)$ is *dense* in A if and only if for every pair $a \leq b$ in A there is some $x \in D$ with $a \leq x \leq b$.

Elements a and b of $(A, \leq)$ are *compatible* if and only if there is some $c \in A$ with $c \leq a$ and $c \leq b$. A subset K of $(A, \leq)$ is an *antichain* if no two distinct elements of K are compatible. $(A, \leq)$ is a *ccc partially ordered set* if and only if every antichain in $(A, \leq)$ is countable.[1] The connection with Souslin's Hypothesis is this: let A be the set of open intervals in a complete line L partially ordered by inclusion. An antichain in A is then simply a disjoint collection of open subsets of L. L is a Souslin line if it has no countable dense subset and $(A, \subseteq)$ is a ccc partially ordered set.

Now for the statement of Martin's Axiom. Let κ be an infinite cardinal. $\mathrm{MA}(\kappa)$ is the assertion that if $(A, \leq)$ is a ccc partially ordered set and D is a collection of at most κ dense subsets of A, then there is a filter F in A such that for all $d \in D$, $F \cap d \neq \emptyset$. Martin's Axiom states that $\mathrm{MA}(\kappa)$ is true for all infinite cardinals κ with κ less than c (the cardinality of the continuum). This axiom has a strong topological flavour, it generalizes the Baire Category Theorem: the intersection of a countable collection of dense open sets of a compact Hausdorff space is non-empty (see the remarks at the end of this subsection for definitions).

The special case $\mathrm{MA}(\aleph_0)$ (the Rasiowa–Sikorski Lemma) is a theorem of ZFC. It is for this reason that Martin's Axiom is of no interest if we assume the Continuum Hypothesis, for in that case there is only one admissible κ, namely $\aleph_0$. At the other extreme, $\mathrm{MA}(c)$ is false in ZFC. There are, just as for the Axiom of Choice, although perhaps not recorded so widely, many different equivalent formulations of Martin's Axiom.

By design, Martin's Axiom implies the Souslin Hypothesis, indeed, continuing to reject the Continuum Hypothesis, $\mathrm{MA}(\aleph_1)$ alone implies the Souslin Hypothesis.[2] Martin's Axiom has many other interesting consequences, for example it implies that c is a regular cardinal and that $2^\kappa = c$ for all infinite cardinals $\kappa < c$; so informally speaking, it says that *all infinite cardinals smaller than c behave like $\aleph_0$*. It is in this sense that Martin's Axiom is a 'weak' Continuum Hypothesis.

REMARKS

1. A topological space is *Hausdorff* if for every distinct pair of points x and y in the space one can find disjoint open sets U and V with $x \in U$ and $y \in V$. All metric spaces are Hausdorff. Non-Hausdorff spaces, being non-metrizable,

---

[1] ccc stands for 'countable chain condition' a misnomer which has become standard, 'countable antichain condition' would have been a sensible alternative.

[2] For a proof see, for example, Just and Weese [**111**] (Chapter 19).

are somewhat counterintuitive, but they do naturally occur in a number of mathematical settings. Recall a topological space X is compact if every open cover of the space has a finite subcover, that is, if $X \subset \cup\{C_\lambda : \lambda \in I\}$, where each C_λ is open, then there is a finite subset F of the index set I such that $X \subset \cup\{C_\lambda : \lambda \in F\}$.

$\mathrm{MA}(\kappa)$ is equivalent to the following: if X is a compact Hausdorff space which satisfies the countable chain condition (i.e. every pairwise disjoint collection of non-empty open subsets of X is countable) then X is not a union of κ or fewer nowhere dense subsets of X.

2. Martin's Axiom is independent of ZFC.

Martin's Axiom is a weakening of the Continuum Hypothesis which implies that the power set of any infinite set of cardinality less than the continuum has cardinality equal to that of the continuum. This is one of many alternative set theoretic axioms that have emerged through a deep study of existing models and techniques. We have often informally mentioned the notion of a model. Before we can properly discuss the phenomenon of independence we need to give a more precise definition.

10

Models

10.1 Satisfaction and restriction

...the grand aim of all science, which is to cover the greatest possible number of empirical facts by logical deduction from the smallest possible number of hypotheses or axioms.

– ALBERT EINSTEIN[1]

In anticipation of the final chapter, and clarifying the earlier casual use of the idea, we make precise the notion of a model of an axiomatic system, restricting our discussion, for the most part, to models of ZF.

10.1.1 Satisfaction

If R is a relation on a class A and ϕ is a well-formed formula of ZF then the sentence '$[A, R]$ satisfies ϕ' is defined recursively as follows:

ϕ	'$[A, R]$ satisfies ϕ' means
$a \in b$	$a, b \in A$ and aRb
$\neg\psi$	$[A, R]$ does not satisfy ψ
$\psi \wedge \eta$	$[A, R]$ satisfies ψ and $[A, R]$ satisfies η
$(\forall x)\psi$	for all $x \in A$, $[A, R]$ satisfies ψ

The notion of satisfaction is clearly eliminable in the sense that '$[A, R]$ satisfies ϕ' is an abbreviation for a well-formed formula. From the equivalence $(\exists x)\phi \leftrightarrow \neg(\forall x)\neg\phi$, the De Morgan Law $\phi \vee \psi \leftrightarrow \neg(\neg\phi \wedge \neg\psi)$ and the equivalence $[\phi \rightarrow \psi] \leftrightarrow [\neg\phi \vee \psi]$ we deduce that:

---

[1] *Relativity and the Problem of Space* (1952) in Einstein [**63**].

ϕ	'$[A, R]$ satisfies ϕ' means
$(\exists x)\psi$	there exists an $x \in A$ such that $[A, R]$ satisfies ψ
$\psi \vee \eta$	$[A, R]$ satisfies ψ or $[A, R]$ satisfies η
$\psi \rightarrow \eta$	if $[A, R]$ satisfies ψ then $[A, R]$ satisfies η
$\psi \leftrightarrow \eta$	$[A, R]$ satisfies ψ if and only if $[A, R]$ satisfies η

If A is non-empty and $[A, R]$ satisfies some sentence ϕ we say that $[A, R]$ is a *model* of ϕ. $[A, R]$ is a model of a collection $\mathcal{C}$ of sentences if $[A, R]$ is a model of ϕ for each ϕ in $\mathcal{C}$. $[A, R]$ is a *standard* structure if R is the usual membership relation $\in$ and the phrase 'A satisfies ϕ', in the absence of any reference to a relation, is understood to mean that $[A, \in]$ satisfies ϕ (where $\in$ is membership restricted to A).

REMARK

Let us take a small example, the finite set

$$A = \{0, 1, 2\} = \{\varnothing, \{\varnothing\}, \{\varnothing, \{\varnothing\}\}\},$$

and ask which of the axioms of ZF are satisfied in this set. Without setting out the fine details (the reader may wish to fill these in and consider what modifications would be required in order to force the set to be a model of a given axiom) we see that Extensionality is satisfied, Pairing is not satisfied (since, for example, $\{1, 2\} \notin A$), Unions is satisfied, Powers is not satisfied (since $P(2) \notin A$), Replacement is not satisfied (see Subsection 4.4.3), Regularity is satisfied and Infinity is not satisfied (since $\omega \notin A$).

To say that a class satisfies some sentence simply means that the sentence is true if the variables are understood to range over elements in the class. We can set this up as a simple recursion on well-formed formulas. For example the sentence $(\forall x)(\forall y)[x \in y \vee y \in x \vee x = y]$ is satisfied in the class of ordinals but is not satisfied in, say, the class of all finite sets. The restriction of a general well-formed formula to an arbitrary class is easily described.

10.1.2   Restriction

The restriction ϕ^A of a well-formed formula ϕ to a class A is defined recursively as follows (here the formula $(\forall x \in A)\phi$ is an abbreviation for $(\forall x)[x \in A \rightarrow \phi]$ and $(\exists x \in A)\phi$ is an abbreviation for $(\exists x)[x \in A \wedge \phi]$):

ϕ	ϕ^A
$a \in b$	$a \in b$
$\neg\psi$	$\neg\psi^A$
$\psi \wedge \eta$	$\psi^A \wedge \eta^A$
$(\forall x)\psi$	$(\forall x \in A)\psi^A$

and by the usual mechanical procedure, we have:

ϕ	ϕ^A
$(\exists x)\psi$	$(\exists x \in A)\psi^A$
$\psi \vee \eta$	$\psi^A \vee \eta^A$
$\psi \rightarrow \eta$	$\psi^A \rightarrow \eta^A$
$\psi \leftrightarrow \eta$	$\psi^A \leftrightarrow \eta^A$

In other words ϕ^A is simply the well-formed formula obtained from ϕ by replacing, for each quantified variable x in ϕ, each occurrence of $\forall x$ by $\forall x \in A$ and each occurrence of $\exists x$ by $\exists x \in A$.

The relativization of a class $\{x : \phi\}$ to a class K, denoted $\{x : \phi\}^K$, is the class $\{x \in K : \phi^K\}$. In particular if a is a set then $a^K = \{x : x \in a\}^K = \{x \in K : [x \in a]^K\} = \{x \in K : x \in a\} = K \cap a$.

REMARK

If ϕ is a sentence, then A satisfies ϕ if and only if ϕ^A holds, so if ϕ is an n-predicate and $a_1, \ldots, a_n \in A$ then A satisfies $\phi(a_1, \ldots, a_n)$ if and only if $\phi(a_1, \ldots, a_n)^A$. It is then a routine matter to prove that if ϕ is a theorem of ZF and if A is a non-empty class that satisfies each non-logical axiom used in a proof of ϕ then ϕ^A.

The assumption is that ϕ $(= \phi_n)$ is the end formula of a string of formulas $\phi_1, \ldots, \phi_n$ where each ϕ_i is either an axiom or can be inferred from earlier formulas in the string. The proof is by induction on n: assuming $\phi_1^A, \ldots, \phi_{k-1}^A$ one proves ϕ_k^A by looking at the three cases: (i) ϕ_k is an axiom; (ii) ϕ_k is inferred by modus ponens from ϕ_i and $\phi_i \rightarrow \phi_k$ for some $i < k$; and (iii) ϕ_k is inferred by generalization from ϕ_i for $i < k$.

It follows that if a proof of a sentence ϕ requires only the logical axioms then *every* non-empty class is a model of ϕ. In particular every non-empty class is a model of the logical axioms of ZF and if two well-formed formulas ϕ and ψ are logically equivalent then a non-empty class is a model of ϕ if and only if it is a model of ψ.

The restriction ϕ^A of a formula ϕ to a class A is the formula obtained by replacing each quantified variable $\forall x$ and $\exists x$ by $\forall x \in A$ and $\exists x \in A$, respectively. If ϕ is a sentence then A satisfies ϕ if and only if ϕ^A. The definition of a model of a general first-order theory is a simple extension of the special case described here.

10.2   General models

Out of nothing I have created a strange new universe.

– JÁNOS BOLYAI[1]

10.2.1   Introducing models

A model of ZF is a class together with a relation which satisfies all of the axioms of ZF in the sense described in Subsection 10.1.1 above. The class is interpreted to be the universe of all sets. The relation models the properties of the membership relation, however, it might not actually be the membership relation; if it *is* the membership relation on the class, then the model is said to be a *standard* model of ZF. Some texts assume at the outset that a model must be a set. We shall not make this assumption, instead using the term 'set model' when we need to emphasize that the model is a set rather than a general class.

Model theory, in combination with first-order logic, has guided set theory from its humble beginnings at the turn of the twentieth century to the mature theory we see today. The informal idea of a model of a set of axioms had already made its mark in the nineteenth century in the context of non-Euclidean geometry, as we saw in Subsection 2.4.2.

From a perspective markedly different from that of set theory, a model of, say, the group axioms *is* a group (a model of the field axioms *is* a field, a model of the axioms of a topological space *is* a topological space and so on); the contrast with set theory being that we know *a priori* that a rich variety of models of the group axioms can be found because these axioms were invented precisely to distill the common properties of a well-known large body of algebraic structures. Set theoretic axioms, on the other hand, are generally drawn from the pulp of intuition and from experience, not from some easily exhibited collection of 'examples of set universes'. In the early pre-model theoretic days, in the absence of any concrete examples, it was perhaps hoped that the axioms of set theory were strong enough to determine only a very limited variety of structures. That they do not is due primarily to the weakness of the underlying first-order logic.

It is possible for a model of ZF to possess a proper subclass, together with some relation, which is also a model of ZF. The theory of *inner models* (a model which contains all ordinals), developed using ZF itself, paves the way to independence results. Later (see Section 11.2), however, we shall see that a new idea is needed to create models which establish the independence of some of the more powerful set theoretic postulates such as the Continuum Hypothesis and the Axiom of Choice.

Of course ZF is just one example of a first-order theory. In the general case a model of a first-order theory T with primitive operators $\{O_i\}$ and relations $\{R_i\}$ is a class A, together with corresponding operators $\{\bar{O}_i\}$ and relations $\{\bar{R}_i\}$ on A, which satisfies all of the axioms of T in the following recursive sense.

---

[1]Exclaimed in a letter to his father in 1823 upon his discovery of non-Euclidean geometry.

(i) If x_i is a variable of T then $\bar{x}_i$ represents an arbitrary element of A, so quantification involving $\bar{x}_i$ is understood to be over the class A.

(ii) If $t = O_i(t_1, \ldots, t_n)$ is a term of T, let $\bar{t} = \bar{O}_i(\bar{t}_1, \ldots, \bar{t}_n)$. (In the case of nullary operators, i.e. constants c of T, this means $\bar{c}$ represents the corresponding element of A.)

(iii) A satisfies $R_i(t_1, \ldots, t_n)$ means $\bar{R}_i(\bar{t}_1, \ldots, \bar{t}_n)$.

(iv) If ψ and η are sentences of T, then:

   (1) A satisfies $\neg\psi$ means A does not satisfy ψ;

   (2) A satisfies $\psi \wedge \eta$ means A satisfies ψ and A satisfies η; and

   (3) A satisfies $(\forall x)\psi(x)$ means $(\forall \bar{x})\psi(\bar{x})$.

REMARK

The case of ZF is noticeably simpler than the general case owing to the absence of primitive operators. For the one primitive relation $\in$, the relation $\bar{\in}$ in a model of ZF need not correspond to set membership.

In the case of PA, unlike ZF, there are several primitive operators, and moreover there is a privileged standard model, namely ω with the von Neumann successor operator. In the above notation we have: $\bar{0} = \varnothing \in \omega$; the unary operator $\bar{s} : \omega \to \omega$ (the von Neumann successor) is defined by $\bar{s}n = n \cup \{n\}$; the binary operators $\bar{+}$ and $\bar{\cdot}$ on ω are defined recursively using $\bar{s}$ (as described by the relevant axioms of PA); and equality is given by standard set equality. The Zermelo model, which is isomorphic to the von Neumann model, is given by the smallest class of sets including $\bar{0} = \varnothing$ which is stable under the Zermelo successor operator $\bar{s}a = \{a\}$, in terms of which one defines the binary operators, and equality is again given by set equality.

A model of a theory is a class together with operations and relations which satisfy all of the axioms of the theory. Model theory makes the study of abstract theories concrete. Through the construction of models we can prove the independence of various set theoretic statements. In a theory we are only able to speak about the provability of a statement; models introduce a clear relative notion of truth via satisfaction. In a sound theory all provable sentences are true, i.e. satisfied in all models, but there may be sentences which are true in some models but not provable.

10.2.2 Completeness and compactness

The Completeness Theorem for first-order logic was proved by Gödel in 1929 and appears in his doctoral dissertation.[1] The terminology is, due to an overuse of the word 'complete' in modern mathematics, the cause of tremendous confusion. One form of the theorem is 'first-order logic is complete', but the word 'complete' is used here in a different sense from 'complete' as it is used in Gödel's First Incompleteness Theorem.

In the sense intended in the Incompleteness Theorem, a theory T is complete, or more precisely, *negation complete*, if for every sentence ϕ in the language of T either ϕ or $\neg\phi$ is provable in T (if both are provable then T is inconsistent).

Completeness of a theory T, as it is used in the Completeness Theorem (i.e. semantic completeness), means that for each sentence ϕ, if ϕ is satisfied in all models of T then ϕ is provable in T. The first-order theory of Peano Arithmetic, for example, is complete in this sense but is not negation complete, while Euclidean Geometry in, say, Tarski's axiomatization, is complete in both senses. The converse, that if ϕ is provable in T then it is satisfied in all models of T, is called the *Soundness Theorem*. One can prove soundness by an easy induction argument on the length of the proof of ϕ. Since one would always presume to be working in a sound theory, meaning that all the axioms are true in some preferred interpretation, it is quite common to use the term 'Completeness Theorem' to refer to the combination of completeness and soundness: a sentence ϕ in the language of T is provable if and only if it is true in all models of T.[2]

Together, completeness and soundness tell us that a first-order theory is consistent if and only if it has a model. If a first-order theory T is inconsistent and has a model M then there is a sentence ϕ such that $\phi \wedge \neg\phi$ is provable in T, but then by soundness, M satisfies $\phi \wedge \neg\phi$. However, 'M satisfies $\phi \wedge \neg\phi$' means 'M satisfies ϕ and M does not satisfy ϕ', a contradiction, which proves that if T has a model then it must be consistent. Conversely, if T has no model then the implication 'for all models M, M satisfies $\phi \wedge \neg\phi$' is vacuously satisfied, so by completeness $\phi \wedge \neg\phi$ is provable in T and hence T is inconsistent.

So, willing as we are to accept that ZF is consistent, we in turn assume the existence of a model of ZF.

The Compactness Theorem states that if T is a collection of sentences in some first-order theory, then T has a model if and only if every finite subset of T has a model. The name 'compactness' comes from an analogy with the

---

[1] Gödel, K., *Über die Vollständigkeit des Logikkalküls*, University of Vienna (1929), later published as 'Die Vollständigkeit der Axiome des logischen Funktionen-kalküls', *Monatshefte für Mathematik und Physik*, **37**, 349–60, (1930). Some of Gödel's results were also discovered earlier by Thoralf Skolem.

[2] In the logical and model theoretic literature the notation for 'A satisfies ϕ' is $A \models \phi$. If T is a theory and every model of T satisfies ϕ then we would write $T \models \phi$. The notation $T \vdash \phi$ means ϕ is provable in T. This useful abbreviation enables us to express the theorems of logic and model theory very efficiently (for instance the Completeness Theorem is often simply stated as: $T \models \phi \leftrightarrow T \vdash \phi$), but I elected to avoid it in this book, not wishing to burden the newcomer with even more potentially unfamiliar symbols.

topological notion.[1]

Compactness is an easy consequence of the equivalence of consistency and model existence. First, if a collection of sentences T has a model then that model is also a model of every finite subset of T. Secondly, to prove the converse, we prove its contrapositive ('if T has no model then T has a finite subset which has no model'): if T has no model then T is inconsistent, so we can derive a contradiction from the sentences in T. Any such proof of contradiction will make use of only a finite subset F of sentences in T. Since we can derive a contradiction from F, it cannot have a model.

REMARKS

1. The Completeness Theorem is used almost unconsciously by mathematicians on a daily basis. Whenever a proof in, say, group theory includes a statement such as 'Let G be a group...' then completeness is being invoked. One is demonstrating that the conjectured sentence is provable by proving that it is true in a generic model.

2. We can use the Compactness Theorem to prove the existence of a model of the theory of real numbers which contains infinitesimal elements. Suppose T is a first-order theory of real numbers. Create a new theory T_ι by adding to the language of T a constant ι and to the theory T the countable schema of axioms $\{\mathfrak{A}_n : n \text{ is a positive integer}\}$ plus the axiom $\iota > 0$, where $\mathfrak{A}_n$ is $\iota < \frac{1}{n}$. In order to show that T_ι has a model, by the Compactness Theorem, we simply need to show that every finite set F of axioms in T_ι has a model. Since F will contain only finitely many instances of the new axioms $\mathfrak{A}_n$, say, $\mathfrak{A}_{n_1}, \ldots, \mathfrak{A}_{n_k}$, by interpreting ι as the real number $\frac{1}{2}\min\{1, \frac{1}{n_1}, \ldots, \frac{1}{n_k}\}$ we see that $\mathbb{R}$ is a model of F (but it is clearly not a model of *all* of the new axioms). As F was an arbitrary finite set of axioms, it follows that T_ι has a model. A similar technique can be used to prove the existence of all sorts of exotic models.

3. Recall that if the set of axioms of a first-order theory is recursively enumerable then the theory is said to be *effective*. In an effective theory we can recursively generate all theorems. By semantic completeness, this means we can recursively generate all statements which are true in all models of the theory (even though the class of models is not recursively enumerable). Completeness fails in unrestricted second-order logic; that is, second-order logic is semantically incomplete. (One can introduce restrictions of second-order logic and retrieve completeness, but we shall not go into that here.)[2]

---

[1]One can interpret it as more than a mere analogy, but the details would take us too far afield here.

[2]There are some subtle issues regarding incompleteness and categoricity in connection with second-order Peano Arithmetic, but these are rather beyond the scope of this text. For a discussion see Shapiro [192] and Chapter 22 of Smith [196].

If a sentence is satisfied in all models of a first-order theory, then it is provable in that theory. This is the Completeness Theorem. The converse, that every provable sentence is satisfied in all models, is the Soundness Theorem. In combination this tells us that a first-order theory has a model if and only if it is consistent. Furthermore, an arbitrary collection of sentences in a first-order theory has a model if and only if every finite subset of that collection has a model. This is the Compactness Theorem. If a theory has a set model then it is natural to ask whether there are any constraints on the model's cardinality. An answer is provided by the Löwenheim–Skolem Theorem.

10.2.3   The Löwenheim–Skolem Theorem

The Löwenheim–Skolem Theorem was proved in full generality by A. I. Maltsev[1] in 1936 following developments by Löwenheim[2], Skolem[3] and Tarski. Maltsev also proved the general version of the Compactness Theorem, Gödel's original only applying to countable languages.

If M is an infinite set which is a model of a collection F of sentences in some first-order theory, then the Löwenheim–Skolem Theorem tells us that there exists a subset of M which is also a model of F (a *submodel*) with cardinality equal to that of F if F is infinite and countable if F is finite.

There are many different variations of the Löwenheim–Skolem Theorem. Sometimes it is named with an allusion to the cardinality of the model whose existence it asserts. The version just given is the Downward Löwenheim–Skolem Theorem. The Upward Löwenheim–Skolem Theorem says that if a theory with α many axioms has an infinite set model, then it has models of arbitrarily large cardinality equal or greater than α. Skolem, who had Finitist sympathies, appears to have been a little perturbed by the attachment of his name to this result.

For example, the Upward Löwenheim–Skolem Theorem tells us that there are uncountable models of Peano Arithmetic.

Remarks

1. We should note that there are also non-standard *countable* models of PA: any such model has order type $\omega + (^*\omega + \omega)\eta$, where η is the order type of $(\mathbb{Q}, <)$. This order type can be pictured as a copy of the natural numbers followed

---

[1] Maltsev, A. I., 'Untersuchungen aus dem Gebeite der mathematischen Logik', *Matematicheskii Sbornik*, **1**, 323–336, (1936).

[2] Löwenheim, L., 'Über Möglichkeiten im Relativkalkül', *Math. Ann.*, **68**, 169–207, (1915).

[3] As set out in Skolem, T., 'Logisch-kombinatorische Untersuchungen über die Erfüllbarkeit und Beweisbarkeit mathematischen Sätze nebst einem Theoreme über dichte Mengen', *Videnskabsakademiet i Kristiania*, Skrifter I, No. 4, 1–36, (1919) and Skolem, T., 'Einige Bemerkung zur axiomatischen Begrundung der Mengenlehre', *Proc. Scand. Math. Congr. Helsinki*, 217–232, (1922).

by a $\mathbb{Q}$-indexed string of copies of $\mathbb{Z}$. So the order structure of countable non-standard models is unique, but we stress that the arithmetic structure is not: there are continuum many non-isomorphic countable models of PA.

2. The Löwenheim–Skolem Theorem does not hold in second-order theories, and quite dramatically so: we have already seen two examples of the failure, one in second-order Peano Arithmetic and the other in the second-order theory of complete ordered fields. Each theory is categorical, it has only *one* model up to isomorphism.

3. If a theory has a model of cardinality κ (some infinite cardinal) and if all models of cardinality κ are isomorphic, then the theory is said to be κ-*categorical*. The theory of algebraically closed fields of characteristic p, for example, is κ-categorical for all κ.

There are many deep and interesting questions connected to κ-categoricity. A celebrated result (Morley's Theorem) tells us that if a first-order theory (in a countable language)[1] is κ-categorical for some uncountable κ then it is κ-categorical for all uncountable κ.[2]

To what degree can categoricity fail? One can ask, for instance, if there is a complete theory which has exactly two non-isomorphic countable models. The answer is that there is not (a result of Robert Vaught). However, for every finite $n \geq 3$ there are complete theories which have exactly n non-isomorphic countable models. Vaught conjectured that if a first-order complete theory has an infinite number ν of non-isomorphic models then ν must be either $\aleph_0$ or $2^{\aleph_0}$. Morley proved that ν must be equal to $\aleph_0$ or $2^{\aleph_0}$ *or* $\aleph_1$. Of course, assuming the Continuum Hypothesis, i.e. $2^{\aleph_0} = \aleph_1$, this proves the conjecture. However, if we assume the Continuum Hypothesis is false, so that $\aleph_0 < \aleph_1 < 2^{\aleph_0}$, there is a chance that there might be a theory with $\aleph_1$ models, providing a counterexample to Vaught's conjecture.

4. Two models M and N of a theory are said to be *elementarily equivalent* if for every sentence ϕ of the theory, M satisfies ϕ if and only if N satisfies ϕ. In other words the two models are indistinguishable in first-order terms. Perhaps the first example of this that springs to mind is the elementary equivalence of $\mathbb{R}$ and the non-standard field $^*\mathbb{R}$ (see Subsection 1.7.11). Clearly any two models of a negation complete theory will be elementarily equivalent, but one can give dramatic examples of elementary equivalence for negation incomplete theories too. For example, as mentioned above, Peano Arithmetic has continuum many non-isomorphic countable models (including the standard model ω), all of them elementarily equivalent.

---

[1]This condition can be weakened. Saharon Shelah proved in 1974 that if the language has cardinality κ and the theory is μ-categorical for some uncountable $\mu \geq \kappa$, then it is μ-categorical for all uncountable $\mu \geq \kappa$.

[2]For more on this subject, see Marker [**145**], Chapter 4.

5. In 1969 Per Lindström published a result,[1] now known as Lindström's Theorem, which gives a kind of characterization of first-order logic. It states that among all logics satisfying certain natural conditions (closure under negation and conjunction), first-order logic is the strongest for which the Downward Löwenheim–Skolem Theorem and the (countable) Compactness Theorem both hold.

The Löwenheim–Skolem Theorem tells us that if a first-order theory with countably many axioms has a set model, then it has a countable model. Indeed, it has models of any infinite cardinality. Since ZF has only countably many axioms, it must, if it has a set model, have a countable model. At first this seems paradoxical.

10.2.4   Skolem's paradox and a route to independence

From the Löwenheim–Skolem Theorem we see that, since ZF has only countably many axioms, if ZF has a set model then *there is a countable model of ZF*. At first the existence of such a model may seem troublesome – how is it that a model of ZF can be countable when it includes uncountable sets such as $P(\omega)$? This is the substance of Skolem's paradox.[2]

Skolem's paradox has, since it was first described, been the subject of an embarrassing number of misunderstandings. It is not a paradox in the sense that, say, Russell's paradox is; it does not lead to contradiction – it is not an antinomy. The difficulty is resolved by closely analyzing what is meant by countability. The statement 'X is uncountable' in some model M is simply the assertion that *in M* there is no bijection $X \to \omega$. In other words, no set of ordered pairs that constitutes a bijection $X \to \omega$ is present in the model. We could say that X is 'externally' countable, meaning that there is another model in which X is equipollent to ω, but in the given model it is 'internally' uncountable.

The statement that a set has a well-ordering and the statement that two sets are equipollent are both assertions about the existence of certain sets of ordered pairs. This suggests that it might be possible to forge, by judicious inclusion or exclusion of sets, models of ZF in which the Axiom of Choice or the Continuum Hypothesis are either true or false. What is needed is a universe of sets built in such a way that it contains or does not contain certain well-orderings or bijections, yet at the same time accommodates all of the axioms of ZF. Gödel and Cohen showed, respectively, that there are models in which Choice and the Continuum Hypothesis are true and models in which they are false.

---

[1] Per Lindström, 'On Extensions of Elementary Logic', *Theoria*, **35**, 1–11, (1969).
[2] Skolem describes the issue in 'Einige Bemerkung zur axiomatischen Begrundung der Mengenlehre', *Proc. Scand. Math. Congr. Helsinki*, 217–232, (1922).

Mathematically Skolem's (non-)paradox is not an issue at all and it is easy to see why this is so. Nevertheless it still seems to generate some debate in some philosophical circles. While these discussions can be interesting, we shall continue to focus on the mathematics.

There is no paradox in the existence of a countable model of ZF. The statement that a set x is uncountable in some model simply means that a set of ordered pairs defining a bijection between x and ω is not present in the model. Viewed externally x may well be countable, but within its own universe it is uncountable, and its precise cardinality is determined by the existence of particular bijections in the model. This indicates that it might be possible to prove independence results by carefully adding sets to existing models, forming a new model in which a statement has been forced to be true or false.

10.3 The Reflection Principle and absoluteness

Mathematics is the most exact science, and its conclusions are capable of absolute proof. But this is so only because mathematics does not attempt to draw absolute conclusions. All mathematical truths are relative, conditional.

– CHARLES P. STEINMETZ[1]

10.3.1 The Reflection Principle

An analogue of the Löwenheim–Skolem Theorem, which also assumes many different forms, is the Reflection Principle. This asserts that for any finite collection F of sentences in ZF there exists a set M which has the property that a sentence ϕ in F is satisfied in M if and only if it is satisfied in the full universe of sets V of ZF, i.e. ϕ^M if and only if ϕ^V. Alternative forms of the Reflection Principle place further restrictions on the set M. Generally speaking, reflection principles posit the existence of sets which, in some restricted sense (i.e. relative to a fixed collection of sentences), are indistinguishable from the class of all sets.

The Reflection Principle and Gödel's Second Incompleteness Theorem together prove that ZF is not finitely axiomatizable, for if it were then the Reflection Principle tells us that in ZF we could prove the existence of a model of ZF, implying ZF can prove its own consistency, contrary to the Second Incompleteness Theorem. Put another way, we cannot escape axiom schemas.

---

[1]Steinmetz [201]. (Lecture IV).

Some reflection principles have been adopted as powerful axioms in alternative set theories. Paul Bernays described one such theory[1] where the reflection axiom asserts that for each class A satisfying some sentence ϕ there exists a transitive set x such that $A \cap x$ also satisfies ϕ. This axiom and axioms like it are closely tied to large cardinals; they imply the existence of certain large cardinals and their consistency with the other axioms of the theory is established by the existence of such cardinals.

REMARK

The Reflection Principle can be proved without the Axiom of Choice, but with Choice one can prove that the model of the given finite set of sentences is countable.[2]

Given any finite collection of sentences in ZF there exists a set M such that each of the sentences is satisfied in M if and only if it is satisfied in the class of all sets V, so that these properties of the entire set theoretic universe are reflected down to the set M. Together with Gödel's Second Incompleteness Theorem this tells us that ZF is not finitely axiomatizable.

10.3.2 Absoluteness

The existence of a model of set theory in which an externally countable set is internally uncountable warns us that we ought to tread carefully when referring to properties of an isolated set in the absence of any concrete set theoretic model. Set theorists express this delicate dependability on the overlying modelling class by saying that countability is not an absolute property. Countability and various other set theoretic notions, as stressed by Skolem, are relative in nature; a set which is uncountable in one model may be countable in another. Countable models play an important role in the metatheory of ZF so we need to get used to this perspective shifting phenomenon.

Given a well-formed formula ϕ and a class A the formula ϕ^A describes the restriction of ϕ to the universe A. It may be the case, countability being an example, that a set has a property ϕ^A but not ϕ^B where A and B are different classes. Another striking example of this dependency is provided by the formula $(\forall x)[x \notin \varnothing]$ characterizing the empty set. It is an immediate consequence of the Axiom of Regularity that every non-empty class A, regarded as a universe of sets, possesses an 'empty set' E in the sense that $(\forall x \in A)[x \notin E]$. From an external vantage point this set may be far from empty ($\aleph_0$ enjoys the role of the 'empty set' in the class of infinite cardinal numbers, for example) and in this setting the label 'minimal set' might be more appropriate.

---

[1]Bernays, P., 'Zur Frage der Unendlichkeitsschemata in der axiomatischen Mengenlehre', in Bar-Hillel [9].
[2]For details, see, for example, Jech [108], Chapter 12.

It is tempting to think that such relativism is an inevitable feature of the theory, however, as the Reflection Principle tells us, for *any* sentence ϕ there exists a set A for which ϕ^A if and only if ϕ^V. In general we shall say that an n-predicate ϕ is *absolute* with respect to a class A if whenever $a_1, \ldots, a_n \in A$ we have $\phi(a_1, \ldots, a_n)^A$ if and only if $\phi(a_1, \ldots, a_n)^V$.

If ϕ and ψ are absolute with respect to A then both $\phi \wedge \psi$ and $\neg\phi$ are absolute with respect to A and consequently so are $\phi \vee \psi$, $\phi \rightarrow \psi$, and $\phi \leftrightarrow \psi$. Establishing absoluteness for formulas including quantifiers requires further (simple) assumptions.[1] Thus we see that for some fixed universe A the set of formulas which are absolute with respect to A is closed under the logical operations.

Absoluteness and satisfaction of generalized well-formed formulas are defined simply by replacing the formula by the formula it abbreviates, i.e. one simply uses ϕ^* (as introduced in Section 3.2) in place of ϕ, i.e. if ϕ is a generalized well-formed formula then: (i) $[A, R]$ satisfies ϕ means $[A, R]$ satisfies ϕ^*; (ii) ϕ^A means $(\phi^*)^A$; and (iii) ϕ is absolute with respect to A means ϕ^* is absolute with respect to A.

Importantly, the irreducible well-formed formula $a \in b$ is (trivially) absolute with respect to any non-empty class, and both $a \subseteq b$ and $a = b$ are absolute with respect to any non-empty *transitive* class. The requirement that the class be transitive in the latter is essential – even in very simple non-transitive cases we find that non-absoluteness introduces some undesirable phenomena. For example, relative to the set $A = \{\emptyset, \{\{\emptyset\}\}\}$, the sets $\emptyset$ and $\{\{\emptyset\}\}$ are indistinguishable in the sense that $(\forall x \in A)[x \in \emptyset \leftrightarrow x \in \{\{\emptyset\}\}]$, so $[\emptyset = \{\{\emptyset\}\}]^A$ is true but $[\emptyset = \{\{\emptyset\}\}]^V$ is false. Note that every non-empty transitive class A contains $\emptyset$, as follows from Regularity: A is non-empty so there exists an x in A such that $x \cap A = \emptyset$, but $x \cap A = x$ by transitivity.

If a class A is non-empty and transitive then both statements 'a is transitive' and 'a is well-ordered by $\in$' are absolute with respect to A. So by restricting the definition of $\mathcal{O}n$ to a non-empty transitive class there is no danger of introducing any new alien objects as ordinals. Consequently, if M is a standard transitive model of ZF, i.e. a transitive class M such that $[M, \in]$ satisfies every axiom of ZF, then the class of ordinals in M is precisely the class $\mathcal{O}n \cap M$.

A version of the Reflection Principle is expressible economically in terms of absoluteness: for any formula ϕ there exists an ordinal number α such that ϕ is absolute with respect to V_α. In the next section, for each axiom ϕ of ZF we shall look at the conditions on α sufficient for V_α to be a model of ϕ.

---

[1] See, for example, Chapter 13 of Takeuti and Zaring [207], which has served as a template for this subsection.

1. Let A be a non-empty class and suppose E and F are sets in A which are minimal in the sense that $(\forall x \in A)[x \notin E]$ and $(\forall x \in A)[x \notin F]$. Then since $[(\forall x)[x \in E \leftrightarrow x \in F]]^A$ is equal to $[(\forall x \in A)[x \in E \leftrightarrow x \in F]]$, a tautology, we have $[E = F]^A$. That is to say, relative to A the 'empty' sets E and F are equal. So in the universe A the uniqueness of the empty set continues to be internally valid, even though E and F may be externally very different. The two element set given in the text, $A = \{\varnothing, \{\{\varnothing\}\}\}$, is then, through the lens of A itself, a one element set comprising the 'empty set' alone! In other words, the sentence $(\forall x)(\forall y)[x = y]$ is true in A.

2. Note in particular that if M is a countable standard transitive model of ZF then the class of all ordinals is faithfully modelled in this model by a class of externally countable ordinals. This admittedly takes a bit of getting used to. The point is that the majority of these ordinals will be 'internally' uncountable, indeed they will form an 'internal proper class'.

A well-formed formula ϕ is absolute with respect to A if $\bar{\phi}^A$ if and only if $\bar{\phi}^V$, where $\bar{\phi}$ is obtained from ϕ by replacing each free variable by some element of A. Since the properties of transitivity and well-ordering are absolute with respect to non-empty transitive classes, the ordinal numbers in a standard transitive model of ZF are simply those elements of $\mathcal{O}n$ that are in the model.

10.4   Standard transitive models of ZF

Questions that pertain to the foundations of mathematics, although treated by many in recent times, still lack a satisfactory solution. Ambiguity of language is philosophy's main source of problems. That is why it is of the utmost importance to examine attentively the very words we use.

– Giuseppe Peano[1]

10.4.1   The Mostowski Collapse Lemma

Suppose R is a relation on a class A with the following two properties:

(i) R is well-founded;

(ii) for all $b, c \in A$, $(\forall a)[aRb \leftrightarrow aRc]$ implies $b = c$.

---

[1] *Arithmetices Principia, nova methodo exposita* [The Principles of Arithmetic, presented by a new method] (1889), in van Heijenoort [**219**].

In other words R behaves like the membership relation. We shall see shortly that, up to isomorphism (i.e. up to the equivalence induced by the existence of a bijection f between relational classes $[A_1, R_1]$ and $[A_2, R_2]$ such that aR_1b if and only if $f(a)R_2f(b)$), the only examples of such structures are given by the membership relation restricted to transitive classes.

For each $x \in A$ define $F(x)$ recursively by $F(x) = \{F(a) : aRx\}$. The class $\mathrm{Mos}(A) = \{F(x) : x \in A\}$, the *Mostowski collapse* of A, is a transitive class with the property that, for any predicate $\phi(x_1, \ldots, x_n)$ and $a_1, \ldots, a_n \in A$, A satisfies $\phi(a_1, \ldots, a_n)$ if and only if $\mathrm{Mos}(A)$ satisfies $\phi(F(a_1), \ldots, F(a_n))$ (with respect to the membership relation). This result, that $F : [A, R] \to [\mathrm{Mos}(A), \in]$ is an isomorphism, is called the *Mostowski Collapse Lemma* (after Andrzej Mostowski).[1] In particular the Mostowski Collapse Lemma tells us that every standard model is $\in$-isomorphic to a standard transitive model, and if A is a submodel of a standard transitive model M then $\mathrm{Mos}(A)$ is a submodel of M.

REMARKS

1. Condition (ii) is sometimes expressed as 'R is extensional' and can be given in the alternative equivalent form: $R^{-1}(x) \neq R^{-1}(y)$ whenever $x \neq y$.

2. The Mostowski collapse of a well-ordered set is the unique ordinal which is order isomorphic to the set. Put another way, the function that maps well-ordered sets to their order types is precisely the restriction of the Mostowski collapse to well-ordered sets.

With each standard model M of ZF we can associate in a canonical way an isomorphism between the model and a standard transitive model $\mathrm{Mos}(M)$ of ZF. $\mathrm{Mos}(M)$ is the Mostowski collapse of M. The Mostowski collapse of a submodel of a standard transitive model is again a submodel of the same standard transitive model.

10.4.2 The axioms of ZF and the sets V_α

The stack of sets V_α, $\alpha \in \mathcal{O}n$, which cumulatively builds the class of all sets V, provides a natural focal point when testing one's intuition about standard transitive models. For which $\alpha > 0$ is V_α a model of a given axiom of ZF? We have the following.[2]

---

[1] Mostowski, A., 'An undecidable arithmetical statement', *Fund. Math.*, **36**, 143–164, (1949).

[2] In a similar vein one can ask what the smallest universe M is which contains $\emptyset$ and which is a (standard) model of a given axiom. For example, if we define $M_0 = \{\emptyset\}$ and for natural n, $M_{n+1} = \{\{a, b\} : a, b \in \cup_{i=0}^{n} M_n\}$, then $M = \cup_{n=0}^{\infty} M_n$ is evidently the smallest (standard) model of the Axiom of Pairing. The reader is invited to examine the remaining axioms.

1. V_α is a standard transitive model of each of the Axiom of Extensionality, the Axiom of Unions and the Axiom of Regularity for all $\alpha > 0$. Every non-empty transitive class is a model of Extensionality and Regularity. A non-empty class M is a model of Unions if it has the obvious closure property $\cup a \in M$ for all $a \in M$.

2. V_α is a standard transitive model of either of the Axiom of Pairing and the Axiom of Powers if and only if α is a limit ordinal. By definition a non-empty class M is a model of Pairing if $\{a, b\} \in M$ whenever $a, b \in M$ and M is a model of Powers if $P(a) \cap M \in M$ for all $a \in M$. The condition for being a model of Powers ultimately follows from the absoluteness of $c \in b \leftrightarrow c \subseteq a$ with respect to the transitive class M. This relativity isn't an issue when considering the classes V_α since, if α is a limit ordinal, $a \in V_\alpha$ implies $P(a) \subset V_\alpha$.

3. As soon as a model of Replacement contains a non-empty set, it must contain infinitely many sets, and as soon as it contains a two-element set it must also satisfy the Axiom of Pairing, and so α is a limit ordinal. However, not all V_α with α a limit ordinal are models of Replacement. V_ω is a model of Replacement, but $V_{\omega+\omega}$, for example, is not. Indeed, suppose $V_{\omega+\omega}$ were a model of Replacement, then the image $\{\omega, \omega+1, \omega+2, \ldots\}$ of the set $\omega \in V_{\omega+\omega}$ under the function $f(n) = \omega + n$ is also a set, implying its union $\omega + \omega$ is an element of $V_{\omega+\omega}$, a contradiction. It soon becomes apparent that Replacement is a powerful axiom that acts in many respects as a strong axiom of infinity. In particular, if a is an element of a model V_α of Replacement, then V_α must be able to accommodate such strongly increasing and possibly transfinitely large cardinal hierarchies as $\mathrm{Card}(a)$, $\mathrm{Card}(P(a))$, $\mathrm{Card}P(P(a))$, $\mathrm{Card}(P(P(P(a))))$,... and their unions. The way points to inaccessibility (see below).[1]

4. V_α is a standard transitive model of the Axiom of Infinity if and only if $\alpha > \omega$.

Using Choice, as mentioned earlier, one can prove that if α is inaccessible then the Grothendieck universe V_α is a standard transitive model of ZF. The statement 'β is inaccessible' is absolute with respect to V_α. Assuming inaccessible cardinals exist and letting α be the smallest such cardinal, V_α is a standard transitive model of ZF which possesses no inaccessible cardinals (any such cardinal, owing to absoluteness, would be inaccessible in V). Consequently, as promised in Section 8.4, we deduce without appealing to Gödel's Second Incompleteness Theorem that the existence of inaccessible cardinals cannot be proved in ZFC.

The results listed above make transparent the independence of some axioms from others. For example, V_ω is a standard transitive model of all of the axioms of ZF except the Axiom of Infinity and $V_{\omega+\omega}$ is a standard transitive model of

---

[1] For more on models of Replacement, see Just and Weese [111], Volume II, Chapter 12.

all axioms except Replacement. In fact, $V_{\omega+\omega}$ is a model of Zermelo's original set theory.

Remark

V_ω is the class of *hereditarily finite sets*. (The *transitive closure* of a set x is the smallest transitive set a such that $x \subseteq a$. A set x is hereditarily finite if and only if its transitive closure is finite.) The theory of hereditarily finite sets ZF_{hf}, obtained from ZF by replacing the Axiom of Infinity by its negation and postulating that every set has a transitive closure, is *bi-interpretable* with PA. This means that the two theories are essentially the same in the following sense. There exists a recursive procedure which transforms any sentence ϕ in ZF_{hf} into a sentence $\hat{\phi}$ in PA, and conversely a procedure which transforms PA-sentences ψ into ZF_{hf}-sentences $\bar{\psi}$, in such a way that ϕ is provable in ZF_{hf} if and only if $\hat{\phi}$ is provable in PA and ψ is provable in PA if and only if $\bar{\psi}$ is provable in ZF_{hf}.[1] Translating arithmetical sentences into the set theoretic language is easy; one takes the usual interpretation of the primitive symbols of PA (von Neumann successor for s, empty set for 0, and so on). In order to translate set theoretic sentences into PA we need a number theoretic relation E which is to play the role of membership. The trick is to define nEm if and only if $b_n = 1$ in the binary representation of m: $b_0 + b_1 2 + b_2 2^2 + \cdots + b_k 2^k + \cdots$ (this relation is recursively definable in PA).

Each set V_α is a model of some of the axioms of ZF. If α is inaccessible then V_α is a model of all of the axioms of ZF. Using the absoluteness of the property 'β is absolute with respect to V_α' and choosing α to be the smallest inaccessible cardinal we conclude, without appealing to Gödel's Second Incompleteness Theorem, that the existence of inaccessible cardinals cannot be proved in ZFC.

---

[1]This result goes back to W. Ackermann. 'Die Widerspruchsfreiheit der allgemeinen Mengenlehre', *Math. Ann.*, **114**, 305–315, (1937).

11

From Gödel to Cohen

11.1   The constructible universe

> *It can be shown that a mathematical web of some kind can be woven about any universe containing several objects. The fact that our universe lends itself to mathematical treatment is not a fact of any great philosophical significance.*
>
> – BERTRAND RUSSELL (attributed)[1]

11.1.1   Constructing a simple universe

Gödel proved the relative consistency of the Axiom of Choice and the Generalized Continuum Hypothesis with Gödel–Bernays set theory by explicitly constructing a model in which both statements hold.

In the introduction to his 1940 Princeton lecture notes *The Consistency of the Continuum Hypothesis*[2] Gödel describes the theory he is using:

> *The system Σ of axioms of set theory which we adopt includes the axiom of substitution [Fraenkel's axiom of substitution] and the axiom of 'Fundierung' [the axiom of foundation]... It is essentially due to P. Bernays... and is equivalent with v. Neumann's system [with a minor modification]... The system Σ has in addition to the ∈-relation two primitive notions, namely 'class' and 'set'.*

The axioms of Σ are listed in the first chapter and are compared with those of Bernays. The translation of Gödel's results from his version of set theory to ZF is not too difficult. I will distort history in this chapter in connection with Gödel's results, as I have done throughout, harmlessly pretending that he worked in ZF.

Gödel's constructible universe L was first defined recursively as the union of sets L_α, $\alpha \in \mathcal{O}n$, where:

---

[1] Quoted in Sullivan [**205**].
[2] Gödel [**81**].

$$L_0 = \emptyset;$$

$$L_{\alpha+1} = \begin{cases} \text{the class of sets definable from } L_\alpha \text{ in the sense} \\ \text{that } x \in L_{\alpha+1} \text{ if and only if there exists an} \\ n+1\text{-predicate } \phi(x_0, x_1, \ldots, x_n), \ n \geq 0, \text{ and} \\ \text{there exist } a_1, \ldots, a_n \in L_\alpha \text{ such that } x = \{y : \\ L_\alpha \text{ satisfies } \phi(y, a_1, \ldots, a_n)\}; \text{ and} \end{cases}$$

$$L_\beta = \bigcup_{\alpha < \beta} L_\alpha \text{ if } \beta \text{ is a limit ordinal.}$$

A set x is then *constructible* if and only if $x \in L$.

Hidden in this construction is a complication that ought to be addressed: how do we express in first-order terms what it means to be definable while avoiding the illegal quantification on well-formed formulas which appears in the definition of $L_{\alpha+1}$? It is possible to overcome this difficulty (via a Gödel numbering – see Section 11.3), but we shall not dwell on it here, for there is an alternative approach to L which avoids the hurdle entirely.

The class of constructible sets is, as Gödel later discovered, the range of a function F on $\mathcal{O}n$ which is defined in terms of the eight basic operations listed below.

Gödel's fundamental operations

$F_1(a, b) = \{a, b\}$,
$F_2(a, b) = a \cap \{(x, y) : x \in y\}$,
$F_3(a, b) = a - b$,
$F_4(a, b)$ is the restriction of a to b,
$F_5(a, b)$ is the intersection of a with the domain of b,
$F_6(a, b) = a \cap b^{-1}$,
$F_7(a, b) = a \cap \{(x, y, z) : (z, x, y) \in b\}$,
$F_8(a, b) = a \cap \{(x, y, z) : (x, z, y) \in b\}$.

Note that the b is redundant in $F_2(a, b)$, but it is convenient in the technical details to treat all F_i as two variable set functions – in particular we avoid picking out F_2 as a special case every time we mention a generic fundamental operation $F_i(a, b)$.

It is easy to prove that if a class M is a standard transitive model of ZF then M is closed under the eight fundamental operations, i.e. if $a, b \in M$ then $F_i(a, b) \in M$ for $i \in \{1, \ldots, 8\}$. Indeed, it is fairly clear that the hypothesis that M is a standard transitive model of ZF is much more than is required. Conversely, if a transitive class M is closed under the eight fundamental operations and is *almost universal*, by which we mean that for all subsets x of M there exists a $y \in M$ with $x \subseteq y$, then M is a standard transitive model of ZF. This extra property of almost universality is strong enough to force $\mathcal{O}n \subseteq M$ and consequently, bearing in mind the Löwenheim–Skolem Theorem, is much more than is generally required of a model of ZF.

We picture L as a growing collection of sets, starting with the empty set, each obtainable by applying the fundamental operations to the sets already formed and, ensuring almost universality, their unions. We describe how to formally capture this recursive hierarchy. The lexicographic ordering $\prec$ of $\mathcal{On} \times \mathcal{On}$ defined by $(\alpha, \beta) \prec (\gamma, \delta)$ if and only if

(i) $\alpha < \delta$; or

(ii) $\alpha = \gamma$ and $\beta < \delta$,

is a well-founded well-ordering and it can be proved that there exists a (unique) order isomorphism

$$\Phi_0 : (\mathcal{On} \times \mathcal{On}, \prec) \to (\mathcal{On}, <).$$

This isomorphism is partly illustrated in the following table:

Φ_0	0	1	2	3	$\cdots$	ω	$\omega + 1$	$\cdots$
0	0	1	4	9	$\cdots$	ω	$\omega 3 + 1$	$\cdots$
1	2	3	5	10	$\cdots$	$\omega + 1$	$\omega 3 + 2$	$\cdots$
2	6	7	8	11	$\cdots$	$\omega + 2$	$\omega 3 + 3$	$\cdots$
3	12	13	14	15	$\cdots$	$\omega + 3$	$\omega 3 + 4$	$\cdots$
$\vdots$	$\vdots$	$\vdots$	$\vdots$	$\vdots$		$\vdots$	$\vdots$	$\vdots$
ω	$\omega 2$	$\omega 2 + 1$	$\omega 2 + 2$	$\omega 2 + 3$	$\cdots$	$\omega 3$	$\omega 4$	$\cdots$
$\omega + 1$	$\omega 4 + 1$	$\omega 4 + 2$	$\omega 4 + 3$	$\omega 4 + 4$	$\cdots$	$\omega 5$	$\omega 5 + 1$	$\cdots$
$\vdots$	$\vdots$	$\vdots$	$\vdots$	$\vdots$	$\vdots$	$\vdots$	$\vdots$	$\ddots$

For finite n and m an explicit formula for $\Phi_0(n, m)$ is easily found:

$$\Phi_0(n, m) = \begin{cases} n^2 + n + m & \text{if} \quad m \le n, \\ m^2 + n & \text{if} \quad m > n. \end{cases}$$

From Φ_0 we obtain the bijection

$$\Phi : \mathcal{On} \times \mathcal{On} \times \{0, 1, 2, 3, 4, 5, 6, 7, 8\} \to \mathcal{On}$$

defined by $\Phi((\alpha, \beta, m)) = 9\Phi_0((\alpha, \beta)) + m$.

The generating function $\mathcal{G} : \mathcal{On} \to L$ of the constructible universe is defined via the bijection Φ as follows. If $\Phi(\alpha, \beta, n) = \gamma$ then

$$\mathcal{G}(\gamma) = \begin{cases} F_n(\mathcal{G}(\alpha), \mathcal{G}(\beta)) & \text{if} \quad n \ne 0, \\ \bigcup_{\delta < \gamma} \mathcal{G}(\delta) & \text{if} \quad n = 0 \end{cases}$$

defines the γth constructible set, where F_n is the nth fundamental operation. Carrying out this construction we find that there is much repetition; the assembly of constructible sets begins: $\emptyset, \{\emptyset\}, \emptyset, \emptyset, \emptyset, \emptyset, \emptyset, \emptyset, \emptyset, \{\emptyset, \{\emptyset\}\}, \{\emptyset, \{\emptyset\}\}, \emptyset, \emptyset, \emptyset, \emptyset, \dots$ By design, the range $L = \{\mathcal{G}(\gamma) : \gamma \in \mathcal{On}\}$ of $\mathcal{G}$ is closed under the application of the fundamental operations, and it is transitive and almost universal, hence it is a model of ZF, that is, we can prove in the assumed consistent theory ZF that the class L is a model of ZF.

If x is constructible then the *order* of x is the ordinal indicating the first stage at which x is constructed (i.e. the smallest α such that $\mathcal{G}(\alpha) = x$).

The constructible universe grows very slowly in comparison with the von Neumann set hierarchy, nevertheless the initial part $\{\mathcal{G}(n) : n \in \omega\}$ of the former coincides with V_ω. Can we assume that the two hierarchies describe the same class of sets? Is every set constructible, or put another way, is L equal to V? The assertion that this is the case, which is expressible as a single sentence in ZF, is called the *Axiom of Constructibility*.

REMARKS

1. One might think of the constructible universe as a more soberly generated version of the von Neumann universe V where the power set used to generate each new successor level of the latter (collecting together *all* subsets of the preceding level) is replaced by the collecting together of all subsets that are *definable* in terms of sets in the preceding level.

2. The inspiration for the constructible universe came from Russell's ramified hierarchy of types and its comparison with axiomatic set theory.

Gödels's constructible universe is a recursively generated class of sets which, assuming ZF is consistent, forms a model of ZF. Each of the sets generated by Gödel's construction is said to be constructible. Comparing the Gödel universe L with the von Neumann universe V we ask if any sets in V are not constructible; does the equality $L = V$ hold? A simple modification of the recursion describes how new sets can be added to L to form a potentially larger model.

11.1.2   Absorbing new sets into L

In a slight extension of constructibility we say that a set is constructible relative to a set a if and only if it can be built up from ω and its elements and the new set a by some possibly transfinite application of the eight fundamental operations and unions of the sets so formed. The collection of all sets obtainable by absorbing such a set a and all of its generated descendants into the full constructible universe L is denoted L_a.

This seeding of L by a is formally described by the following simple mutation $\mathcal{G}_a$ of the function $\mathcal{G}$ on $\mathcal{O}n$. If $\Phi(\alpha, \beta, n) = \gamma$ then the γth a-constructible set is:

$$\mathcal{G}_a(\gamma) = \begin{cases} \gamma & \text{if} \quad \gamma < \omega + 1, \\ a & \text{if} \quad \gamma = \omega + 1, \\ F_n(\mathcal{G}_a(\alpha), \mathcal{G}_a(\beta)) & \text{if} \quad n \neq 0 \text{ and } \gamma > \omega + 1, \\ \bigcup_{\delta < \gamma} \mathcal{G}_a(\delta) & \text{if} \quad n = 0 \text{ and } \gamma > \omega + 1. \end{cases}$$

The first $\omega + 1$ a-constructible sets are $0, 1, 2, 3, \ldots, \omega, a$. Since $\Phi(0, \omega, 2) = \omega + 2$ we see that the $\omega + 2$th a-constructible set is $\mathcal{G}_a(\omega + 2) = F_2(\mathcal{G}_a(0), \mathcal{G}_a(\omega)) = F_2(0, \omega) = 0 \cap \{(x, y) : x \in y\} = 0$, and so on. Like L, L_a is a standard transitive model of ZF and $\mathcal{O}n \subseteq L_a$.

Gödel's class of constructible sets and simple modifications of it together with the models implied by the existence of large cardinals were, for a long time, the main examples of standard models of ZF.

REMARK

It is easy to modify the definition to introduce a possibly infinite collection of new sets, the new sets a_1, a_2, a_3, and so on being the $\omega + 1$th, $\omega + 2$th, $\omega + 3$th, etc. constructible sets.

One can naturally enlarge the constructible universe to incorporate any number of new sets. The resulting structure is, like the constructible universe itself, a standard transitive model of ZF which includes all ordinals. The hypothesis that all sets are constructible has some strong corollaries.

11.1.3 Consequences of constructibility

Gödel's constructible universe occupies an important position among the standard transitive models of ZF. If M is a standard transitive model of ZF with $\mathcal{O}n \subseteq M$ then $L \subseteq M$ (and if $a \in M$, then $L_a \subseteq M$), in other words L *is the smallest standard transitive model of ZF that contains all ordinals*.

Because L is well-ordered (a precedes b in L if the order of a is less than the order of b) it follows that every constructible set, itself an aggregate of constructible sets, has a well-ordering and hence the Axiom of Choice holds in L. Evidently this does much more than prove that the Axiom of Choice holds in L – it exhibits an explicit *universal* choice function, and consequently the result proved is significantly stronger than the *strong* form of the Axiom of Choice, which merely asserts the existence of such a function.

The proof that Constructibility implies the Generalized Continuum Hypothesis is more difficult. It is typically split into two subarguments (we do not give details): first, the cardinality of $\{F(\beta) : \beta < \aleph_\alpha\}$ in L is $\aleph_\alpha$; secondly, every subset of $\{F(\beta) : \beta < \aleph_\alpha\}$ has order less than $\aleph_{\alpha+1}$, together forcing $2^{\aleph_\alpha} = \aleph_{\alpha+1}$. The relative consistency of the Generalized Continuum Hypothesis and the Axiom of Choice with ZF is then established by proving the relative consistency of the Axiom of Constructibility with ZF. A model of Constructibility is provided by L itself. This is perhaps not quite as obvious as it may sound at first. That L is a model of Constructibility means that the relativization of $V = L$ to L is a theorem of ZF, a fact which ultimately derives from the absoluteness of '$x = \mathcal{G}(\alpha)$' with respect to L.

Besides the 'headline' consequences of the Axiom of Choice and the Generalized
Continuum Hypothesis, the axiom of constructibility ($V = L$) also has a number
of interesting consequences in descriptive set theory, it proves the negation of
Souslin's Hypothesis and also the non-existence of measurable cardinals.

*The constructible universe is not only a model of ZF, it is also a model of a
strong Axiom of Choice and of the Generalized Continuum Hypothesis. It is
the smallest standard transitive model of ZF which includes the class $\mathcal{O}n$ of all
ordinal numbers. The assumption that there is a set which is a model of ZF has
some interesting consequences.*

11.1.4   The minimal model

Assuming the existence of a *set* which is a standard model of ZF, the *Standard
Model Hypothesis* (a statement which can be expressed as a single sentence in
ZF without fear of self-reference – see Section 11.3), it is possible to prove the
existence of a model which is minimal among all standard transitive models of
ZF. This model, which is countable, is a model of Constructibility and hence
also of the Axiom of Choice and the Generalized Continuum Hypothesis.

An outline of the proof is as follows: the Standard Model Hypothesis, together with the Mostowski Collapse Lemma, implies the existence of a set m
which is a standard *transitive* model of ZF. As the property of being an ordinal
is absolute with respect to standard transitive models, the ordinal numbers in
m are just those elements of $\mathcal{O}n$ which happen to be in m. Furthermore, if some
ordinal α is in m then all ordinals smaller than α (i.e. all elements of α) are also
in m. Let us say that an ordinal α is 'admissible' if there is a standard transitive
model M of ZF such that $\alpha = \mathcal{O}n \cap M$, then the class of admissible ordinals,
which we have just deduced is not empty, must have a smallest member α_0.

Let N be a standard transitive model of ZF with $\alpha_0 = \mathcal{O}n \cap N$. $L =
\{x : (\exists \alpha \in \mathcal{O}n)[x = \mathcal{G}(\alpha)]\}$, so using the mechanical rules of relativization (see
Section 10.1) and the fact that $\alpha \in N$ implies $\mathcal{G}(\alpha) \in N$ (as follows by induction
on α, N being closed under the fundamental operations) we have:

$$
\begin{aligned}
L^N &= \{x \in N : (\exists \alpha \in N)[x = \mathcal{G}(\alpha)]^N\} \\
&= \{x \in N : (\exists \alpha \in N)[x = \mathcal{G}(\alpha)]\} \\
&= \{x : (\exists \alpha \in N)[x = \mathcal{G}(\alpha)]\} \\
&= \{x : (\exists \alpha < \alpha_0)[x = \mathcal{G}(\alpha)]\} \\
&= \{\mathcal{G}(\alpha) : \alpha < \alpha_0\}.
\end{aligned}
$$

Let M_0 be the set $L^N = \{\mathcal{G}(\alpha) : \alpha < \alpha_0\}$. As L and N are models of ZF, so
is M_0. Put another way, the minimal model M_0 is equal to L_κ, where κ is the

smallest ordinal such that L_κ is a model of ZF, assuming that such an ordinal exists.

M_0 is a model of Constructibility: since $x = \mathcal{G}(\alpha)$ is absolute with respect to all standard transitive models of ZF, the relativization $(V = L)^{M_0}$, i.e. $[(\forall x)(\exists \alpha)[x = \mathcal{G}(\alpha)]]^{M_0}$, is simply $(\forall x \in M_0)(\exists \alpha \in M_0)[x = \mathcal{G}(\alpha)]$. As the set of ordinals in M_0 is α_0 this is just a restatement of the definition of M_0. Note also that, by minimality, the Standard Model Hypothesis is *false* in M_0.

By construction, M_0 is a subset of all standard transitive models of ZF. The countability of M_0 is a consequence of the existence of a countable standard transitive model of ZF, or indeed, via the Mostowski Collapse Lemma, of a countable standard model of ZF. The existence of the latter follows by a combination of the Downward Löwenheim–Skolem Theorem and the Standard Model Hypothesis.

The set $P = \{x \in M_0 : x \subseteq \omega\}$ assuming the role of $P(\omega)$ in M_0 is not countable *in* M_0 in the sense that there does not exist a bijection $\omega \leftrightarrow P(\omega)$ in M_0, as follows from Cantor's Theorem, but P is countable from the point of view of some larger models of ZF. The fact that we are calling M_0 'countable' betrays our natural vantage point – we cannot help but view M_0 externally. The minimal model M_0 is used as an initial building block for many interesting models of ZF.

REMARK

We should perhaps stress that M_0 is minimal only among standard transitive models of ZF. If a set in M_0 is a model of ZF then it will necessarily be a non-standard one. If no set is a standard model of ZF then the constructible universe (a proper class) is the minimal standard model of ZF.

Assuming the existence of a set model of ZF, there exists an initial segment of the constructible universe which is a countable standard transitive model of ZF and which is the smallest such model. This minimal model is the initial seed from which new interesting models of ZF grow.

11.2   Limitations of inner models

> *Civilization is the process of reducing the infinite to the finite.*
> – OLIVER WENDELL HOLMES[1]

The Axiom of Constructibility is relatively consistent with ZF, but this says nothing of the *negation* of Constructibility. If the negation of Constructibility is

---

[1]Letter to Frederick Pollock, November 19, 1922.

relatively consistent with ZF, and if we are willing to accept that ZF itself is consistent, then there must exist a model of ZF which possesses a non-constructible set.

Given the success of the technique elsewhere, it is natural to ask if we can find a property ϕ such that, assuming the Standard Model Hypothesis, it is provable in ZF that $\{x : \phi\}$ is a model of ZF and the negation of Constructibility. Unfortunately, as described in Subsection 11.1.4, the Standard Model Hypothesis entails the existence of the minimal model M_0, which, as we shall see below, forces us to concede that no such ϕ exists. This is a result of J. C. Shepherdson,[1] who was the first to investigate the properties of the minimal model. Cohen only became aware of this older work after independently finding many of the results himself.

If we could prove from ZF and the Standard Model Hypothesis the existence of a predicate ϕ for which $A = \{x : \phi\}$ is a model of ZF and the negation of Constructibility then $A^{M_0} = \{x \in M_0 : \phi^{M_0}(x)\}$ too would be a model of ZF and the negation of Constructibility ($[A \text{ satisfies } \psi]^{M_0} \to [A^{M_0} \text{ satisfies } \psi]$). Since M_0 is a model of Constructibility and A^{M_0} is not, A^{M_0} is a proper submodel of M_0. But then $\mathrm{Mos}(A^{M_0})$ is a standard transitive proper submodel of M_0, contradicting minimality. The same obstacle applies when we seek models of ZF plus the negation of Choice and of ZFC plus the negation of the Generalized Continuum Hypothesis.

The method clearly has its limitations. Unable to use such means to prove the independence of Constructibility, the Axiom of Choice or the Generalized Continuum Hypothesis from ZF, the mathematician is forced to seek an alternative strategy. The technique invented by Cohen is to dilate the minimal model M_0 by introducing certain new sets and then forming the class of all sets constructible from these newcomers and M_0, an idea anticipated in our introduction of L_a. As all finite sets (and indeed ω itself) are constructible, a non-constructible set, if it exists, must be infinite.

If the class $M_0(a)$ (to be formally defined in Section 11.5) comprising all sets constructible from the minimal model and a non-constructible set a were a model of ZF, it would be one in which the Axiom of Constructibility fails. However, the existence of such a model is not quite as easy to establish as this suggests; $M_0(a)$ is not guaranteed to be a model of ZF for arbitrary a. Nevertheless it is possible to prove, via the so-called method of forcing, the existence of a non-constructible set a satisfying the requisite properties and it can be shown further, using arguments similar to those used in the much larger model L_a, that an a exists such that both the Axiom of Choice and the Generalized Continuum Hypothesis hold in $M_0(a)$, yet Constructibility does not. A *very* rough sketch of some of Cohen's original ideas is set out in next few sections.

The Axiom of Constructibility is clearly a very powerful assumption, but it has consequences which some regard as discordant with any intuitive picture of sets, and it can seem too restrictive. One of its most famous implications, the

---

[1]Shepherdson, J. C., 'Inner Models for Set Theory, Part III', *J. Symb. Logic*, **18**, 145–167, (1953).

Continuum Hypothesis, was once regarded by Cantor as 'self-evidently true' but has in modern times gone out of favour among some Platonists.

With hindsight, from the point of view of the global development of mathematics, the failure of the method of inner models to prove certain independence results was beneficial. The new techniques invented by Cohen to reach independence were far reaching and amenable to substantial generalization. The stronger the barriers that mathematics encounters, the greater the force that builds up behind them.

REMARK

If M is a model of ZF then we can form the model L^M, which is the smallest submodel of M containing all the ordinals of M. L^M is a model of Constructibility and hence is a model of the Axiom of Choice and the Generalized Continuum Hypothesis – this is the case regardless of the nature of the original model M.

The basic idea used to prove the consistency of Constructibility, the Axiom of Choice and the Generalized Continuum Hypothesis from ZF cannot be used to prove their independence. Instead a new technique called 'forcing' is used. This powerful method, invented by Paul Cohen in the early 1960s, carefully adds new elements to existing models, proving the existence of new sets that give the resulting augmented model the desired properties. One of the tools used in this work is the far-ranging notion of a Gödel numbering.

11.3   Gödel numbering

My desire and wish is that the things I start with should be so obvious that you wonder why I spend my time stating them. This is what I aim at because the point of philosophy is to start with something so simple as not to seem worth stating, and to end with something so paradoxical that no one will believe it.

– BERTRAND RUSSELL[1]

11.3.1   An example of a Gödel numbering of ZF

To express the Standard Model Hypothesis and other such metaprinciples as single sentences in ZF we employ an ingenious yet simple device: so-called *Gödel numbers* are assigned to the well-formed formulas of ZF.

The term 'Gödel numbering' is applied to any consistent numerical labelling of the formulas of a given logical theory, many variations of which can be found

---

[1]Russell [**183**].

in the literature, each one sculpted according to the context it is intended to be used in. This arithmetization of logical statements, used in combination with diagonalization, is a key component in Gödel's Incompleteness Theorems (see Appendix C); with the aid of a Gödel numbering any theory which possesses the modest resources needed to describe basic arithmetic is able to talk about itself without violating the constraints of first-order logic and without fear of running into self-referential paradoxes. Indeed, it can use semantic paradoxes as a source of inspiration, yielding profound incompleteness results.

As an example let us invent a Gödel numbering for ZF which in some ways imitates the original type of numbering employed by Gödel in his incompleteness proofs. Gödel's numbering applied to a different system, but the basic idea is the same. We first need to number the basic symbols of ZF. Although we could restrict ourselves to a smaller alphabet without any loss of generality, we'll be generous and label all the logical symbols and variables.

symbol	number
x_n	$2n \ (n \in \{0, 1, 2, \ldots\})$
$\in$	1
$\neg$	3
$\vee$	5
$\wedge$	7
$\rightarrow$	9
$\leftrightarrow$	11
$\forall$	13
$\exists$	15
(	17
)	19
[	21
]	23

This numbering is arbitrary; it has been invented only to illustrate the method. There are several different approaches available – for theoretical (not practical) reasons we choose a numbering which is easily codable and decodable and which can be described by a clearly defined terminating recursion. The numbering of Peano Arithmetic used in, say, Raymond Smullyan's *Gödel's Incompleteness Theorems* (which is a modification of a numbering due to Quine) is perhaps the most natural way of generating an injection from the class of formulas of Peano Arithmetic to the set of integers. There the language has thirteen symbols, labelled by the integers 0 through 12, and any given formula is mapped to the number in base 13 represented by the concatenation of its component symbols' labels. Thus if the symbol ϕ_i has label n_i then the formula '$\phi_m \phi_{m-1} \ldots \phi_1 \phi_0$' has Gödel number $13^m n_m + \cdots + 13^2 n_2 + 13 n_1 + n_0$. This numbering has several convenient properties (one can also exploit the fact that 13 is prime). The trick Smullyan uses to reduce the number of basic symbols to a finite set is to employ a generic variable symbol v and an auxiliary symbol $'$ so that the variables are given by the strings $v, v', v'', v''', v'''', \ldots$.

Making use of our labelling, we shall assign a number both to the formulas of ZF and to finite sequences (i.e. ordered n-tuples) of formulas of ZF. The same construction will be used to encode both objects. Following Gödel we make use of the uniqueness of prime factorization. Suppose a formula ϕ of ZF is the concatenation of symbols $s_1 s_2 \dots s_n$ and that symbol s_i has numerical label g_i, then the Gödel number of ϕ is

$$2^{g_1} \cdot 3^{g_2} \cdot 5^{g_3} \cdot 7^{g_4} \dots p_n^{g_n},$$

where p_n is the nth prime. In practice this is going to be an enormous number, however, this characteristic surfeit of magnitude is of no consequence for there is generally no need to explicitly calculate Gödel numbers. It is the fact of the existence of a useful Gödel numbering which is important, not our ability to mechanically find the Gödel number of some given formula.

Using this numbering we can encode any formula as a Gödel number and, conversely, given any Gödel number we can calculate its prime factorization and from that read off the sequence of exponents yielding the formula to which it corresponds. For finite sequences of formulas we repeat the procedure. If $(\phi_1, \dots, \phi_n)$ is an n-tuple of formulas of ZF, first calculate their Gödel numbers, yielding an n-tuple $(G_1, G_2, \dots, G_n)$ and from this form what we shall call the *Gödel sequence number* of $(\phi_1, \dots, \phi_n)$:

$$2^{G_1} \cdot 3^{G_2} \cdot 5^{G_3} \dots p_n^{G_n}.$$

Thus we can encode any finite sequence of formulas, yielding a unique Gödel sequence number, and conversely to each Gödel sequence number corresponds a unique finite sequence of formulas.

REMARK

Although a Gödel numbering typically assigns a natural number to each formula of a theory, with a little imagination one can see that alternatives might be useful. We shall refer to an ordinal-valued numbering shortly, but in general the objects needn't be numbers at all – they might be algebraic structures (say, a group-valued Gödel 'numbering') or perhaps formulas in another language.

By assigning a different number to each of the basic symbols of a theory one can associate with each well-formed formula, and with each finite sequence of well-formed formulas, another unique number. There are many ways of doing this in practice. These numbers are respectively the Gödel numbers and Gödel sequence numbers of the theory. If a theory has sufficient arithmetical resources to recursively identify certain classes of Gödel numbers and Gödel sequence numbers then it will be able to indirectly talk about its own theorems and proofs.

11.3.2   Recursively identifying types of formula

In practice the algorithms we consider below are computationally very time consuming, however, it is the fact that one can describe the algorithm that is important, we are not interested in actually carrying out the procedure. By building up a library of recursively decidable predicates we can determine, for example, whether a given Gödel number is that of a well-formed formula. The details are as follows.

Formulas are built by concatenation of symbols, so an arithmetization of this key operation should be our first undertaking. If n is the Gödel number of a string of symbols S and m is the Gödel number of a string of symbols T, then we denote by $n * m$ the Gödel number of ST, i.e. the string comprising all the symbols of S followed by all the symbols of T. One can recursively define $n * m$: if $n = 2^{a_1} 3^{a_2} 5^{a_3} \cdots p_s^{a_s}$ and $m = 2^{b_1} 3^{b_2} 5^{b_3} \cdots p_t^{b_t}$ then $n * m = 2^{a_1} 3^{a_2} 5^{a_3} \cdots p_s^{a_s} p_{s+1}^{b_1} p_{s+2}^{b_2} p_{s+3}^{b_3} \cdots p_{s+t}^{b_t}$, where p_k is the kth prime.

Note that one of the advantages of the Quine–Smullyan style of numbering is that this concatenation operation is a simple arithmetic combination of the operands. In the Gödel style of numbering concatenation is certainly describable by a recursive procedure but the recursion is not quite as elementary.

Since the variable x_i has symbol number $2i$ we can easily specify an algorithm, a terminating recursion, which returns true or false for the predicate $\mathrm{var}(n)$: 'n is the Gödel number of a variable'. This algorithm simply determines whether or not n is of the form 2^{2i}.

Next we define the predicate $\mathrm{iwff}(n)$: 'n is the Gödel number of an irreducible well-formed formula'. Making use of the predicate var and since the label of $\in$ is 1, we have $\mathrm{iwff}(n) = (\exists s < n)(\exists t < n)[\mathrm{var}(s) \wedge \mathrm{var}(t) \wedge n = s * 1 * t]$. In other words, given a natural number n we can determine, by a well-defined algorithm, whether it is the Gödel number of an irreducible well-formed formula, i.e. n is of the form $2^{2i} \cdot 3 \cdot 5^{2j}$.

We think of a well-formed formula as the end term of a finite sequence of formulas where the first formula is irreducible and where all subsequent formulas are either irreducible or obtainable from its predecessors by application of the usual rules for well-formed construction; let us call such a sequence a well-formed string. The predicate $\mathrm{wfstring}(n, m)$ means 'n is the Gödel sequence number of a well-formed string terminating in a formula with Gödel number m'. $\mathrm{wfstring}(n, m)$ is true if and only if the following holds, where $n = 2^{G_1} 3^{G_2} \cdots p_k^{G_k}$.

(i) $\mathrm{iwff}(G_1)$.

(ii) $m = G_k$.

(iii) For $i = 2$ to k, either $\mathrm{iwff}(G_i)$ or one of the following holds for some $s, t < i$:

   1. $G_i = 3 * G_s$ (negation),

   2. $G_i = G_s * 5 * G_t$ (disjunction),

3. $G_i = G_s * 7 * G_t$ (conjunction),

4. $G_i = G_s * 9 * G_t$ (implication),

5. $G_i = G_s * 11 * G_t$ (equivalence),

6. $G_i = 17 * 13 * v * 19 * G_t$ where $\mathtt{var}(v)$ (universal quantification),

7. $G_i = 17 * 15 * v * 19 * G_t$ where $\mathtt{var}(v)$ (existential quantification).

Finally we define $\mathtt{wff}(m) = (\exists n)\mathtt{wfstring}(n, m)$, '$m$ is the Gödel number of a well-formed formula'.

A recursive procedure can only claim to be an algorithm if it is guaranteed to terminate. If m is not the Gödel number of a well-formed formula then none of $\mathtt{wfstring}(1, m), \mathtt{wfstring}(2, m), \mathtt{wfstring}(3, m), \mathtt{wfstring}(4, m), \ldots$ will be true, so in this case we need a little more information which will tell us when we can stop the search and conclude $\neg\mathtt{wff}(m)$. It suffices to identify, assuming the truth of $\mathtt{wff}(m)$, an upper bound of the set $\{n : \mathtt{wfstring}(n, m)\}$ which can be recursively determined from m. Among all well-formed strings terminating in a given well-formed formula, there will be one which has the following properties:

(i) the Gödel numbers of the terms in the string increase to a maximum attained by the Gödel number of the terminal formula; and

(ii) there are no redundant terms in the string, i.e. every term in the string is a component of the terminal formula.

The number of terms in such a well-formed string cannot exceed the number of symbols of the final formula (its *length*). The length $l(m)$ of the formula with Gödel number m can be recursively determined; it is precisely the number of distinct primes in the prime factorization of m. It follows that $n \leq p_{l(m)}^m$, where $p_{l(m)}$ is the $l(m)$th prime. Of course we could do a lot better than this crude bound, and we could greatly improve the algorithm in many ways, but we stress that it is the existence of the algorithm that is important, not its elegance or efficiency. We run n from 1 through to $p_{l(m)}^m$ and either we hit an n for which $\mathtt{wfstring}(n, m)$ holds, concluding $\mathtt{wff}(m)$, or we don't, concluding $\neg\mathtt{wff}(m)$.

Having determined a procedure for identifying Gödel numbers of well-formed formulas we go on to build algorithms to recognize Gödel numbers of sentences, axioms and, one of the most powerful uses of Gödel sequence numbers, *proofs*. This, of course, includes all of the logical axiom schemas in addition to the non-logical axioms and axiom schema. For example, the first logical axiom schema is: if ϕ and ψ are well-formed formulas then $\phi \rightarrow (\psi \rightarrow \phi)$. So the predicate '$n$ is the Gödel number of an instance of Logical Axiom 1' is given by $(\exists s < n)(\exists t < n)[\mathtt{wff}(s) \wedge \mathtt{wff}(t) \wedge n = s * 9 * 17 * t * 9 * s * 19]$. Other axioms introduce more complications, asking us to identify free and bound variables.

More precisely, it is possible to define (in terms of the other recursive functions) the predicate $\mathtt{proof}(n, m)$: 'n is the Gödel sequence number of a finite sequence of formulas each of which is either an axiom or can be inferred from earlier formulas by the rules of inference of ZF, the last formula having Gödel number m', or more colloquially 'n is the Gödel sequence number of a proof of the formula with Gödel number m'.

Having 'Gödelized' the language we note that the rules of inference can now be viewed as number theoretic relations. Modus ponens can be formulated as follows: $\mathtt{modpon}(m, n, p)$ means $\mathtt{wff}(m) \wedge \mathtt{wff}(p) \wedge n = m * 9 * p$, i.e. 'the formula with Gödel number p follows by modus ponens from the formulas with Gödel numbers m and n'; and generalization $\mathtt{gen}(m, v, n)$ is $\mathtt{wff}(m) \wedge \mathtt{var}(v) \wedge n = 17 * 13 * v * 19 * m$, i.e. 'the formula with Gödel number n follows by generalization, via the variable with Gödel number v, from the formula with Gödel number m'. In determining $\mathtt{proof}(n, m)$ we would first find the prime factorization of n, the exponents yielding the sequence of Gödel numbers $G_1, G_2, \ldots, G_k$, we verify that G_1 is the Gödel number of an axiom, $G_k = m$, and for $2 \leq i < k$, G_i is either the Gödel number of an axiom or we have one of the following for some $s, t < i$: $\mathtt{modpon}(G_s, G_t, G_i)$ or $(\exists v < G_i)\mathtt{gen}(G_s, v, G_i)$.

This is a sketch. There is some benefit in filling in all the details at least once in a lifetime, or at least reading through an account which does so.

Since ZF has the available machinery to define the above recursions we begin to see how it might be able to talk about matters of the provability of its own sentences. For the purposes of our sketch of forcing, however, we need not delve very deeply into the finer points of the arithmetization of logic.[1]

REMARK

Gödel's original paper carefully constructs forty-five recursive predicates, topping these off with the non-recursive $(\exists n)\mathtt{proof}(n, m)$ asserting that m is the Gödel number of a provable formula (Gödel denoted the latter by Bew(m), for *Beweisbar* – 'provable'). Many texts (including this one) are content instead to give a reasonable set of directions, focusing on the gist of the argument rather than the minutiae of primitive recursive functions.

Once a Gödel numbering of a theory has been fixed we can, by building up a sequence of simple recursive procedures, define an explicit algorithm which decides, given natural numbers n and m, whether m is the Gödel sequence number of a proof of the sentence ϕ with Gödel number n. The statement that there exists such an m is then an encoded statement that ϕ is provable. Thus a theory which is able to describe this recursion will be able to talk about the provability of its own sentences. The Gödel numbering used in the version of forcing sketched below is ordinal-valued and is defined on a simple extension of ZF.

---

[1] However, we do make use of this material in Appendix C.

11.3.3 An extension of ZF

With Cohen's original method of forcing in mind we work with an extension of ZF. For a fixed ordinal α_0 we denote by $\mathrm{ZF}[\alpha_0]$ the theory which adjoins to ZF a collection $\{k_\alpha\}_{\alpha \in \alpha_0}$ of constants. (The α_0 we will be using shortly is the very same α_0, the minimal 'admissible' ordinal, that was described in Subsection 11.1.4, i.e. the smallest ordinal not in the minimal model, however the construction is perfectly general.) This collection of atoms labels a class of sets that is as yet indeterminate but will later be imbued with sufficient structure to form a model of ZF satisfying some given property – the negation of Constructibility, for example.

The well-formed formulas in these extended theories are defined as follows.

(i) If a and b are variables or constants then $a \in b$ is a well-formed formula. These are the *prime* formulas.

(ii) If ϕ and ψ are well-formed formulas then so are $\neg\phi$, $\phi \vee \psi$, $\phi \wedge \psi$, $\phi \rightarrow \psi$ and $\phi \leftrightarrow \psi$.

(iii) If ϕ is a well-formed formula and x_i is a free variable in ϕ then $(\forall x_i)\phi$ and $(\exists x_i)\phi$ are well-formed formulas.

(iv) A string of symbols is a well-formed formula if and only if it can be formed using the three rules above.

The definition is much the same as for ZF, except that constants are now allowed to appear in the irreducible formulas. We have also taken the opportunity to banish multiple quantification over the same variable by insisting that the quantified variable is free in the scope of the quantifier. For example, the formula $(\exists x_i)(\exists x_i)\phi$ is no longer included in our class of well-formed formulas. This has several technical advantages.

The axioms and rules of inference of $\mathrm{ZF}[\alpha_0]$ are identical to those of ZF. One can define in a fairly natural way a Gödel numbering on $\mathrm{ZF}[\alpha_0]$, such a numbering assigning an ordinal number to each formula. It is the fact of the existence of such a well-defined numbering which is important, not its specification, so we'll suppress all details of calculation throughout.[1]

Using only the resources available in $\mathrm{ZF}[\alpha_0]$ we can define the class of Gödel numbers of the axioms of $\mathrm{ZF}[\alpha_0]$, that is to say, there is a predicate **ax** such that $\mathbf{ax}(\beta)$ if and only if β is the Gödel number of an axiom of $\mathrm{ZF}[\alpha_0]$. There are some tedious technicalities in setting this up, for one we need to deal with matters of bound and free variables, and with substitution of variables, but let us take on trust that all of this has been done: there is a predicate $\mathbf{ax}(\alpha)$ *definable in* $\mathrm{ZF}[\alpha_0]$ which tells us that a given ordinal α is the Gödel number of an axiom.

Working in the first-order theory ZF (and its extensions $\mathrm{ZF}[\alpha_0]$) we cannot quantify over well-formed formulas, a fact which obstructs any naive attempt

---

[1]See Takeuti and Zaring [**207**] (Chapter 16) for an example of a numbering of this extended theory.

to formulate, say, the Axiom Schema of Replacement or the Standard Model
Hypothesis as a single sentence. However, we *can* quantify over Gödel numbers
and consequently we can legitimately form a single well-formed formula of the
form $(\forall\alpha)[\mathtt{ax}(\alpha) \to \phi(\alpha)]$, where $\phi(\alpha)$ is the well-formed formula asserting for
some set b that the well-formed formula with Gödel number α is satisfied in b.

A Gödel numbering is clearly an 'external' construction; being an injection
from the class of well-formed formulas into $\mathcal{O}n$, its definition quantifies on well-
formed formulas. The point is that it gives us a precise means of expressing, in
the language of ZF, albeit in an indirect encripted form, such statements as 'the
set b is a model of ZF' and 'this sentence is not provable'. This arithmetization
of satisfaction is our next task.

REMARK

I should stress that the type of forcing we are about to very loosely sketch is the
earliest version of the technique and that the alternative streamlined versions
which have appeared since the early 1960s are now rather different from this
slightly cumbersome looking construction (although the inspiration provided by
the old technique can still be read between the lines). I want to try to convey
something of the struggle that was involved in the original creation of these
models of ZF rather than appealing to new slick alternatives.

---

*ZF is augmented by adding a collection of constants and an ordinal-valued Gödel
numbering is defined on this extended theory. Via this numbering the extended
theory is able to recursively identify ordinal numbers which are Gödel numbers
of axioms. If we can arithmetize the notion of satisfaction then the theory will
be able to speak indirectly about its own models.*

11.4   Arithmetization

Salviati: ...Now you see how easy it is to understand.

*Sagredo: So are all truths, once they are discovered; the point is in
being able to discover them.*

– GALILEO GALILEI[1]

In this section we indicate how the sentence 'b is a model of ZF' can be 'arithmetized', that is, how it can be encoded within ZF via the tool of Gödel numbering.
In fact, without any extra effort we can obtain a more general arithmetization,
encoding 'b is a model of ZF$[\alpha_0]$' within ZF$[\alpha_0]$. The set a which is to determine

---

[1] Galilei [**76**] (The Second Day) (translation by Stillman Drake). This dialogue is sometimes
paraphrased as 'All truths are easy to understand once they are discovered; the point is to
discover them'.

the membership connections between the constants k_α of ZF$[\alpha_0]$ is treated here as a generic set (forcing, to be roughly outlined soon, proves the existence of a set a which ensures that the class $\{\mathcal{G}_a(\alpha) : \alpha < \alpha_0\}$ models certain target properties).

Let us assume that a Gödel numbering of ZF$[\alpha_0]$ has been fixed which assigns to each well-formed formula ϕ an ordinal Gdl(ϕ) so that we can express, in the language of ZF$[\alpha_0]$, whether a given ordinal α is the Gödel number of various types of well-formed formula, in particular we have the means to express the predicate $\mathbf{ax}(\alpha)$: 'α is the Gödel number of an axiom of ZF$[\alpha_0]$'.

For a set b we regard a sequence $f : \omega \to b$ as an assignment of sets in b to the variables $x_0, x_1, x_2, \ldots$. Interpreting each free x_i as the set $f(i)$ and k_α as the αth a-constructible set $\mathcal{G}_a(\alpha)$, the following table gives us enough information, once a has been specified, to recursively determine whether b f-satisfies Gdl(ψ), i.e. we first identify which of the seven classes our Gödel number belongs to (identified in the left column), reducing the satisfaction to the corresponding condition(s) described in the right column, and after a finite number of steps the condition is resolved either positively or negatively.

Gdl(ψ)	b f-satisfies Gdl(ψ) if and only if
Gdl$(x_i \in x_j)$	$f(i) \in f(j)$
Gdl$(x_i \in k_\alpha)$	$f(i) \in \mathcal{G}_a(\alpha)$
Gdl$(k_\alpha \in x_i)$	$\mathcal{G}_a(\alpha) \in f(i)$
Gdl$(k_\alpha \in k_\beta)$	$\mathcal{G}_a(\alpha) \in \mathcal{G}_a(\beta)$
Gdl$((\exists x_i)\phi)$	there is a $g : \omega \to b$ with $f(j) = g(j)$, for all $j \neq i$, such that b g-satisfies Gdl(ϕ)
Gdl$(\neg\phi)$	b does not f-satisfy Gdl(ϕ)
Gdl$(\phi \wedge \eta)$	b f-satisfies Gdl(ϕ) and b f-satisfies Gdl(η)

This simple recursion is definable within ZF$[\alpha_0]$. Denote by $\mathbf{sat}(f, b, \alpha)$ the predicate 'α is the Gödel number of a well-formed formula and b f-satisfies α'. We say b satisfies Gdl(ψ), $\mathbf{Sat}(b, \text{Gdl}(\psi))$, if there exists an assignment f such that b f-satisfies Gdl(ψ), i.e. $(\exists f)[f : \omega \to b \wedge \mathbf{sat}(f, b, \text{Gdl}(\psi))]$. If ψ is a sentence then $\mathbf{Sat}(b, \text{Gdl}(\psi))$ if and only if ψ^b, in other words we have arithmetized the restriction of ψ to b. This retrospectively motivates what is perhaps the only mysterious entry in the table, the condition for 'b f-satisfies Gdl$((\exists x_i)\phi)$'. If we were being very formal we would have to set up $\mathbf{Sat}$ more carefully based on an induction on degree (see Subsection 11.5.2).

The statement 'b is a model of ZF$[\alpha_0]$' is now expressible in the language of ZF$[\alpha_0]$ as $(\forall \alpha)[\mathbf{ax}(\alpha) \to \mathbf{Sat}(b, \alpha)]$.

REMARK

If ZF is consistent then it has a model (indeed it has a model of any given cardinality $\geq \aleph_0$), and vice versa, so an arithmetization of 'b is a model of ZF' is just an arithmetization of 'ZF is consistent' – it is, assuming consistency, a true but unprovable statement in ZF.

Interpreting the constants of the extended theory $ZF[\alpha_0]$ as the first α_0 a-constructible sets and considering assignments of sets in a set b to the variables of the theory we can arithmetize the statement 'b is a model of $ZF[\alpha_0]$' so that $ZF[\alpha_0]$ can indirectly talk about its own models. Working in this extended theory one can prove, by the method of forcing, the existence of a set a (or a collection of sets) for which this initial part of the a-constructible universe is a model of ZF together with some desired properties.

11.5   A sketch of forcing

Set theory could never be the same after Cohen, and there is simply no comparison whatsoever in the sophistication of our knowledge about models of set theory today as contrasted to the pre-Cohen era.

– Dana S. Scott[1]

11.5.1   The strategy

The independence of such set theoretic postulates as Constructibility, the Axiom of Choice or the Continuum Hypothesis, if provable, must be established by some novel method quite different from the technique of inner models which was employed to prove their consistency with ZF. At the heart of Cohen's general approach is the absorption into a ground model M (the minimal model M_0, say) of a collection G of new so-called generic sets, these additions crafted so that the augmented model $M(G)$ has certain target properties. The technique used to prove the existence of models by this process of careful extension is known as 'forcing' and is widely applicable.

Models in which the Axiom of Choice holds but the Continuum Hypothesis fails are obtained by adding to the base model M_0 a set K of distinct subsets a_β of ω, $\beta < \alpha$, the extension controlled in such a way that $\alpha \geq \aleph_2$ in $M_0(K)$. In order to construct a model in which the Axiom of Choice fails, infinitely many sets are added to M_0 so that, relative to any set in the augmented model, it is only possible to distinguish between finitely many of the new sets. The new sets form a set G and for any set a in the augmented model $M_0(G)$ of ZF we

---

[1]In Bell [15].

have $[g = h]^a$ for all but finitely many pairs of sets $g, h \in G$. G is an infinite set in $M_0(G)$ but no subset of G is countable in $M_0(G)$, contradicting the Axiom of Choice.

We will focus on one of the simplest applications of forcing, the problem of the independence of Constructibility, where only one new subset of ω is absorbed into the base model M_0. As M_0 is countable there is an abundance of subsets of ω to choose from; the difficulty lies in proving that at least one of these has the right characteristics.

From here on α_0 is the smallest ordinal not in M_0. For $a \subseteq \omega$ we define $M_0(a)$ to be the initial part $\{\mathcal{G}_a(\alpha) : \alpha < \alpha_0\}$ of the a-constructible universe L_a introduced in Subsection 11.1.2. $M_0(a)$ is a 'thin' extension of M_0; it does not introduce any new ordinals, i.e. the class of ordinals in $M_0(a)$ is equal to α_0. This inadmission of new ordinals is one of the key controlling features of forcing.

$M_0(a)$ is non-empty and transitive and is closed under the fundamental operations, but it need not be a model of ZF (it cannot be almost universal because it doesn't contain all ordinals). Sufficient additional properties which ensure that $M_0(a)$ is a model of ZF are that it is a model of the Axiom of Powers and of Replacement, the two 'large set' axioms.

Working in the extension ZF[α_0], the constants k_α interpreted as the sets $\mathcal{G}_a(\alpha)$ comprising $M_0(a)$, it is proved that a non-constructible a exists such that $M_0(a)$ is a model of ZF. Together with the example of Gödel's constructible universe, this establishes the independence of the Axiom of Constructibility from ZF.

The method used to label the elements of $M_0(a)$ is not important. Cohen draws an analogy with the process of extending a field F by adjoining a root α of an irreducible polynomial f over F. All elements of the extended field are of the form $p(\alpha)$, where p is a polynomial, and the only restriction imposed is that $p(\alpha) = q(\alpha)$ if and only if the difference $p(\alpha) - q(\alpha)$ is divisible by f. α is just a convenient formal marker that facilitates the algebraic manipulations; there are several alternative concrete descriptions of the extended field.[1]

What must a look like? Given a well-formed formula ϕ of ZF, the statement '$M_0(a)$ satisfies Gdl(ϕ)' reduces by the recursion described in the previous section to a collection of membership connections of the form $\mathcal{G}_a(\alpha) \in \mathcal{G}_a(\beta)$ or $\mathcal{G}_a(\alpha) \notin \mathcal{G}_a(\beta)$, $\alpha, \beta < \alpha_0$, and so, by enforcing these membership connections, we impose restrictions on the set a itself. Put another way, the set a that we seek is sculpted by the properties that we wish $M_0(a)$ to satisfy.

On the face of it, it is remarkable that forcing can be made to work at all; there seems to be no way to begin the construction – a completely determines the model $M_0(a)$, yet at the same time by insisting that $M_0(a)$ satisfies some collection of properties we impose restrictions on a. Where do we start?[2]

---

[1] See, for example, our extension of the field $\mathbb{Q}$ by the root $\sqrt{2}$ of the irreducible polynomial $x^2 - 2$ in Subsection 1.6.2.

[2] Cohen describes the conceptual difficulties that he managed to tame in April 1963 in his interesting article 'The discovery of forcing', (*Rocky Mountain Journal of Mathematics*, **32**, no. 4, 1071–1100, (2002)).

REMARKS

1. In attempting to prove the independence of the Axiom of Choice and the Continuum Hypothesis from ZF, Cohen made an early decision to seek *standard* models. Given how unnatural the negation of the Axiom of Choice and (arguably) the negation of the Continuum Hypothesis are, Cohen's choice might seem to be unusually conservative, and perhaps self-defeating. One might expect that models of unnatural statements may have to be non-standard. In fact, the limitation actually helped to suggest possibilities for an attack on the problem. Assuming the standard model hypothesis, one cannot prove the existence of an *uncountable* standard model in which Choice is true and in which the Continuum Hypothesis is false, nor can one find a model in which Choice is true and which contains non-constructible real numbers. This therefore forced Cohen to consider countable models. Nevertheless, we should stress that there is a healthy liberation in considering non-standard models, if only to gain an understanding of how varied the models of ZF can be, and to appreciate that we ought not to marry ourselves too readily to the usual interpretation of the symbol '$\in$'.

2. It is instructive to give an example of a subset a of ω such that $M_0(a)$ is *not* a model of ZF. The ordinal α_0 is countable, so it corresponds to a well-ordering of ω. That is, α_0 corresponds to a subset of $\omega \times \omega$. But $\omega \times \omega$ is equipollent to ω and so, via such a bijection, α_0 corresponds to a subset a of ω. Now if $M_0(a)$ were a model of ZF then it would include these correspondences and so it would tell us that the ordinal α_0 is in $M_0(a)$, a contradiction. In particular this tells us that our choice of a cannot tell us any information about M_0 which is strictly 'external', such as the countability of α_0; the existence of ordinals beyond α_0 must be unattainable in $M_0(a)$. The set a must therefore behave like a typical (or 'generic') set in M_0. An example of a property which cannot give such external information is '$n \in a$' or '$n \notin a$', where $n \in \omega$. Collecting together a consistent finite number of such statements we form the notion of a forcing condition (see Subsection 11.5.3), and one can ask which statements are forced to be true by such a forcing condition.

3. In his excellent article *A Beginner's Guide to Forcing*,[1] Timothy Chow accurately describes forcing as an 'open exposition problem'. It seems that no matter how one approaches forcing there is always a point where some idea or structure is introduced as if taken out of a hat – one can see that it works and that it is logically sound, but not where the idea came from. The big idea is very difficult to reduce to a series of small digestible steps.

The task of constructing a model $M_0(a)$ of ZF comprising an initial part of the a-constructible universe for some set a poses dual problems. The set a

[1] *Contemporary Mathematics*, **479**, 25–40, (2009).

determines the set $M_0(a)$, but by insisting that the set $M_0(a)$ is a model of ZF and certain other properties we in turn place constraints on the set a. Some technical ingenuity is required to control the simultaneous construction of these mutually influential elements.

11.5.2   A partial order on sentences

Forcing is defined recursively via a natural partial order $\prec$ on the class of Gödel numbers of sentences of $\mathrm{ZF}[\alpha_0]$. The idea is that $\mathrm{Gdl}(\phi) \prec \mathrm{Gdl}(\psi)$ if ϕ is 'less complex' than ψ so that one can determine whether a formula is 'forced' by appealing to ever more simple subformulas. This measure of complexity is set up by isolating certain properties of well-formed formulas and using them to define an order as follows.

Limited quantification It is convenient to add to the theory a new set of symbols, the limited quantifiers $\exists^\beta$, where $0 < \beta < \alpha_0$. The formula $(\exists^\beta x_i)\phi$ is an abbreviation for $(\exists x_i)(\exists \alpha < \beta)[x_i = k_\alpha \wedge \phi(k_\alpha)]$ (which is itself a substantial abbreviation). Here $\phi(k_\alpha)$ is the formula obtained by replacing each occurrence of x_i (all necessarily free) in ϕ with k_α. It makes no material difference to the theory if we accommodate this new abbreviation in an extra rule of well-formed construction: if ϕ is a well-formed formula, $\beta < \alpha_0$ and x_i is free in ϕ, then $(\exists^\beta x_i)\phi$ is a well-formed formula (we could also add $\forall^\beta$ to the theory, with the obvious interpretation, however we will not need this here). Limited quantifiers are given their own Gödel numbers in a way which does not intrude on the existing numbering (so each Gödel number still determines a unique formula). A Gödel number $\mathrm{Gdl}(\phi)$ is *limited* if all the quantifiers in ϕ are limited, otherwise $\mathrm{Gdl}(\phi)$ is *unlimited*.

Degree Each well-formed formula ϕ is the terminal formula of a string of subformulas of ϕ where each term in the string is either a prime formula or obtainable from earlier formulas using the rules of well-formed construction. There may be several different strings which terminate in ϕ, some longer than others – the length of the *longest* such string is the *degree* of $\mathrm{Gdl}(\phi)$.

Type The *type* of a limited $\mathrm{Gdl}(\phi)$, where ϕ possesses limited quantifiers $\exists^{\beta_1}, \exists^{\beta_2}, \ldots$ and constants $k_{\alpha_1}, k_{\alpha_2}, \ldots$, is the maximum of the finite set

$$\{\beta_1, \beta_2, \ldots, \alpha_1 + 1, \alpha_2 + 1, \ldots\}.$$

Parity A limited $\mathrm{Gdl}(\phi)$ has *parity* 0 if β is less than the type of $\mathrm{Gdl}(\phi)$ whenever $(\exists^\beta x_i)\psi$ is a subformula of ϕ and $\alpha + 1$ is less than the type of $\mathrm{Gdl}(\phi)$ whenever $k_\alpha \in a$ is a subformula of ϕ, where a is a constant or variable; otherwise $\mathrm{Gdl}(\phi)$ has parity 1. To clarify: in a parity 0 formula of type α the α only appears in a subformula of the form $a \in k_\alpha$ where a is either a variable or a constant k_β with $\beta < \alpha$; it does not appear in a limited quantifier $\exists^\alpha$ and it does not appear in a subformula of the form $k_\alpha \in a$.

We then define $\alpha \prec \beta$ if:

  (i) α, β are unlimited and the degree of α is strictly less than the degree of β;

 (ii) α is limited and β is unlimited; or

(iii) α, β are both limited and:

  (1) the type of α is strictly less than the type of β; or

  (2) α and β have the same type, α has parity 0 and β has parity 1; or

  (3) α and β have equal parity and type and the degree of α is strictly less than the degree of β.

REMARK

$\prec$ is a well-founded relation (as is easily seen) and so it is sufficiently structured to allow induction.

A partial ordering of the (Gödel numbers of) sentences of ZF[α_0] is described which arranges the sentences in order of complexity. Forcing is defined recursively on this partially ordered set.

11.5.3 Forcing conditions

A *forcing condition* (in the present simple context) is a disjoint ordered pair of finite subsets of ω. Alternatively, one can think of a forcing condition (P, Q) as a pair of finite collections of statements $P = \{n_1 \in a, n_2 \in a, \ldots\}$ and $Q = \{n_t \notin a, n_{t+1} \notin a, \ldots\}$, but nothing is lost by replacing these sets with the bare essential information they contain, i.e. the sets of integers $\{n_1, n_2, \ldots\}$ and $\{n_t, n_{t+1}, \ldots\}$. That P and Q must be disjoint simply reflects our desire for consistency, an integer cannot be both in and not in a.

The set of all forcing conditions is partially ordered by defining $(P_1, Q_1) \sqsubseteq (P_2, Q_2)$ if $P_1 \subseteq P_2$ and $Q_1 \subseteq Q_2$. We detail below a set of conditions which determine whether or not a forcing condition $\mathbf{P} = (P, Q)$ *forces* Gdl(ϕ), where ϕ is a sentence in ZF[α_0]. Keeping in mind the interpretation of k_α as $\mathcal{G}_a(\alpha)$, the variables x_i ranging over the set of all constants $\{k_\alpha : \alpha < \alpha_0\}$ that is to form a model of ZF, P a set of n for which $n \in a$, Q a set of n for which $n \notin a$, and remembering that $\mathcal{G}_a(\omega) = \omega$ and $\mathcal{G}_a(\omega + 1) = a$, the conditions are fairly natural. The negation property (property 2 below), for example, ensures that as soon as $\mathbf{P}$ fails to force Gdl(ϕ) (i.e. either P contains an integer not in a or Q contains an integer that is in a) then all larger $\mathbf{P}$ also fail to force Gdl(ϕ) (the 'illegal' member still being present in the larger forcing condition).

1. $\mathbf{P}$ forces $\mathrm{Gdl}((\exists x_i)\phi(x_i))$ if there exists a $\beta < \alpha_0$ such that $\mathbf{P}$ forces $\mathrm{Gdl}(\phi(k_\beta))$.

2. $\mathbf{P}$ forces $\mathrm{Gdl}(\neg\phi)$ if, for all $\mathbf{Q} \sqsupseteq \mathbf{P}$, $\mathbf{Q}$ does not force $\mathrm{Gdl}(\phi)$.

3. $\mathbf{P}$ forces $\mathrm{Gdl}(\phi \wedge \psi)$ if $\mathbf{P}$ forces $\mathrm{Gdl}(\phi)$ and $\mathbf{P}$ forces $\mathrm{Gdl}(\psi)$.

4. $\mathbf{P}$ forces $\mathrm{Gdl}((\exists^\beta x_i)\phi(x_i))$ if there exists an $\alpha < \beta$ such that $\mathbf{P}$ forces $\mathrm{Gdl}(\phi(k_\alpha))$.

5. If $\beta < \alpha$ then $\mathbf{P}$ forces $\mathrm{Gdl}(k_\alpha \in k_\beta)$ if $\mathbf{P}$ forces $\mathrm{Gdl}((\exists^\beta x_i)[k_\alpha = x_i \wedge x_i \in k_\beta])$.

6. Suppose $\alpha < \beta$, $\omega + 1 < \beta$ and $\Phi(\gamma, \delta, n) = \beta$. Our principal interpretation leads us to the following reductions of '$\mathbf{P}$ forces $\mathrm{Gdl}(k_\alpha \in k_\beta)$'. Since $\alpha < \beta$ the case $n = 0$ is settled immediately: $\mathbf{P}$ forces $\mathrm{Gdl}(k_\alpha \in k_\beta)$. For $n \neq 0$ we have '$\mathbf{P}$ forces $\mathrm{Gdl}(k_\alpha \in F_n(k_\gamma, k_\delta))$', where F_n is the nth fundamental operation and '$k_\alpha \in F_n(k_\gamma, k_\delta)$' is naturally interpreted as follows:

n	'$k_\alpha \in F_n(k_\gamma, k_\delta)$'
1	$k_\alpha = k_\gamma \vee k_\alpha = k_\delta$
2	$(\exists^\beta x_i)(\exists^\beta x_j)[k_\gamma = (x_i, x_j) \wedge x_i \in x_j \wedge k_\alpha \in k_\gamma]$
3	$k_\alpha \in k_\gamma \wedge k_\alpha \notin k_\delta$
4	$(\exists^\beta x_i)(\exists^\beta x_j)[k_\alpha = (x_i, x_j) \wedge k_\alpha \in k_\gamma \wedge x_i \in k_\delta]$
5	$k_\alpha \in k_\delta \wedge (\exists^\beta x_j)[(k_\alpha, x_j) \in k_\delta]$
6	$(\exists^\beta x_j)(\exists^\beta x_i)[k_\alpha \in k_\gamma \wedge k_\alpha = (x_j, x_i) \wedge (x_i, x_j) \in k_\delta]$
7	$(\exists^\beta x_i)(\exists^\beta x_j)(\exists^\beta x_k)[k_\alpha \in k_\gamma \wedge k_\alpha = (x_i, x_j, x_k) \wedge (x_k, x_i, x_j) \in k_\delta]$
8	$(\exists^\beta x_i)(\exists^\beta x_j)(\exists^\beta x_k)[k_\alpha \in k_\gamma \wedge k_\alpha = (x_i, x_j, x_k) \wedge (x_i, x_k, x_j) \in k_\delta]$

7. If $\alpha \in P$ then (P, Q) forces $\mathrm{Gdl}(k_\alpha \in k_{\omega+1})$.

8. If $\alpha \in Q$ then (P, Q) does not force $\mathrm{Gdl}(k_\alpha \in k_{\omega+1})$.

9. $\mathbf{P}$ does not force $\mathrm{Gdl}(k_\omega \in k_{\omega+1})$.

10. For all α, $\mathbf{P}$ does not force $\mathrm{Gdl}(k_\alpha \in k_\alpha)$.

11. If $\alpha, \beta \leq \omega$ then $\mathbf{P}$ forces $\mathrm{Gdl}(k_\alpha \in k_\beta)$ if and only if $\alpha < \beta$.

The equalities appearing in conditions 5 and 6 are, of course, abbreviations for a sentence which includes an unlimited quantifier. As each application of the rules is supposed to reduce the complexity of the formulas in question (in the sense of $\prec$) these hidden quantifiers ought to be limited. This is easy enough to remedy: the equality in 5, for example, is in the scope of the limited quantifier $\exists^\beta$, and $\beta < \alpha$, so nothing is lost by replacing the hidden $\exists$ by $\exists^\alpha$. All equalities appearing in subsequent limited cases are treated in the same fashion by replacing each hidden quantified expression $(\exists x)\psi$ by the limited $(\exists^\delta x)\psi$ where δ is no smaller than the maximum of all ordinals appearing in ψ. Components involving membership relations between constants and ordered pairs or triples must similarly be unravelled with care.

By successively applying these rules the statement 'P forces $\mathrm{Gdl}(\phi)$' is reduced to the consideration of statements of the form 'P forces $\mathrm{Gdl}(\psi)$' with $\mathrm{Gdl}(\psi) \prec \mathrm{Gdl}(\phi)$, each component – possibly in the scope of some (limited) quantifier – eventually decided either positively (satisfying one of case $n = 0$ of 6 above, case 7 or, with $\alpha < \beta$, case 11) or negatively (cases 8, 9, 10 or, with $\alpha > \beta$, case 11).

A concrete example is needed. Let us choose an arbitrary sentence, say $(\forall x_0)[x_0 \in k_0]$, and unravel the statement P forces $\mathrm{Gdl}((\forall x_0)[x_0 \in k_0])$. First we need to write the sentence in a form to which our rules apply, i.e. as $\neg(\exists x_0)\neg[x_0 \in k_0]$. We will use the abbreviation $\mathbf{PF}\phi$ for 'P forces $\mathrm{Gdl}(\phi)$'.

If $\mathbf{PF}\neg(\exists x_0)\neg[x_0 \in k_0]$, then by rule 2, $(\forall \mathbf{Q} \sqsupseteq \mathbf{P})\neg[\mathbf{QF}(\exists x_0)\neg[x_0 \in k_0]]$. By rule 1 we have $(\forall \mathbf{Q} \sqsupseteq \mathbf{P})\neg[(\exists \beta < \alpha_0)\mathbf{QF}\neg[k_\beta \in k_0]]$ and, using 2 again, $(\forall \mathbf{Q} \sqsupseteq \mathbf{P})\neg[(\exists \beta < \alpha_0)\neg[\mathbf{QF}k_\beta \in k_0]]$. In other words, $(\forall \mathbf{Q} \sqsupseteq \mathbf{P})(\forall \beta < \alpha_0)\mathbf{QF}k_\beta \in k_0$, but if we take $\beta = 0$, we have $\mathbf{QF}k_0 \in k_0$, contradicting rule 10. So for no forcing condition P is this sentence forced. This is as we would expect; the sentence proposes that all sets are elements of the set labelled k_0, i.e. $\varnothing$.

The key result in forcing is:

> *For every sentence ϕ and every forcing condition P there exists a forcing condition $\mathbf{Q} \sqsupseteq \mathbf{P}$ such that $\mathbf{Q}$ forces $\mathrm{Gdl}(\phi)$ or $\mathbf{Q}$ forces $\mathrm{Gdl}(\neg\phi)$.*

This follows immediately from the negation condition alone, for if P does not force $\mathrm{Gdl}(\neg\phi)$ then there must exist a $\mathbf{Q} \sqsupseteq \mathbf{P}$ such that $\mathbf{Q}$ forces $\mathrm{Gdl}(\phi)$. From this it is easy to construct a *complete sequence* of forcing conditions, this being a chain $(\mathbf{P}_n)$ of forcing conditions ($\mathbf{P}_n \sqsubseteq \mathbf{P}_{n+1}$ for all n) such that for every sentence ϕ there exists an n for which either $\mathbf{P}_n$ forces $\mathrm{Gdl}(\phi)$ or $\mathbf{P}_n$ forces $\mathrm{Gdl}(\neg\phi)$. Fix an enumeration $(\mathrm{Gdl}(\phi_n))$ of the class of Gödel numbers of sentences of $\mathrm{ZF}[\alpha_0]$ (here we are making use of the fact that α_0 is a countable ordinal) and an enumeration $(\mathbf{P}_n)$ of all forcing conditions. A complete sequence of forcing conditions $(\mathbf{C}_n)$ is generated by setting $\mathbf{C}_0 = (\varnothing, \varnothing)$ and defining $\mathbf{C}_{n+1}$ to be the first $\mathbf{P}_k$ in our enumeration with $\mathbf{C}_n \sqsubseteq \mathbf{P}_k$ ($\mathbf{C}_n \neq \mathbf{P}_k$) and which has the property that $\mathbf{P}_k$ forces $\mathrm{Gdl}(\phi_n)$ or $\mathbf{P}_k$ forces $\mathrm{Gdl}(\neg\phi_n)$.

If $\mathbf{P}_n = (P_n, Q_n)$ and $(\mathbf{P}_n)$ is a complete sequence of forcing conditions then, the union $\cup P_n$ is a *forcing set*. A sentence ϕ is satisfied in $M_0(a)$ if and only if $\mathbf{P}_n$ forces ϕ for some n.

The remaining part of the proof of the independence of Constructibility comprises two technical arguments. First, it is shown that if a is a forcing set then $M_0(a)$ satisfies the Axiom of Powers and each instance of the Axiom Schema of Replacement, completing the requirements for $M_0(a)$ to be a standard transitive model of ZF. Secondly it is shown via an enumeration of the sentences of $\mathrm{ZF}[\alpha_0]$, α_0 and the set of forcing conditions that there exists a forcing set a not in M_0; a is not constructible in $M_0(a)$.

REMARK

By the same argument used for L_a, in addition to being a standard transitive model of ZF and the negation of Constructibility, $M_0(a)$ is also a model of Choice and the Generalized Continuum Hypothesis.

A forcing set a is built as the union of a sequence of finite sets designed to ensure that $M_0(a)$ is a model of ZF. It is shown that there exists a non-constructible forcing set a. For this a, $M_0(a)$ is a model of ZF, Choice, the Generalized Continuum Hypothesis and the negation of Constructibility. Together with Gödel's constructible universe this proves the independence of Constructibility from ZF. Cohen's original method, sketched here, has since developed into a wide-ranging technique.

11.6   The evolution of forcing

> *The mathematician's patterns, like the painter's or the poet's must be beautiful; the ideas like the colours or the words, must fit together in a harmonious way. Beauty is the first test: there is no permanent place in the world for ugly mathematics.*
>
> – G. H. HARDY[1]

Some independence results had been obtained before Cohen, albeit in a more restricted setting. In 1922 Fraenkel proved the independence of Choice from certain theories that admitted atoms.[2] These techniques were further developed by others[3] but could not be directly extended to ZF. Cohen was the first to prove independence theorems in the atomless ZF.

The technique of forcing has been significantly refined and extended and is now one of the basic tools used by modern set theorists. The new methods are much slicker than the so-called 'ramified forcing' that has just been sketched, but it is perhaps easier to see in the early stages of ramified forcing where the motivation comes from, even if the technical machinery needed to make the basic idea work is difficult. There seems, in any case, to be a healthy ebb and flow in the literature between a preference for an abstract approach on one hand and something closer to the original forcing on the other.

Solomon Feferman[4] was among the pioneers who obtained new results with forcing and who developed variants of the technique. These early days also

---

[1] Hardy [93].

[2] Fraenkel, A. A., 'Über den Begriff "definit" und die Unabhängigkeit des Auswahlaxioms', *Sitzungsber. Preuss. Akad. Wiss., Phys.-math. Kl.*, 253–257, (1922).

[3] Mostowski, A., 'Axiom of choice for finite sets', *Fund. Math.*, **33**, 137–68, (1945), for example.

[4] See, for example, Feferman, S., 'Some applications of the notions of forcing and generic sets', *Fundamenta Mathematicae*, **56**, 325–345, (1965).

saw William Easton's[1] treatment of forcing come to fruition in his results on powers of regular cardinals. Feferman and Azriel Levy[2] exploited the method to collapse cardinals, i.e. to add a generic set constituting a bijection from a cardinal to another cardinal which is strictly smaller in the ground model, ultimately arriving at the Feferman–Levy model – a model in which the continuum is a countable union of countable sets. Further developments were presented by Robert Solovay and Dana Scott, independently, who realized that the use of the constructible universe was unnecessary and interpreted forcing instead in terms of Boolean algebras. At the same time Petr Vopěnka[3] developed a theory of forcing using open sets in a topological space as forcing conditions, an approach which eventually converged to a Boolean-valued version of forcing more or less identical to the Scott–Solovay version of the same year. Joseph Shoenfield[4] was to further clarify 'unramified forcing'.

Iteration of forcing was first used by Solovay and Stanley Tennenbaum[5] in their proof of the independence of Souslin's Hypothesis from ZF. Martin went on to observe that many properties of models obtained by iteration of forcing follow from a single axiom, namely Martin's Axiom (see Subsection 9.3.5), the relative consistency of which is also proved by iterated forcing.

Cohen's proof of the independence of the Continuum Hypothesis and the Axiom of Choice from ZF rank amongst the most important developments in the foundations of mathematics. Cohen's methods, since generalized, have unlocked a large number of additional important results in set theory. The independence theorems uncovered by these techniques exhibit in startling clarity the intuitive incompleteness of the standard axioms of set theory. This gap between the wildly underdetermined reality of set theory and the completionist hopes of the early set theorists is much larger than any of them could have feared. Future developments in set theory, perhaps viewed through the broader lens of topos theory, may lead to new axiomatizations of set theory which surpass ZF in elegance, intuitive plausibility and power, but I suspect the 'multiverse' view of set theory[6] will increasingly be embraced, each theory being just one of infinitely many alternatives in a larger ensemble (independent statements, proved to be so by forcing, providing the passage from one universe to its augmented partner).[7]

---

[1] The abridged PhD. thesis eventually published as 'Powers of regular cardinals', *Annals of Mathematical Logic*, **1**, (1970).

[2] Feferman, S. and Levy, A., 'Independence results in set theory by Cohen's method', *Notices of the American Mathematical Society*, **10**, 593, (1963).

[3] Vopěnka, P., 'The general theory of ∇-models', *Commentationes Mathematicae Universitatis Carolinae*, **8**, 145–170, (1967).

[4] Shoenfield, J. R., 'Unramified forcing', *Axiomatic Set Theory* (Dana S. Scott, ed.), *Proceedings of Symposia in Pure Mathematics*, vol. 13, American Mathematical Society, Providence, 357–381, (1971).

[5] See Solovay, R. M. and Tennenbaum, S., 'Iterated Cohen extensions and Souslin's problem', *Annals of Mathematics*, **94**, 201–245, (1971).

[6] See Hamkins, J. D., 'The set-theoretic multiverse', *Review of Symbolic Logic*, **5**, 416–449, (2012).

[7] For a very readable account of the development and influence of forcing see Akihiro Kanamori's article 'Cohen and set theory', *Bulletin of Symbolic Logic*, **14**, 3, Sept. (2008).

Remarks

1. Fraenkel's proof of the independence of Choice in an atomic theory is critically tied to the fact that atoms cannot be distinguished from one another by pure set theoretic statements (he produced models formed by permuting sets of atoms and proved that such sets needn't have choice functions).

2. The prototypical example of a Boolean algebra is $P(X)$ for a non-empty set X with addition symmetric difference Δ (i.e. $A\Delta B = (A-B)\cup(B-A)$) and multiplication $\cap$; the general definition being an abstraction of this structure.

   The use of Boolean algebra in forcing is akin to multi-valued logic. Roughly speaking, one defines a Boolean algebra-valued function on the set of sentences of ZF (the language augmented by some constant terms) which assigns the maximal element '1' to true sentences (i.e. sentences we want to be true, such as the axioms and any other target properties), the minimal element '0' to false sentences, and intermediate Boolean values to the remaining sentences in a way that is compatible with the logical operations. The notion of a Boolean-valued set is defined recursively: the empty set is a Boolean-valued set and given any set of Boolean-valued sets a function from that set to the Boolean algebra is another Boolean-valued set. Given a model M of ZF one defines the corresponding Boolean-valued model to be the set of Boolean-valued sets in M, and to the language of ZF one adds a constant for each such Boolean-valued set. With an appropriate choice of Boolean algebra and notion of satisfaction a model of ZF is then formed by a natural quotient construction on this collection of Boolean-valued sets.[1]

3. Here we are forced to stop. One can see how forcing was an explosive event in set theory and how sophisticated mathematical ideas enter into the frame, using and interpreting forcing in different ways. A detailed account of these topics is far beyond the scope of this book – and even to dare to try to describe forcing in the present context was already pushing the boundaries of what was feasible!

The version of forcing we have just sketched is Cohen's original 'ramified forcing' applied to one of the simplest cases. Almost as soon as Cohen shared his proof various improvements and modifications started to appear. Forcing is now a mature subject, providing the set theorist with a means of exploring independence, opening portals in the set theoretic multiverse.

---

[1]For an account of the Boolean approach to forcing, and much more, see Jech [**108**] and Bell [**15**]. For a short summary of the technique seek out Timothy Chow's article mentioned in the remarks to Subsection 11.5.1, which provided the raw material for this remark.

Appendix A

Peano Arithmetic

A.1 The axioms

Peano Arithmetic is a first-order theory with four primitive operators 0 (constant), s (unary), $+$ and $\cdot$ (both binary) and one primitive binary relation $=$. There are countably many variables. In the following the variables will be denoted by a, b, c and x.

A *term* is defined recursively as follows: (i) 0 is a term; (ii) every variable is a term; (iii) if t_1 and t_2 are terms then st_1, $t_1 + t_2$ and $t_1 \cdot t_2$ are terms.

0 is a *numeral*, and if n is a numeral so is sn. Every numeral is clearly a term. Intuitively the numerals collectively form the set of natural numbers.

A *formula* is defined recursively as follows: (i) if t_1 and t_2 are terms then $t_1 = t_2$ is a formula; (ii) if f_1 and f_2 are formulas then $\neg f_1$, $f_1 \wedge f_2$, $f_1 \vee f_2$, $f_1 \rightarrow f_2$ and $f_1 \leftrightarrow f_2$ are formulas; (iii) if a is a variable and f is a formula then $(\forall a)f$ and $(\exists a)f$ are formulas.

Bound and free variables are defined in the usual way. A formula with no free variables is a sentence, otherwise it is a predicate.

The relation $=$ is interpreted as the relation of identity, i.e. an equivalence relation with substitutivity.

(E1) $(\forall a)[a = a]$.
(E2) $(\forall a)(\forall b)[a = b \rightarrow b = a]$.
(E3) $(\forall a)(\forall b)(\forall c)[a = b \wedge b = c \rightarrow a = c]$.
(E4) If $\phi(x)$ is a predicate then $(\forall a)(\forall b)[a = b \rightarrow [\phi(a) \leftrightarrow \phi(b)]]$.

The operator s is governed by the two axioms

(S1) $(\forall a)(\forall b)[sa = sb \rightarrow a = b]$,
(S2) $\neg(\exists a)[sa = 0]$.

The operators $+$ and $\cdot$ are governed by the following axioms.

411

(A1) $(\forall a)[a + 0 = a]$.
(A2) $(\forall a)(\forall b)[a + sb = s(a + b)]$.
(M1) $(\forall a)[a \cdot 0 = 0]$.
(M2) $(\forall a)(\forall b)[a \cdot sb = a \cdot b + a]$.

Some treatments of Peano Arithmetic admit another binary primitive relation $\leq$ with axioms (N1), (N2) and (N3) below or alternatively treat $a \leq b$ as an abbreviation for $(\exists x)[a + x = b]$, the three axioms below then being easy theorems:

(N1) $(\forall a)[a \leq 0 \rightarrow a = 0]$.
(N2) $(\forall a)(\forall b)[a \leq sb \leftrightarrow [a \leq b \lor a = sb]]$.
(N3) $(\forall a)(\forall b)[a \leq b \lor b \leq a]$.

The final axiom is the induction schema (an infinite collection $\{I_\phi\}$ of axioms indexed by predicates $\phi(x)$):

$$(I_\phi) \qquad [\phi(0) \land (\forall a)[\phi(a) \rightarrow \phi(sa)]] \rightarrow (\forall a)\phi(a).$$

A.2   Some familiar results

We illustrate this formalized arithmetic with a few simple results. In all cases below the $\Diamond$ indicating the end of the proof can be read as 'the result follows by induction'.

Associativity of addition
$$(\forall a)(\forall b)(\forall c)[(a + b) + c = a + (b + c)]$$

Proof. Fix a and b. Let $\phi(c)$ be the formula $(a + b) + c = a + (b + c)$. Since $(a + b) + 0 = a + b = a + (b + 0)$ we have $\phi(0)$. Assuming $\phi(c)$, we have $(a + b) + sc = s((a + b) + c) = s(a + (b + c)) = a + s(b + c) = a + (b + sc)$.$\Diamond$

Distributivity
$$(\forall a)(\forall b)(\forall c)[a \cdot (b + c) = a \cdot b + a \cdot c]$$

Proof. Fix a and b. Let $\phi(c)$ be the formula $a \cdot (b + c) = a \cdot b + a \cdot c$. Since $a(b + 0) = a \cdot b = a \cdot b + 0 = a \cdot b + a \cdot 0$ we have $\phi(0)$. Assuming $\phi(c)$, we have $a \cdot (b + sc) = a \cdot s(b + c) = a \cdot (b + c) + a = (a \cdot b + a \cdot c) + a$, which by associativity of addition equals $a \cdot b + (a \cdot c + a) = a \cdot b + a \cdot sc$.$\Diamond$

Associativity of multiplication
$$(\forall a)(\forall b)(\forall c)[(a \cdot b) \cdot c = a \cdot (b \cdot c)]$$

Proof. Fix a and b. Let $\phi(c)$ be the formula $(a \cdot b) \cdot c = a \cdot (b \cdot c)$. Since $(a \cdot b) \cdot 0 = 0 = a \cdot 0 = a \cdot (b \cdot 0)$ we have $\phi(0)$. Assuming $\phi(c)$, we have

$(a \cdot b) \cdot sc = (a \cdot b) \cdot c + a \cdot b = a \cdot (b \cdot c) + a \cdot b$, which by distributivity is equal to
$a \cdot (b \cdot c + b) = a \cdot (b \cdot sc).\diamondsuit$

Commutativity of addition
$$(\forall a)(\forall b)[a + b = b + a]$$

(i) $(\forall a)[0 + a = a + 0]$.

Proof. Let $\phi(a)$ be the formula $0 + a = a + 0$. Since $0 + 0 = 0 + 0$ we have
$\phi(0)$. Assuming $\phi(a)$, we have $0 + sa = s(0 + a) = s(a + 0) = sa = sa + 0.\diamondsuit$

(ii) $(\forall a)[a + s0 = sa + 0]$.

Proof. Let $\phi(a)$ be the formula $a + s0 = sa + 0$. Since $0 + s0 = s(0 + 0) = s0 = s0 + 0$ we have $\phi(0)$. Assume $\phi(a)$, then $sa + s0 = s(sa + 0) = ssa = ssa + 0.\diamondsuit$

(iii) $(\forall a)(\forall b)[a + sb = sa + b]$.

Proof. Fix a. Let $\phi(b)$ be the formula $a + sb = sa + b$. By (ii) $\phi(0)$. Assume
$\phi(b)$, then $a + ssb = s(a + sb) = s(sa + b) = sa + sb.\diamondsuit$

(iv) $(\forall a)(\forall b)[a + b = b + a]$.

Proof. Fix b. Let $\phi(a)$ be the formula $a + b = b + a$. By (i) $\phi(0)$. Assume
$\phi(a)$, then $b + sa = s(b + a) = s(a + b) = a + sb$, which by (iii) is equal to
$sa + b.\diamondsuit$

Commutativity of multiplication
$$(\forall a)(\forall b)[a \cdot b = b \cdot a]$$

(i) $(\forall a)[a \cdot 0 = 0 \cdot a]$.

Proof. Let $\phi(a)$ be the formula $a \cdot 0 = 0 \cdot a$. Since $0 \cdot 0 = 0 \cdot 0$ we have $\phi(0)$.
Assume $\phi(a)$, then $0 \cdot sa = 0 \cdot a + 0 = 0 \cdot a = a \cdot 0 = 0 = sa \cdot 0.\diamondsuit$

(ii) $(\forall a)(\forall b)[sa \cdot b = a \cdot b + b]$.

Proof. Fix a. Let $\phi(b)$ be the formula $sa \cdot b = a \cdot b + b$. Since $sa \cdot 0 = 0 = a \cdot 0 + 0$
we have $\phi(0)$. Assuming $\phi(b)$ we have, making use of associativity twice in the
fourth equality, $sa \cdot sb = sa \cdot b + sa = s(sa \cdot b + a) = s(a \cdot b + b + a) = s(a \cdot b + a + b) = s(a \cdot sb + b) = a \cdot sb + sb.\diamondsuit$

(iii) $(\forall a)(\forall b)[a \cdot b = b \cdot a]$.
Proof. Fix a. Let $\phi(b)$ be the formula $ab = ba$. By (i) $\phi(0)$. Assume $\phi(b)$,
then $a \cdot sb = a \cdot b + a = b \cdot a + a$, which by (ii) is equal to $sb \cdot a.\diamondsuit$

A.3   Exponentiation

We could introduce a primitive binary operator $\uparrow$ for exponentiation subject to axioms $(\forall a)[a \uparrow 0 = s0]$ and $(\forall a)[a \uparrow sn = (a \uparrow n) \cdot a]$, however, it turns out that the theory is not strengthened by incorporating $\uparrow$; the relation $a \uparrow b = c$ is definable from $+$ and $\cdot$ alone.[1]

A.4   Second-order Peano Arithmetic

In the introduction (Subsection 1.2.2) we defined induction as follows: if A is a set containing 0 and $n+1$ whenever $n \in A$, then A contains all natural numbers. The induction schema of first-order Peano Arithmetic is substantially weaker than this, for it only applies to sets of the form $\{x : \phi(x)\}$, where ϕ is some predicate, so it applies to only countably many sets. Once again this boils down to a simple cardinal inequality: no countable language can describe all subsets of the natural numbers.

In the full second-order version of Peano Arithmetic (or even just the monadic extension – see Subsection 2.3.4) the axiom schema of induction is replaced by a much more powerful single axiom which allows quantification over *arbitrary* subsets, not just first-order describable subsets. We should stress that this extension is much more powerful than the meagre second-order extension which replaces the induction schema by a single sentence

$$(\forall \phi)[\phi(0) \wedge (\forall a)[\phi(a) \to \phi(sa)]] \to (\forall x)\phi(x),$$

where the initial quantification is interpreted to range over all first-order predicates.[2]

The difference between first- and second-order Peano Arithmetic (and indeed between any first- and second-order theory) is considerable. Most notably the second-order theory is *categorical*, meaning that any two models of it are isomorphic. First-order Peano Arithmetic on the other hand, in common with all first-order theories possessing an infinite model, has a wealth of non-isomorphic models. The 'standard model' of Peano Arithmetic is ω with von Neumann successor $sa = a \cup \{a\}$.

The *full theory* of ω, also called *true arithmetic* and denoted $\mathrm{Th}(\omega)$, is the set of all sentences in PA which are satisfied in ω. All models of $\mathrm{Th}(\omega)$ will therefore be elementarily equivalent to ω. The full theory of ω, although rendered negation complete by definition, has the defect that its set of provable sentences, which coincides with its set of true sentences and its set of axioms, is not recursively enumerable, i.e. there is no algorithm which, when presented with a PA sentence ϕ, will tell us whether $\phi \in \mathrm{Th}(\omega)$ or $\phi \notin \mathrm{Th}(\omega)$.

---

[1]Indeed, all primitive recursive functions are definable in Peano Arithmetic. This is a result of Gödel. For a precise statement of the elimination of exponentiation, and a proof, see Smullyan [198], Chapter IV.

[2]This much overlooked point is stressed in Chapter 22 of Smith [196].

A.5 Weaker theories

Although clearly of limited expressive power relative to ZF, Peano Arithmetic is nevertheless capable of producing highly non-trivial results and is sufficiently complex to act as an interesting test case for many metamathematical questions. Weaker relatives of Peano Arithmetic have also been studied. The theory without + or · (the elementary theory of the successor function) serves as a useful example of a negation complete theory (but is otherwise of little practical use). If we augment this limited structure by including + and its axioms we obtain another negation complete theory known as *Presburger Arithmetic* (after Mojsesz Presburger). This theory is too weak, however, to define multiplication. There is also an elementary theory of multiplication (without s or $+$), known as *Skolem Arithmetic*, which is also negation complete.

Appendix B

Zermelo–Fraenkel set theory

Zermelo–Fraenkel set theory (ZF) is a first-order theory with one primitive binary relation $\in$ and no primitive operators together with the following non-logical axioms. Here the axioms are given in a semi-colloquial form making use of some of the notation and terminology which is discussed in more detail in the main text.

Axiom 1: Axiom of Extensionality
For all sets a and b, $a =_o b$ if and only if $a = b$.

Axiom 2: Axiom of Pairing
For all sets a and b, $\{a, b\}$ is a set.

Axiom 3: Axiom of Unions
For all sets a, $\cup a$ is a set.

Axiom 4: Axiom of Powers
For all sets a, $P(a)$ is a set.

Axiom 5: Axiom Schema of Replacement
If a predicate $\phi(x, y)$ induces a function then for all sets a,
$\{y : x \in a \text{ and } \phi(x, y)\}$ is a set.

Axiom 6: Axiom of Regularity
If $a \neq \emptyset$ then there exists an $x \in a$ such that $x \cap a = \emptyset$.

Axiom 7: Axiom of Infinity
ω is a set.

The Axiom of Pairing is redundant (i.e. it is a consequence of the other axioms).

Neither provable nor disprovable in ZF is the following, which is also assumed by most mathematicians.

Axiom of Choice
For every set a there exists a function f such that,
for all $x \in a$, if $x \neq \varnothing$ then $f(x) \in x$.

Among other statements the Axiom of Choice is equivalent to the Well-Ordering Theorem (every set has a well-ordering) and trichotomy (given any pair of sets a and b there either exists an injection $a \to b$ or an injection $b \to a$).

ZF is sometimes augmented by the Continuum Hypothesis, which, like the Axiom of Choice, is neither provable nor disprovable in ZF.

Continuum Hypothesis
If there exists an injection $\omega \to b$ and an injection $b \to P(\omega)$,
then b is either equipollent to ω or equipollent to $P(\omega)$.

The Generalized Continuum Hypothesis, also neither provable nor disprovable in ZF, implies the Axiom of Choice.

Generalized Continuum Hypothesis
If there exists an injection $\omega \to a$, an injection $a \to b$ and an injection $b \to P(a)$,
then b is either equipollent to a or equipollent to $P(a)$.

Appendix C

Gödel's Incompleteness Theorems

Gödel's Incompleteness Theorems are now justly regarded both as one of the most profound discoveries in mathematical logic and as one of the gems of twentieth century mathematics.

Like all major scientific and mathematical landmarks (see also the Heisenberg Uncertainty Principle – in fact all of quantum mechanics – and the theory of complex dynamical systems ('chaos')), Gödel's theorems have captured the imagination of an excitable horde of pseudointellectuals who, not bothering to trouble themselves with actually studying the subject or understanding what the theorems say, instead abuse them, treating Gödel's deep and precise work as a springboard for flowery nonsense (generally hollow wordplay and embarrassingly vague analogies designed to obscure a chronic lack of content – a species of 'mathematics envy'). For an account of how the Incompleteness Theorems have been misunderstood and mishandled see Torkel Franzen's *Gödel's Incompleteness Theorem: an incomplete guide to its use and abuse.*[1] To fully appreciate Gödel's Theorems there is no better alternative than to carefully read a modern technical account which provides all the details. I recommend both Peter Smith's *Gödel's Theorems*[2] and Raymond Smullyan's *Gödel's Incompleteness Theorems.*[3] The latter is the main source for the condensed abstract sketch presented here.

Gödel's famous paper[4] of 1931 begins by stating that large tracts of mathematics have become formalized, allowing proofs to be carried out by mechanical rules, the systems described by Russell and Whitehead's *Principia Mathemat-*

---

[1] Franzen [**73**].

[2] Smith [**196**].

[3] Smullyan [**198**].

[4] 'Über formal unentscheidbare Sätze der Principia Mathematica und verwandter Systeme I', *Monatschefte für Mathematik und Physik*, **38**, 173–198, (1931). The English translation *On formally undecidable propositions of Principia Mathematica and related systems* is published by Dover, see Gödel [**83**].

ica (PM) and Zermelo–Fraenkel set theory (ZF) being two notable and very far-ranging examples. Given the success and wide scope of PM and ZF one might be led to believe that *all* questions which are expressible in the underlying code of these formal systems can be decided in them. This is not so, and more surprisingly it can be shown that in both systems there are propositions about the *natural numbers* which cannot be proved true or false: these systems are 'incomplete'. Furthermore, this is not a peculiarity of the given systems but a phenomenon which holds for a very broad class of formal systems.

Below is a brief account of Gödel's theorems, following Smullyan, which is intended only to highlight some of the main ideas. It is, although fairly detailed in parts, no substitute for a proper course on incompleteness, and in the case of the First Incompleteness Theorem we restrict attention to the semantic version, i.e. 'soundness implies negation incompleteness', rather than the more technically demanding 'consistency implies negation completeness'. As we have already sketched in Section 11.3 some of the necessary preliminaries, we will first describe the incompleteness theorems in the concrete setting of ZF. Since it is only the 'arithmetic part' of a theory that is needed, most accounts prove the incompleteness theorems for PA. The approach is the same as described here for ZF.

C.1   The basic idea

We indicated in Section 11.3 how to construct the predicate $\mathtt{proof}(n, m)$: 'n is the Gödel sequence number of a proof of the sentence with Gödel number m'. If n is the Gödel number of $\phi(y)$ then the *diagonalization* $\mathrm{diag}(n)$ of n is equal to the Gödel number of the sentence $\phi(\bar{n})$, where $\bar{n}$ denotes the set that represents the natural number n (i.e. one of $\varnothing$, $\{\varnothing\}$, $\{\varnothing, \{\varnothing\}\}$, etc.). There exists a 2-predicate $\pi(x, y)$ which has the property that $\mathtt{proof}(n, \mathrm{diag}(m))$ if and only if $\pi(\bar{n}, \bar{m})$ (built into π is the condition $x, y \in \omega$).

Proving that such a well-formed formula π exists forms the bulk of the technical part of the proof. The usual approach is to prove a much more general result: every primitive recursive function is expressible by a Σ_1-formula (see the remarks to Subsection 2.3.5).

Let $F(y)$ be the formula $(\forall x)\neg\pi(x, y)$, let g be the Gödel number of $F(y)$ and let $G = F(\bar{g})$, so that G has Gödel number $\mathrm{diag}(g)$. In considering the truth of G we unravel the following sequence of equivalences:

 (i) G is true;

 (ii) $F(\bar{g})$ is true;

 (iii) $(\forall x)\neg\pi(x, \bar{g})$ is true;

 (iv) $\pi(x, \bar{g})$ is false for all x;

 (v) $\mathtt{proof}(n, \mathrm{diag}(g))$ is false for all natural numbers n;

(vi) the sentence with Gödel number diag(g) is not provable;

(v) G is not provable.

So we see that G *is true if and only if G is not provable!*

Assuming ZF is *sound* (i.e. every provable sentence is true), if G is provable in ZF then it is true, and hence by the above equivalence is unprovable, a contradiction. So G is not provable, which by the above equivalence again means G is true, so $\neg G$ is false, meaning, by soundness, $\neg G$ too is unprovable in ZF. Neither G nor $\neg G$ is provable in ZF, or in other words *if ZF is sound then it is negation incomplete.*

Gödel's original incompleteness results were based on the intuitively rather obscure assumption of ω-consistency (sometimes called *arithmetic consistency*), a condition weaker than soundness but stronger than consistency.[1] PA is ω-consistent if whenever $(\exists n)\phi(n)$ is provable then at least one of the statements $\phi(0)$, $\phi(1)$, $\phi(2)$,... is not refutable. The notion of ω-consistency is definable in any theory in which one can isolate an arithmetic fragment, i.e. a first-order theory in which it is possible to define the objects of PA: the constant '0', equality $=$, the successor operator s, addition $+$ and multiplication $\cdot$, in such a way that all the axioms of PA are satisfied (ZF is one such system) – such a theory is an *arithmetic theory*. An ω-inconsistent theory can be consistent! The strengthening of Gödel's results from ω-consistency to the more intuitively clear consistency is due to J. Barkley Rosser.[2] The grand result usually referred to as the First Incompleteness Theorem for ZF is: *if ZF is consistent then it is negation incomplete.*

The true but not provable sentence constructed in the proof of the First Incompleteness Theorem is a *Gödel–Rosser sentence* for ZF. Rosser's strengthening of Gödel's result involved the construction of a new sentence, more complex than the undecidable sentence of Gödel's original, which was proved to be undecidable based on the much weaker assumption of consistency.

If ZF is consistent then it cannot prove $0 = 1$ and since an inconsistent theory can prove all its sentences we also have the converse; thus the statement 'ZF is consistent' is entirely captured by the statement '$0 = 1$ is not provable in ZF', something that is expressible in ZF, i.e. there is a sentence Con of ZF which is true if and only if ZF is consistent. Making use of the predicate π introduced earlier, if n is the Gödel number of the sentence $0 = 1$ then we can take Con to be the sentence $\neg(\exists x)\pi(x, \bar{n})$. Let G be the Gödel–Rosser sentence constructed in the proof of the First Incompleteness Theorem. That proof asserts the implication Con $\rightarrow G$. However, the implication Con $\rightarrow G$ can be derived *within ZF*, so if it were possible to prove in ZF that ZF is consistent then we would have by modus ponens a proof of G in ZF, but G is not provable,

---

[1] Consistency is much weaker than soundness: it is obvious that every sound theory is consistent, but there exist theories which are consistent but not sound, such as PA+¬Con(PA).

[2] Rosser, J. Barkley, 'Extensions of some theorems of Gödel and Church', *J. Symbolic Logic*, 1, 87–91, (1936).

a contradiction. Therefore *if ZF is consistent then it cannot prove its own consistency.* This is the Second Incompleteness Theorem for ZF.

At the heart of the matter is the fact that the underlying theory ZF has sufficient expressive resources to capture arithmetic statements of provability and consistency (these resources are already present in the much weaker theory PA). We outline below, in abstract form, those generic features which are sufficient for a theory to be negation incomplete and be unable to prove its own consistency.

C.2   Negation incompleteness

Extracting (and abstracting) the basic ingredients of Gödel's argument, the First Incompleteness Theorem applies to a theory **L** which has the following features. To help dilute the abstractness of this presentation we will describe in each case the intended interpretation in ZF.

E A countable set of *expressions*. In ZF, **E** is the set of all well-formed formulas.

S *Sentences*, a subset of **E**. In ZF, **S** is the set of well-formed formulas with no free variables.

K *Class names*, a subset of **E** (the term 'class name' is a tribute to Gödel's original terminology). In ZF, **K** is the set of 1-predicates, the 'class name' of a predicate ϕ referring to the class symbol $\{x : \phi\}$.

P *Provable sentences*, a subset of **S**. In ZF, **P** is the set of provable sentences.

D *Disprovable sentences*, a subset of **S**. In ZF, **D** the set of disprovable sentences.

T TRUE *sentences*, a subset of **S**. In ZF the truth of a sentence is determined by some prior fixed model of ZF.[1]

Φ A function $\mathbf{E} \times \mathbb{N} \to \mathbf{E}$, the image of (E, n) to be denoted $E(n)$, with the property that $K(n) \in \mathbf{S}$ for all $K \in \mathbf{K}$. In ZF, the function Φ could be taken to map (ϕ, n) to the sentence obtained by replacing all free variables of ϕ by the set $\bar{n}$ representing the natural number n, or to the well-formed formula obtained by replacing the first appearing free variable by $\bar{n}$. It is only its action on 1-predicates that concerns us.

With each class name is associated a subset of $\mathbb{N}$: a natural number n satisfies a class name K if $K(n)$ is TRUE. The set *expressed* by a class name K is the set of all natural numbers satisfying K. By Cantor's Theorem, since **K** is countable, a cardinal majority of subsets of $\mathbb{N}$ will have no class name. A subset of $\mathbb{N}$ is

---

[1] In PA we have the advantage of a privileged 'standard' model $\mathbb{N}$ so that the truth of a sentence in PA is often measured against this standard. ZF has no such privileged model.

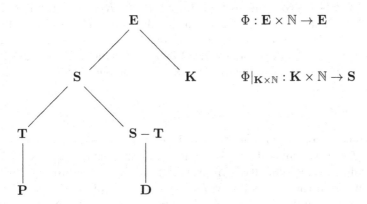

$$\Phi : \mathbf{E} \times \mathbb{N} \to \mathbf{E}$$

$$\Phi|_{\mathbf{K} \times \mathbb{N}} : \mathbf{K} \times \mathbb{N} \to \mathbf{S}$$

Figure C.1 A diagram depicting, for a sound theory **L**, the basic framework to which the First Incompleteness Theorem applies, the lines indicating inclusion, the lower class being a subset of the upper. See the text for an interpretation of the symbols.

expressible in **L** if it is expressed by some class name in **L**. The sentence $K(n)$ is interpreted as 'n is an element of the set associated with K'.

In ZF, the subset of natural numbers associated with a 1-predicate ϕ is precisely $\{n \in \mathbb{N} : \phi(\bar{n})\}$, and '$n$ satisfies ϕ' simply means that n is an element of the latter set, which is the set expressed by ϕ; a subset A of $\mathbb{N}$ is expressible if there is a 1-predicate ϕ such that $A = \{n \in \mathbb{N} : \phi(\bar{n})\}$.

The theory **L** is *sound* if all provable sentences are TRUE.

A *Gödel numbering* is an injection $g : \mathbf{E} \to \mathbb{N}$. We assume here, for convenience – it is not necessary but shortens the argument – that a Gödel numbering g has been fixed which is, in addition, surjective. The *Gödel number* of an expression E is the natural number $g(E)$. E_n denotes the unique expression with Gödel number n, i.e. $E_n = g^{-1}(n)$.

The Gödel numbering we defined for ZF in Section 11.3 is injective, as required, but is not surjective. It can be modified to be so but at the expense of a more complicated encoding.

The *diagonalization* of E_n is the expression $E_n(n)$. If E_n is a class name then its diagonalization is a sentence which is TRUE if and only if E_n is satisfied by its own Gödel number n. The *diagonal function* of **L** is the function $d : \mathbb{N} \to \mathbb{N}$ mapping n to the Gödel number of $E_n(n)$.

Let $\mathcal{P}rov$ be the set of Gödel numbers of all provable sentences.

An Abstract Incompleteness Theorem[1]

*If $d^{-1}(\mathbb{N} - \mathcal{P}rov)$ is expressible in a sound theory **L** then there is a TRUE sentence of **L** which is not provable in **L**.*

---

[1]See Smullyan [**198**], p. 7, 'Theorem (GT)—After Gödel with shades of Tarski'.

The proof is as follows. Suppose $\mathbf{L}$ is sound and a class name K expresses $d^{-1}(\mathbb{N} - \mathcal{P}rov)$. Let k be the Gödel number of K. The statement that K expresses $d^{-1}(\mathbb{N} - \mathcal{P}rov)$ means that, for all $n \in \mathbb{N}$, $K(n)$ is TRUE if and only if $n \in d^{-1}(\mathbb{N} - \mathcal{P}rov)$. In particular the diagonalization $K(k)$ of K is TRUE if and only if $k \in d^{-1}(\mathbb{N} - \mathcal{P}rov)$, i.e. if $d(k) \notin \mathcal{P}rov$. But $d(k)$ is the Gödel number of $K(k)$ so $d(k) \notin \mathcal{P}rov$ if and only if $K(k)$ is not provable in $\mathbf{L}$. Since $\mathbf{L}$ is sound (so that every provable statement is TRUE), we conclude that $K(k)$ is TRUE and not provable in $\mathbf{L}$.

Of course, for a specific theory $\mathbf{L}$ one must establish the condition '$d^{-1}(\mathbb{N} - \mathcal{P}rov)$ is expressible in $\mathbf{L}$', which in practice is not easy (or, of course, may be false!). The difficulty is in proving that $\mathcal{P}rov$ is expressible in $\mathbf{L}$ – the rest follows from the expressibility, for expressible A, of $\mathbb{N} - A$ and $d^{-1}(A)$. In our concrete example of ZF this is the part of the proof that involves firstly generating the predicate **proof** and then proving that it is expressible by a well-formed formula of ZF.

Incompleteness via Gödel sentences

A *Gödel sentence* for $A \subseteq \mathbb{N}$ is a sentence E_n with the property that either E_n is TRUE and $n \in A$ or E_n is not true and $n \notin A$. A Gödel sentence for A is guaranteed if $d^{-1}(A)$ is expressible in $\mathbf{L}$, for if K expresses $d^{-1}(A)$ and k is the Gödel number of K, then $K(k)$ is TRUE if and only if $d(k) \in A$, and since $d(k)$ is the Gödel number of $K(k)$, $K(k)$ is a Gödel sentence for A.

In terms of Gödel sentences the proof of the incompleteness theorem is as follows. If $d^{-1}(\mathbb{N} - \mathcal{P}rov)$ is expressible in $\mathbf{L}$, then there is a Gödel sentence for $\mathbb{N} - \mathcal{P}rov$. Such a sentence is TRUE if and only if it is not provable, and assuming $\mathbf{L}$ is sound, it is TRUE but not provable in $\mathbf{L}$.

We cannot replace $\mathcal{P}rov$ with the set $\mathcal{T}rue$ of Gödel numbers of the TRUE sentences of $\mathbf{L}$; $d^{-1}(\mathbb{N} - \mathcal{T}rue)$ is not expressible in $\mathbf{L}$. This is known as *Tarski's Theorem*. The proof is straightforward. If $d^{-1}(\mathbb{N} - \mathcal{T}rue)$ were expressible in $\mathbf{L}$ then there would be a Gödel sentence for $\mathbb{N} - \mathcal{T}rue$, but such a sentence would be TRUE if and only if its Gödel number is not the Gödel number of a TRUE sentence, a contradiction.

Undecidable sentences

$\mathbf{L}$ is *consistent* if $\mathbf{P} \cap \mathbf{D} = \emptyset$ and is *inconsistent* otherwise. If $\mathbf{L}$ is sound then it is automatically consistent, but not conversely. A sentence is *decidable* if it is in the union $\mathbf{P} \cup \mathbf{D}$ and is *undecidable* otherwise. $\mathbf{L}$ is *negation complete* if $\mathbf{S} = \mathbf{P} \cup \mathbf{D}$ and is *negation incomplete* otherwise. Our Incompleteness Theorem then says that if $\mathbf{L}$ is sound and if $d^{-1}(\mathbb{N} - \mathcal{P}rov)$ is expressible in $\mathbf{L}$ then $\mathbf{L}$ is negation incomplete.

C.3 Consistency

We fix a Gödel numbering g on an arithmetic theory S.

A formula $p(x)$ is a *provability predicate* if it satisfies the following, where we use $p\phi$ as an abbreviation for $p(\overline{g(\phi)})$, $\overline{g(\phi)}$ being the term in the underlying theory which represents the natural number $g(\phi)$. The properties of a provability predicate are modelled on those of a formula which expresses the set $\mathcal{P}\text{rov}$ in an ω-consistent theory. In ZF the formula $(\exists x)\pi(x, y)$ is a provability predicate.

1. ϕ is provable in S if and only if $p\phi$ is provable in S.
2. $p(\phi \to \psi) \to (p\phi \to p\psi)$ is provable in S.
3. $p\phi \to pp\phi$ is provable in S.

Applying the rules mechanically to the provable $\phi \to (p\phi \to \psi)$, we soon deduce that $p\phi \to (pp\phi \to p\psi)$ is provable. Assuming $p\phi$ we have: (i) by the latter implication $pp\phi \to p\psi$; and (ii) by property 3, $pp\phi$. So by modus ponens we deduce $p\psi$, hence $p\phi \to p\psi$ is provable. In summary:

4. If $\phi \to (p\phi \to \psi)$ is provable in S then so is $p\phi \to p\psi$.

In addition we posit the existence of a sentence G with the property that $G \leftrightarrow \neg pG$ is provable in S. This is modelled on the property of a Gödel sentence for $\mathbb{N} - \mathcal{P}\text{rov}$, i.e. a Gödel–Rosser sentence for the theory S. Let us fix some disprovable sentence $\mathbf{F}$ (the choice is not important – we could let $\mathbf{F}$ be the sentence $\neg 0 = 0$, for example) and let $\text{Con}(S)$ be the sentence $\neg p\mathbf{F}$. If p expresses $\mathcal{P}\text{rov}$, then $\text{Con}(S)$ is an S-encoded statement of the consistency of S: it is true if and only if S is consistent. We now have all the ingredients for the following abstract result.

A Second Incompleteness Theorem[1]
*If a consistent arithmetic theory S has a provability predicate p
and there exists a sentence G such that $G \leftrightarrow \neg pG$ is provable in S,
then $\text{Con}(S)$ is not provable in S.*

Proof.

(i) *G is not provable in S*
If G were provable in S then, by property 1 and the assumption that $G \leftrightarrow \neg pG$ is provable, both $\neg pG$ and pG would be provable, contradicting the consistency of S.

(ii) *$\text{Con}(S) \to G$ is provable in S*
Since $\mathbf{F}$ is disprovable, the sentence $\neg pG \to (pG \to \mathbf{F})$ is provable in S, and by the assumption that $G \leftrightarrow \neg pG$ is provable it follows that $G \to (pG \to \mathbf{F})$ is provable and so, by property 4, $pG \to p\mathbf{F}$ is provable. Thus $\neg p\mathbf{F} \to \neg pG$ is

---

[1]See Smullyan **[198]**, Chapter IX, Section 2.

provable, and, again using the provability of $G \leftrightarrow \neg pG$, we have that $\neg p\mathbf{F} \to G$ is provable. In other words $\mathrm{Con}(S) \to G$ is provable.

If $\mathrm{Con}(S)$ were provable in S, then by (ii) G would be provable in S, contradicting (i). This completes the proof.

C.4   A further abstraction

Further abstracting the above system, we can state in great generality a skeletal version of both Incompleteness Theorems. Some of the assumptions we make can be weakened for the purpose of proving one or other theorem and related results, but for brevity we keep all assumptions at full strength.

Our theory M comprises a set S of sentences, including a sentence $\mathbf{F}$, interpreted as a falsehood, a binary operator $\to$ from $S \times S$ to S, a subset $\mathbf{P}$ of S interpreted as the set of provable sentences, and a map p from S to S mapping ϕ to $p\phi$. We interpret $p\phi$ as 'ϕ is provable'.

A *valuation set* V is a subset of S satisfying the properties:

(i) $\mathbf{F} \notin V$;

(ii) $(\phi \to \psi) \in V$ if and only if either $\phi \notin V$ or $\psi \in V$.

ϕ is a *tautology* if it belongs to every valuation set. We define $\neg\phi$ to be $\phi \to \mathbf{F}$ and the other logical operations $\wedge, \vee, \leftrightarrow$ are defined in terms of $\neg$ and $\to$ in the usual way.

$\mathbf{P}$ has the following properties:

(i) $\phi \in \mathbf{P}$ for all tautologies ϕ;

(ii) if $\phi \in \mathbf{P}$ and $(\phi \to \psi) \in \mathbf{P}$ then $\psi \in \mathbf{P}$, i.e. $\mathbf{P}$ is closed under modus ponens.

p satisfies the following direct analogues of the properties given earlier.

1. $\phi \in \mathbf{P} \leftrightarrow p\phi \in \mathbf{P}$.
2. $(p(\phi \to \psi) \to (p\phi \to p\psi)) \in \mathbf{P}$.
3. $(p\phi \to pp\phi) \in \mathbf{P}$.

M is *consistent* if $\mathbf{F} \notin \mathbf{P}$. $\mathrm{Con}(M)$ is the sentence $\neg p\mathbf{F}$.

Skeletal Incompleteness Theorems[1]

If M is consistent and possesses a sentence G such that $(G \leftrightarrow \neg pG) \in \mathbf{P}$ then:

(i) *(First Incompleteness Theorem) neither G nor $\neg G$ is provable in M.*

---

[1] See Smullyan [198], Chapter XI, Part II.

(ii) *(Second Incompleteness Theorem)* $\mathrm{Con}(M)$ *is not provable in* M.

The Incompleteness Theorems are usually summarized in the following vague form: *if a consistent effective theory is capable of describing 'a certain amount of arithmetic' then there is an arithmetic sentence expressible in the underlying language which is neither provable nor refutable in the theory, and furthermore the theory cannot prove its own consistency.* What is meant by 'a certain amount of arithmetic' can be answered in different ways, as indicated in the preceding abstract discussion. We must not confuse the theorem with the false statement that 'sufficiently complex' theories are negation incomplete; there are very simple negation incomplete systems and very complex negation complete systems.

Bibliography

[1] Abel, N. H. *Oeuvres Complètes*. Imprimerie de Groundhal & Son, 1881.

The complete works of Niels Henrik Abel.

[2] Aczel, P. *Non-Well-Founded Sets*. Stanford University Center for the Study of Language and Information Lecture Notes no. 14, 1988.

Non-well-founded sets seen in a positive light. Add to ZF the Axiom of Choice, remove the Axiom of Regularity and replace it with a new axiom, the anti-foundation axiom: *'every graph has a unique decoration by sets'*. Consequences, variations and applications are discussed and an historical survey is provided in an Appendix. No deep set theoretic prerequisites are assumed.

[3] Annas, J. *Aristotle's Metaphysics: Books M and N*. Clarendon Press, 1976.

Books M and N are the last two books of Aristotle's *Metaphysics*. Here, in his only lengthy foray into the philosophy of mathematics, Aristotle criticizes Plato and suggests his own alternatives. Annas counters criticism that M–N is difficult and mysterious 'mystical nonsense', presenting Aristotle's ancient work as still very much alive.

[4] Archimedes. *The Works of Archimedes*. Rough Draft Printing, 2007. (First edition 1897.)

Thomas Heath's 186 page introduction followed by a translation of Archimedes' works: on the sphere and cylinder, measurement of a circle, on conoids and spheroids, on spirals, on the equilibrium of planes, the sand-reckoner, quadrature of the parabola, on floating bodies, book of lemmas, and the cattle-problem.

[5] Aristotle. *Metaphysics: Books* Γ, Δ *and E.* Translated with notes by C. Kirwan. Clarendon Press, 1971.

A close translation and critical commentary of three (of fourteen) books of Aristotle's *Metaphysics*, including an examination of the logical principles of contradiction and excluded middle.

[6] Aristotle. *The Complete Works of Aristotle.* Barnes, J. (ed.) Princeton University Press, 1984.

The standard English version of everything that has survived of Aristotle's writing (and also some material that he probably didn't write). Nearly two and a half thousand pages in total.

[7] Bacon, R. *The Opus Majus of Roger Bacon.* Translated by R. B. Burke. Two Volumes. Kessinger Publishing, 2002.

(The source of the quote at the beginning of Section 1.2.)

[8] Balaguer, M. *Platonism and Anti-Platonism in Mathematics.* Oxford University Press, 1998.

An argument that there can be no rational reason for believing or disbelieving in abstract objects by studying mathematical theory and practice.

[9] Bar-Hillel, Y., Poznanski, E. I. J., Rabin, M. O. and Robinson, A. *Essays on the Foundations of Mathematics: dedicated to A. A. Fraenkel on his seventieth anniversary.* Magnes Press, Hebrew University, 1966. (Original edition 1962.)

A much referenced collection of articles including contributions by, among others, P. Bernays (Zur Frage der Unendlichkeitsschemata in der axiomatischen Mengenlehre), P. Erdos and A. Tarski (On some problems involving inaccessible cardinals), A. Lévy (Comparing the axioms of local and universal choice), R. Montague (Frankel's addition to the axioms of Zermelo), D. Scott (More on the axiom of extensionality), A. Robinson (On the construction of models), T. Skolem (Interpretation of mathematical theories in the first-order predicate calculus) and E. Mendelson (On nonstandard models for number theory).

[10] Bartoszynski, T. and Judah, H. *Set Theory. On the Structure of the Real Line.* A. K. Peters Ltd., 1995.

A research level monograph detailing progress on measure and category in set theory in the period 1980 to 1995, including some background on forcing.

[11] Barwise, J. and Etchemendy, J. *The Liar: an essay on truth and circularity*. Oxford University Press, 1987.

The ancient paradox of the liar is analysed using modern tools (based in part on Peter Aczel's theory of non-well-founded sets).

[12] Bekkali, M. *Topics in Set Theory: Lebesgue measurability, large cardinals, forcing axioms, rho-functions*. Vol. 1476 of Lecture Notes in Mathematics. Springer, 1991.

Research level lecture notes based on a 1987 course delivered by S. Todorcevic at the University of Colorado. The topics covered are adequately described in the title!

[13] Bell, E. T. *The Queen of the Sciences*. The Williams & Wilkins Company, 1931.

Although the sciences often give rise to new mathematics, the reverse is also true. Interesting pure mathematics is created without applications in mind, yet after decades, or centuries, it often finds applications to fields that no one could have foreseen at the time it was created. (The source of the quote at the beginning of Section 6.1.)

[14] Bell, E. T. *The Development of Mathematics*. Courier Dover Publications, 1945. (Original 1940.)

A survey of the role of mathematics in civilization. (The source of the quote at the beginning of Section 1.11.)

[15] Bell, J. L. *Set Theory: Boolean-Valued Models and Independence Proofs in Set Theory*. Clarendon Press, 2005. (Follow-up to the 1977 original.)

A research level monograph on independence proofs, including, among other things, proofs of the independence of the continuum hypothesis and the axiom of choice from ZF.

[16] Benacerraf, P. and Putnam, H. (eds.) *Philosophy of Mathematics: selected readings*. Prentice Hall Inc., 1983. (First edition 1964.)

Twenty-eight selected writings on the foundations of mathematics, the existence of mathematical objects, mathematical truth and the concept of set.

[17] Berkeley, G. *The Analyst, or, A Discourse Addressed to an Infidel Mathematician*. J. Tonson, 1734.

George Berkeley's famous criticism of the lack of rigour to be found in the calculus of the early eighteenth century *'wherein it is examined whether the Object, Principles, and Inferences of the modern Analysis are more distinctly conceived, or more evidently deduced, than Religious Mysteries and Points of Faith'*. A full copy

is freely available online via the Michigan University archive. Also
included in Ewald (2000).

[18] Bernays, P. *Axiomatic Set Theory*, with an historical introduction
by Abraham A. Fraenkel. Courier Dover Publications, 1991. (First
edition 1958.)

A book of two parts. In part I Fraenkel introduces the original
Zermelo–Fraenkel set theory. In part II (four fifths of the book)
Bernays introduces his class theory.

[19] Bishop, E. *Foundations of Constructive Analysis*. McGraw-Hill,
1967.

A constructivist (in the mathematical sense) manifesto. This now
famous book countered the pessimism of earlier mathematicians
by demonstrating that analysis can be developed in a constructivist way. A revised version, *Constructive Analysis*, with D.
Bridges, was completed in 1985.

[20] Black, M. *The Nature of Mathematics. A Critical Survey*. Routledge, 2001. (Original 1933.)

A critical exposition of *Principia Mathematica* and accounts of
formalism and intuitionism intended to introduce the reader to
the literature of the time.

[21] Bolzano, B. *Paradoxien des Unendlichen*. Translated by Fr. Pri-
honsky with an historical introduction by D. A. Steele as 'Paradoxes of the Infinite'. Routledge and Kegan Paul, 1950. (Original
1851.)

Bolzano's posthumously published work on some of the peculiarities of infinite sets. Examples of bijections between sets and their
proper subsets are given. The full original is freely available online.

[22] Boole, G. *An Investigation of the Laws of Thought on Which are
Founded the Mathematical Theories of Logic and Probabilities*.
Macmillan, 1854. Reprinted with corrections by Dover Publica-
tions, 1958.

Boole's highly influential monograph on algebraic logic (there was
a predecessor, *Mathematical Analysis of Logic*, but Boole wanted
the second volume to be regarded as the maturation of his ideas
– '*its methods are more general, and its range of applications far
wider. It exhibits the results, matured by some years of study and
reflection, of a principle of investigation relating to the intellectual
operations, the previous exposition of which was written within
a few weeks after its idea had been conceived*'). A copy is freely
available online.

[23] Boolos, G. *The Unprovability of Consistency: an essay in modal logic*. Cambridge University Press, 1979.

[24] Boolos, G. *The Logic of Provability*. Cambridge University Press, 1995.

The second book is a fully updated revision of the first. Provability is examined from the novel perspective of modal logic, an ancient logic centered on necessity and possibility. Once dismissed and ignored by philosophers and mathematicians alike, modal logic makes its debut here as a respectable science. Aimed both at specialists and interested higher level undergraduates.

[25] Borges, J. L. *Fictions*. Translated with an afterword by Andrew Hurley. Penguin, 2000.

[26] Borges, J. L. *Labyrinths*. Translated by James E. Irby. New Directions, 2007.

[27] Borges, J. L. *The Total Library: non-fiction 1922-1986*. Edited by E. Weinberger; translated by E. Allen, S. J. Levine and E. Weinberger. Allen Lane, 2000.

Mathematical ideas permeate Borges' writings. One short story alone, *The Library of Babel*, has triggered many a mathematical and philosophical discussion.

[28] Bourbaki, N. *Elements of Mathematics: Theory of Sets*. Translated from the 1968 *Théorie des Ensembles*. Springer, 2004.

Set theory receives the highly abstract Bourbaki treatment. The first volume in an influential series.

[29] Cantor, G. *Contributions to the Founding of the Theory of Transfinite Numbers*. Translated and introduced by P. E. B. Jourdain. Open Court, 1915.

Philip E. Jourdain translates and introduces Cantor's two most famous papers of 1895 and 1897. These papers are the birthplace of the new theory of ordinal and cardinal numbers.

[30] Carruccio, E. *Mathematics and Logic in History and in Contemporary Thought*. Translated from the Italian by I. Quigly. Faber, 1964.

A history of mathematics focusing on the development of the concepts central to the evolution of logic, culminating with a look at the lively problems of contemporary logic.

[31] Chang, C. C. and Keisler, H. J. *Model Theory*. North-Holland, 1990. (First edition 1973.)

A graduate level text on model theory, the branch of logic dealing
with the relation between a formal language and its interpretations
(that is, its models). The focus is on classical model theory, the
model theory of first-order predicate logic.

[32] Changeux, J-.P. and Connes, A. *Conversations on Mind, Matter, and Mathematics.* Translated by M. B. DeBevoise. Princeton
University Press, 1995.

Dialogues between the neurobiologist Jean-Pierre Changeux and
mathematician Alain Connes. The chapter titles give a good impression of the content: Mathematics and the Brain, Plato as Materialist?, Nature Made to Order, The Neuronal Mathematician,
Darwin among the Mathematicians, Thinking Machines, The Real
and the Rational and Ethical Questions.

[33] Ciesielski, K. *Set Theory for the Working Mathematician.* Cambridge University Press, 1997.

Aimed at the lower level graduate or advanced undergraduate
level. The course covers such topics as transfinite induction, Zorn's
Lemma, the continuum hypothesis, Martin's Axiom, the diamond
principle, and elements of forcing. These topics are chosen as being
useful tools in the application of set theory to geometry, analysis,
topology and algebra.

[34] Cohen, P. *Set Theory and the Continuum Hypothesis.* W. A. Benjamin, 1966.

Much has been written on the subject since, but here is Cohen's
original book detailing his proof of the independence of the continuum hypothesis. Based on a course given at Harvard in 1965,
the exposition is self-contained, the first part of the book covering
the logical preliminaries for the later forcing argument.

[35] Connes, A., Lichnerowicz, A. and Schützenberger, M. P. *Triangle of Thoughts.* Translated by J. Gage. American Mathematical
Society, 2001.

Conversations on logic and reality, the nature of mathematical
objects, physics and mathematics, fundamental theory and real
calculation, mathematics and the description of the world, cosmology and grand unification, interpreting quantum mechanics
and reflections on time.

[36] Conway, J. B. *Functions of One Complex Variable.* Springer,
Graduate Texts in Mathematics 11, 1978.

A well-respected course on the basics of complex analysis.

[37] Conway, J. H. *On Numbers and Games.* A. K. Peters, 2001.

This is the famous guide to game theory in which Conway introduced what are now called the surreal numbers, at once extending
the class of ordinal numbers and the field of real numbers.

[38] Conway, J. H. and Smith D. A. *On Quaternions and Octonions*.
A. K. Peters, Ltd., 2003.

A modern investigation of the geometry of the quaternions and
octonions, starting with a discussion of the properties of 3- and
4-dimensional Euclidean spaces via the quaternions.

[39] Dales, H. G. and Oliveri, G. (eds.) *Truth in Mathematics*. Clarendon Press, 1998.

What is truth in mathematics? Different answers give rise to different and mutually incompatible theories and philosophical positions. Here we find a compilation of articles on the subject of
mathematical truth, authored by a number of notable mathematicians, which are collected under four broad headings: knowability,
constructivity, and truth; formalism and naturalism; realism in
mathematics; and sets, undecidability, and the natural numbers.

[40] Darwin, C. (Edited by F. Darwin) *The Autobiography of Charles
Darwin, and selected letters*. Dover Press 1958. First published
in 1892 as *Charles Darwin, His Life Told in an Autobiographical
Chapter and in a Selected Series of his Published Letters*.

(The source of the quote at the beginning of Section 1.14.)

[41] Dauben, J. W. *Georg Cantor: His Mathematics and Philosophy of
the Infinite*. Harvard University Press, 1979.

An account of Cantor's mathematics and of the man behind it.

[42] Davis, M. (ed.) *The Undecidable: basic papers on undecidable propositions, unsolvable problems and computable functions*.
Raven Press, 1965.

An anthology of papers on undecidability and unsolvability, starting with Kurt Gödel's landmark paper of 1931. Includes papers
of Alonzo Church, Alan Turing, J. B. Rosser, Stephen Kleene and
Emil Post.

[43] Dedekind, R. *Was Sind und Was Sollen die Zahlen?* (What are
numbers and what should they be?) Revised, edited and translated by H. Pogorzelski, W. Ryan and W. Sinder. Research Institute for Mathematics, 1995. (Original 1888.)

Richard Dedekind's influential work in which he discusses set theory, the definition of finiteness and infiniteness, definition by recursion and the formal arithmetic of natural numbers.

[44] De Morgan, A. *Formal Logic, or the Calculus of Inference, Necessary and Probable.* Taylor and Walton, 1847.

Augustus De Morgan's pioneering work on mathematical logic.

[45] Dennett, D. C. *Elbow Room: the varieties of free will worth wanting.* Clarendon Press, 1984.

If determinism is true, and every decision is the outcome of physical forces that happen to be acting at that moment, how could we be free?

[46] Dennett, D. C. *Consciousness Explained.* The Penguin Press, 1992.

Dennett tries to change the mind of those who regard the demystification of consciousness as 'intellectual vandalism'.

[47] Dennett, D. C. *Darwin's Dangerous Idea.* Penguin, 1995.

Dennett places Darwin's great idea ahead of those of Newton, Einstein and everyone else. It is a unifying idea, says Dennett, but at the same time it is a dangerous idea.

[48] Dennett, D. C. *Freedom Evolves.* Viking, 2003.

Free will re-examined from an evolutionary perspective.

[49] Descartes, R. *Discourse on Method.* Translated by L. J. LaFleur. Macmillan, 1956. (Original French edition 1637.)

Descartes' famous philosophical treatise, including in the appendix the first appearance of cartesian coordinates. (The source of the quote at the beginning of Section 2.2.)

[50] Deutsch, D. *The Fabric of Reality.* Penguin Books, 1998.

A stimulating account of the explanatory power of science which weaves together the four main strands of epistemology, quantum physics, the theory of computation and evolution.

[51] Devlin, K. *Constructibility.* Springer, 1984.

The constructible universe and its properties.

[52] Devlin, K. *The Joy of Sets: Fundamentals of Contemporary Set Theory.* Springer, 1993.

An accessible course in Zermelo–Fraenkel set theory at the advanced undergraduate/beginning graduate level. The statement of the axioms is sensibly delayed until an intuitive appreciation of naive set theory has been developed.

[53] Dickman, M. A. *Large Infinitary Languages: Model Theory.* North-Holland Publishing Co., 1975.

A first-order language traditionally limits its sentences to finite strings of symbols. One can nevertheless extend such languages to allow sentences of infinite length. The result is an infinitary language.

[54] Dieudonné, J. A. E. *Foundations of Modern Analysis.* Academic Press, 1960.

A broad course in modern analysis.

[55] Dodd, A. J. *The Core Model.* Cambridge University Press, 1982.

Graduate level lecture notes on the 'Core Model', a generalization of Gödel's constructible universe.

[56] Drake, F. R. *Set Theory: an introduction to large cardinals.* Elsevier Science Publishing Co. Inc., 1974.

A graduate course in Zermelo–Fraenkel set theory, including material on forcing.

[57] Drake, F. R. and Singh, D. *Intermediate Set Theory.* Wiley, 1996.

An intermediate undergraduate/graduate level text introducing first-order logic and set theory including a chapter on constructibility and forcing.

[58] Dummett, M. *Frege: Philosophy of Mathematics.* Duckworth, 1991.

A detailed discussion of Frege's philosophy of mathematics as found in his *The Foundations of Arithmetic* and *Basic Laws of Arithmetic.*

[59] Ebbinghaus, H.-D., Hermes, H., Hirzebruch, F. *et al. Numbers.* Graduate Texts in Mathematics 123. Springer, 1990.

An accessible tour through the conceptual development of number. Includes chapters on natural numbers, integers, rational numbers, real numbers, complex numbers, π, p-adic numbers, quaternions, Cayley numbers, nonstandard analysis and Conway numbers.

[60] Eddington, A. S. *The Nature of the Physical World.* Nabu Press, 2011. (Original 1928.)

Based on lectures delivered in 1927. '*It treats of the philosophical outcome of the great changes in scientific thought which have recently come about.*' (The source of the quote at the beginning of the Synopsis.)

[61] Edwards, H. M. *Galois Theory*. Graduate Texts in Mathematics 101. Springer, 1984.

A course in Galois theory including in the appendix an English translation of Galois' *Memoir on the Conditions for Solvability of Equations by Radicals*.

[62] Einstein, A. *Sidelights on Relativity*. Kessinger Publishing, 2004. (Original 1922.)

Two addresses: 'Ether and the theory of relativity' and 'Geometry and experience'. (The source of the quote at the beginning of Section 7.2.)

[63] Einstein, A. *Ideas and Opinions*. Based on *Mein Weltbild*. Random House, 1997. (Original 1954.)

A collection of Einstein's writing covering a diverse range of topics. (Includes *Relativity and the Problem of Space*, the source of the quote at the beginning of Section 10.1.)

[64] Enderton, H. B. *Elements of Set Theory*. Academic Press, 1977.

An undergraduate course on Zermelo–Fraenkel set theory, up to and including inaccessible cardinals.

[65] Euclid. *Euclid's Elements: all thirteen books complete in one volume*. Densmore, D. (ed.), T. L. Heath (translation). Green Lion Press, 2002.

The world's most influential scientific text. Geometry and number theory developed axiomatically circa 300BC.

[66] Ewald, W. B. (ed.) *From Kant to Hilbert. A Source Book in the Foundations of Mathematics*. Clarendon Press, 2000.

A useful two volume compilation of articles on the foundations of mathematics, including writings of Berkeley, MacLaurin, D'Alembert, Kant, Lambert, Bolzano, Gauss, Gregory, De Morgan, Hamilton, Boole, Sylvester, Clifford, Cayley, Peirce, Riemann, Helmholtz, Dedekind, Cantor, Kronecker, Klein, Poincaré, Hilbert, Brouwer, Zermelo, Hardy and the Bourbaki group.

[67] Feferman, S. *The Number Systems: foundations of algebra and analysis*. Addison-Wesley, 1964.

A rigorous introduction to the basic number systems (natural numbers, integers, rational numbers, real numbers, complex numbers and algebraic numbers), their structure (integral domains, polynomial rings and fields) and their properties.

[68] Ferreirós, J. *Labyrinth of Thought: a history of set theory and its role in modern mathematics*. Birkhäuser, 1999.

A panoramic history of set theory from its nineteenth century mathematical origins, through Cantor to twentieth century axiomatics.

[69] Fitting, M. *Intuitionistic Logic, Model Theory and Forcing.* North-Holland Publishing Co., 1969.

A technical introduction to forcing from the point of view of intuitionistic logic, classical results reworked and shown in a new light.

[70] Forster, T. E. *Set Theory with a Universal Set: Exploring an Untyped Universe.* Clarendon Press, 1995.

The assumption that Russell's paradox shows that the universe cannot be a set is false; one can only draw this conclusion if every subclass of a set is a set. Forster explores alternative systems for which the universe *is* a set and for which, adopting a suitable notion of smallness, the small sets form a model of ZF.

[71] Fraenkel, A. *Abstract Set Theory.* North-Holland Publishing Co., 1953.

A translated, revised, and expanded edition of the first three chapters of Fraenkel's *Einleitung in die Mengenlehre* (appearing in three editions 1919–1928). Fraenkel gives a complete philosophical and mathematical treatment of the following topics without resorting to cold axiomatics: the concept of set (and examples); finiteness and infiniteness; denumerable sets; the continuum; transfinite cardinal numbers; ordering of cardinals; cardinal arithmetic; infinitesimals; ordered sets; ordinals and the well-ordering theorem. The notation is a little archaic and unfamiliar. Also features a huge bibliography.

[72] Fraenkel, A. A. (posthumously), Bar-Hillel, Y., Levy, A., van Dalen, D. *Foundations of Set Theory.* North-Holland Publishing Co., 1976. (First edition 1958.)

A survey of the issues central to set theory: the paradoxes; axiomatic foundations; type theory; intuitionistic mathematics; and metamathematical and semantical approaches.

[73] Franzén, T. *Gödel's Theorem: an incomplete guide to its use and abuse.* A. K. Peters, 2005.

When the incompleteness theorems are invoked outside the field of logic, the results are often nonsensical and embarrassing misunderstandings or at best the result of misjudged free association. Franzén presents an accessible account of the limits of the theorems and of the abuse that they receive in the hands of non-logicians.

[74] Frege, F. L. G. *The Foundations of Arithmetic: a logico-mathematical enquiry into the concept of number (Die Grundlagen der Arithmetick: eine logische mathematische Untersuchung über den Begriff der Zahl)*. Blackwell, 1953.

Frege's 1884 investigation into the philosophical foundations of arithmetic. Although not widely read at the time, Frege's work later proved to be highly influential, via Russell and Wittgenstein, in analytical philosophy circles.

[75] Galilei, G. *The Assayer*. Translation of *Il Saggiatore* by S. Drake and C. D. O'Malley in *The Controversy on the Comets of 1618*. University of Pennsylvania Press, 1960.

(The source of the quote at the beginning of Section 2.1.)

[76] Galilei, G. *Dialogue Concerning the Two Chief World Systems*. Translated by S. Drake. Foreword by Albert Einstein. University of California Press, 1967. (Original 1632.)

A series of dialogues over four days between Salviati (supporting the Copernican system), Sagredo (a neutral layman) and Simplicio (supporting the Ptolemaic system).

[77] Galilei, G. *Two New Sciences*. University of Wisconsin Press, 1974. S. Drake's translation of *Discorsi e dimostrazioni matematiche, intorno a due nuove scienze*, 1638.

Four dialogues. Day 1: first new science, treating of the resistance which solid bodies offer to fracture; Day 2: concerning the cause of cohesion; Day 3: second new science, treating of motion (uniform motion and naturally accelerated motion); Day 4: violent motions – projectiles; and an appendix on theorems and demonstrations concerning the centres of gravity of solids. Includes the observation that the set of natural numbers is equipollent with the set of squares.

[78] Garciadiego, A. R. *Bertrand Russell and the Origins of the Set Theoretic 'Paradoxes'*. Birkhäuser, 1992.

The history and prehistory of the set theoretic paradoxes, shattering some myths and correcting the chronology on the way.

[79] Geach, P. and Black, M. (eds.) *Translations from the Philosophical Writings of Gottlob Frege*. Blackwell, 1952.

Extracts from Frege's work including selections from *Begriffsschrift, Grundgesetze der Arithmetik*, Frege's review of Husserl's *Algebra der Logik* and other writings.

[80] George, A. (ed.) *Mathematics and Mind.* Oxford University Press, 1994.

A compilation of articles on the nature of mathematics. Contributions from Michael Dummett (What is Mathematics About?), George Boolos (The Advantages of Honest Toil over Theft), W. W. Tait (The Law of Excluded Middle and the Axiom of Choice), Wilfried Sieg (Mechanical Procedures and Mathematical Experience), Daniel Isaacson (Mathematical Intuition and Objectivity), Charles Parsons (Intuition and Number) and Michael Hallett (Hilbert's Axiomatic Method and the Laws of Thought).

[81] Gödel, K. *The Consistency of the Continuum Hypothesis.* Princeton University Press, 1940. (Published in 1968: notes by George Brown of Gödel's lectures 1938-1939.)

Gödel's account of his proof of the consistency of the Continuum Hypothesis.

[82] Gödel, K. *Collected Works.* Solomon Feferman, John W. Dawson, Warren Goldfarb, Charles Parsons and Wilfred Sieg (eds.). Five volumes. Clarendon Press, 1986-2003.

Gödel's writings. Vol. I: Publications 1929–1936; Vol. II: Publications 1938–1974; Vol. III: Unpublished essays and lectures; Vol. IV: Correspondence A–G; Vol. V: Correspondence H–Z.

[83] Gödel, K. *On Formally Undecidable Propositions of Principia Mathematica and Related Systems.* Translated by B. Meltzer. Dover, 1992.

An English translation of Gödel's paper *Über formal unentschiedbare Sätze der Principia Mathematica und verwandter Systeme I* with an introduction by R. B. Braithwaite.

[84] Grattan-Guinness, I. (ed.) *From the Calculus to Set Theory 1630–1910: An Introductory History.* Duckworth, 1980.

Techniques of the calculus, 1630–1660 (Kirsti Møller Pedersen); Newton, Leibniz and the Leibnizian tradition (H. J. M. Bos); The emergence of mathematical analysis and its foundational progress, 1780–1880 (I. Grattan-Guinness); The origins of modern theories of integration (Thomas Hawkins); The development of Cantorian set theory (Joseph W. Dauben); and Developments in the foundations of mathematics, 1870–1910 (R. Bunn).

[85] Gray, J. *Henri Poincaré: A Scientific Biography.* Princeton University Press, 2012.

Jeremy Gray describes Henri Poincaré's influence, from his interpretation of non-Euclidean geometry, his pioneering work in topology, his disovery of chaotic motion via celestial mechanics,

his insights into physics which anticipated relativity, and his criticism of the newly developing theory of sets and the reduction of
mathematics to logic.

[86] Grünbaum, A. *Modern Science and Zeno's Paradoxes*. Wesleyan
University Press, 1967.

A modern discussion of Zeno's paradoxes (partly reproduced in
Salmon (2001)).

[87] Hadamard, J. *An Essay on the Psychology of Invention in the
Mathematical Field*. Dover, 1954. (Original 1945.)

How creativity is tapped in science; the unconscious mind and
discovery; intuition versus verbal reasoning; Poincaré's forgetting
hypothesis; and creative techniques of Einstein, Pascal, Wiener
and others.

[88] Hajnal, A. and Hamburger, P. *Set Theory*. Cambridge University
Press, 1999.

An introduction to set theory in two parts. The first part presents
an intuitive course in the theory of cardinal and ordinal numbers
(no previous knowledge of set theory is assumed). An appendix
presents the Zermelo–Fraenkel axiomatization. The second part
('topics in combinatorial set theory') takes a look at, among other
things, inaccessible cardinals and powers of singular cardinals.

[89] Hallett, M. *Cantorian Set Theory and Limitation of Size*. Clarendon Press, 1984.

Cantor's theory of infinity; the ordinal theory of powers; Cantor's
theory of number; the origin of the limitation of size idea; the
limitation of size argument; the completability of sets; the Zermelo
system; and von Neumann's reinstatement of the ordinal theory
of size.

[90] Halmos, P. R. *Naive Set Theory*. D. Van Nostrand Co., 1960.

A much praised textbook on set theory covering the basics up to
ordinal and cardinal arithmetic.

[91] Halmos, P. R. *A Hilbert Space Problem Book*. Graduate Texts in
Mathematics 19. Springer, 1982.

A course in the theory of Hilbert spaces in the form of 250 problems, hints and solutions.

[92] Hamilton, N. T. and Landin, J. *Set Theory and the Structure of
Arithmetic*. Prentice-Hall International, 1962.

The elements of the theory of sets; the natural numbers; the integers and the rational numbers; the real numbers; and the deeper
study of the real numbers.

[93] Hardy, G. H. *A Mathematician's Apology*. Cambridge University Press, 1940.

A defence of pure mathematics (not that it needs one), spoken from the point of view of the Oxbridge culture of the early twentieth century. (The source of the quote at the beginning of Section 11.6.)

[94] Hartshorne, C. and Weiss, P. (eds.) *Collected Papers of Charles Sanders Peirce*. Harvard University Press, 1932. Reprinted in two volumes in 1974.

Peirce's writings in general philosophy, logic (deductive, inductive, and symbolic), pragmatism and metaphysics.

[95] Hatcher, W. S. *The Logical Foundations of Mathematics*. Pergamon, 1981.

An extension of Hatcher's earlier *Foundations of Mathematics*. Topics covered are: first-order logic, the origin of modern foundational studies, Frege's system and the paradoxes, the theory of types, Zermelo–Fraenkel set theory, Hilbert's program and Gödel's incompleteness theorems, the foundational systems of W. V. O. Quine, and categorical algebra.

[96] Hausdorff, F. *Set Theory*. Translation of *Mengenlehre*. American Mathematical Soc., 1957. (Original 1937.)

An introduction to set theory by one of the most influential mathematicians of the first half of the twentieth century.

[97] Henkin, L., Smith, W. N., Varineau, V. J. and Walsh, M. J. *Retracing Elementary Mathematics*. Macmillan Co., 1962.

Starting with a variant of Peano's postulates the natural numbers, integers, rational numbers and real numbers are constructed, stressing throughout the underlying logic and the axiomatic treatment of each structure.

[98] Henle, J. M. *An Outline of Set Theory*. Springer, 1986.

A problem-oriented undergraduate level introduction to set theory. Comprises problems, hints and solutions.

[99] Heyting, A. *Intuitionism. An Introduction*. North-Holland Publishing Co., 1971.

A mature condensed introduction to intuitionistic mathematics presented as a dialogue between a classical mathematician, a formalist, an intuitionist, a finitistic nominalist, a pragmatist and a significist. The intuitionist first defends intuitionistic mathematics and then presents mathematics for the others to judge.

[100] Hilbert, D. *The Foundations of Geometry*. Open Court Publishing Co. 10th revised edition, 1977. (Extended by Paul Bernays.)

David Hilbert's axiomatisation of Euclidean geometry. Freely available online.

[101] Hintikka, J. *Language, Truth and Logic in Mathematics*. Kluwer Academic, 1998.

Twelve essays on the foundations of mathematics with a focus on the logical and linguistic tools used by mathematicians.

[102] Hodges, W. *Model Theory*. Cambridge University Press, 1993.

A thorough graduate level introduction to model theory. Assumes a familiarity with first-order logic.

[103] Hofstadter, D. *Gödel, Escher, Bach: an eternal golden braid*. Harvester Press, 1979.

'*A metaphorical fugue on minds and machines in the spirit of Lewis Carroll.*' Pulitzer prize-winning fantasia on the emergence of cognition via self-reference.

[104] Hrbacek, K. and Jech, T. *Introduction to Set Theory*. Marcel Dekker, 1999.

A course in set theory covering the basics together with material on filters and ultrafilters and combinatorial set theory.

[105] Huxley, A. *Themes and Variations*. Books for Libraries Press, 1970. (Original 1950.)

A collection of essays from Aldous Huxley. (The source of the quote at the beginning of Section 7.1.)

[106] Jacquette, D. *David Hume's Critique of Infinity*. Brill's Studies in Intellectual History. Brill Academic Publishing, 2000.

The eighteenth century empiricist and sceptic David Hume argued that since we have no physical experience of infinite sets we cannot argue about their properties. Jacquette's book assesses Hume's ideas both in their historical context and in contemporary philosophy.

[107] Jech, T. *The Axiom of Choice*. North-Holland Publishing Company, 1973.

An analysis of what can and what cannot be proved assuming the axiom of choice.

[108] Jech, T. *Set Theory*. The Third Millenium edition, revised and expanded. Monographs in Mathematics. Springer, 2003.

A well regarded graduate text on set theory. Covers forcing, inner models, large cardinals and descriptive set theory.

[109] Johnstone, P. T. *Notes on Logic and Set Theory*. Cambridge University Press, 1987.

The language and techniques of first-order logic, including the Completeness Theorem, recursion theory, Zermelo–Fraenkel set theory up to a traditional treatment of ordinal and cardinal arithmetic and a proof of Gödel's Incompleteness Theorems, plus an informal discussion of independence proofs.

[110] Jourdain, P. E. B. *Selected Essays on the History of Set Theory and Logics (1906-1918)*. Ivor Grattan-Guinness (ed.). Editrice CLUEB, 1991.

A reprinting of three of Jourdain's essays: The Development of the Theory of Transfinite Numbers, The Development of Theories of Mathematical Logic and the Principles of Mathematics, and The Philosophy of Mr. B*rtr*nd R*ss*ll.

[111] Just, W. and Weese, M. *Discovering Modern Set Theory*. American Mathematical Society, 1997.

A two volume graduate text on set theory. Volume I (The Basics), aimed at beginning graduate students, introduces formal languages, models, the axiomatic method, infinite arithmetic, and other topics. Volume II (Set-Theoretic Tools for Every Mathematician) continues with more advanced material on trees, partition calculus, cardinal invariants of the continuum, Martin's Axiom, closed unbounded and stationary sets, the Diamond Principle and elementary submodels.

[112] Kac, M. and Ulam, S. M. *Mathematics and Logic: retrospect and prospects*. Praeger, 1968.

A discussion of the origins and development of mathematics, and of the relation between mathematics and the empirical sciences.

[113] Kamke, E. *Theory of Sets*. Dover Publications, 1950.

A short course on set theory, covering the rudiments of set theory, arbitrary sets and their cardinal numbers, ordered sets and their order types and well-ordered sets and their ordinal numbers.

[114] Kanamori, A. *The Higher Infinite: Large cardinals in set theory from their beginnings*. Springer, 1994.

A comprehensive account of the theory of large cardinals tracing their historical beginnings through to current research. The first of a projected multi-volume series.

[115] Kasner, E. and Newman, J. *Mathematics and the Imagination*. With diagrams by R. Isaacs. Penguin, 1979.

A popular account of the art and science of mathematics and how it continues to 'lead the creative faculties beyond even imagination and intuition'. (The source of the quote at the beginning of Section 3.1.)

[116] Kaufmann, F. *The Infinite in Mathematics: logicao-mathematical writings*. Reidel, 1978. (Original edition 1930.)

An account of Husserl's phenomenology of finitism in mathematics. As this was written before Kurt Gödel's earth-shattering paper of 1931, some of the material is now clearly mistaken. Neverthless it still offers food for thought.

[117] Kaye, R. *Models of Peano Arithmetic*. Oxford University Press, 1991.

A graduate level introduction to non-standard models of arithmetic.

[118] Kechris, A. S. *Classical Descriptive Set Theory*. Springer, 1994.

A graduate level introduction to descriptive set theory covering the basic theory of Polish spaces, Borel sets, analytic and co-analytic sets and the theory of projective sets.

[119] Keene, G. B. *Abstract Sets and Finite Ordinals. An Introduction to the Study of Set Theory*. Pergamon Press, 1961.

An elementary treatment of the elements of set theory and the Bernays theory of finite classes and finite sets.

[120] Keisler, H. J. *Model Theory for Infinitary Logic: logic with countable conjunctions and finite quantifiers*. North-Holland Publishing Company, 1971.

Thirty-four lectures on the infinitary logic $L_{\omega_1,\omega}$.

[121] Kelley, J. L. *General Topology*. D. Van Nostrand Co., 1955.

An account of general topology intended as a background for analysis (Kelley had wanted to call the book 'What Every Young Analyst Should Know'). Includes in the appendix an account of what was to become Morse–Kelley set theory.

[122] Kennedy, H. C. *Peano. The life and work of Giuseppe Peano*. Reidel, 1980.

A biography of one of the pioneers of logic and set theory.

[123] Kershner, R. B. and Wilcox, L. R. *The Anatomy of Mathematics*. Ronald Press, 1950.

A treatise on the axiomatic method in mathematics.

[124] Khinchin, A. Ya. *Continued Fractions*. Dover Publications, 1997. (First Russian edition 1935.)

A short elementary introduction to continued fractions in three chapters: Properties of the Apparatus; The Representation of Numbers by Continued Fractions; and The Measure Theory of Continued Fractions.

[125] Kleene, S. C. *Introduction to Metamathematics*. North-Holland Publishing Co., 1952.

A highly influential introduction to logic and model theory as it was understood in the 1950s.

[126] Kline, M. *Mathematical Thought from Ancient to Modern Times*. Oxford University Press, 1990. (Three volume paperback.)

A work tracing the key themes of mathematics from its Babylonian origins through to the early twentieth century.

[127] Kneebone, G. T. *Mathematical Logic and the Foundations of Mathematics. An introductory survey*. D. Van Nostrand Co., 1963.

A detailed historical study of mathematical logic, the foundations of mathematics, and the philosophy of mathematics.

[128] Knuth, D. E. *Surreal Numbers*. Addison-Wesley Professional, 1974.

A mathematical novelette: *How Two Ex-Students Turned on to Pure Mathematics and Found Total Happiness*. Knuth's short novel introducing Conway's new numbers.

[129] Kossak, R. and Schmerl, J. H. *The Structure of Models of Peano Arithmetic*. Oxford University Press, 2006.

A research level monograph detailing four decades worth of research on the theory of non-standard models of Peano Arithmetic.

[130] Krivine, J.-L. *Introduction to Axiomatic Set Theory*. Translated from the French by David Miller. Reidel, 1968.

The Zermelo–Fraenkel axioms of set theory; ordinals and cardinals; the axiom of foundation; the reflection principle; the set of expressions; ordinal definable sets – relative consistency of the axiom of choice; Frankel–Mostowski models – relative consistency of the negation of the axiom of choice (without the axiom of foundation); and constructible sets – relative consistency of the generalized continuum hypothesis.

[131] Kunen, K. *Set Theory: An Introduction to Independence Proofs.* North-Holland, 1980.

The foundations of set theory; infinitary combinatorics; the well-founded sets; easy consistency proofs; defining definability; the constructible sets; forcing; and iterated forcing.

[132] Kuratowski, K. and Mostowski, A. *Set Theory.* Translated from the Polish by M. Maczyński. North-Holland, 1968.

A course in set theory combining both 'naive' set theory (providing intuitive motivation) and the formal axiomatic theory. The second edition features some substantial changes from the first (ten chapters versus the original six). A 1976 edition appeared with the title *Set theory: with an introduction to descriptive set theory.*

[133] Lakatos, I. *Proofs and Refutations: The Logic of Mathematical Discovery.* Cambridge University Press, 1976.

Essays on the philosophy and practice of mathematics, largely in the form of a dialogue between teacher and students. Lakatos observes that the philosophy of mathematics in the twentieth century, triggered by developments in metamathematics, followed the trend often observed in the history of thought whereby a newly discovered method, having attracted the limelight, saturates the field while the rest is unduly ignored or forgotten.

[134] Landau, E. *Foundations of Analysis.* Second edition. Chelsea, 1960.

Subtitled *the arithmetic of whole, rational, irrational and complex numbers: a supplement to text-books on the differential and integral calculus.* Starting with the axiomatization of the natural numbers Landau's book builds, via the rational field, the fields of real and complex numbers.

[135] Lavine, S. *Understanding the Infinite.* Harvard University Press, 1994.

Lavine presents set theory as an idealization of human experience, no more remote than the similar idealizations associated with numbers and geometric objects, painting a picture of mathematical knowledge that is much closer to scientific knowlege than is so often portrayed.

[136] Lebesgue, H. *Leçons sur l'Intégration et la Recherche des Fonctions Primitives.* Gauthier-Villars, 1904.

Lebesgue introduces the integral that now bears his name.

[137] Leng, M. *Mathematics and Reality*. Oxford University Press, 2010.

An absorbing defense of mathematical fictionalism (the thesis that we can have no reason to believe that there are any mathematical objects) and an examination of the indispensability argument for the existence of mathematical objects.

[138] Leng, M., Paseau, A. and Potter, M. D. *Mathematical Knowledge*. Oxford University Press, 2007.

Eight essays on the nature of mathematical knowledge. Contributions by Michael Potter, W. T. Gowers, Alan Baker, Marinella Cappelletti, Valeria Giardino, Mary Leng, Mark Colyvan, Alexander Paseau and Crispin Wright.

[139] Levy, A. *Basic Set Theory*. Dover Publications, 2002.

A course on set theory plus some applications to topology and combinatorics.

[140] Machover, M. *Set Theory, Logic and their Limitations*. Cambridge University Press, 1996.

An accessible introduction to set theory and logic aimed at a mathematical audience with a keen interest in philosophy.

[141] Mackay, A. L. *The Harvest of a Quiet Eye: a selection of scientific quotations*. Institute of Physics, 1977.

A personal selection of scientific quotations from which the quote at the beginning of Section 4.1 was taken.

[142] Mac Lane, S. *Mathematics: Form and Function*. Springer, 1986.

Mac Lane's survey of mathematics, its origins and structure.

[143] Mac Lane, S. *Categories for the Working Mathematician*. Graduate Texts in Mathematics 5. Springer, 1998.

A graduate introduction to category theory authored by one of its inventors.

[144] Maddy, P. *Realism in Mathematics*. Clarendon Press, 1990.

A simplistic definition of realism in mathematics is that it is the science of mathematical objects (sets, functions, numbers, groups etc.) in just the same way that physics, say, is the study of subatomic particles. Maddy examines this view and finds that it is rather fragile.

[145] Marker, D. *Model Theory: an introduction*, Graduate Texts in Mathematics 217. Springer, 2002.

An introductory text in model theory which both develops the basics and also presents the interplay between the abstract theory and its applications.

[146] Martin, G. *The Foundations of Geometry and the Non-Euclidean plane*. Undergraduate Texts in Mathematics. Springer, 1998.

An introduction to non-Euclidean geometry.

[147] Mayberry, J. P. *The Foundations of Mathematics in the Theory of Sets*. Cambridge University Press, 2000.

Mayberry proposes that, equipped with the techniques of modern set theory, we might re-examine the roots of mathematics and form what he calls *Euclidean set theory*, which replaces the assumption that the collection of natural numbers forms a set with the traditional notion that every set is strictly larger than each of its proper subsets.

[148] Monk, J. D. *Introduction to Set Theory*. McGraw-Hill Book Co., 1969.

An introduction to Morse–Kelley set theory.

[149] Monk, J. D. *Mathematical Logic*. Graduate Texts in Mathematics 37. Springer, 1976.

Part I covers elements of recursive function theory. Part II is a short course in elementary logic including the completeness and compactness theorems. Part III discusses decidability. Part IV examines the relationship between semantic properties of languages and their formal characteristics. Part V looks at alternative languages (type theory and infinitary logic, for example).

[150] Moore, A. W. *The Infinite*. Routledge, 1990.

Moore examines the notion of the infinite via an understanding of how it has been approached by other thinkers over the last 2500 years.

[151] Moore, E. H. *Introduction to a Form of General Analysis*. Yale University Press, 1910.

Moore highlights the analogies between (to use modern terminology) $\mathbb{R}^n$, the space of real continuous functions on an interval and certain functions of infinitely many variables. (The source of the quote at the beginning of Section 3.3.)

[152] Moore, E. and Robin, R. (eds.) *Studies in the Philosophy of Charles S. Peirce*. University of Massachusetts Press, 1933.

The philosophy of a man who Bertrand Russell once described as one of the most original minds of the later nineteenth century.

[153] Moore, G. H. *Zermelo's Axiom of Choice: Its origins, development and influence*. Springer, 1982.

A history of the Axiom of Choice and an analysis of some developments inspired by Zermelo's research.

[154] Moore, R. L. *Foundations of Point Set Theory*, American Mathematical Society Colloquium Publications vol. 13, 1970.

A self-contained axiomatic treatment of the foundations of continuity (topology). The primitive notions are 'point' and 'region'.

[155] Morse, A. P. *A Theory of Sets*. Academic Press, 1965.

A course on set theory in three parts: language and inference; logic; and set theory.

[156] Moschovakis, Y. N. *Descriptive Set Theory*. North-Holland, 1980.

An account of descriptive set theory with an emphasis on infinite games and determinacy.

[157] Moschovakis, Y. N. *Notes on Set Theory*. Springer, 1994.

Equinumerosity; paradoxes and axioms; are sets all there is?; the natural numbers; fixed points; well-ordered sets; choices; choice's consequences; Baire spaces; replacement and other axioms; and ordinal numbers.

[158] Mostowski, A. *Sentences Undecidable in Formalized Arithmetic: An Exposition of the Theory of Kurt Gödel*. North-Holland Publishing Co., 1952.

An exposition of Gödel's Incompleteness Theorems.

[159] Nagel, E. and Newman, J. R. *Gödel's Proof*. Routledge and Kegan Paul, 1959.

A non-technical and accessible introduction to the ideas behind Gödel's Incompleteness Theorems.

[160] Needham, T. *Visual Complex Analysis*. Oxford University Press, 1997.

An undergraduate course in complex analysis which focuses on the abundant geometry of the subject, not on the kind of dry symbolic manipulation that one sometimes finds in other treatments.

[161] Oxtoby, J. *Measure and Category*. Graduate Texts in Mathematics 2. Springer, 1971.

A survey of the analogies between topological and measure spaces. Oxtoby's book explores the Baire category theorem as a method of proving existence and also examines the duality between measure and Baire category.

[162] Painlevé, P. *Analyse des Travaux Scientifiques*. Gauthier-Villars, 1900.

(The source of the quote at the beginning of Section 1.8.)

[163] Parsons, C. *Mathematics in Philosophy: Selected Essays.* Cornell University Press, 1983.

Eleven essays in three parts: mathematics, logic, and ontology; interpretations; and sets, classes, and truth.

[164] Penrose, R. *The Road to Reality: a complete guide to the laws of the universe.* Jonathan Cape, 2004.

A very ambitious attempt to describe the fundamentals of current physics in a single volume. The first 382 pages cover pure mathematical topics and the remaining seven hundred pages embark on a discussion of physics.

[165] Pinter, C. C. *Set Theory.* Addison-Wesley, 1971.

An introduction to set theory.

[166] Poincaré, J. H. *Science and Hypothesis.* Cosimo, Inc., 2007. (Original 1901.)

Thirteen chapters divided into four parts: Number and Magnitude; Space; Force; and Nature. (The source of the quote at the beginning of Section 2.5.)

[167] Poincaré, J. H. *Science and Method.* English translation by Francis Maitland. Dover, 1914. (Original French edition 1908.)

Four chapters on The Scientist and Science; Mathematical Reasoning; The New Mechanics; and Astronomical Science. (The source of the quotes at the beginning of Sections 1.4 and 4.5.)

[168] Potter, M. *Reason's Nearest Kin: philosophies of arithmetic from Kant to Carnap.* Oxford University Press, 2000.

Progress in the philosophical investigation of the properties of the natural numbers, and all that goes with it, from the 1880s to the 1930s.

[169] Potter, M. *Sets: an introduction.* Clarendon Press, 1990.

[170] Potter, M. *Set Theory and its Philosophy: a critical introduction.* Oxford University Press, 2004.

The second book is a substantial revision of the first, written for a broad philosophical readership. Introducing a theory ZU (urelements are included but the schema of replacement is not) this volume covers all the main themes of modern set theory.

[171] Quine, W. V. O. *Mathematical Logic.* Norton, 1940.

Quine shows that Whitehead and Russell's *Principia Mathematica* can be considerably condensed. Seven chapters: statements; quantification; terms; extended theory of classes; relations; number; and syntax. The *Principia* style notation may be off-putting.

[172] Quine, W. V. O. *From a Logical Point of View: 9 logico-philosophical essays.* Harvard University Press, 1953 (fourth printing 2003).

Essays on the problem of meaning and on ontological commitment. (The source of the quote at the beginning of Section 1.1.)

[173] Quine, W. V. O. *Set Theory and its Logic.* Belknap Press, 1969.

The elements of logic; transfiniteness; and axiom systems.

[174] Quine, W. V. O. *Philosophy of Logic.* Prentice-Hall, 1970.

Truth, grammar and the boundaries of logic.

[175] Quine, W. V. O. *Quiddities: An intermittently philosophical dictionary.* Harvard University Press, 1987.

Quine's musings, anecdotes and general discussions on a wide variety of topics ranging from *Alphabet* to *Zero*.

[176] Robinson, A. *Non-Standard Analysis.* North-Holland, 1966.

An introduction to non-standard analysis written by its inventor. Kurt Gödel described this approach as 'the analysis of the future'.

[177] Roitman, J. *Introduction to Modern Set Theory.* Wiley, 1990.

A self contained course on set theory covering the axioms of ZF, infinite arithmetic, the constructible universe and models of the form V_κ (κ inaccessible), and infinite combinatorics.

[178] Rota, G.-C. *Indiscrete Thoughts.* Birkhäuser, 1997.

Remembrances of Princeton and Yale in the early 50s; the phenomenology of mathematics; book reviews and more.

[179] Rubin, H. and Rubin, J. *Equivalents of the Axiom of Choice.* North-Holland, 1985.

A substantial collection of statements equivalent to the axiom of choice (in Gödel–Bernays set theory).

[180] Rucker, R. v. B. *Infinity and the Mind: the science and philosophy of the infinite.* Birkhäuser, 1982.

A popular discussion of, among other things, paradoxes, parallel worlds, physical infinities and Gödel and his theorems.

[181] Russell, B. *The Principles of Mathematics.* Cambridge University Press, 1903.

The prequel to *Principia Mathematica.* The work has two goals. Firstly it is proposed that all of pure mathematics can be reduced to concepts definable in terms of a very small number of logical concepts and secondly that all of its propositions are deducible from a small number of rules of inference.

[182] Russell, B. *Mysticism and Logic, and Other Essays*. Longmans, Green and Company, 1918.

Ten essays on various subjects. (The source of the quotes at the beginning of Sections 1.7 and 6.3.)

[183] Russell, B. *The Philosophy of Logical Atomism*. Open Court, 1998. (Original 1918.)

Russell discusses some ideas learnt from his former pupil Ludwig Wittgenstein. (The source of the quotes at the beginning of Sections 5.3 and 11.3.)

[184] Russell, B. *Introduction to Mathematical Philosophy*. Allen & Unwin, 1919.

Written in the summer of 1918 while Russell was imprisoned for making a statement 'insulting to a war-time ally of Great Britain' this is an accessible philosophical account of the definition of number, finitude, order, relations, the classical number systems, transfinite numbers, continuity, the axiom of choice ('multiplicative axiom'), and logic.

[185] Russell, B. *ABC of Relativity*. Taylor & Francis, 2009. (Original 1925.)

Russell gives a readable exposition of Einstein's theories of special and general relativity. (The source of the quote at the beginning of Section 8.1.)

[186] Russell, B. *Sceptical Essays*. Routledge, 2004. (Original 1928.)

Russell offers a few 'mild but revolutionary' propositions. (The source of the quote at the beginning of Section 9.2.)

[187] Russell, B. *Unpopular Essays*. Routledge, 1996. (Original 1950.)

Bertand Russell tries to combat the growth of dogmatism. (The source of the quote at the beginning of Section 6.5.)

[188] Russell, B. *The Impact of Science on Society*. Routledge, 1985. (Original 1952.)

(The source of the quote at the beginning of Section 1.12.)

[189] Russell, B. *The Autobiography of Bertrand Russell*. George Allen and Unwin Ltd, 1967.

Three volumes: I (1872–1914); II (1914–1944); III (1944–1967). (The source of the quote at the beginning of Section 5.1.)

[190] Sainsbury, R. M. *Paradoxes*. Cambridge University Press, 1995.

A discussion of paradoxes and their resolutions, ranging from Zeno's paradoxes, the paradox of the heap, the prisoner's dilemma, the unexpected examination through to the paradox of the liar and its relation to Russell's paradox.

[191] Salmon, W. C. (ed.) *Zeno's Paradoxes*. Hackett Publishing Company, 2001.

Selected essays on Zeno's paradoxes by Abner Shimony, Wesley C. Salmon, Bertrand Russell, Henri Bergson, Max Black, J. O. Wisdom, James Thomson, Paul Benacerraf, G. E. L. Owen and Adolf Grünbaum.

[192] Shapiro, S. *Foundations without Foundationalism: a case for second-order logic*. Oxford Logic Guides 17. Clarendon Press, 1991.

A formal development of higher-order logic and an argument in favour of higher-order logic in the foundations of mathematics and in philosophy.

[193] Shelah, S. *Cardinal Arithmetic*. Clarendon Press, 1994.

A research-level monograph focusing in particular on cardinal arithmetic associated with singular cardinals.

[194] Shen, A. and Vereshchagin, N. K. *Basic Set Theory*. American Mathematical Society, 2002.

A leisurely treatment of the main notions of set theory.

[195] Sierpiński, W. *Cardinal and Ordinal Numbers*. Monografie Matematyczne, vol. 34. Państwowe Wydawnictwo Naukowe, 1958.

A leisurely and very illuminating exposition of the theory cardinal and ordinal numbers. Although the discussion is largely in terms of naive set theory, the role of the Axiom of Choice is covered in detail.

[196] Smith, P. *An Introduction to Gödel's Theorems*. Cambridge introductions to philosophy. Cambridge University Press, 2008.

Aimed at both philosophy students interested in mathematical logic and mathematicians seeking an accessible exposition of Gödel's incompleteness theorems.

[197] Smith, D. E. *A Source Book in Mathematics*. Courier Dover Publications, 1959.

A collection of 125 treatises and articles from the Renaissance to the end of the nineteenth century including work of Newton, Liebniz, Pascal, Riemann, Bernoulli, and others.

[198] Smullyan, R. M. *Gödel's Incompleteness Theorems*. Oxford University Press, 1992.

Written for the general mathematician, philosopher, computer scientist and anyone else who might be interested. Assumes a light acquaintance with first-order logic.

[199] Smullyan, R. M. *First-order Logic*. Dover, 1995.

An introduction to first-order logic with an emphasis on tableaux. Part I introduces analytic tableaux, Part II covers first-order logic, and Part III covers Gentzen systems, elimination theorems, prenex tableaux, Craig's interpolation lemma and Beth's definability theorem, symmetric completeness theorems and systems of linear reasoning.

[200] Smullyan, R. M. and Fitting, M. *Set Theory and the Continuum Problem*. Clarendon Press, 1996.

A self-contained course in set theory including a proof of the independence of the continuum hypothesis and the axiom of choice. Includes some novel results not found in similar textbooks.

[201] Steinmetz, C. P. *Four Lectures on Relativity and Space*. Kessinger Publishing, 2005. (Original 1923.)

A non-technical sketch of relativity and its consequences. (The source of the quote at the beginning of Section 10.3.)

[202] Stewart, I. *Letters to a Young Mathematician*. Basic Books, 2007.

Intended as an updating of G. H. Hardy's *A Mathematician's Apology*, Stewart presents an inside view of what it is like to be a mathematician.

[203] Stillwell, J. *Roads to Infinity: The mathematics of truth and proof*. A. K. Peters, 2010.

A popular account of the diagonal argument, ordinals, computability and proof, logic, arithmetic, unprovability and axioms of infinity.

[204] Stoll, R. R. *Set Theory and Logic*. Freeman, 1963.

Sets and relations; the natural number sequence and its generalizations; the extension of the natural numbers to the real numbers; logic; informal axiomatic mathematics; Boolean algebras; informal axiomatic set theory; several algebraic theories (groups, rings, integral domains, fields); and first-order theories.

[205] Sullivan, J. W. N. *The Limitations of Science*. The Viking Press, 1933.

A broad overview of science and its nature from the perspective of an early twentieth century commentator. (The source of the quote attributed to Bertand Russell at the beginning of Section 11.1.)

[206] Suppes, P. *Axiomatic Set Theory*. Dover, 1972.

A course in Zermelo–Fraenkel set theory which assumes no prerequisites in logic or set theory.

[207] Takeuti, G. and Zaring, W. M. *Introduction to Axiomatic Set Theory*. Graduate Texts in Mathematics 1. Springer, 1971.

[208] Takeuti, G. and Zaring, W. M. *Axiomatic Set Theory*. Graduate Texts in Mathematics 8. Springer, 1973.

The first volume is a very formal course on Zermelo–Fraenkel set theory ending with a proof of the independence of the axiom of constructibility. The second volume covers relative constructibility, general forcing and their relationship.

[209] Tarski, A., Mostowski, A. and Robinson, R. M. *Undecidable Theories*. North-Holland Publishing Co., 1953.

An account of the undecidability of various mathematical systems, including group theory, the theory of lattices and abstract projective geometry.

[210] Tarski, A. *Ordinal Algebras*. North-Holland Publishing Co., 1956.

Following his work on cardinal algebras (a model of which is provided by the arithmetic of cardinal numbers), Tarski introduces in this book ordinal algebras, modelled by the arithmetic of order types.

[211] Tarski, A. *The Collected Papers of Alfred Tarski*. Givant, S. R. and McKenzie, R. N. (eds.). Four volumes. Birkhäuser, 1986.

The collected works of one of the giants of twentieth century logic. 110 papers and monographs, 96 abstracts, 18 problems, reviews and discussions.

[212] Tiles, M. *The Philosophy of Set Theory: an historical introduction to Cantor's paradise*. Basil Blackwell, 1989.

Aimed at philosophy (and mathematics) students, but also at the interested general reader, Tiles' book emphasizes that the philosophy of mathematics is not isolated from mainstream philosophy, but on the contrary is tightly intertwined with it.

[213] Titchmarsh, E. C. *Mathematics for the General Reader*. Hutchinson's University Library, 1948.

(The source of the quote at the beginning of Section 4.4.)

[214] Tourlakis, G. *Lectures in Logic and Set Theory. Volume 1. Mathematical Logic*. Cambridge University Press, 2003.

[215] Tourlakis, G. *Lectures in Logic and Set Theory. Volume 2. Set Theory*. Cambridge University Press, 2003.

Volume 1 covers completeness, compactness and applications to analysis, incompleteness including a complete proof of Gödel's second incompleteness thoerem, and recursion theory. Volume 2 is a course on Zermelo–Fraenkel set theory including a chapter on forcing.

[216] Truss, J. K. *Foundations of Mathematical Analysis*. Clarendon Press, 1997.

An introduction to the set theoretic underpinnings of analysis beginning with the classical number systems through to metric spaces, topology, measure and category, the continuum hypothesis and ending with a discussion of constructive analysis.

[217] Ulam, S. M. *Adventures of a Mathematician*. Charles Scribner's Sons, 1976.

Stanisław Ulam's autobiography.

[218] van Dalen, Dirk. *L. E. J. Brouwer – Topologist, Intuitionist, Philosopher: how mathematics is rooted in life*. Springer, 2012.

A biography of L. E. J. Brouwer (1881–1966), the founder of Intuitionism.

[219] van Heijenoort, J. (ed.) *From Frege to Gödel: a source book in mathematical logic, 1879–1931*. Harvard University Press, 1967.

From Frege's *Begriffsschrift* to Gödel's work on incompleteness, a treasure trove of pioneering articles in mathematical logic.

[220] Vaught, R. L. *Set Theory. An Introduction*. Birkhäuser, 1985.

An introduction to set theory which largely focuses on the intuitive theory before presenting the axiomatic approach in the last two chapters.

[221] von Neumann, J. *Collected Works*. Six Volumes. Pergamon Press, 1961.

I: Logic, Theory of Sets and Quantum Mechanics; II: Operators, Ergodic Theory and Almost Periodic Functions in a Group; III:

Rings of Operators; IV: Continuous Geometry and Other Topics; V: Design of Computers, Theory of Automata and Numerical Analysis; and VI: Theory of Games, Astrophysics, Hydrodynamics and Meteorology.

[222] Wagon, S. *The Banach–Tarski Paradox*. Cambridge University Press, 1985.

A detailed account of what some regard as one of the most surprising results of modern mathematics.

[223] Wang, H. *From Mathematics to Philosophy*. Routledge & Kegan Paul, 1974.

Mathematical logic and philosophy of mathematics; characterization of general mathematical concepts; Russell's logic and some general issues; logical truth; metalogic; the concept of set; theory and practice in mathematics; necessity, analyticity, and apriority; mathematics and computers; minds and machines; notes on knowledge and life; themes and approaches; and an appendix: exercises in criticism.

[224] Weyl, H. *Philosophy of Mathematics and Natural Science*. Princeton University Press, 1949.

Philosophy as led by scientific discovery, and science via philosophy. A book of two parts, one on mathematics and the other on the physical sciences.

[225] Whitehead, A. N. and Russell, B. *Principia Mathematica*. Second Edition. Cambridge University Press, Vol I (1925), Vol II (1927), Vol III (1927).

Whitehead and Russell avoid the set theoretic paradoxes by building an elaborate theory of types. This monster of a book is regarded as a seminal work in mathematical logic, but is (because of future simplifications, corrections and better alternatives) seldom read. The full three volume *Principia Mathematica*, which runs to nearly 2000 pages, seems to go in and out of print. Fortunately the appearance of 'print on demand' services and the existence of various online copies make it easier to find than in the past. It has a reputation as an unreadable and obsolete, even flawed, work and hence is regarded by publishers as a financial non-starter. Even when it was written there seemed to be tremendous difficulty in funding its publication. An affordable but heavily truncated edition (just under two-thirds of volume I) is also available ('Principia Mathematica to *56' (Cambridge Mathematical Library), Cambridge University Press; second edition (11 Sep 1997)).

[226] Whitehead, A. N. *Dialogues of Alfred North Whitehead* (recorded by Lucien Price). David R. Godine Publisher, 2001. (Original 1954.)

Conversations recorded in Whitehead's home. (The source of the quote at the beginning of Section 1.5.)

[227] Wilder, R. L. *Introduction to the Foundations of Mathematics.* John Wiley & Sons, 1952.

Set theory, the construction of the real numbers, group theory and applications from the point of view of foundational research. Discussion of different approaches to the foundations of mathematics, and mathematics in its cultural setting. (The source of the quote at the beginning of Section 2.4.)

[228] Wittgenstein, L. *Lectures on the Foundations of Mathematics.* Cambridge 1939: from the notes of R. G. Bosanquet, N. Malcolm, R. Rhees and Y. Smythies. C. Diamond (ed.) Harvester Press, 1976.

Thirty-one of Ludwig Wittgenstein's 1939 Cambridge lectures on the philosophical foundations of mathematics, based on notes taken by four of the students present.

[229] Wittgenstein, L. *Remarks on the Foundations of Mathematics.* G. H. von Wright, R. Rhees, G. E. M. Anscombe, (ed.) Translated from the German by G. E. M. Anscombe. Blackwell, 1978.

Wittgenstein offers some thoughts, occasionally insightful, occasionally opaque, on the foundations of mathematics: Logical compulsion and mathematical conviction; calculation as experiment; mathematical surprise, discovery, and invention; Russell's logic, Gödel's theorem, Cantor's diagonal procedure, Dedekind's cuts; the nature of proof and contradiction; and the role of mathematical propositions in the forming of concepts.

[230] Wright, C. *Wittgenstein on the Foundations of Mathematics.* Duckworth, 1980.

An examination of Wittgenstein's *Remarks on the Foundations of Mathematics*, Wittgenstein's later philosophy of mathematics and links with his philosophy of language.

[231] Wright, C. *Frege's Conception of Numbers as Objects.* Aberdeen University Press, 1983.

An exposition of Frege's conception of number as presented in his *Foundations of Arithmetic* and support for the revival of Frege's theory via the removal of the problematic principal of unrestricted comprehension.

[232] Yandell, B. *The Honors Class: Hilbert's problems and their solvers*. A. K. Peters, 2001.

An account of Hilbert's problems, who worked on them, who solved them, how they were solved, which problems remain open and some twentieth century developments that Hilbert had not predicted.

Index

𝔸, 127
Abel sum, 81
abelian group, 222
absolute, 376, 377, 380
    infinite, 244
    truth, 202, 232
    value, 59, 82
absolutely convergent, 80
absoluteness, 377, 380
abstract nonsense, 182
abundant number, 63
acceleration, 95
addition
    cardinal, 151, 325
    Cauchy class, 60
    complex, 107
    integer, 40
    natural, 29
    order type, 136
    ordinal, 303, 305
    polynomial, 127
    $\mathbb{Q}[\sqrt{2}]$, 52
    rational, 45
    real, 56
additive
    identity, 30, 84
    inverse, 41
affine plane geometry, 231
aleph, 153, 172, 326, 356
algebraic
    closure, 129
    number, 127, 129, 131
algebraically closed field, 129
aliquot sequence, 64
almost universal, 386
alternating series, 80
alternative algebra, 126

amicable pair, 64
analytic function, 112
antichain, 363
Anti-foundation Axiom, 270
antilexicographic order, 138, 139
antisymmetric relation, 37
Archimedean field, 105
argument, 113
arithmetic, 76, 181, 412
    consistency, 421
    theory, 421
arithmetization, 75, 392
arrow paradox, 95, 167
associative laws, 190
associativity of addition
    cardinal, 151
    natural, 29
    order type, 137
    PA, 412
associativity of multiplication
    cardinal, 151
    natural, 30
    order type, 138
    PA, 412
atom, 15, 197, 250
atomic formula, 204, 210
Axiom
    of Abstraction, 209, 240
    of Choice, 12, 13, 20, 85, 151, 171,
        173, 176, 179, 197, 209, 281,
        335–337, 339, 340, 342, 343,
        346, 347, 349, 350, 352, 354,
        374, 383, 387, 388, 390, 400,
        408, 418
    of Choice (strong form), 335, 387
    of Constructibility, 334, 386–390,
        397, 401

of Countable Choice, 337
of Countable Dependent Choice,
    338
of Definable Determinacy, 362
of Determinacy, 261, 347, 348, 361
of Extensionality, 209, 236, 380,
    417
of Foundation, 267
of Infinity, 289, 380, 417
of Pairing, 256, 266, 380, 417
of Powers, 260, 261, 380, 401, 417
of Regularity, 267–270, 274, 290,
    298, 301, 380, 417
of Separation, 262
of Symmetry (Freiling), 357
of Unions, 259, 380, 417
of Weak Extensionality, 235
Schema of Replacement, 246, 263–
    266, 290, 295, 380, 381, 398,
    401, 417
axiom schema, 179, 213, 263, 375
axiomatic method, 181, 232
axioms of infinity, 334

back-and-forth, 140
Baire
    Class, 147
    line, 362
Banach–Tarski paradox, 343, 344, 346–
    348, 361
    generalized, 345
barber paradox, 178
Basel problem, 96
basis, 84
Bell number, 281
Benacerraf's problem
    epistemological, 257
    identification, 257
Bernoulli number, 25
Bernstein decomposition, 358
Berry's paradox, 243
best approximation, 66
betweenness, 229
bi-interpretable, 381
bijection, 10
bijective, 10

Boolean algebra, 409
Borel set, 358
bound variable, 204, 205
bounded
    above, 57
    sequence, 77
Bunyakovsky conjecture, 17
Burali-Forti paradox, 148, 241, 285, 322

$\mathbb{C}$, 107
calculus, 15, 75, 77, 88–90, 92, 94, 95,
    100, 102, 168, 169
Cantor normal form, 312, 313, 316
Cantor's
    paradox, 322
    ternary set, 358, 359
    Theorem, 134, 153, 322, 353, 389
Cantor–Bendixson Theorem, 358
Cantor–Bernstein Theorem, 11, 70, 71,
    133, 319, 342
cardinal number, 149–153, 319, 325
cardinality, 10, 325
cartesian product, 34, 273
categorical theory, 373
category theory, 182
Cauchy
    complete, 77
    sequence, 59, 60, 77, 87, 88
Cauchy–Riemann equations, 113
Cayley's Theorem, 220
Cayley–Dickson
    algebra, 125
    process, 125
ccc partially ordered set, 363
Cèsaro
    mean, 80
    summable, 80
chain, 350
Chaitin's Theorem, 243
character of cofinality, 328
choice function, 335
Church–Kleene ordinal, 309
class, 7
    name, 422
closed curve, 118
codomain, 10

cofinal, 328
commutativity of addition
    cardinal, 151
    natural, 29
    PA, 413
commutativity of multiplication
    cardinal, 151
    natural, 30
    PA, 413
Compactness Theorem, 370, 371
compatible, 363
complete
    Cauchy, 77
    Dedekind, 87, 88
    induction, 23, 292
    line, 359
    logically complete, 370
    negation, 370
    normal form, 312, 314, 316
    ordered field, 57, 59, 60
    sequence (of forcing conditions), 406
    theory, 356, 370
Completeness Theorem, 370
completion, 78
complex
    analysis, 107, 108, 111
    conjugation, 111
    differentiable, 111
    dynamics, 123
    integration, 118
    logarithm, 120
    number, 18, 107, 108, 111, 114
    plane, 114
composite, 275
computable number, 231
computable set, 216
conditionally convergent, 80
conformal
    equivalence, 115
    mapping, 115
congruent by finite decomposition, 345
conjunction, 188
connected set, 111
conservative extension, 179
consistent, 4, 20, 215, 339, 424, 426
constructible

real number, 362
    set, 384–387, 390
    universe, 383, 386, 401
Constructivist, 200
continued fraction, 65
continuity, 82
continuous, 147
    function, 81, 82, 146
    nowhere differentiable function, 99
    ordinal function, 308
continuum, 54
Continuum Hypothesis, 20, 104, 144,
        171, 172, 176, 179, 326, 334,
        353, 354, 356–358, 362, 374,
        400, 408
contradiction, 191, 290
contrapositive, 12
convergent
    sequence, 77
    series, 80
countable, 41
    chain condition, 363
cover, 333
cubic equation, 108
cumulative hierarchy, 254
curve, 118
cut, 54

Darboux integral, 92
De Morgan's Laws, 190, 207
de Moivre's formula, 113
decidable sentence, 424
decimal representation, 61
Dedekind
    complete, 57
    cut, 54, 105
    finite, 12, 323
    set, 55
deficient number, 63
definable
    number, 231
    set, 231
definite integral, 88, 90
definiteness, 262
definition, 189, 195
degree

of a formula, 403
of a polynomial, 128
Δ_n-formula, 217
dense, 359, 363
    ordered set, 140
derivative, 88, 90, 95
derived set, 145, 357
descriptive set theory, 357
detachment, 207
determined set, 361
diagonal
    function, 423
    proof, 73, 242
diagonalization, 73, 392, 420, 423
dichotomy paradox, 164
differentiation, 88, 90, 94, 95
dimension, 84, 85
Diophantine set, 216
direct unity, 110
directed
    graph, 270
    set, 308
discontinuity, 82
disjoint, 35
disjunction, 187, 255
disprovable sentence, 422
distributive laws, 191
distributivity
    natural, 30
    PA, 412
divergent
    sequence, 77
    series, 80
divisible group, 222
domain, 10, 275
dyadic normal form, 312

effective theory, 211, 371
element, 6
elementarily equivalent, 373
elementary theory of the successor function, 415
elliptic geometry, 225
elucidation, 200
empty set, 14, 21, 249, 250, 254, 265, 376

ensembles extraordinaires, 268
Entscheidungsproblem, 215
epsilon-delta, 90, 92
epsilon number, 309
equality, 34, 233
equipollence, 11, 319
    of $[0, \infty)$ and $[0, 1)$, 69
    of $[0, 1]$ and $[0, 1)$, 69
    of $[0, 1) \times [0, 1)$ and $[0, 1)$, 71
    of all intervals, 68
    of $\mathbb{N}$ and $\mathbb{A}$, 131
    of $\mathbb{N}$ and $\mathbb{Q}$, 47
    of $\mathbb{N}$ and $\mathbb{Z}$, 41
    of $\mathbb{N}$ and $\mathbb{Z} \times \mathbb{Z}$, 42
    of $\mathbb{N}$ and $\mathbb{Z}^n$, 42
    of $\mathbb{R}$ and $(-1, 1)$, 68
    of $\mathbb{R}$ and $P(\mathbb{N})$, 133
    of $\mathbb{R}$ and $\mathbb{R}^n$, 71
equivalence, 188, 208
    class, 35
    relation, 34–36
essential singularity, 116
Euclid's proof (infinitely many primes), 26
Euclid–Mullin sequence, 26
Euclidean geometry, 102, 200, 225, 229, 230, 356
Euler–Mascheroni constant, 129
Euler's
    formula, 112
    number, 87, 93
eventually
    null sequence, 127
    periodic representation, 63
    positive sequence, 60
everywhere discontinuous integrable function, 100
exclusive or, 187
existential quantifier, 189
exponential function, 93
exponentiation
    cardinal, 152, 325
    complex, 112
    integer, 45
    naive Peano Arithmetic, 30
    PA, 414

real, 93
expressed (by a class name), 422
expressible set, 423
expression, 422
extended real line, 78, 334
extensional relation, 379

field, 18, 45, 52, 84, 128, 129
    Archimedean, 105
    extension, 18, 52, 129
    isomorphism, 57
field of constructible points, 128
filter, 363
finite
    character, 352
    plane, 232
    set, 286
    simple group, 223
finiteness, 6, 12, 13, 322
Finitist, 160, 161, 163, 176
First Principle of Generation, 143
first uncountable ordinal, 309
first-order
    completeness axiom schema, 58
    logic, 368, 370
    theory, 210
forcing, 390, 400, 404, 408
    condition, 404, 408
    set, 406
Formalism, 215
Formalist, 16, 20, 202
foundational relation, 277
Fourier
    analysis, 80
    coefficient, 95
    series, 96, 97
free variable, 204
Freiling's Axiom of Symmetry, 357
full theory, 414
function, 9, 273
fundamental operation, 385, 386, 401
Fundamental Theorem
    of Algebra, 107
    of Calculus, 90
fusion, 14, 265
fuzzy set, 236

Galois theory, 128
game theory, 105, 360
Gelfond–Schneider Theorem, 130
General Convergence Principle, 87
General Principle of Limitation, 144
Generalized Continuum Hypothesis, 134,
    171, 172, 331, 353–355, 383,
    387, 388, 390
generalized well-formed formula, 248
geometry, 53, 75, 76, 94, 111, 181, 202,
    224, 229, 230
Gibbs phenomenon, 98
Gödel
    number, 391, 393, 398, 423
    number (limited), 403
    number (unlimited), 403
    numbering, 391, 393, 423
    sentence, 424
    sequence number, 393
Gödel–Bernays set theory, 179, 180
Gödel–Rosser sentence, 421
Gödel's Incompleteness Theorem, 73,
    215, 231, 356, 370, 392, 419
    first, 423
    second, 293, 331, 375, 380
Goldbach conjecture, 198
    ternary, 198
Goldbach's comet, 200
golden ratio, 65
Goodstein sequence, 314–316
Goodstein's Theorem, 316
graph, 81
Great Picard Theorem, 116
Grelling–Nelson paradox, 243
Grothendieck universe, 329
group, 219, 368
    abelian, 222
    divisible, 222
    first-order theory, 220
    theory, 219
    torsion, 223
    torsion-free, 223

$\mathbb{H}$, 125
Hartogs number, 340, 354
Hartogs' Theorem, 340

Hausdorff
    dimension, 85
    space, 363
Hausdorff's maximal principle, 351
hereditarily finite sets, 381
Hessenberg sum, 313
heterologicality, 243
Hilbert's
    Hotel, 13
    tenth problem, 216
holomorphic, 111
homeomorphism, 85
hyperbolic geometry, 225, 226
hyper-inaccessible cardinal, 331
hyperreals, 103

ideal, 103
image, 275
imaginary part, 111
implication, 188
inaccessible cardinal, 329, 331, 333, 380
incidence, 229
inclusion, 14
incommensurable, 48, 49
inconsistent, 370, 424
independence, 176, 180, 230, 380, 390,
        400, 401, 408
index, 119
individuals, 15
induction, 22–24, 27, 28, 48, 268, 286,
        292, 293, 412
inductive set, 57, 290
infinite
    cardinal, 323, 325, 330
    dimensional vector space, 84
    element, 103
    ordinal product, 317
    ordinal sum, 317
    set, 286
infinitesimal, 90, 102
    element, 103
injective, 10
    class, 275
inner
    equality, 235
    model, 368, 400

inside, 118
integer, 38–41, 46
integrable, 90
integration, 88, 90, 94
interior point, 345
intersection, 35
interval, 67
Intuitionism, 170, 198
Intuitionist, 170
Intuitionistic logic, 198
inverse unity, 110
inversion, 124
irrational number, 18, 55, 63
irreducible formula, 204
isolated singularity, 116
isometry, 345
isomorphism, 28, 57, 281

Jordan curve, 118
Jordan Curve Theorem, 118
Jourdain's paradox, 244

κ-additive measure, 333
Khinchin's constant, 67
Kleene–Mostowski arithmetical hierar-
        chy, 217
Klein model, 225, 226
Kuratowski's principle, 351

large cardinal axiom, 332, 361
lateral unity, 110
Laurent expansion, 120
least
    number principle, 23, 24
    upper bound, 57
Lebesgue
    integral, 90
    measurable set, 333
    measure, 333
left
    continuous, 83
    distributivity (order type), 139
Leibniz's Law, 235
lexicographic order, 139, 385
limit, 61, 77, 81, 87, 111
    ordinal, 144, 285, 308

limitation of size, 252, 253
limited
    Gödel number, 403
    quantifier, 403
Lindström's Theorem, 374
line, 47, 53, 54, 68, 100, 200, 225, 226,
    229, 359
linearly independent, 84
Liouville number, 130
Liouville's Theorem, 124
logical
    axiom, 206
    paradox, 241, 245
logically
    complete, 370
    equivalent, 207, 367
long line, 172
Löwenheim–Skolem Theorem, 372, 374,
    375
    Downward, 372
    Upward, 372
lower
    integral, 90
    sum, 90, 91

Mandelbrot set, 121
manifold, 53
many-valued logic, 199
Martin's Axiom, 362, 363, 408
mathematical paradox, 241
Matiyasevich's Theorem, 216
maximal element, 350
measurable, 333, 346, 347
    cardinal, 333, 388
measure, 332, 333
    theory, 78, 332–334, 338
mechanics, 95
membership, 14, 185
    chain, 290
    loop, 267, 268, 270
mereology, 15
Mersenne prime, 63
metric, 78
    space, 77
minimal model, 389, 390, 400
model, 366, 368

theory, 213, 368
modulus, 111
modus ponens, 207
monadic second-order logic, 213
Morley's Theorem, 373
Morse–Kelley set theory, 179
Mostowski Collapse
    Lemma, 379, 388
Mostowski collapse, 379
multi-valued mapping, 120
multiplication
    cardinal, 151, 325
    Cauchy class, 60
    complex, 107
    integer, 40
    natural, 30
    ordered set, 138
    ordinal, 305
    polynomial, 128
    $\mathbb{Q}[\sqrt{2}]$, 52
    rational, 45
    real, 57
multiplicative
    axiom, 336
    identity, 30
    inverse, 45
musical intervals, 50

$\mathbb{N}$, 29
naive set, 16
    theory, 16
NAND, 192
natural
    logarithm, 93
    number, 13, 19, 21, 22, 24, 27, 28,
        32, 72, 285–287
    product, 313, 314
    sum, 313, 314
negation, 187, 197
    complete, 370
    complete theory, 424
    incomplete theory, 424
negative number, 18, 38
net, 308
n-gonal number, 25
Nichomachus' Theorem, 25

no-class theory, 252
Noetherian induction, 301
non-constructible set, 390
non-Euclidean geometry, 225, 229, 368
non-logical axiom, 206, 207
non-standard
    analysis, 102, 104
    model, 58, 103
non-well-founded set, 270
    theory, 270
NOR, 191
norm, 82
normal form, 312
null sequence, 59, 77
number base, 62
Number Class, 143, 144, 153

$\mathbb{O}$, 125
object language, 245
octonion, 125
ω-consistent, 421
$\mathcal{O}n$, 142
one-to-one, 10
onto, 10
open
    interval, 67, 359
    set, 82, 111, 358
order
    equivalent, 135
    isomorphism, 135, 142
    of constructibility, 386
    topology, 308
    type, 135, 136, 140, 142, 283
ordered
    pair, 34, 256, 257
    product, 138
    set, 135, 136, 138, 140, 142, 283
    sum, 136
ordinal, 142
    class, 284
    division, 305
    limit, 308
    number, 96, 135, 142, 143, 145,
        283–285, 312
    subtraction, 305
ordinary finite, 12

Ostrowski's Theorem, 78
outer
    equality, 235
    measure, 333
outside, 118

PA, 28
p-adic
    absolute value, 78
    analysis, 78
    integer, 79
    number, 78, 79
paradox, 148, 163–169, 177, 185, 209,
        240–244, 248, 267, 285, 343,
        344, 374
    Achilles and the tortoise, 164
    arrow, 95, 167
    Banach–Tarski, 343, 344, 346–348,
        361
    Banach–Tarski (generalized), 345
    barber, 178
    Berry's, 243
    Burali-Forti, 148, 241, 285, 322
    Cantor's, 322
    dichotomy, 164
    Epimenides the liar, 241
    Grelling–Nelson, 243
    Jourdain's, 244
    logical, 241, 245
    mathematical, 241
    Quine's, 245
    Richard's, 242
    Russell's, 148, 177, 185, 241, 244,
        248
    semantic, 241–243, 245
    Skolem's, 374
    Smale's, 346
    sorites, 155
    stadium, 166
    Zeno's, 163, 168, 169
paradoxical decomposition, 346
parallel postulate, 202, 224–226, 335,
    356
parity of a formula, 403
partial
    order, 363

sum, 80
partition, 35
path, 118
pathological function, 98
PCF theory, 331
Peano
    Arithmetic, 28, 58, 103, 293, 314,
        370, 372, 392, 411, 412, 414,
        415
    curve, 85
Peano's Postulates, 27–29, 31, 285
perfect
    number, 63
    set, 357, 358, 361
Π_0-formula, 217
Π_1-formula, 217
Π_n-formula, 217
plane, 53, 100, 111, 200, 226, 229
Platonism, 16, 246
Platonist, 16, 20, 200
Playfair's Axiom, 225
point, 200, 226, 229
pointwise limit, 146
pole, 116
Polish notation, 192
polynomial, 107, 127, 128
power set, 133, 134, 260, 261, 353
predicate, 205
    n-predicate, 205
predicative
    extension, 180
    functions, 252
Presburger Arithmetic, 415
prime
    formula, 397
    number theorem, 300
primes of the form $n^2 + 1$, 17
primitive, 2
    n-ary operator, 210
    n-ary relation symbol, 210
    recursive arithmetic, 293
principal argument, 113
principle
    of finite induction, 23, 286, 292
    of finite recursion, 294, 295
    of purity, 197

projective
    determinacy, 362
    plane geometry, 228, 232
    set, 348, 362
proper
    class, 7
    subset, 11
provability predicate, 425
provable sentence, 422
punctured neighbourhood, 145
Pythagoras' Theorem, 48, 49

$\mathbb{Q}$, 44
quadratic
    algebra, 126
    equation, 108
quartic equation, 108
quaternion, 125
Quine atom, 270
Quine's paradox, 245
quintic equation, 109

$\mathbb{R}$, 55
ramified theory of types, 252
Ramsey Theorem (finite), 316
range, 275
rank, 254, 298, 301
Rasiowa–Sikorski Lemma, 363
rational
    function, 104
    number, 44–48, 63
real
    algebra, 126
    closed field, 230, 231
    number, 18, 32, 49, 52–55, 59, 61
    part, 111
real projective plane, 228
rearrangement of series, 80
rectifiable curve, 118
recursion with a parameter, 294
recursive
    ordinal, 309
    sequences, 294
recursively enumerable set, 216
Reflection Principle, 375, 377
reflexive relation, 34

reflexivity of equality, 233
region, 111
regular cardinal, 329
relation, 34, 274
    antisymmetric, 37
    equivalence, 34, 35
    foundational, 277
    reflexive, 34
    symmetric, 34
    transitive, 34
relatively constructible set, 386, 387
relativization, 367
removable singularity, 116
residue, 120
restriction, 275
reverse order type, 136
Richard's paradox, 242
Riemann
    Mapping Theorem, 115
    integrable, 90, 91
    integral, 90
    sphere, 121
    surface, 120
right continuous, 83
ring, 37, 103
R-initial segment, 277
R-minimal element, 277
root, 128
R-predecessor, 277
rules of inference, 207
Russell's paradox, 148, 177, 185, 241,
    244, 248

salient ordinal, 312
satisfaction, 365
sawtooth wave, 97
schema
    of comprehension, 246
    of separation, 246
Second Principle of Generation, 143,
    144, 148
second-order
    Peano Arithmetic, 414
    completeness axiom, 58
    logic, 212, 213
sedenions, 125

self-reference, 244
semantic paradox, 241–243, 245
sentence, 205, 422
    of Goldbach type, 217
sequence, 59
sequential convergence, 77
sequentially compact, 309
set, 7
Sheffer stroke, 191
sigma algebra, 333, 358
Σ_1-formula, 217
Σ_n-formula, 217
Silver's Theorem, 330
singleton, 255
single-valued class, 275
singular cardinal, 329
Singular Cardinals Hypothesis, 330
singularity, 116
    essential, 116
    isolated, 116
    pole, 116
    removable, 116
Skolem Arithmetic, 415
Skolem's paradox, 374
Smale's paradox, 346
sociable number, 64
sorites paradox, 155
sound theory, 421, 423
Soundness Theorem, 370
Souslin line, 359
Souslin's Hypothesis, 359, 363, 388, 408
space-filling curve, 85
span, 84
sphere eversion, 347
square wave, 97
stadium paradox, 165, 166
standard
    model, 368
    structure, 366
    transitive model, 377, 380, 384, 387,
        388
Standard Model Hypothesis, 388, 390,
    398
strategy, 361
strong induction principle, 23
subset, 11

substitution schema, 234
substitutivity of equality, 234
subtraction (integer), 40
successor, 27, 29, 286, 287, 290
    ordinal, 143, 285, 298
superbase, 313, 316
surjective, 10
surreal numbers, 105
syllogism, 196, 197
symmetric relation, 34
symmetry of equality, 233

Tarski finite, 13, 323
Tarski's Theorem, 424
tautology, 191, 426
Teichmüller–Tukey Lemma, 352
term, 210
terminating decimal, 62, 63
ternary Goldbach conjecture, 198
theory of types, 213
topological space, 82
topology, 82
topos theory, 182
torsion group, 223
torsion-free group, 223
transcendental number, 127, 129–131
transfinite
    cardinal, 323
    induction, 147, 292, 293, 296, 349
    recursion, 147, 295, 296, 303
transitive
    class, 284, 377
    closure, 381
    law, 191
    relation, 34
    set, 284
transitivity of equality, 234
triangle inequality, 78
trichotomy, 173, 350
trigonometric series, 145
true arithmetic, 414
true sentence, 422
truth table, 207
tuple, 255, 257, 259
type, 213
    of a formula, 403

theory, 178, 197, 252

Ultrafinitist, 160
uncountability of $\mathbb{R}$, 73
uncountable, 73, 374
undecidable sentence, 424
uniform convergence, 148
union, 35
unique, 274
universal
    choice function, 335, 387
    quantifier, 189
unlimited Gödel number, 403
unordered pair, 255
upper
    integral, 90
    sum, 90, 91
urelements, 15

valuation set, 426
variable, 186, 204, 210
Vaught's conjecture, 373
vector space, 84
velocity, 95, 168
vibrating string, 95
Vitali set, 358
von Neumann successor, 286, 287

weak extensionality, 235
weak second-order logic, 213
weakly inaccessible cardinal, 329
Weierstrass Approximation Theorem,
    83
well-formed formula, 203, 204, 210, 397
    generalized, 248
well-founded
    relation, 280, 281
    set, 298
well-ordered set, 142, 284
well-ordering, 135, 142, 173, 174, 279,
    281, 336, 340, 349
Well-Ordering Theorem, 173, 174, 197,
    281, 336, 349, 350

$\mathbb{Z}$, 39
Zeno's paradoxes, 163, 168, 169

Zermelo–Fraenkel set theory, 7, 15, 20,
        179, 417
zero, 14, 18, 21
    divisor, 30, 45, 125
    of a function, 128
ZF, xxi, 7
ZFC, 190
Zig-Zag theory, 252
Zorn's Lemma, 104, 350, 351